STUDY GUIDE AND WORKBOOK
AN INTERACTIVE APPROACH

for Starr and Taggart's

BIOLOGY
The Unity and Diversity of Life

SIXTH EDITION

STUDY GUIDE AND WORKBOOK
AN INTERACTIVE APPROACH

for Starr and Taggart's
B I O L O G Y
The Unity and Diversity of Life
SIXTH EDITION

JANE B. TAYLOR
Northern Virginia Community College

JOHN D. JACKSON
North Hennepin Community College

Wadsworth Publishing Company
Belmont, California
A Division of Wadsworth, Inc.

Biology Editor: Jack Carey
Production Editor: Angela Mann
Designers: Carolyn Deacy, Stephen Rapley, and Alan Noyes
Print Buyer: Diana Spence
Art Editor: Donna Kalal
Permissions: Peggy Meehan
Copy Editor: Mary Roybal
Technical Illustrator: Alexander Teshin Associates
Compositor: Omegatype Typography, Inc.
Printer: Malloy

Credits: Photographs from C. Starr and R. Taggart, *Biology,* Sixth Edition, Wadsworth, 1992:
p. 238 (34), Hans Paerl; p. 238 (35), Tony Brain/Science Photo Library/Photo Researchers;
p. 238 (37), Tony Brain/Science Photo Library/ Photo Researchers; p. 238 (38), G. Shih and
R. G. Kessel, *Living Images,* Jones and Bartlett Publishers, Inc., Boston, © 1982; p. 238 (41), John
Clegg/Ardea, London; p. 239 (42), Gary W. Grimes and Steven L'Hernault; p. 239 (43), C. McLaren
and F. Siegel/ Burroughs Wellcome Co.; p. 253 (12), W. Merrill; p. 253 (14), Victor Duran; p. 253 (15),
Ken Davis/Tom Stack & Associates; p. 253 (16), Gary T. Cole, University of Texas, Austin/BPS;
p. 254 (17), Victor Duran; p. 254 (18), John E. Hodgin; p. 254 (20), Jane Burton/Bruce Coleman Ltd.;
p. 254 (21), G. L. Barron, University of Guelph; p. 262, D. J. Patterson/Seaphot Ltd.: Planet Earth
Pictures; p. 263, Jane Burton/Bruce Coleman Ltd.; p. 265, Anthony and Elizabeth Bomford/Ardea,
London; p. 267, Edward S. Ross; p. 286 (a), Ian Took/Biofotos; p. 286 (b), John Mason/Ardea,
London; p. 286 (c), Kjell B. Sandved; p. 286 (d), Douglas Faulkner/Sally Faulkner Collection; p. 295,
C. R. Wyttenbach, University of Kansas/BPS; p. 302 (6), Christopher Crowley; p. 302 (7), Bill
Wood/Bruce Coleman Ltd.; p. 303 (9), Reinhard/ZEFA; p. 303 (10), Peter Scoones/Seaphot Limited:
Planet Earth Pictures; p. 303 (11), Erwin Christian/ ZEFA; p. 304 (12), Peter Scoones/Seaphot
Limited: Planet Earth Pictures; p. 304 (13), Allan Power/Bruce Coleman Ltd.; p. 304 (14), Rick M.
Harbo; p. 304 (15), Hervé Chaumeton/Agence Nature; p. 313, Biophoto Associates; p. 315, Chuck
Brown; p. 342 (a, b), Patricia Schulz; p. 342 (c, d), Ray F. Evert; p. 342 (e, f), Ripon Microslides, Inc.;
p. 362 (47), Lennart Nilsson from *Behold Man,* © 1974 Albert Bonniers Forlag and Little, Brown and
Company, Boston; p. 394, Manfred Kage/Peter Arnold, Inc.; p. 396, C. Yokochi and J. Rohen,
Photographic Anatomy of the Human Body, Second Edition, Igaku-Shoin Ltd., 1979; p. 415, Ed Reschke;
p. 428, D. Fawcett, *The Cell,* Philadelphia: W. B. Saunders Co., 1966; p. 497 (left), Lennart Nilsson,
A Child Is Born, © 1966, 1977 Dell Publishing Company, Inc.; p. 497 (right), Lennart Nilsson, *A Child
Is Born,* © 1966, 1977 Dell Publishing Company, Inc.

*This book is printed on acid-free paper that meets Environmental Protection Agency standards for
recycled paper.*

2 3 4 5 6 7 8 9 10—96 95 94 93 92

ISBN 0-534-16569-9

CONTENTS

PREFACE

Tell me and I will forget, show me and I might remember,
involve me and I will understand
Chinese Proverb

The proverb outlines three levels of learning, each successively more effective than the method before it. The writer of the proverb understood that humans learn most efficiently when they *involve* themselves in the material to be learned. This study guide is like a tutor; when properly used it increases the efficiency of your study periods. The interactive exercises actively *involve* you in the most important terms and central ideas of your text. Specific tasks ask you to recall key concepts and terms and apply them to life; they test your understanding of the facts and indicate items to reexamine or clarify. They also estimate your next test score based on specific material. Most important, though, the study guide and text together help you make informed decisions about matters that affect your own well-being and the well-being of your environment. In the years to come our survival will require administrative and managerial decisions based on an informed biological background.

HOW TO USE THIS STUDY GUIDE

Following this preface, you will find an outline that will show you how the study guide is organized and will help you use it efficiently. Each chapter begins with a title and a list of the 1-, 2- and 3-level headings in that chapter. The *Interactive Exercises* follow, wherein each main (1-level) heading is labeled 1-I, 1-II, and so on. A variety of interactive exercise types are presented under each main (1-level) heading. These exercises include completion/short-answer, fill-in-the-blanks, matching, label-match, crossword puzzles, problems, labeling, sequencing, multiple-choice, and completion of tables. Interactive Exercises are followed by a list of page-referenced *Chapter Terms*; space is provided by each term for

you to formulate a definition in your own words. Immediately following is a *Self-Quiz* composed primarily of multiple-choice questions although sometimes we may present another examination device. Any wrong answers in the Self-Quiz indicate which portions of the text you need to reexamine. A series of *Chapter Objectives/Review Questions* follows each Self-Quiz section. These are tasks that you should be able to accomplish if you have understood the assigned reading in the text. Some objectives require you to compose a short answer or long essay while others may require a sketch or supplying correct words. *Integrating and Applying Key Concepts* invites you to try your hand at applying major concepts to situations in which there is not necessarily a single pat answer and so none is provided in the chapter answer section (except for problems in Chapters 11, 12, and 22). Your text generally will provide enough clues to get you started on an answer, but these sections are intended to stimulate your thought and provoke group discussions. This is followed by a section called *Critical Thinking*. This section presents you with problem situations that concentrate on the critical and higher-level thinking skills used by scientists. Solving these problems requires application of chapter information to new perspectives, analysis of data, drawing conclusions, making predictions, and identification of basic assumptions. *Answers to Interactive Exercises* follows Critical Thinking. Here the answers to all interactive exercises may be conveniently and quickly checked at the end of each chapter.

A person's mind, once stretched by a new idea, can never
return to its original dimension.
Oliver Wendell Holmes

We would like to thank Jim Maxwell and the staff at Wadsworth, especially Jack Carey, Angela Mann, Carolyn Deacy, Donna Kalal, Stephen Rapley, and Alan Noyes, for their help and support.

STRUCTURE OF THE STUDY GUIDE

The outline below indicates how each chapter in this study guide is organized.

Chapter Number ————————————→   **10**

Chapter Title ————————————→   **A CLOSER LOOK AT MEIOSIS**

Chapter Outline ————————————→   **ON ASEXUAL AND SEXUAL REPRODUCTION**
OVERVIEW OF MEIOSIS
    Think "Homologues"
    Overview of the Two Divisions
    Prophase I Activities
    Separating the Homologues
    Separating the Sister Chromatids
MEIOSIS AND THE LIFE CYCLES
    Gamete Formation
        Gamete Formation in Animals
        Gamete Formation in Plants
    More Gene Shufflings at Fertilization
MEIOSIS COMPARED WITH MITOSIS

Interactive Exercises ————————→   Divided into sections by main (1-level) headings

Main (1-Level) Heading ————————→   10-1 ON ASEXUAL AND SEXUAL REPRODUCTION
Groups of **interactive exercises** that vary in type and require constant interaction with the important chapter information

Chapter Terms ————————————→   The page-referenced important terms from the chapter

Self-Quiz ————————————————→   Usually a set of multiple-choice questions that sample important blocks of text information

Chapter Objectives/Review Questions ——→   Tasks consisting of combinations of relative objectives and questions to be answered

Integrating and Applying Key Concepts ——→   Applications of text materials to questions for which there may be more than one correct answer

Critical Thinking ————————————→   Problem situations that concentrate on thinking skills used by scientists

Answers to Interactive Exercises ————→   Answers for all interactive exercises found under main (1-level) headings in each chapter

1

METHODS AND CONCEPTS
IN BIOLOGY

SHARED CHARACTERISTICS OF LIFE
    DNA and Biological Organization
    Metabolism
    Interdependency Among Organisms
    Homeostasis
    Reproduction
    Mutation and Adapting to Change
LIFE'S DIVERSITY
    Five Kingdoms, Millions of Species
    An Evolutionary View of Diversity

THE NATURE OF BIOLOGICAL INQUIRY
    On Scientific Methods
    Testing Alternative Hypotheses
    The Role of Experiments
    About the Word "Theory"
    Uncertainty in Science
    The Limits of Science

Interactive Exercises

SHARED CHARACTERISTICS OF LIFE (1-I, pp. 3–9)

True-False

If false, explain why. *Cells are the smallest living unit*

F 1. Cells and rocks are composed of the same fundamental chemical particles and types of molecules. *DNA molecules are not found in non living objects ie a rock* *[yes]*

T 2. Converting sunlight energy into chemical energy is an example of energy conversion. *Photsynthesis energy chloro phyl* *[no]*

T 3. To stay alive, an organism must obtain energy from someplace else.

F 4. Green plants absorb ~~sugar,~~ water, and minerals from the soil. ~~The sugar absorbed in this manner is used to do cellular work.~~ *Plants produce sugar* *All*

T 5. Trapping light energy in the chemical bonds of ~~sugar is an example of an energy transformation.~~ *it transfers chem. energy to other molecules* *Phosphate to a certain molecule. Together called ATP*

T 6. Photosynthesis and aerobic respiration involve energy transfers. *(Stored chemical energy can be trapped for use)*

F 7. Instructions that result in particular developmental and maintenance patterns being followed are encoded in the ~~ATP~~ molecule. *[DNA]*

F 8. Mutations are never advantageous to the organism in which they occur.

T 9. Sweating is part of a homeostatic control system that keeps body temperature more or less constant in mammals.

T 10. The instructions for production of each insect developmental stage exist before eggs are produced. *DNA*

Matching

Choose the most appropriate answer to match with each term.

11. _F_ organ system
12. _E_ cell
13. _H_ community
14. _J_ ecosystem
15. _G_ molecule
16. _C_ organelle
17. _L_ population
18. _B_ subatomic particle
19. _M_ tissue
20. _D_ biosphere
21. _K_ multicellular organism
22. _A_ organ
23. _I_ atom

A. One or more tissues interacting as a unit
B. A proton, neutron, or electron
C. A well-defined structure within a cell, performing a particular function
D. regions of the earth where organisms can live
E. The smallest unit of life
F. Two or more organs whose separate functions are integrated to perform a specific task
G. Two or more atoms bonded together
H. All of the populations interacting in a given area
I. The smallest unit of a pure substance that has the properties of that substance
J. A community interacting with its nonliving environment
K. An individual composed of cells arranged in tissues, organs, and often organ systems
L. A group of individuals of the same species in a particular place at a particular time
M. A group of cells that work together to carry out a particular function

Fill-in-the-Blanks

(24) _Photosynthesis_ is the chemical process whereby some organisms are able to trap and store sunlight energy in energy-rich molecules. Organisms release energy from energy-rich molecules for cellular work through a chemical process known as aerobic (25) _respiration_. (26) _Metabolism_ refers to the cell's capacity to extract and transform energy from its environment and use energy to maintain itself, grow, and reproduce. (27) _ATP_ is a molecule that transfers energy to other molecules. Organisms show (28) _interdependance_ in that they depend directly or indirectly on one another for energy and raw materials. Molecules and structures called (29) _receptors_ permit organisms to detect specific information about the environment. (30) _Homeostasis_ is the capacity to maintain internal conditions within some tolerable range, even when external conditions vary. The production of offspring by one or more parents is known as (31) _reproduction_. The four stages in the moth life cycle, in sequence, are (32) _egg_, (33) _larvae_, (34) _pupa_, and (35) _adult_. (36) _DNA_ is the molecule of inheritance. (37) _Mutations_ are changes that occur in the structure or number of DNA molecules. An (38) _adaptive_ trait is one that improves the survival and reproduction of an organism in a certain environment.

LIFE'S DIVERSITY (1-II, pp. 9–12)

Fill-in-the-Blanks

The proposal that characteristics of organisms change through natural selection is primarily linked with the name of (1) _Darwin_. Multicelled consumers are classified in the kingdom (2) _Animalia_. Single-celled producers or consumers of considerable internal complexity are placed in the kingdom (3) _protista_. Organisms that are, for the most part, multicelled decomposers that digest their food externally and then absorb it belong to the kingdom (4) _Fungi_. Kingdom (5) _Monera_ holds relatively simple colonies or single cells that can be producers or decomposers. Organisms that are mostly multicelled producers are classified in the kingdom (6) _Plantae_

Sequence

Arrange in correct hierarchical order with the largest, most inclusive category first and the smallest, most exclusive category last.

7. _D_   A. Class
8. _F_   B. Family
              C. Genus
9. _A_   D. Kingdom
10. _E_   E. Order
              F. Phylum
11. _B_   G. Species
12. _C_
13. _G_

True-False

If false, explain why.

_F_ 14. The most inclusive (largest) taxonomic category is the ~~phylum.~~ _Kingdom_

_T_ 15. There is a larger number of different species in a class than in an order.

_F_ 16. If some organisms in a population inherit traits that lend them a survival advantage, they will be ~~less likely to produce offspring.~~ → _that's not survival_

_T_ 17. Darwin described the selection occurring in pigeons as natural selection.

_F_ 18. All bacteria are single-celled and are therefore ~~protistans.~~ _Monera + decomposers; not complex[ed]_

THE NATURE OF BIOLOGICAL INQUIRY (1-III, pp. 12–15)

True-False

If false, explain why.

_F_ 1. Scientists develop only ~~one best~~ hypothesis as to what the solution to a problem might be. _as many that can make reasonable sense_

_T_ 2. Making observations, developing models, and performing experiments are all means of testing the accuracy of predictions drawn from the hypothesis.

___ 3. A control group is used to evaluate possible side effects of the manipulation of the experimental group.

___ 4. Members of a control group should be identical to those of an experimental group in all respects.

Sequence

Arrange the following steps of the scientific method in a possible chronological sequence from first to last (A through G).

5. ___   A. Develop one or more hypotheses about what the solution to a problem might be.

6. ___   B. Devise ways to test the accuracy of predictions drawn from the hypothesis (use of observations, models, and experiments).

7. ___   C. Repeat or devise new tests (different tests might support the same hypothesis).

8. ___   D. Think about what predictability will occur or be observed if the hypothesis is correct.

9. ___   E. If the tests do not provide the expected results, check to see what might have gone wrong.

10. ___   F. Objectively report the results from tests and the conclusions drawn.

11. ___   G. Identify a problem or ask a question of nature.

Labeling

Assume that you have to determine what object is inside a sealed, opaque box. Your only tools to test the contents are a bar magnet and a triple-beam balance. Label each of the following with an *O* (for observation) or a *C* (for conclusion).

___ 12. The object has two flat surfaces.

___ 13. The object is composed of nonmagnetic metal.

___ 14. The object is not a quarter, a half-dollar, or a silver dollar.

___ 15. The object weighs *x* grams.

___ 16. The object is a penny.

Chapter Terms

The following page-referenced terms are important; most were in boldface type in the chapter. Compose a *written* definition for each term *in your own words* without looking at the text. Next, compare your definition with that given in the chapter or in the text glossary. If your definition seems to lack accuracy, allow some time to pass and repeat this procedure until you can define each term rather quickly (rapidity of answering is a gauge of the effectiveness of your learning).

energy (3)
DNA (3)
metabolism (4)
ATP (5)
receptors (6)
homeostasis (7)

mutations (8)
adaptive trait (9)
species (9)
genus (9)
Monera (9)
Protista (9)

Fungi (9)
Plantae (9)
Animalia (9)
evolution (11)
natural selection (12)

hypotheses (13)
experiments (13)
control group (14)
variable (14)
theory (14)

Crossword Puzzle: Methods and Concepts in Biology

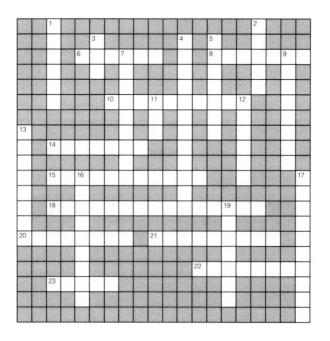

Across

6. the capacity to make things happen, to do work
8. kingdom including mostly multicelled plants
10. extract, transform, and use energy
14. group differing from the experimental group by the variable under study
15. maintaining tolerable conditions in the internal environment
18. the need that organisms have for each other to supply energy and raw materials
20. a trait that helps an organism survive and reproduce
21. changing character of a population through successive generations
22. key factor peculiar to the control group
23. category including all species with perceived similarities

Down

1. broad explanation formed by a related set of hypotheses
2. kingdom including mostly multicelled decomposers
3. molecule containing instructions for assembling new organisms
4. production of offspring by one or more parents
5. category including different "kinds" of organisms
7. molecules and structures that detect environmental information
9. kingdom including mostly multicelled consumers
11. a molecule capable of transferring energy to other molecules
12. kingdom including all bacteria
13. kingdom including single cells of considerable complexity
16. changes in the structure or number of DNA molecules
17. educated guesses as to the solution to scientific problems
19. selection resulting from survival/reproduction differences

Self-Quiz

___ 1. About 12 to 24 hours after the last meal, a person's blood-sugar level normally varies from about 60 to 90 mg per 100 ml of blood, though it may attain 130 mg/100 ml after meals high in carbohydrates. That the blood-sugar level is maintained within a fairly narrow range despite uneven intake of sugar is due to the body's ability to carry out _____.
   a. prediction
   b. inheritance
   c. metabolism
   d. homeostasis

___ 2. As an eel migrates from saltwater to freshwater, the salt concentration in its environment decreases from as much as 35 parts of salt per 1,000 parts of seawater to less than 1 part of salt per 1,000 parts of freshwater. The eel stays in the freshwater environment for many weeks because of its body's ability to carry out _____ .
   a. adaptation
   b. inheritance
   c. puberty
   d. homeostasis

___ 3. A boy is color-blind just as his grandfather was, even though his mother had normal vision. This situation is the result of _____.
   a. adaptation
   b. inheritance
   c. metabolism
   d. homeostasis

___ 4. The digestion of food, the production of ATP by respiration, the construction of the body's proteins, cellular reproduction by cell division, and the contraction of a muscle are all part of _____.
   a. adaptation
   b. inheritance
   c. metabolism
   d. homeostasis

___ 5. Which of the following does *not* involve using energy to do work?
   a. atoms being bound together to form molecules
   b. the division of one cell into two cells
   c. the digestion of food
   d. none of these

____ 6. The experimental group and control group are identical except for _____.
   a. the number of variables studied
   b. the variable under study
   c. two variables under study
   d. the number of experiments performed on each group

____ 7. A hypothesis should *not* be accepted as valid if _____.
   a. the sample studied is determined to be representative of the entire group
   b. a variety of different tools and experimental designs yield similar observations and results
   c. other investigators can obtain similar results when they conduct the experiment under similar conditions
   d. several different experiments, each without a control group, systematically eliminate each of the variables except one

____ 8. The principal point of evolution by natural selection was that _____.
   a. it measures the difference in survival and reproduction that has occurred among individuals that differ from one another in one or more traits
   b. even bad mutations can improve survival and reproduction of organisms in a population
   c. evolution does not occur when some forms of traits increase in frequency and others decrease or disappear with time
   d. individuals lacking adaptive traits make up more of the reproductive base for each new generation

____ 9. The purpose of experiments is to _____.
   a. identify a problem
   b. develop a hypothesis
   c. test predictions
   d. provide conclusions
   e. reject as many hypotheses as possible

____ 10. The least inclusive of the taxonomic categories listed is _____.
   a. family
   b. phylum
   c. class
   d. order
   e. genus

Chapter Objectives/Review Questions

This section lists general and detailed chapter objectives that can be used as review questions. You can make maximum use of these items by writing answers on a separate sheet of paper. Fill in answers where blanks are provided. To check for accuracy, compare your answers with information given in the chapter or glossary.

Page	Objectives/Questions
(3)	1. _____ interactions among molecules bind the parts of a rock and the parts of a frog together.
(3)	2. Name the special molecule that sets living things apart from the nonliving world; explain the functions of this molecule in general terms.
(4)	3. A _____ is the basic living unit.
(4)	4. Distinguish between single-celled organisms and multicelled organisms.
(4–5)	5. Arrange in order, from smallest to largest, the levels of organization that occur in nature. Define each as you list it.
(4–5)	6. _____ means "energy transfers" within the cell.
(5)	7. Contrast the general functions of the processes known as photosynthesis and aerobic respiration.
(5)	8. Organisms use a molecule known as _____ to transfer chemical energy from one molecule to another.
(5–6)	9. Explain how the basic processes of energy capture and energy release create interdependencies among organisms.
(5–6)	10. Describe the general pattern of energy flow through Earth's life forms and explain how Earth's resources are used again and again (cycled).
(6–7)	11. Fully describe the role of receptors in the life of organisms.

(7)   12. _____ is defined as a state in which the conditions of an organism's internal environment are maintained within tolerable limits.

(8)   13. _____ is the means by which each new organism arises.

(8–9)   14. Explain the origin of trait variations that function in inheritance.

(9)   15. A trait that assists an organism in survival and reproduction in a certain environment is said to be _____.

(9)   16. Explain the use of genus and species names by considering your Latin name, *Homo sapiens*.

(9–10)   17. List the five kingdoms of life that are currently recognized by most scientists; tell generally what kinds of organisms are classified in each kingdom.

(9)   18. Arrange in order, from greater to fewer organisms included, the following categories of classification: class, family, genus, kingdom, order, phylum, and species.

(11)   19. As organisms move through time by successive generations, the character of populations changes; this is called _____.

(9–11)   20. Explain what is meant by the term *diversity* and speculate about what caused the great diversity of life forms on Earth.

(11)   21. Darwin used _____ selection as a model for natural selection.

(12)   22. Define natural selection and briefly describe what is occurring when a population is said to evolve.

(12–13) 23. Outline a set of steps that might be used in the scientific method of investigating a problem.

(13)   24. Tests performed to reveal nature's secrets are called _____.

(13–14) 25. Explain why a control group is used in an experiment.

(14)   26. Generally, members of a control group should be identical to those of the experimental group except for the key factor under study, the _____.

(14)   27. Define what is meant by a theory; cite an actual example.

(14–15) 28. Explain the advantages of the "uncertainty" related to scientific endeavors.

(15)   29. Explain how the methods of science differ from answering questions by using subjective thinking and systems of belief.

Interpreting and Applying Key Concepts

1. Humans have the ability to maintain body temperature very close to 37°C.
   a. What conditions would tend to make the body temperature drop?
   b. What measures do you think your body takes to raise body temperature when it drops?
   c. What conditions would cause body temperature to rise?
   d. What measures do you think your body takes to lower body temperature when it rises?
2. Do you think that all humans on Earth today should be grouped in the same species?
3. What sorts of topics are usually regarded by scientists as untestable by the methods that scientists generally use?

Critical Thinking Exercises

1. One day you are watching a wasp drag a grasshopper down a hole in the ground. A friend sees you and asks what you are doing. You reply, "I am watching that wasp store grasshoppers in her nest to feed her offspring." Which of the following is the best word to describe your statement?

   a. observation b. hypothesis c. theory d. assumption e. prediction

ANALYSIS

   a. Observations are the only basis for evidence in science. They are perceptions of actual properties of the external world. They are the things we can see, hear, smell, taste, or feel, sometimes aided by instruments. Your statement contains an observation—that you are watching a wasp—but the most important part of your statement is an explanation of what the wasp is doing. This part of the statement is from inside your own brain, not from the external world; hence, it is not an observation. A statement of your observation of the wasp would be limited to only what you could see her do.

b. Hypotheses are possible explanations of observations or tentative answers to questions. They go beyond the available facts. Your words "... *store* grasshoppers in *her nest to feed her offspring*" explain your observation. You have observed no wasp offspring, you have no evidence that the hole is the wasp's nest, and *store* implies a future time that you have not observed. Most of your statement is a possible, tentative answer to the question "Why is the wasp doing that?"

c. A theory is a collection of related hypotheses that together explain a broad range of observations. Your statement explains only this one behavior in this one species, or perhaps this one individual organism. A theory of animal behavior would have to include hypotheses that explain all the things that all animals do.

d. An assumption is a statement that is accepted without evidence in order to proceed in science. Assumptions are always added to observations in order to make hypotheses, make predictions, and interpret experiments. They do not explain observations, but they are necessary in order to create explanations, because we never have *all* the facts. For example, in order to make your stated hypothesis, you had to assume that the behavior of the wasp was useful to the wasp. Without that assumption, you would also have to accept the hypothesis that the grasshopper had provided a stimulus that made the wasp carry the grasshopper into a sheltered location.

e. A prediction is a statement of expected observations given that some hypothesis is true. It always has the "if-then" form and must contain a verb in the future tense. Your statement has none of these characteristics.

2. In order to test your hypothesis, you plan an experiment. Which of the following experimental results would provide the best support for your hypothesis?

   a. Follow the wasp and see that she digs a similar hole in another place.
   b. Watch the first hole for a while and see that the wasp brings in another grasshopper.
   c. Dig into the hole and see the wasp eating the grasshopper.
   d. Dig into the hole and find a grasshopper buried with wasp eggs on its side.
   e. Come back to the hole three weeks later and see several adult wasps come out of it.

ANALYSIS

In order for the results of an experiment—new observations—to support your hypothesis, they have to be predicted by the hypothesis *and* not be predicted by possible alternative hypotheses.

   a. This observation is consistent with your hypothesis but is not predicted by it. The wasp could be storing food for her young in a hole she had taken over from some other organism, or she might be able to store enough food for all her possible offspring in a single hole. This observation neither supports nor contradicts your hypothesis.

   b. Repeating observations is necessary in order for you to have confidence that they are real, normal events and not artifacts caused by the process of observing. But confidence in the facts does not affect the possible explanations of those facts. Your hypothesis is neither supported nor contradicted by repeated observations.

   c. Your hypothesis generates the prediction that the wasp will *not* eat the grasshopper but will leave it for her young. This observation contradicts your hypothesis and supports the alternative hypothesis that the behavior is simple predation.

   d. Your hypothesis predicts that wasp offspring will at some time be present on the grasshopper, and this observation shows that they are. Therefore, this observation supports your hypothesis. Note, however, that an alternative hypothesis—that the grasshoppers will decay in the nest and produce heat that incubates the wasp eggs—also makes this same prediction. This observation supports but does not *prove* your hypothesis. Also note that your hypothesis could still be valid even if no eggs were found. They might be buried near the grasshopper, and the young might come to the grasshopper only in a later stage of development.

   e. This observation, like choice (a), is consistent with your hypothesis but is not necessarily predicted by it. The emerging adult wasps might simply have crawled into the hole while you were not watching. This observation neither supports nor contradicts your hypothesis.

3. You and your friend wait until the wasp flies away; then you dig into the hole. You find three grasshoppers, each with a small white egg on its side. You claim that this observation supports your hypothesis. Which of the following assumptions are you most likely making?

 a. The wasp stored the grasshopper to feed her offspring.
 b. The eggs are grasshopper eggs.
 c. The wasp laid the eggs.
 d. The wasp dug the hole.
 e. The wasp will not return.

ANALYSIS

 a. An assumption is something you accept as true without evidence in order to make your conclusion. This choice is a statement of the conclusion itself, not something that underlies the conclusion.
 b. If you assumed this statement, you would conclude that your hypothesis was *not* valid, because it predicts that the wasp's own offspring will be present on the grasshopper.
 c. This statement must be true if you are to conclude that the presence of the eggs supports your hypothesis, yet you did not see her lay them, nor do you even know that they are wasp eggs. Many organisms lay small, white eggs, and some of them are parasites of grasshoppers and of wasps. This assumption could become the hypothesis of your next experiment—take the eggs back to the laboratory and wait to see what hatches from them.
 d. While you do not have any evidence to support this statement, it does not have to be true in order for your hypothesis to be valid. The wasp may have taken over a hole dug by some other organism.
 e. You probably are assuming this when you decide to dig into the hole, but it does not have to be true in order for the observation to support your hypothesis.

Answers

Answers to Interactive Exercises

SHARED CHARACTERISTICS OF LIFE (1-I)
1. F; 2. T; 3. T; 4. F; 5. T; 6. T; 7. F; 8. F; 9. T; 10. T; 11. F; 12. E; 13. H; 14. J; 15. G; 16. C; 17. L; 18. B; 19. M; 20. D; 21. K; 22. A; 23. I; 24. Photosynthesis; 25. respiration; 26. Metabolism; 27. ATP; 28. interdependency; 29. receptors; 30. Homeostasis; 31. reproduction; 32. egg; 33. larva; 34. pupa; 35. adult; 36. DNA; 37. Mutations; 38. adaptive.

LIFE'S DIVERSITY (1-II)
1. Darwin; 2. Animalia; 3. Protista; 4. Fungi; 5. Monera; 6. Plantae; 7. D; 8. F; 9. A; 10. E; 11. B; 12. C; 13. G; 14. F; 15. T; 16. F; 17. F; 18. F.

THE NATURE OF BIOLOGICAL INQUIRY (1-III)
1. F; 2. T; 3. T; 4. F; 5. G; 6. A; 7. D; 8. B; 9. E; 10. C; 11. F; 12. O; 13. O; 14. C; 15. O; 16. C.

Answers to Crossword Puzzle: Methods and Concepts in Biology

Answers to Self-Quiz

1. d; 2. a; 3. b; 4. c; 5. d; 6. b; 7. d; 8. a; 9. c; 10. e.

2

CHEMICAL FOUNDATIONS
FOR CELLS

ORGANIZATION OF MATTER
    The Structure of Atoms
    Commentary: Dating Fossils, Tracking
    Chemicals, and Saving Lives—Some Uses
    of Radioisotopes
    Isotopes

BONDS BETWEEN ATOMS
    The Nature of Chemical Bonds
    Ionic Bonding
    Covalent Bonding
    Hydrogen Bonding

PROPERTIES OF WATER
    Polarity of the Water Molecule
    Temperature-Stabilizing Effects
    Cohesive Properties
    Solvent Properties
ACIDS, BASES, AND SALTS
    Acids and Bases
    The pH Scale
    Buffers
    Dissolved Salts
WATER AND BIOLOGICAL ORGANIZATION

Interactive Exercises

ORGANIZATION OF MATTER (2-I, pp. 19–22)

1. Define *element* and *compound.*

    a. element_____

    b. compound_____

2. Compose a sentence that correctly defines the terms *atom* and *molecule.*

    _____

    _____

3. Describe the particles that make up the atom by giving their electric charge and their location in an atom.

    a. protons_____

    b. electrons_____

    c. neutrons_____

4. $^{12}_{6}C$ is a shorthand method for recording the symbol of the element, the atomic mass, and the atomic

    number. C is the symbol for the element _____ ; 12 is the atomic _____ ; there are

    (number) _____ neutrons and (number) _____ protons; 6 is the atomic _____ ;

    there are (number) _____ protons; there are (number)_____electrons in the atom.

5. Only four kinds of elements compose most of the human body. List them in any order: _____ ,

_____ , _____ , and _____ .

6. Complete the following table (refer to Tables 2.1 and 2.2 in the text).

Element	Symbol	Atomic Number	Mass Number	Electron Distribution First Shell	Second Shell	Third Shell
a. Hydrogen	_____	_____	_____	_____	_____	_____
b. Carbon	_____	_____	_____	_____	_____	_____
c. Nitrogen	_____	_____	_____	_____	_____	_____
d. Oxygen	_____	_____	_____	_____	_____	_____
e. Sodium	_____	_____	_____	_____	_____	_____
f. Magnesium	_____	_____	_____	_____	_____	_____
g. Phosphorus	_____	_____	_____	_____	_____	_____
h. Sulfur	_____	_____	_____	_____	_____	_____
i. Chlorine	_____	_____	_____	_____	_____	_____

7. Following the model below (number of protons and neutrons shown in the nucleus), construct the indicated atoms of the elements illustrated below. Show electrons in the form (2e⁻).

MODEL:

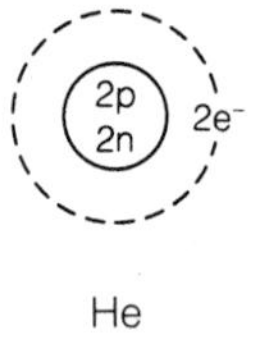

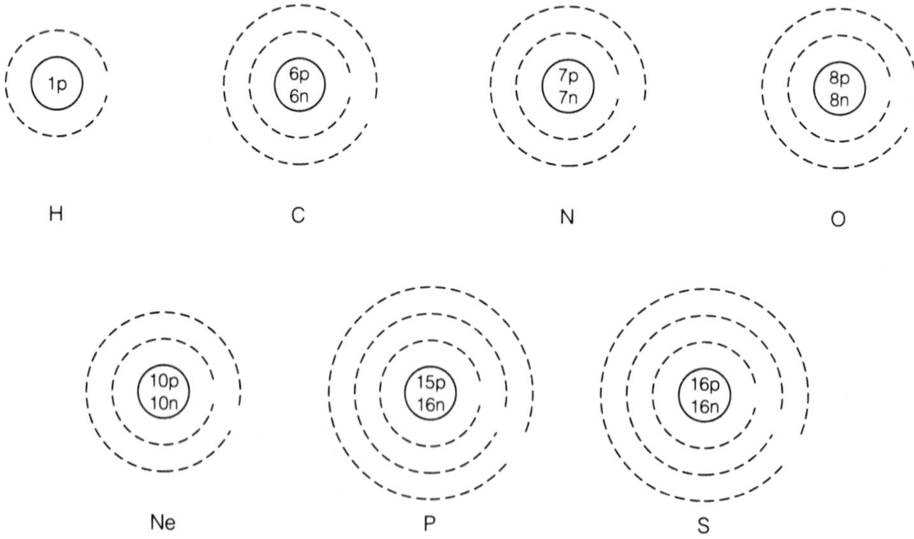

He

H   C   N   O

Ne   P   S

8. Which of the elements above have all shells filled and are considered inert because they do not combine with atoms of other elements? _____

9. Define *isotopes* and *radioisotopes*.

  a. isotopes_____

  b. radioisotopes_____

10. List several uses of radioisotopes._____

_____

_____

_____

True-False

If false, explain why.

___ 11. A molecule is a unit of two or more atoms of different elements bonded together.

___ 12. The number of protons in the nucleus of an atom is equal to the number of electrons outside the nucleus.

___ 13. The protons in a nucleus directly determine how that atom reacts with other atoms.

___ 14. If two particles bear the same electric charge, they are attracted to each other.

___ 15. Carbon 12 and carbon 14 are identical except for the number of electrons in each atom.

Matching

Choose the one most appropriate answer for each. Not every letter may be matched with a number.

16. ___ atom

17. ___ atomic number

18. ___ electric charge

19. ___ electron

20. ___ element

21. ___ energy

22. ___ isotope

23. ___ matter

24. ___ molecule

25. ___ neutron

26. ___ proton

27. ___ radioisotopes

A. An uncharged subatomic particle
B. The number of protons in the nucleus of one atom of an element
C. The smallest neutral unit of an element that shows the chemical and physical properties of that element
D. That which occupies space and has mass
E. Two or more atoms linked together by one or more chemical bonds
F. Contributes the capacity to do work—that is, to move a certain amount of material across a specific distance
G. A positively charged subatomic particle
H. A term applied to unstable isotopes
I. The sum of protons and neutrons in an atom
J. A negatively charged subatomic particle
K. Fundamental substance of which ninety-two different types occur in nature
L. A form of an element, the atoms of which contain a different number of neutrons from other forms of the same element
M. Pushes away similar particles but attracts oppositely charged particles

BONDS BETWEEN ATOMS (2-II, pp. 23–26)

1. What is meant by a chemical bond?

_____

_____

2. Explain the relationship between electrons, orbitals, and shells (energy levels).

_____

_____

3. In an atom, where are the electrons that are said to be at the lowest energy level located? the highest

energy level?_____

_____

4. State how many electrons are required to fill the one (and only) shell of the hydrogen atom and how

many electrons are required to fill the outermost shell of all other atoms._____

_____

5. State the conditions of atomic structure that exist when atoms tend to react with other atoms.

_____

_____

_____

6. Define *ion* and *ionic bond*.

_____

_____

7. Following the model below, sketch the transfer of electron(s) (by arrows) to show how positive and
   negative ions form ionic bonds to create a molecule of $MgCl_2$ (magnesium chloride).

   MODEL:

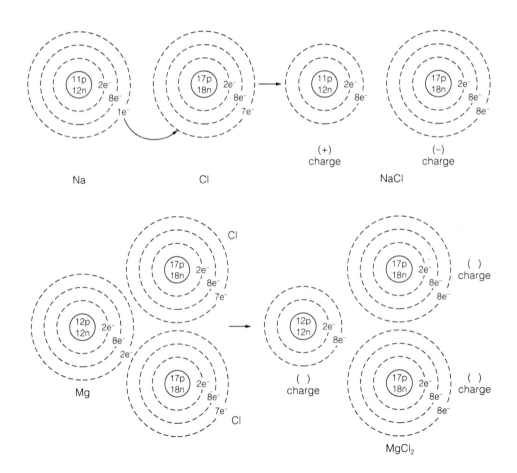

8. Define covalent bond; distinguish between a nonpolar covalent bond and a polar covalent bond.

_____

_____

_____

9. Following the model of hydrogen gas below, sketch electrons (as dots) in the outer shells to illustrate the nonpolar covalent bonding to form oxygen gas; similarly illustrate polar covalent bonds by completing electron structures to form a water molecule.

MODEL:

10. Define hydrogen bond; cite one example of a large molecule within which such bonds exist.

_____

_____

True-False

If false, explain why.

____ 11. In an atom, the volume of space that can accommodate (at most) two electrons is called a *shell*.

____ 12. Electrons moving outside the nucleus of an atom spend as much time as they can away from the nucleus but as close to other electrons as possible.

____ 13. For an atom to react with another atom, all orbitals of the first atom must already contain two electrons.

____ 14. Electrons that are farther away from the nucleus of an atom are said to be at higher energy levels.

____ 15. An ionic chemical bond is formed when two atoms share a pair of electrons, and a covalent chemical bond is formed when atoms lose or gain one or more electrons.

____ 16. Covalent bonds typically occur within molecules; weaker hydrogen bonds usually form between two or more different molecules but can also form within the same molecule.

____ 17. The atoms in table salt (sodium chloride) are linked together with covalent bonds.

____ 18. In order to form a single covalent bond, a pair of electrons is shared between the two nuclei.

____ 19. A polar covalent bond is formed when two atoms share a pair of electrons equally.

____ 20. Water molecules are polar; that is, due to the formation of polar covalent bonds, the oxygen end becomes positive and the hydrogen end becomes negative.

PROPERTIES OF WATER (2-III, pp. 26–29)

True-False

If false, explain why.

____ 1. The polarity of water molecules allows them to interact with each other.

____ 2. Hydrophobic substances generally dissolve in water.

____ 3. Hydrophilic substances are nonpolar.

____ 4. Water changes its temperature more slowly than air because of the great amount of heat required to break the hydrogen bonds between water molecules.

____ 5. The escape of water molecules from a fluid surface to the surrounding air is known as evaporation.

____ 6. Below 0°C, water molecules hydrogen-bond in a bonding pattern similar to that of water molecules at room temperature.

____ 7. Cohesion is the property of water that explains how insects walk on water and how long, narrow water columns rise to the tops of tall trees.

____ 8. "Spheres of hydration" form around ions and prevent them from interacting chemically.

____ 9. A solute is any fluid in which one or more substances can be dissolved.

____ 10. The formation of spheres of hydration around individual ions or molecules means that a substance is dissolved.

Fill-in-the-Blanks

As water absorbs heat and the molecules move more quickly, (11)_____ bonds between neighboring water molecules are more readily broken down and reformed. The (12)_____ of a substance is a measure of how fast its molecules are moving. When molecules at the surface of a liquid begin moving fast enough to escape into the atmosphere, the process is called (13)_____. The surface of the liquid left behind drops in temperature because much of its (14)_____ has been released by the escaping molecules. In pure water, each molecule is attracted to and hydrogen-bonded with other water molecules; such an attraction between like molecules is called (15)_____. Water is an outstanding (16)_____ because it dissolves many different types of (17)_____.

ACIDS, BASES, AND SALTS (2-IV, pp. 30–31)

True-False

If false, explain why.

____ 1. Some substances dissolve in water and release hydrogen ions, OH⁻.

____ 2. A pH scale is used to express the concentration of hydrogen ions in different solutions.

____ 3. A pH of 4 is ten times more acid than a pH of 3.

____ 4. Most living cells maintain a pH of about 4.

____ 5. Hydrochloric acid serves as a major buffer in cells.

Fill-in-the-Blanks

When acids dissolve in water, they release (6)_____ ions; when bases dissolve in water, they release (7)_____ ions. The (8)_____ scale is used to express the (9)_____ ion concentration of solutions. Most living cells maintain an H^+ concentration close to pH (10)_____. A pH of 3 has an H^+ concentration ten times higher than a pH of (11)_____. (12)_____ are molecules that combine with or release hydrogen ions to prevent rapid shifts in pH. A (13)_____ is formed when an acid reacts with a base.

14. Consult Figure 2.13 in the text and fill in the missing information in the following table.

Solution	pH Value	Acid/Base
a. Household ammonia	_____	_____
b. Blood	_____	_____
c. Saliva	_____	_____
d. Urine	_____	_____
e. Stomach acid	_____	_____
f. Hydrochloric acid	_____	_____

Matching

Choose the most appropriate answer for each.

15. ___ acid
16. ___ base
17. ___ ions with vital roles in cells
18. ___ hydrogen ion
19. ___ hydroxide ion
20. ___ neutral pH
21. ___ salt

A. OH^-
B. Ca^{++}, K^+
C. H^+
D. NaCl
E. A substance that releases OH^- in water
F. A substance that releases H^+ in water
G. 7

WATER AND BIOLOGICAL ORGANIZATION (2-V, p. 32)

1. Explain why proteins remain dispersed in watery cellular fluid.

_____

_____

True-False

If false, explain why.

___ 2. Cellular fluid is a formless substance with little structure.

___ 3. Protein surfaces can be positively or negatively charged overall.

___ 4. Water molecules are attracted to charged regions and polar groups on protein surfaces.

___ 5. Through interactions with water and ions, proteins randomly settle out in the cellular fluid.

___ 6. Proteins have specific molecular regions on which many important cellular chemical reactions occur.

Chapter Terms

The following page-referenced terms are important—those that were in boldface type in the chapter. Refer to the instructions given in Chapter 1, p. 4.

elements (19)
trace elements (19)
compounds (20)
molecule (20)
atom (20)
atomic number (20)

mass number (20)
isotopes (21)
radioisotopes (21)
chemical bond (23)
ion (25)
ionic bond (25)

covalent bond (25)
hydrogen bond (26)
hydrophilic (27)
hydrophobic (27)
evaporation (28)
hydrogen ions (30)

acid (30)
hydroxide ion (30)
base (30)
pH scale (30)
buffer (31)
salt (31)

Self-Quiz

_____ 1. A molecule is _____.
  a. a combination of two or more atoms
  b. less stable than its constituent atoms separated
  c. electrically charged
  d. a carrier of one or more extra neutrons

_____ 2. If lithium has an atomic number of 3 and an atomic mass of 7, it has _____ neutron(s) in its nucleus.
  a. one
  b. two
  c. three
  d. four
  e. seven

_____ 3. An ionic bond is one in which _____.
  a. electrons are shared equally
  b. electrically neutral atoms have a mutual attraction
  c. two charged atoms have a mutual attraction due to electron transfer
  d. electrons are shared unequally

_____ 4. A hydrogen bond is _____.
  a. a sharing of a pair of electrons between a hydrogen nucleus and an oxygen nucleus
  b. a sharing of a pair of electrons between a hydrogen nucleus and either an oxygen or a nitrogen nucleus
  c. an attractive force that involves a hydrogen atom and an oxygen or a nitrogen atom that are either in two different molecules or within the same molecule
  d. none of the above

_____ 5. A mixture of sugar and water is an example of a(n) _____.
  a. compound
  b. solution
  c. suspension
  d. colloid
  e. ion

_____ 6. Radioactive isotopes have _____.
  a. excess electrons
  b. excess protons
  c. excess neutrons
  d. insufficient neutrons
  e. insufficient protons

_____ 7. The shapes of large molecules are controlled by _____ bonds.
  a. hydrogen
  b. ionic
  c. covalent
  d. inert
  e. single

_____ 8. A pH solution of 10 is _____ times as basic as a pH of 7.
  a. 2
  b. 3
  c. 10
  d. 100
  e. 1,000

_____ 9. Substances that are nonpolar and are repelled by water are _____.
  a. hydrolyzed
  b. nonpolar
  c. hydrophilic
  d. hydrophobic

___ 10. Any molecule that combines with or
releases hydrogen ions, or both, and
helps stabilize pH is known as a(n)
_____.

a. neutral molecule
b. salt
c. base
d. acid
e. buffer

Chapter Objectives/Review Questions

This section lists general and detailed chapter objectives that can be used as review questions. You can make maximum use of these items by writing answers on a separate sheet of paper. Fill in answers where blanks are provided. To check for accuracy, compare your answers with information given in the chapter or glossary.

Page   Objectives/Questions

(19)   1. All forms of matter are composed of one or more _____.
(19)   2. _____ elements represent less than 0.01 percent of all the atoms in any organism.
(20)   3. Know the symbols for the elements listed in Table 2.1 in the text.
(20, 25)   4. Explain how protons, electrons, and neutrons are arranged into atoms and ions.
(20)   5. Define atomic number and atomic mass.
(21)   6. Atoms with the same atomic number but a different mass number are _____.
(21)   7. Define radioisotopes and list three ways they are useful.
(23)   8. The union between the electron structures of atoms is known as the chemical _____.
(23–24)   9. Describe the distribution of electrons in the space around the nucleus of an atom.
(23)   10. Electrons farthest away from the nucleus of an atom are said to be at [choose one] ( ) lower,
               ( ) higher energy levels.
(24)   11. An atom tends to react with other atoms when its outermost shell is only partly filled with
               _____.
(25)   12. Define *ion* and describe the conditions under which ionic bonds form between positive and
               negative ions.
(25)   13. In a _____ bond, atoms share electrons.
(25–26)   14. Distinguish between a nonpolar covalent bond and a polar covalent bond; give an example of
               each.
(26)   15. Define hydrogen bond; describe conditions under which hydrogen bonds form and cite one
               example.
(27)   16. Explain what is meant by the polarity of the water molecule.
(27)   17. Describe how the polarity of water molecules allows them to interact with one another.
(27)   18. Define *hydrophilic* and *hydrophobic*; relate these terms to different types of substances that
               contact water.
(28)   19. Explain why water cools and warms more slowly than air.
(28)   20. The escape of water molecules from fluid surfaces to the surrounding air is called _____.
(28)   21. Explain why ice floats on water in terms of the bonding between adjacent water molecules.
(28–29)   22. The surface tension of water and the movement of long water columns to the tops of trees are
               explained by a property of water known as _____.
(29)   23. Dissolved substances are called _____ ; a fluid in which one or more substances can
               dissolve is called a _____.
(29)   24. Substances are dissolved in water when spheres of _____ form around their individual
               ions or molecules.
(30)   25. Define *acid* and *base;* cite an example of each.
(30)   26. The concentration of free hydrogen ions in solutions is measured by the _____ scale.
(30)   27. The pH of hair remover is 13; it is a(n) [choose one] ( ) base, ( ) acid. The pH of vinegar is 3; it
               is a(n) [choose one] ( ) base, ( ) acid.
(30–31)   28. Describe the pH existing within the chemistry of living systems.
(31)   29. Define *buffer;* cite an example.
(31)   30. Explain how a salt is formed.
(32)   31. Organization of nearly all large biological molecules is influenced by their interactions with
               water; describe this interaction as it exists with protein molecules.

Integrating and Applying Key Concepts

1. Explain what would happen if water were a nonpolar molecule instead of a polar molecule. Would water be a good solvent for the same kinds of substances? Would the nonpolar molecule's specific heat likely be higher or lower than that of water? surface tension? cohesive nature? ability to form hydrogen bonds? Is it likely that the nonpolar molecules could form unbroken columns of liquid? What implications would that hold for trees?
2. If the ways that atoms bond affect molecular shapes, do the ways that molecules behave toward one another influence the shapes of cellular organelles?

Critical Thinking Exercises

1. The text says that radioisotopes can be used to identify the pathway or destination of a substance that has been introduced into an organism. Which of the following assumptions is most important for such an experiment?

   a. Each radioisotope decays spontaneously into a different isotope.
   b. Molecules that contain radioactive atoms are not changed into different compounds.
   c. Ionizing radiation damages cells and can kill them.
   d. Instruments can detect the presence and location of radioisotopes.
   e. Cells act upon molecules that contain radioactive atoms in exactly the same way they act upon molecules that do not contain radioactive atoms.

ANALYSIS

   a. This statement is true, but in order to interpret the results of a tracer experiment, you must assume the opposite. You must assume that most atoms of the radioisotope do not decay during the time of the experiment. If they did, you would end up studying the pathway or destination of the decay product, not the original element. Furthermore, if the product element was not radioactive, you would have no way to detect it. Some radioisotopes decay in seconds and are not suitable for tracer experiments. Fortunately, biologically important elements like carbon, hydrogen, and sulfur decay slowly.
   b. You may or may not have to assume this, depending on what kind of experiment you are doing. If you want to study the location or destination of a substance, you must assume this; otherwise, you will gather data about the movements of some other substance. On the other hand, if you want to study the series of compounds in a metabolic pathway, you must assume the opposite—that the tracer undergoes normal biochemical conversions.
   c. Radiation does damage cells, but again you must assume the opposite during the course of the experiment. You are interested in the normal course of events. You can't study normal events if your method disrupts them.
   d. This must be true in order to do the experiment; however, it is not an assumption. A variety of techniques are available that do detect radioactivity.
   e. In order to draw conclusions about the processing of normal, nonradioactive molecules, you must use data from molecules that undergo the same processes. If the cell handles the tracer by unusual pathways or reactions, your conclusions will be invalid.

2. Suppose you dissolve a little acid in water and determine that the pH of the solution is 5.2. Now you add 100 times as much acid to the solution. Which of the following is the best estimate of the new pH?

   a. 6.0 b. 5.2 c. 4.0 d. 3.0 e. 2.0

ANALYSIS

You are adding some acid to the unbuffered solution, so the pH will change: Eliminate choice (b).

The pH scale is inverted; the more acid, the lower the pH. You are adding acid, so the pH will drop: Eliminate choice (a).

The pH scale changes one unit every time the concentration of hydrogen ion changes tenfold. You are changing the hydrogen ion concentration about one-hundredfold, so the pH will drop about two units, to approximately 3.2. Choice (d) is closest.

3. The text says that desert oasis plants use evaporative water cooling. Suppose a student tested this hypothesis by weighing some plants before and after a very hot day; the weights did not change, and the student concluded that the hypothesis was not valid. Which of the following would be the best criticism of the student?

    a. The plants replace the evaporated water through their roots as quickly as it is lost, keeping their weight constant.
    b. Because of the high heat of evaporation of water, only a little water loss provides sufficient cooling.
    c. The plants don't cool by water evaporation; they live in the shade of other plants.
    d. The plants use evaporative water cooling.
    e. Organic acids and bases, produced by photosynthesis, react and replace the evaporated water, keeping the weights constant.

ANALYSIS

    a. If this is true, the observation is explained and the hypothesis could still be valid. Furthermore, this hypothesis is consistent with the idea that open desert plants do not use evaporative cooling. This is a good criticism.
    b. If this is true, the failure to measure a change in weight is explained, but only if we also assume that the weighing technique is not very sensitive. This is a criticism, but it is not as good as (a) because it involves an extra assumption and casts doubt on the data as a whole.
    c. This is not a criticism; it is an alternative to the hypothesis the student rejects. If you say this, you are accepting the student's conclusion.
    d. This is the original hypothesis. Repeating it does not affect the student's data or reasoning. Saying this would be arguing dogmatically, not scientifically.
    e. This is a criticism; it would explain the observations and leave the original hypothesis still possibly valid. However, this would require a great deal of metabolic effort from the plants, and many assumptions must be made in order to accept this mechanism. Thus, this is a weak criticism, even though we have no observations to contradict it.

Answers

Answers to Interactive Exercises

ORGANIZATION OF MATTER (2-I)
1. a. Matter is made up of elements; elements cannot be broken down to different substances by ordinary means.
    b. Compounds are substances composed of two or more elements combined in fixed proportions.
2. An atom is the smallest unit of matter unique to a particular element; a molecule is a unit of two or more atoms bonded together.
3. a. Protons are positively charged particles in the nucleus of the atom.

    b. Electrons are negatively charged particles that move outside the nucleus of the atom.
    c. Neutrons are electrically neutral particles in the nucleus of the atom.
4. carbon; mass; six; six; number; six; six.
5. oxygen, carbon, hydrogen, nitrogen.
6. a. H, 1, 1 (1) (0) (0); b. C, 6, 12 (2) (4) (0);
c. N, 7, 14 (2) (5) (0); d. O, 8, 16 (2) (6) (0);
e. Na, 11, 23 (2) (8) (1); f. Mg, 12, 24 (2) (8) (2);
g. P, 15, 31 (2) (8) (5); h. S, 16, 32 (2) (8) (6); i. Cl, 17, 35 (2) (8) (7).

7.

    H   C   N   O   Ne   P   S

8. He and Ne.

9. a. Isotopes are atoms of the same element that have the same atomic number but a different mass number; they vary slightly in the number of neutrons they have.

  b. Radioisotopes are radioactive isotopes that are unstable; they decay into more stable atoms.

10. Radioactive dating, tracking chemicals in living systems, saving lives by diagnosis and disease treatment.

11. F; 12. T; 13. F; 14. F; 15. F; 16. C; 17. B; 18. M; 19. J; 20. K; 21. F; 22. L; 23. D; 24. E; 25. A; 26. G; 27. H.

BONDS BETWEEN ATOMS (2-II)

1. A chemical bond is a union between the electron structures of atoms.

2. Electrons move in various orbitals outside the nucleus, and each orbital can have no more than two electrons; electron orbitals occupy spaces called shells.

3. Electrons in orbitals closest to the nucleus are at the lowest energy level; electrons farther from the nucleus are at higher energy levels.

4. The only shell of H is filled with two electrons; all other atoms require eight electrons to fill their outermost shell.

5. An atom tends to react with other atoms when its outermost shell is only partly filled with electrons.

6. An ion is an atom that becomes positively or negatively charged by loss or gain of electrons; an ionic bond is an association of oppositely charged particles.

7.

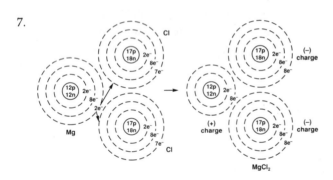

8. In a covalent bond, atoms share electrons to fill their outermost shells. In a nonpolar covalent bond, atoms attract shared electrons equally. In a polar covalent bond, atoms do not share electrons equally, and the bond is positive at one end, negative at the other (for example, the water molecule).

9.

10. In a hydrogen bond, an atom or molecule interacts weakly with a neighboring hydrogen atom that is already participating in a polar covalent bond; in the DNA molecule, the two nucleotide chains are held together by hydrogen bonds.

11. F; 12. F; 13. F; 14. T; 15. F; 16. T; 17. F; 18. T; 19. F; 20. F.

PROPERTIES OF WATER (2-III)

1. T; 2. F; 3. F; 4. T; 5. T; 6. F; 7. T; 8. T; 9. F; 10. T; 11. hydrogen; 12. temperature; 13. evaporation; 14. energy; 15. cohesion; 16. solvent; 17. solutes.

ACIDS, BASES, AND SALTS (2-IV)

1. F; 2. T; 3. F; 4. F; 5. F; 6. hydrogen; 7. hydroxide; 8. pH; 9. hydrogen; 10. 7 or neutral; 11. 4; 12. Buffers; 13. salt.

14. a. household ammonia: 10.5–11.9, base; b. blood: 7.3–7.5, slightly basic; c. saliva: 6.2–7.4, slightly acid; d. urine: 5–7, slightly acid or neutral; e. stomach acid: 1.0–3.0, acid; f. hydrochloric acid: 0, strongly acid.

15. F; 16. E; 17. B; 18. C; 19. A; 20. G; 21. D.

WATER AND BIOLOGICAL ORGANIZATION (2-V)

1. The charged surfaces of proteins interact with water and ions and remain dispersed as chemical reaction sites.

2. F; 3. T; 4. T; 5. F; 6. T.

Answers to Self-Quiz

1. a; 2. d; 3. c; 4. c; 5. b; 6. c; 7. a; 8. e; 9. d; 10. e.

3

CARBON COMPOUNDS IN CELLS

Interactive Exercises

PROPERTIES OF CARBON COMPOUNDS (3-I, pp. 35–37)

1. Describe the bonding properties that are responsible for carbon's central role in the molecules of life.

    _____

2. List the four main families of small molecules._____

    _____

3. List the main macromolecules of life's chemistry._____

    _____

4. Define *functional group* and cite an example; give the importance of functional groups to organic

    compounds._____

    _____

5. Define *enzyme;* state the general role of enzymes as they relate to organic compounds.

    _____

    _____

6. Define *condensation reaction* and *hydrolysis reaction*.

a. condensation reaction_____

_____

b. hydrolysis reaction_____

_____

7. The structural formulas of two adjacent amino acids are shown below. Show how enzyme action causes formation of a covalent bond and a water molecule by circling an H atom from one amino acid and an —OH group from the other amino acid. Circle the covalent bond that formed the dipeptide.

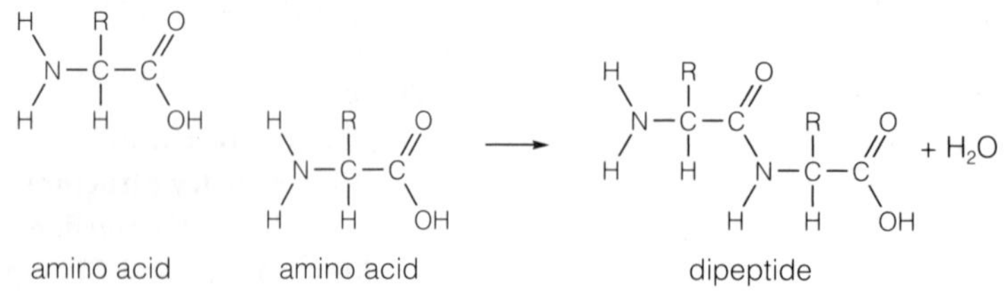

amino acid	amino acid		dipeptide

8. Describe hydrolysis through enzyme action for the molecules in exercise 7._____

_____

9. By reference to Figure 3.3 in the text, complete the missing information in the table below.

Name of Functional Group	Functional Group Formula	Compounds of Occurrence
a. Phosphate		phosphate compounds
b. _____	$-CH_3 \left(-\overset{H}{\underset{H}{C}}-H \right)$	fats, oils, waxes
c. Amino		amino acids, proteins
d. _____	$-OH\ (-OH)$	sugars
e. Carboxyl	$-COOH \left(\overset{O}{\underset{-C-OH}{\|}} \right)$	
f. Aldehyde		sugars
g. _____	$\left(\overset{O}{\underset{-C-C-C}{\|}} \right)$	sugars

Identifying

Study the following structural formulas of organic compounds; identify the circled functional groups (sometimes repeated) by entering the correct name in the blanks with matching numbers below the sketches.

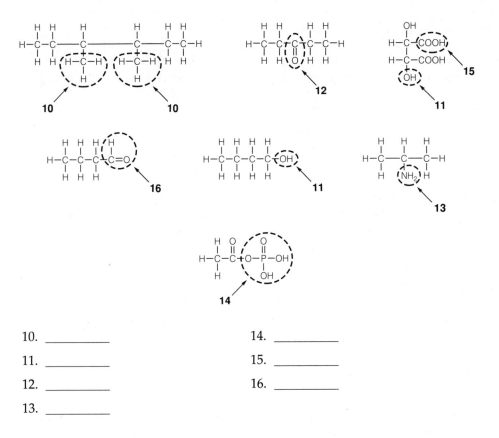

10. _____	14. _____
11. _____	15. _____
12. _____	16. _____
13. _____	

Fill-in-the-Blanks

A(n) (17)_____ group contains an oxygen atom and a hydrogen atom. A(n) (18)_____ group contains two oxygen atoms, a carbon atom, and a hydrogen atom. A(n) (19)_____ group can contain four oxygen atoms, two hydrogen atoms, and a phosphorus atom. A(n) (20)_____ group may contain one nitrogen atom and two hydrogen atoms. A(n) (21)_____ group contains one carbon atom and three hydrogen atoms.

True-False

If false, explain why.

___ 22. Molecules are assembled into polymers by means of hydrolysis.

___ 23. Carbon atoms are part of so many different substances because in one of their electron configurations there are four unpaired electrons in the outermost occupied energy level. Thus, each unpaired electron can form a covalent bond with a variety of other atoms, including other carbon atoms.

___ 24. All organic molecules contain carbon.

___ 25. When a ring like ⬡ is shown, it is understood that a carbon atom occurs at every corner and that no other atoms are attached. (Think!)

___ 26. In a branched chain, each carbon atom is never bonded to more than one other carbon atom. (Think!)

CARBOHYDRATES (3-II, pp. 38–40)

1. Define *carbohydrate*._____

   _____

2. List the general functions of carbohydrates._____

   _____

3. Distinguish between the following terms: *monosaccharide, oligosaccharide, disaccharide,* and *polysaccharide.*

   _____

   _____

   _____

   _____

4. In the diagram below, illustrate condensation reaction sites between the two glucose molecules by circling the components of the water removed that allow a covalent bond to form between the glucose molecules. Note that the reverse reaction is hydrolysis and that both condensation and hydrolysis reactions require enzymes in order to proceed.

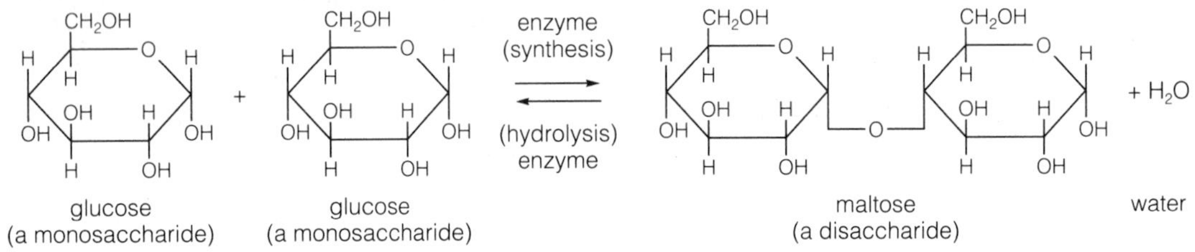

glucose (a monosaccharide) + glucose (a monosaccharide) enzyme (synthesis) ⇌ (hydrolysis) enzyme maltose (a disaccharide) + H_2O water

5. In the table below, enter the name of the carbohydrate class as monosaccharide, oligosaccharide (disaccharide), or polysaccharide; also cite the function of the listed carbohydrates.

Carbohydrate	Carbohydrate Class	Function
a. Sucrose	_____	_____
b. Ribose	_____	_____
c. Maltose	_____	_____
d. Cellulose	_____	_____
e. Deoxyribose	_____	_____
f. Lactose	_____	_____
g. Chitin	_____	_____
h. Glycogen	_____	_____

True-False

If false, explain why.

____ 6. Glucose, a crystalline sugar, has the molecular formula $C_6H_{12}O_6$. The mass numbers of hydrogen, carbon, and oxygen are 1, 12, and 16, respectively. The molecular weight of glucose is 29.

____ 7. Common polysaccharides such as starch, cellulose, glycogen, and chitin are all polymers of glucose.

____ 8. In living organisms, carbohydrates serve as structural supports and as food reserves.

___ 9. Polysaccharides can be broken down into simple sugars if the appropriate enzymes and water molecules are present.

___ 10. Condensation is the process that assembles simple sugars into starches.

LIPIDS (3-III, pp. 40–42)

1. Define *lipids* and list their general functions._____

   _____

2. Describe a "fatty acid."_____

   _____

3. Define *glyceride.*_____

   _____

4. Combine glycerol with three fatty acids below to form a triglyceride by circling the participating atoms that will form three covalent bonds; circle the covalent bonds in the triglyceride.

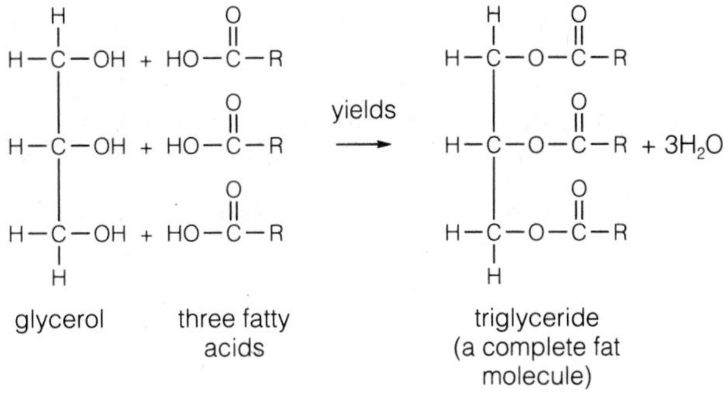

| glycerol | three fatty acids | | triglyceride (a complete fat molecule) |

5. Distinguish saturated fats from unsaturated fats._____

   _____

6. In the appropriate blanks, label the molecules shown below as saturated or unsaturated.

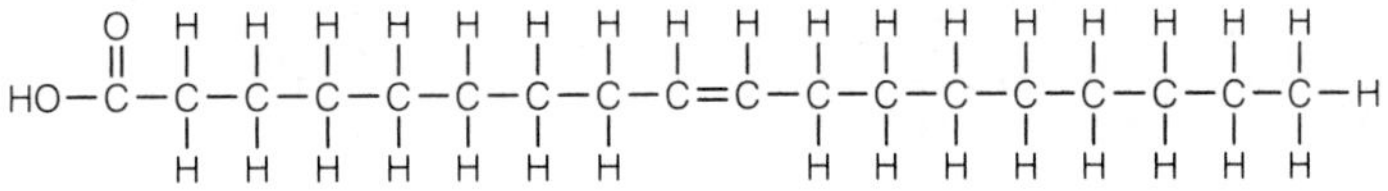

oleic acid

a

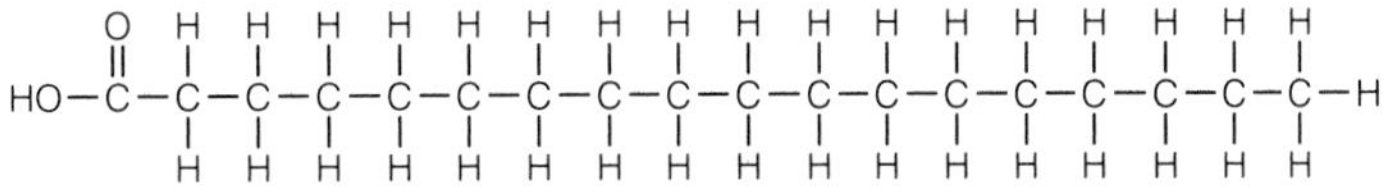

stearic acid

b

   a. _____   b. _____

7. Define *phospholipid;* describe the structure and biological functions of these molecules._____

   _____

   _____

   _____

8. Describe the chemical structure of waxes; list several important functions of waxes in plants and animals._____

_____

_____

9. Some lipids lack fatty acids; steroids are lipids without fatty acids. Describe the chemical structure of steroids and list the importance of the steroids known as cholesterol and hormones._____

_____

_____

_____

True-False

If false, explain why.

____ 10. Lipids contain oxygen, but carbohydrates lack oxygen.

____ 11. A true fat is broken down by hydrolysis into three fatty acid molecules and one molecule of glycerol.

____ 12. Saturated fats contain the maximum possible number of hydrogen atoms that can be covalently bonded to the carbon skeleton of their fatty acid tails.

____ 13. A polyunsaturated fat is usually an oil at room temperature and generally contains many double bonds in the fatty acid component.

____ 14. Waxes are long-chain alcohols combined with long-chain fatty acids. They help organisms prevent water loss as well as protect and lubricate.

____ 15. Some of the male and female sex hormones belong to the group of lipids known as *terpenes*.

Matching

Choose the appropriate answer for each.

16. ____ cholesterol
17. ____ cutin
18. ____ phospholipid
19. ____ saturated fat
20. ____ unsaturated fat

A. Basic fabric for all membranes
B. Butter and bacon
C. Vegetable oil
D. Wax
E. Steroid

PROTEINS (3-IV, pp. 43–46)

1. Define *proteins*._____

_____

2. Describe the basic structure of amino acids._____

_____

3. List the major functions of proteins._____

_____

_____

4. In the illustration of four amino acids in cellular solution (ionized state) below, circle the atoms and ions that form water to allow formation of covalent (peptide) bonds between adjacent amino acids to form a polypeptide. On the completed polypeptide below, circle the newly formed peptide bonds.

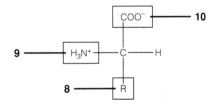

5. Describe the primary, secondary, tertiary, and quaternary structure that results in the three-dimensional shape of proteins._____

_____

_____

_____

_____

6. Distinguish lipoproteins from glycoproteins._____

_____

_____

7. Describe protein denaturation._____

_____

Matching

For exercises 8, 9, and 10, match the major parts of *every* amino acid by entering the letter of the part in the blank corresponding to the number on the molecule.

8. ___   A. R group (a symbol for a characteristic group of atoms that differ in number and arrangement from one amino acid to another)
9. ___   B. Carboxyl group
10. ___   C. Amino group

True-False

If false, explain why.

___ 11. Amino acids are linked by hydrolysis, a process that splits molecules of water as the amino acid subunits are linked together.

___ 12. Bone and cartilage are constructed, in part, of specific proteins.

___ 13. R groups projecting from the main carbon skeleton determine how a long-chain protein interacts chemically with other substances.

___ 14. The primary structure of a protein is formed principally by hydrogen bonds linking various amino acids.

___ 15. An amino group contains a nitrogen atom and two hydrogen atoms; a carboxyl group contains two oxygen atoms, a carbon atom, and a hydrogen atom.

NUCLEOTIDES AND NUCLEIC ACIDS (3-V, pp. 46–47)

1. Define *nucleotide*._____

   _____

   _____

2. Define and give examples of adenosine phosphates, nucleotide enzymes, and nucleic acids.

   a. adenosine phosphates_____

   _____

   b. nucleotide enzymes_____

   _____

   c. nucleic acids_____

   _____

Matching

For exercises 3, 4, and 5, match the following answers to the diagram below depicting the parts of a nucleotide.

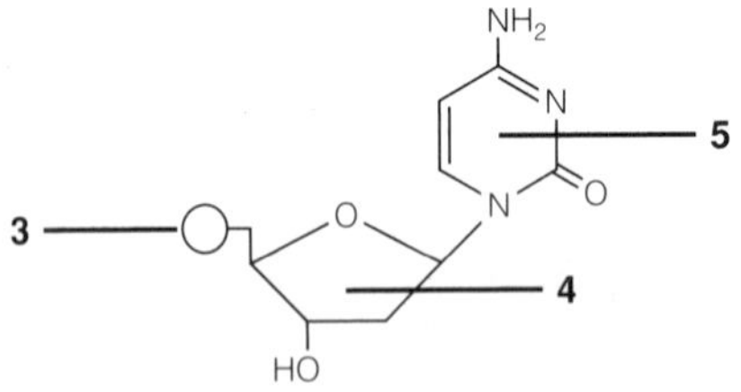

3. ___   A. A five-carbon sugar (ribose or deoxyribose)
4. ___   B. Phosphate group
5. ___   C. A nitrogen-containing base that has either a single-ring or double-ring structure

6. In the diagram of a single-stranded nucleic acid molecule below, completely circle each nucleotide component composed of a phosphate, a sugar, and a nitrogen-containing base.

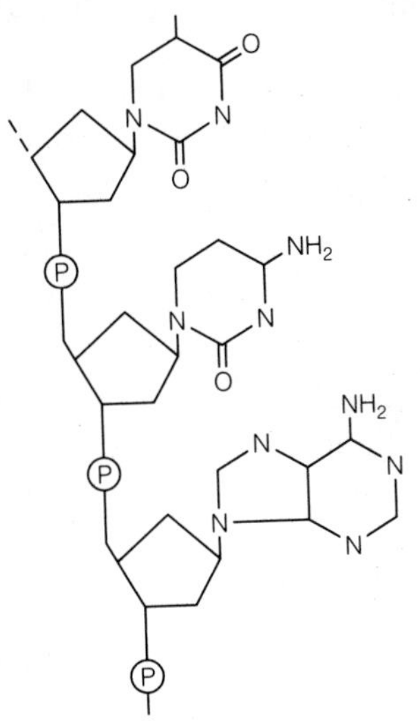

True-False

If false, explain why.

___ 7. ATP is a temporary energy-storage molecule.

___ 8. Nitrogen-containing bases are either single-ring or double-ring structures.

___ 9. Adenosine phosphates act as chemical messengers between cells and as energy carriers.

___ 10. Nucleic acids are long chains of nucleotides strung together, with nitrogenous bases connecting the phosphates and with sugars sticking out to the side.

___ 11. DNA is usually a single-stranded molecule that twists to form a helix.

___ 12. RNA molecules function in processes that utilize genetic instructions to construct proteins.

Matching

Choose *all* the appropriate answers for each.

13. ___ adenosine phosphates
14. ___ nucleotide coenzymes
15. ___ nucleic acid

A. Long; single-stranded or double-stranded
B. ATP
C. Transport hydrogen ions and their associated electrons
D. cAMP, a chemical messenger
E. NAD^+ and FAD
F. RNA and DNA

Chapter Terms

The following page-referenced terms are important—those that were in boldface type in the chapter. Refer to the instructions given in Chapter 1, p. 4.

functional groups (36)   oligosaccharide (38)   waxes (41)   lipoproteins (46)
condensation reaction (37) polysaccharide (38)   steroids (42)   glycoproteins (46)
hydrolysis (37)   lipids (40)   proteins (43)   denaturation (46)
carbohydrate (38)   glyceride (40)   amino acid (43)   nucleotide (46)
monosaccharide (38)   phospholipid (41)   polypeptide chain (43)

Self-Quiz

____ 1. Carbon is part of so many different
substances because _____.
a. carbon generally forms two bonds with
a variety of other atoms
b. carbon generally forms four bonds
with a variety of atoms
c. carbon ionizes easily
d. carbon is a polar compound

____ 2. _____ are compounds used by cells
as transportable packets of quick energy,
storage forms of energy, and structural
materials.
a. Lipids
b. Nucleic acids
c. Carbohydrates
d. Proteins

____ 3. Proteins_____.
a. include all hormones
b. are composed of nucleotide subunits
c. are not very diverse in structure and
function
d. include all enzymes

____ 4. Glucose dissolves in water because it
_____.
a. ionizes
b. is a polysaccharide
c. forms many hydrogen bonds with the
water molecules
d. has a very reactive primary structure

____ 5. Hydrolysis could be correctly described as
the _____.
a. heating of a compound in order to
drive off its excess water and
concentrate its volume
b. breaking of a long-chain compound into
its subunits by adding water molecules
to its structure between the subunits
c. linking of two or more molecules by
the removal of one or more water
molecules
d. constant removal of hydrogen atoms
from the surface of a carbohydrate

____ 6. DNA _____.
a. is one of the adenosine phosphates
b. is one of the nucleotide coenzymes
c. contains protein-building instructions
d. translates protein-building instructions
into actual protein structures

____ 7. Amino acids are linked by _____
bonds to form the primary structure of a
protein.
a. disulfide
b. hydrogen
c. ionic
d. peptide

____ 8. Lipids _____.
a. serve as food reserves in many
organisms
b. include cartilage and chitin
c. include fats that are broken down into
one fatty acid molecule and three
glycerol molecules
d. are composed of monosaccharides

____ 9. Most of the chemical reactions in cells
must have _____ present before they
proceed.
a. RNA
b. salt
c. enzymes
d. fats

____ 10. Genetic instructions are encoded in
the bases of _____; molecules
of _____ function in processes
using genetic instructions to construct
proteins.
a. DNA; DNA
b. DNA; RNA
c. RNA; DNA
d. RNA; RNA

Chapter Objectives/Review Questions

This section lists general and detailed chapter objectives that can be used as review questions. You can make
maximum use of these items by writing answers on a separate sheet of paper. Fill in answers where blanks
are provided. To check for accuracy, compare your answers with information given in the chapter or glossary.

(34)   1. _____ is the legacy of photosynthesis in ancient swamp forests.
(35)   2. Name the three most abundant elements in living things.
(35)   3. Each carbon atom can form as many as _____ covalent bonds with other carbon atoms as well as with other elements.
(36)   4. List the four main families of small organic molecules.
(36–37)   5. Define *functional group;* be able to recognize the major functional groups of organic compounds as shown in Figure 3.3 of the text.
(37)   6. _____ are a special class of proteins that speed up chemical reactions in cells.
(37)   7. _____ reactions result in the formation of covalent bonds between small molecules to form larger organic molecules.
(37)   8. Describe what occurs during hydrolysis reactions.
(38)   9. Define *carbohydrates* and list their functions.
(38)   10. The simplest carbohydrates are sugar monomers, the _____ ; be able to give examples and their functions.
(38)   11. An _____ is a short chain of two or more sugar monomers; be able to give examples of well-known disaccharides and their functions.
(38–39)   12. A _____ is a straight or branched chain of hundreds or thousands of sugar monomers, of the same or different kinds; be able to give common examples and their functions.
(40)   13. Define *lipids* and list their functions.
(40)   14. Describe a "fatty acid."
(40–41)   15. A _____ molecule has one to three fatty acid tails attached to a backbone of glycerol.
(41)   16. Distinguish a saturated fat from an unsaturated fat.
(41)   17. A _____ has two fatty acid tails attached to a glycerol backbone; cite the importance of these molecules.
(41)   18. _____ have long-chain fatty acids linked to long-chain alcohols or to carbon rings; list functions of these molecules.
(41–42)   19. Describe general functions of lipids without fatty acids.
(42)   20. Define *steroids* and describe their chemical structure; cite the importance of the steroids known as cholesterol, phytosterols, and hormones.
(43)   21. Describe proteins and cite their general functions.
(43)   22. Be able to sketch the three parts of every amino acid.
(43–44)   23. Describe the complex structure of a protein through its primary, secondary, tertiary, and quaternary structure; relate this to the three-dimensional structure of proteins.
(46)   24. Describe lipoproteins and glycoproteins.
(46)   25. _____ refers to the loss of a molecule's three-dimensional shape through disruption of the weak bonds responsible for it.
(46–47)   26. Describe the three parts of every nucleotide; give examples of adenosine phosphates, nucleotide enzymes, and nucleic acids.
(47)   27. Give the general functions of DNA and RNA molecules.

Integrating and Applying Key Concepts

1. Humans can obtain energy from many different food sources. Do you think this ability is an advantage or a disadvantage in terms of long-term survival? Why?
2. If the ways that atoms bond affect molecular shapes, do the ways that molecules behave toward one another influence the shapes of organelles? Do the ways that organelles behave toward one another influence the structure and function of the cells?
3. If proteins have the most diverse shapes and the most complex structure of all molecules, why do you suppose that proteins are not the code molecules used to construct new proteins?

Critical Thinking Exercises

1. The glucose content of intact muscle tissue is low, but when the cells are homogenized in a blender, glucose appears in high concentration in the homogenate. You formulate the hypothesis that breaking the muscle cells in the blender releases or activates enzymes that increase the rate of hydrolysis of glycogen. Which of the following would be the best test of this hypothesis?

   a. Measure the glucose content of muscle tissue before and after homogenization.
   b. Measure the protein content of muscle tissue before and after homogenization.
   c. Add glycogen to muscle homogenate and measure glucose concentration after a period of time.
   d. Add an enzyme that does increase the rate of glycogen hydrolysis and measure glucose concentration after a period of time.
   e. Homogenize muscle tissue for various periods of time and measure the glucose concentrations in the homogenates.

ANALYSIS

First think of alternative hypotheses: The forces involved in the homogenization might have broken the glycogen polymer into monomers. Some other molecule is converted to glucose as a result of homogenization. Glycogen is normally broken down by an enzyme in muscle tissue but is replaced at the same rate; homogenization stops the replacement mechanism. Now, for each proposed experiment, think of what outcome would be predicted by each hypothesis.

   a. This observation has already been made and has generated the hypothesis. It cannot be used to test the hypothesis.
   b. The hypothesis predicts the same protein content before and after homogenization, but so do alternative hypotheses. Thus, this experiment will not allow you to conclude that any hypothesis has greater validity than any other.
   c. If enzymes are responsible for the hydrolysis of glycogen in the homogenate, they would also break down added glycogen, and the glucose concentration would rise. On the other hand, if homogenization breaks down glycogen mechanically or if some other substance is the precursor, glycogen added after homogenization would not contribute to the glucose concentration. This experiment would allow elimination of two hypotheses, although it could not distinguish between the original and the third alternative hypothesis.
   d. If you can assume that there is some glycogen left after homogenization, added enzyme would cause it to break down to glucose. This would be predicted by all the hypotheses, and thus this experiment is not worthwhile.
   e. If homogenization mechanically releases glucose from glycogen, then longer periods of homogenization should release more glucose. But they might release more enzyme, convert more of something else to glucose, or more effectively inhibit glycogen condensation. If all the hypotheses predict the same outcome, there is no experiment.

2. A protein chemist studied a protein that had no regions of helically coiled chain. Because protein molecules made only of the amino acid glutamate (Figure 3.15) are totally helical, the chemist concluded that the protein contained no glutamate. Which of the following would be the best criticism of this conclusion?

   a. Other amino acids besides glutamate can form helically coiled chains.
   b. Helically coiled chains are stabilized by hydrogen bonds along the backbone of the string of amino acids.
   c. No natural protein contains only glutamate.
   d. Glutamate can participate in other types of secondary structure, depending on the kinds of amino acids next to it.
   e. Peptide bonds are always identical.

ANALYSIS

   a. This is true, but it does not criticize the conclusion. If the conclusion was that a protein that *did* contain helical coil must contain glutamate, this statement would be a criticism.
   b. This is also true, but it does not depend on the types of amino acids present. Any segment of helical coil, with or without glutamate, would have backbone hydrogen bonds.

c. This is true, too, but it is not a criticism of the conclusion. Polyglutamate can be made and does form only the helically coiled structure. The question is whether glutamate can be present in a mixed polymer when helical coil is not present.

d. This is true and is a criticism. The protein might contain glutamate in an amino acid sequence that forms other types of secondary structure.

e. This is true in some senses, but peptide bonds can twist in various ways and lead to various secondary structures, depending on which amino acids are participating. If all peptide bonds were completely identical, all proteins would have the same secondary structure.

3. The text says that one of the "lipid culprits" in the formation of atherosclerotic plaques is cholesterol. From this hypothesis, you might predict that atherosclerotic plaques would contain more and more cholesterol as they developed from fatty streaks to deeper fatty nodules, then became fibrous plaques, and finally calcified. Developing atherosclerotic lesions were studied this way, and the following data were gathered:

	Normal Artery Wall	Fatty Streak	Stage of Plaque Formation Fatty Nodule	Fibrous Plaque	Calcified Plaque
Amount of Lipid (mg/dl)	11	28	61	47	50
Percent Cholesterol	55%	72%	79%	72%	78%

Calculate the cholesterol concentration (in mg/dl) in each kind of specimen. Do these data support or contradict the text's hypothesis?

ANALYSIS

Specimen	Cholesterol Concentration
Normal wall	6.1 mg/dl
Fatty streak	20.2 mg/dl
Fatty nodule	48.2 mg/dl
Fibrous plaque	33.8 mg/dl
Calcified plaque	39.0 mg/dl

Clearly, all the stages of plaque formation do contain more cholesterol than normal artery wall, as predicted. This by itself, however, does not lead to the conclusion that cholesterol causes atherosclerosis. The high level of cholesterol might form as a result of some other process that causes the plaque. Similarly, the second stage of plaque formation contains a greater concentration of cholesterol than the first stage, as predicted. But again, this could be a secondary effect of plaque formation and is not conclusive support for the hypothesis. In the later stages, plaques actually contain less cholesterol, which is not predicted. On the other hand, the early accumulation of cholesterol might trigger plaque formation, and later other damage processes might begin that are independent of cholesterol. Formation of fibers and deposition of calcium in the plaques could even be dependent on prior deposition of cholesterol but result in the replacement of cholesterol by fiber and calcium salts. These observations are consistent with the hypothesis but do not give it strong support.

Reference: Smith *et al., J. Atherosclerosis Res.* 5:231 (1965) and 7:177 (1967).

Answers

Answers to Interactive Exercises

PROPERTIES OF CARBON COMPOUNDS (3-I)
1. One carbon atom is able to form as many as four covalent bonds with other carbon atoms as well as with atoms of other elements; carbon atoms may bond to form organic compounds in chains, branched chains, and rings.
2. simple sugars, fatty acids, amino acids, and nucleotides.
3. polysaccharides (one of three classes of carbohydrates), lipids, proteins, and nucleic acids.

4. A functional group is a group of atoms covalently bonded to the carbon backbone of an organic molecule. Examples of functional groups are methyl, hydroxyl, aldehyde, ketone, carboxyl, amino, and phosphate. Functional groups are important because they determine the structure and chemical behavior of organic compounds.

5. Enzymes belong to a special class of proteins that speed up reactions between specific substances; when enzymes work on target molecules involved in chemical reactions, they are recognizing specific functional groups and cause specific changes in their structure.

6. a. Condensation reactions are chemical reactions that form covalent bonds between small molecules to form larger molecules; this is accompanied by the formation of water molecules.

b. Hydrolysis reactions reverse the chemistry of condensation reactions; in the presence of water, larger molecules are split to their component smaller molecules. Both condensation and hydrolysis require the presence of enzymes specific to the particular molecules involved.

7.

amino acid   amino acid   dipeptide

8. A water molecule and the appropriate enzyme break the dipeptide into the two amino acids shown.

9.

Name of Functional Group	Functional Group Formula	Compounds of Occurrence
a. Phosphate	$-PO_4(p_i)$	phosphate compounds
b. Methyl	$-CH_3$	fats, oils, waxes
c. Amino	$-N; -NH; -NH2$	amino acids, proteins
d. Hydroxyl	$-OH (-OH)$	sugars
e. Carboxyl	$-COOH$	sugars, fats, amino acids
f. Aldehyde	$-CHO$	sugars
g. Ketone		sugars

10. methyl group; 11. hydroxyl group; 12. ketone group; 13. amino group; 14. phosphate group; 15. carboxyl group; 16. aldehyde group; 17. hydroxyl; 18. carboxyl; 19. phosphate; 20. amino; 21. methyl; 22. F; 23. T; 24. T; 25. F; 26. F.

CARBOHYDRATES (3-II)

1. A carbohydrate is a simple sugar or a polymer assembled from a number of sugar units; carbohydrates are probably the most abundant organic molecules in the world of life.

2. Carbohydrates serve cells as quick sources of energy, storage forms of energy, and structural materials.

3. Monosaccharides are the simplest type of carbohydrate, the sugar monomer; oligosaccharides are short chains of two or more sugar monomers; disaccharides are oligosaccharides consisting of two covalently bonded monomers; polysaccharides are straight or branched chains of hundreds or thousands of sugar monomers, of the same or different kinds.

4. See page 38 for the answer to this question.

5. a. sucrose: oligosaccharide (disaccharide), most plentiful sugar in nature, transport form of carbohydrates in leafy plants; b. ribose: monosaccharide, five-carbon sugar occurring in RNA; c. maltose: oligosaccharide (disaccharide), present in germinating seeds; d. cellulose: polysaccharide, structural material of plant cell walls; e. deoxyribose: monosaccharide; five-carbon sugar occurring in DNA; f. lactose: oligosaccharide (disaccharide), sugar present in milk; g. chitin: polysaccharide, main structural material in external skeletons and other hard body parts, also the main structural material in fungal cell walls; h. glycogen: branched polysaccharide, animal starch.

6. F; 7. T; 8. T; 9. T; 10. T.

LIPIDS (3-III)

1. Lipids are greasy or oily compounds that show little tendency to dissolve in water but can dissolve in nonpolar solvents like ether. Lipids can be hydrolyzed into subunits. Lipids function in energy storage, cell membrane structure, cell coatings, and other cell structures.

2. Fatty acids are long, water-insoluble chains of mostly carbon and hydrogen, with a —COOH (carboxyl) group on one end.

3. Glycerides have one to three (mono-, di-, and triglycerides) fatty acid tails attached to a backbone of glycerol. Fats and oils are glycerides.

4.

glycerol   three fatty acids   triglyceride (a complete fat molecule)

5. Saturated fats such as butter and lard tend to be solid at room temperature; all carbon atoms in the fatty acid tails have single C—C bonds and as much bonded hydrogen as possible. Unsaturated fats or oils tend to be liquid at room temperature; one or more double bonds occur between carbon atoms in the fatty acid tails, which means there is less bonded hydrogen.

6. a. unsaturated; b. saturated.

7. Phospholipids have two fatty acid tails attached to a glycerol backbone; they have hydrophilic heads that dissolve in water. Phospholipids are the main structural materials of cell membranes.

8. Waxes are lipids that have long-chain fatty acids linked to long-chain alcohols or to carbon rings; wax secretions form protective and water-repellant coatings on skin, hair, feathers, and leaves.

9. All steroids have the same backbone of four carbon rings, but they vary in the number, position, and type of attached functional groups. Cholesterol is an important component of animal cell membranes and is used in vitamin D synthesis. Hormones help regulate body growth, development, and reproduction.

10. F; 11. T; 12. T; 13. T; 14. T; 15. F; 16. E; 17. D; 18. A; 19. B; 20. C.

PROTEINS (3-IV)

1. Proteins are polymers of twenty or so different kinds of amino acids; they are the most diverse of all biological molecules.

2. Amino acids are small organic molecules with three parts: an amino group, a carboxyl group, and one or more atoms termed the R group. In cellular solutions, the amino and carboxyl groups ionize.

3. Enzymes are proteins that speed up the chemical reactions occurring in life's chemistry. Proteins are the molecules of cell movements and transport of cell substances. Many hormones and antibodies are proteins, as are the major structural materials of organisms.

4.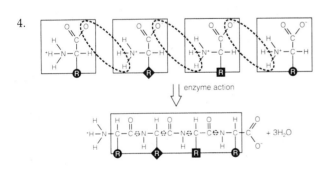

5. The specific sequence of amino acids in a polypeptide chain is the protein's primary structure. The secondary structure of proteins refers to the helical or extended pattern brought about by hydrogen bonds at regular intervals along a polypeptide chain. The tertiary structure of proteins refers to folding that arises through interactions among the R groups of a polypeptide chain. The quaternary structure of proteins means that the protein incorporates two or more polypeptide chains. Proteins with quaternary structure are hemoglobin, keratin, and collagen.

6. Lipoproteins have both lipid and protein components; they circulate in the blood, where they transport fats and cholesterol. Glycoproteins are proteins to which oligosaccharides are covalently bonded; they make up the outer surface of animal cells, most cellular protein secretion, and most of the blood proteins.

7. Protein denaturation refers to the loss of a molecule's three-dimensional shape through disruption of the weak bonds responsible for it; high temperatures and solutions of strong pH can denature proteins.

8. A; 9. C; 10. B; 11. F; 12. T; 13. T; 14. F; 15. T.

NUCLEOTIDES AND NUCLEIC ACIDS (3-V)

1. Nucleotides are essential to life; each nucleotide has three parts: a five-carbon sugar (ribose or deoxyribose), a nitrogen-containing base that has either single-ring or double-ring structure, and a phosphate group.

2. a. Adenosine phosphates are small molecules that serve as chemical messengers within and between cells and as energy carriers; cAMP is a chemical messenger, and ATP carries chemical energy between cellular reaction sites.

b. Nucleotide enzymes transport hydrogen atoms and electrons between cellular reaction sites; NAD$^+$ and FAD are two major coenzymes.

c. Nucleic acids are large single or double strands of nucleotides; DNA is a double-stranded molecule whose bases are encoded with genetic instructions; RNA is a single-stranded molecule that functions in the process of using genetic instructions to build proteins.

3. B; 4. A; 5. C.

6.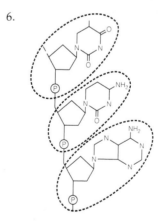

7. T; 8. T; 9. T; 10. F; 11. F; 12. T; 13. B, D; 14. C, E; 15. A, F.

Answers to Self-Quiz

1. b; 2. c; 3. d; 4. c; 5. b; 6. c; 7. d; 8. a; 9. c; 10. b.

CARBOHYDRATES (3-II) *continued*

4. Below is the answer to question 4 from page 36.

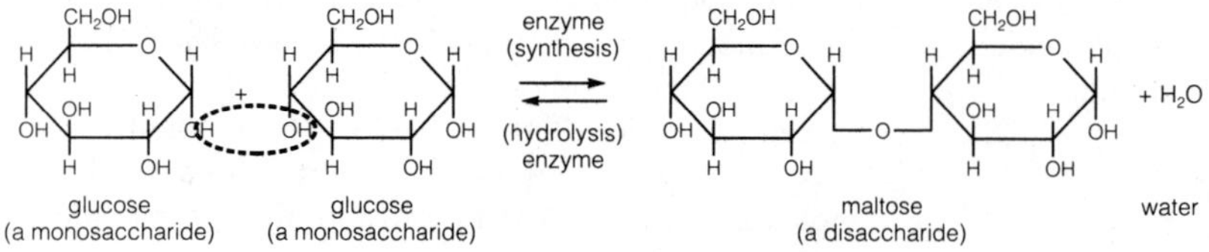

glucose
(a monosaccharide)

glucose
(a monosaccharide)

enzyme
(synthesis)

(hydrolysis)
enzyme

maltose
(a disaccharide)

water

4

CELL STRUCTURE
AND FUNCTION:
AN OVERVIEW

Interactive Exercises

GENERALIZED PICTURE OF THE CELL (4-I, pp. 51–56)

1. Summarize important contributions to the emergence of the cell theory by completing the following table.

Contributor	Years	Contribution
a. Galileo Galilei	(1564–1642)	_____
b. Robert Hooke	(1635–1703)	_____
c. Antony van Leeuwenhoek	(1632–1723)	_____
d. Robert Brown	(1773–1858)	_____
e. Matthias Schleiden	(1804–1881)	_____

f. Theodor Schwann   (1810–1882)   _____

g. Rudolf Virchow   (1821–1902)   _____

Label-Match

Although cells vary in many specific ways, they are all alike in a few basic respects. Identify each part of the illustration below. Complete the exercise by matching and entering the letter of the proper description in the parentheses following each label.

2. _____ ( )

3. _____ ( )

4. _____ ( )

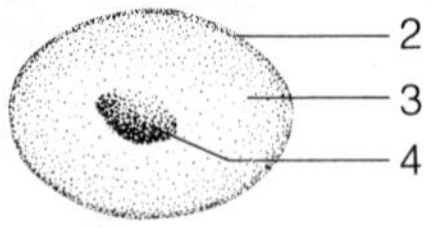

A. Everything in the cell that is enclosed by the plasma membrane, except the nucleus (in all but bacterial cells, specific compartments exist where different metabolic reactions occur)
B. A membrane-bound compartment that contains the hereditary instruction (DNA)
C. Outermost membrane of the cell; separates the internal events from the external environment

5. Bacterial cells are an exception to the concept illustrated above. Describe why this is so._____

_____

True-False

If false, explain why.

____ 6. The cell is the smallest independent living unit.

____ 7. The average cell is approximately as large as your thumbnail.

____ 8. A plasma membrane permits some substances to cross it and prevents other substances from crossing.

____ 9. From the standpoint of resources and wastes, the reason an elephant is not composed of one large cell is that the surface area of that cell would be far too small to service the volume within.

____ 10. All cells have a plasma membrane, an inner cytoplasm, and DNA molecules enclosed within a distinct structure, the nucleus.

Fill-in-the-Blanks

A(n) (11)_____ is a photograph of an image formed with a microscope. If you want to observe a living cell with a compound microscope, the cell must be small or thin enough for (12)_____ to pass through. (13)_____ is the property that dictates whether small objects close together can be seen as separate things. A(n) (14)_____ is one-billionth of a meter. With a (15)_____ electron microscope, a beam of electrons is transmitted through a prepared, thinly sliced section of a cell or organism. With a (16)_____ electron microscope, a narrow beam of electrons is played back and forth across a specimen's surface, which has been coated with a thin metal layer. The (17)_____ is a membrane-bound zone of hereditary control.

PROKARYOTIC CELLS—THE BACTERIA (4-II, p. 57)

Fill-in-the-Blanks

(1)_____ are the smallest and most structurally simple cells. Such prokaryotic cells may have a wall that surrounds the (2)_____ membrane; this membrane controls movement of substances into and out of the (3)_____. (4)_____ are composed of two molecular subunits that serve as workbenches for making proteins. The DNA in bacterial cells is located not in a nucleus but in an irregularly shaped region of cytoplasm called the (5)_____. The word *prokaryotic* means "before the (6)_____."

EUKARYOTIC CELLS (4-III, pp. 57–60)

Matching

Select the single best answer.

1. ___ nucleus
2. ___ endoplasmic reticulum
3. ___ Golgi bodies
4. ___ lysosomes
5. ___ transport vesicles
6. ___ mitochondria
7. ___ ribosomes
8. ___ cytoskeleton
9. ___ chloroplasts
10. ___ central vacuoles

A. ATP formation
B. Free in the cytoplasm or attached to certain membranes; protein synthesis
C. Internal cellular network of protein filaments
D. Modifies polypeptide chains into mature proteins; lipid synthesis
E. Further modifies, sorts, and ships proteins and lipids for secretion or use in the cell
F. Occupy most of the space in fungal and plant cells
G. Convey a variety of materials to and from organelles and the plasma membrane
H. Place where photosynthesis occurs
I. Physically isolates and organizes DNA
J. Cellular digestion

THE NUCLEUS (4-IV, pp. 61–62)

True-False

If false, explain why.

___ 1. The nuclear envelope has a one-membrane structure dotted with ribosomes on the cytoplasm side.

___ 2. Pores for exchange of substances between nucleus and cytoplasm occur at regular intervals over the entire nuclear envelope.

___ 3. Nucleoli are sites where the protein and RNA subunits of Golgi bodies are assembled.

___ 4. In eukaryotic cells, the DNA with its attached proteins is referred to as a chromosome.

___ 5. Chromosomes have an identical appearance throughout the life of a cell.

THE CYTOMEMBRANE SYSTEM (4-V, pp. 62–65)

Label-Match

Identify each indicated part of the accompanying illustration. Complete the exercise by matching and entering the letter of the proper function description in the parentheses following each label.

1. _____ _____ ( )
2. _____ _____ ( )
3. _____ _____ ( )
4. _____ _____ _____ ( )
5. _____ _____ ( )
6. _____ ( )
7. _____ ( )
8. _____ _____ _____ ( )
9. _____ ( )

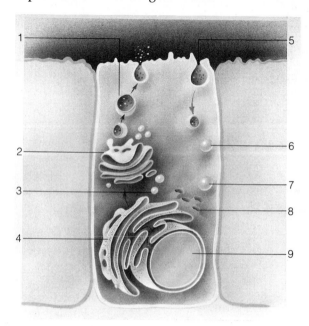

A. Import extracellular substances into cytoplasm. Buds form from plasma membrane and fuse with lysosomes.
B. Formed as buds from the ER; contain enzymes that degrade hydrogen peroxide, fats, oils.
C. Formed as buds from Golgi membranes; carry proteins and other substances to the plasma membrane and fuse with it, releasing contents to the outside.
D. Proteins are constructed according to DNA instructions located here; RNA molecules are assembled here in protein-coding regions of DNA, then move to ribosomes in cytoplasm where proteins entering the cytomembrane system are assembled.
E. Bud from the ER, deliver proteins and lipids to Golgi bodies.
F. Further modify and sort proteins and lipids arriving from ER; ship these chemicals to specific organelles or plasma membrane.
G. Formed as buds from Golgi membranes; contain enzymes for intracellular digestion.
H. New proteins initially modified in this are incorporated into membranes, enclosed in organelles, or secreted.
I. Lipids synthesized in this are incorporated into membranes, enclosed in organelles, or secreted.

MITOCHONDRIA (4-VI, p. 65)

True-False

If false, explain why.

___ 1. The mitochondrion is an organelle in eukaryotic cells that specializes in liberating energy stored in sugars and using it to form many DNA molecules.

___ 2. Mitochondria extract far more energy from sugars than can be extracted by any other means; they do this with the help of carbon dioxide.

___ 3. Muscle cells and other cells that demand high-energy output generally have many more mitochondria than less active cells have.

___ 4. Mitochondria may represent bacteria that have been simplified through evolution.

___ 5. Mitochondria have a double-membrane system that creates one compartment.

SPECIALIZED PLANT ORGANELLES (4-VII, pp. 65–66)

Matching

Choose the one best answer for each.

1. ___ chromoplast
2. ___ amyloplast
3. ___ central vacuole
4. ___ stroma
5. ___ chloroplast
6. ___ grana

A. Colorless; accumulates starch grains
B. Disk-shaped compartments organized in stacks within the chloroplast
C. Stores plant pigments
D. In most living plant cells, a large storage compartment (amino acids, sugars, ions, and toxic wastes) that is generally fluid-filled
E. Double-membraned organelle that contains pigments, stores starch, and functions in photosynthesis
F. Semifluid substance surrounding the grana within the chloroplast

THE CYTOSKELETON (4-VIII, pp. 67–70)

Fill-in-the-Blanks

Cellular shapes and internal cell organization are made possible by the (1)_____. This system of bundled fiber, slender threads, and lattices extends from the nucleus to the plasma membrane; its main components are (2)_____ , (3)_____ , and intermediate filaments. Some parts of the cytoskeleton, such as chromosome spindle fibers, are (4)_____ in that they appear and disappear at different times in a cell's life. Other elements of the cytoskeleton, such as filaments in skeletal muscle cells, flagella, and cilia, which remain are said to be (5)_____. Microtubules consist of globular protein subunits called (6)_____. The protein subunit of microfilaments is (7)_____. Flagella and cilia are both organelles that function in moving cells through their (8)_____. Cilia are shorter than flagella, but the internal microtubule organization of both is known as the (9)_____ array. Within cells, the microtubules are organized by masses of proteins and other substances in the cytoplasm known as (10)_____. In most animal cells, prominent MTOCs near the nucleus include a pair of (11)_____. In ciliated or flagellated cells that are forming, a centriole migrates from the area of the nucleus to the plasma membrane; there it acts as a pattern for assembly of one or more (12)_____ bodies. These bodies give rise to the (13)_____ forming the core structure of cilia and flagella. Although centrioles do not seem to be necessary for cell division, they do appear to govern the (14)_____ of cell division.

CELL SURFACE SPECIALIZATIONS (4-IX, pp. 70–71)

Matching

Choose the appropriate answer for each.

1. ___ gap junction
2. ___ cell walls
3. ___ plasmodesmata
4. ___ extracellular matrix
5. ___ tight junction
6. ___ adhering junction
7. ___ middle lamella

A. Layer of pectins cementing adjacent primary cell walls in plants
B. General meshwork of macromolecules holding the cells of many tissues together
C. Like spot welds at the plasma membranes of two adjacent epithelial cells in animals
D. Small, open channels that directly link the cytoplasm of adjacent animal cells
E. Cellulose; primary and sometimes secondary layers
F. Cytoplasmic channels extending across adjacent plant cell walls that transport substances
G. Rows of cytoskeletal strands at the plasma membranes of animal cells matching up and forming seals between adjacent epithelial cells

---

Chapter Terms

The following page-referenced terms are important—those that were in boldface type in the chapter. Refer to the instructions given in Chapter 1, p. 4.

cell theory (52)
plasma membrane (53)
nucleus, -i (53)
cytoplasm (53)
surface-to-volume ratio (56)
ribosome (57)
organelles (57)

nuclear envelope (61)
nucleolus, -oli (62)
chromosomes (62)
cytomembrane system (62)
endoplasmic reticulum, ER (63)
Golgi bodies (64)
lysosomes (64)

peroxisomes (65)
glyoxysome (65)
mitochondrion, -ria (65)
chloroplast (66)
central vacuole (66)
cytoskeleton (67)
flagellum, -a (69)

cilium, -a (69)
microtubule organizing center, MTOC (69)
centrioles (69)
basal body (70)
cell walls (70)

Self-Quiz

Label-Match

Identify each indicated part of the accompanying illustrations. Complete the exercise by matching and entering the letter of the proper function description in the parentheses following each label. Some letter choices must be used more than once.

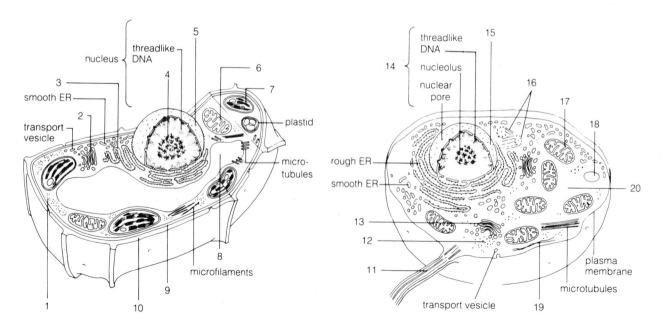

1. _____ ( )
2. _____ _____ ( )
3. _____ _____ _____ ( )
4. _____ ( )
5. _____ _____ ( )
6. _____ ( )
7. _____ ( )
8. _____ _____ ( )
9. _____ _____ ( )
10. _____ _____ ( )
11. _____ ( )
12. _____ ( )
13. _____ _____ ( )
14. _____ ( )
15. _____ _____ ( )
16. _____ ( )
17. _____ ( )
18. _____ ( )
19. _____ ( )
20. _____ ( )

A. Two-membrane structure; outermost part of the nucleus
B. Protection and structural support
C. Increasing cell surface area and storage
D. Everything enclosed by the plasma membrane, except the nucleus
E. Formed as buds from Golgi membranes; contain enzymes for intracellular digestion
F. Small cylinders composed of triplet microtubules; act as a pattern for assembling basal bodies
G. Microtubular structures for propelling eukaryotic cells; longer than cilia but with similar microtubular structure
H. Site of protein synthesis
I. A membrane-bound compartment that houses DNA in eukaryotic cells
J. Further modification, sorting, and shipping of proteins and lipids for secretion or for use in the cell
K. Sites where the protein and RNA subunits of ribosomes are assembled
L. A major component of the cytoskeleton
M. Photosynthesis and some starch storage
N. Control of material exchanges; mediates cell-environment interactions
O. Site of aerobic respiration
P. Isolation, modification, and transport of proteins and other substances

Multiple Choice

___ 21. _____ distilled the meaning of the new microscopic observations by himself, _____ , and earlier investigators into the first two generalizations of the cell theory.
   a. van Leeuwenhoek; Hooke
   b. Brown; Schleiden
   c. Schwann; Schleiden
   d. Schwann; Brown

___ 22. Which of the following is *not* found as a part of prokaryotic cells?
   a. ribosomes
   b. DNA
   c. nucleus
   d. cytoplasm
   e. cell wall

___ 23. Which of the following statements most correctly describes the relationship between cell surface area and cell volume?

   a. As a cell expands in volume, its diameter increases at a rate faster than its surface area does.
   b. Volume increases with the square of the diameter, but surface area increases with the cube.
   c. If a cell were to grow four times in diameter, its volume of cytoplasm would increase sixteen times and its surface area would increase sixty-four times.
   d. Volume increases with the cube of the diameter, but surface area increases with the square.

___ 24. Animal cells dismantle and dispose of waste materials by _____.
   a. using centrally located vacuoles
   b. several lysosomes fusing with a vesicle that encloses the wastes
   c. microvilli packaging and exporting the wastes
   d. mitochondrial breakdown of the wastes

25. The nucleolus is a dense region of _____ where the subunits that later will be constructed into _____ are made.
   a. nucleoplasm; ribosomes
   b. cytoplasm; vesicles
   c. cytoplasm; chromosomes
   d. nucleoplasm; chromatin

26. The _____ is free of ribosomes, curves through the cytoplasm, and is the main site of lipid synthesis.
   a. lysosome
   b. Golgi body
   c. smooth ER
   d. rough ER

27. Which of the following is *not* present in all cells?
   a. cell wall
   b. plasma membrane
   c. ribosomes
   d. DNA molecules

28. A nanometer is _____ of a meter.
   a. one-ninth
   b. one-tenth
   c. one-hundredth
   d. one-billionth

29. Mitochondria convert energy stored in _____ to forms that the cell can use, principally ATP.
   a. water
   b. carbon compounds
   c. $NADPH_2$
   d. carbon dioxide

30. _____ are vesicles that bud from ER; they contain enzymes that use oxygen to break down fatty acids and amino acids.
   a. Lysosomes
   b. Glyoxysomes
   c. Golgi bodies
   d. Peroxisomes

Fill-in-the-Blanks

If the cell structure is present in all or most members of a given group, put a check (√) on its line; if the cell structure is present in some of the group, put a cross (+) on its line. If it is not present, leave the line blank.

	Monera	Protista	Fungi	Plantae	Animalia
Cell wall	31. ___	32. ___	33. ___	34. ___	35. ___
Plasma membrane	36. ___	37. ___	38. ___	39. ___	40. ___
Photosynthetic pigments	41. ___	42. ___	43. ___	44. ___	45. ___
Chloroplasts	46. ___	47. ___	48. ___	49. ___	50. ___
Mitochondria	51. ___	52. ___	53. ___	54. ___	55. ___
Ribosomes	56. ___	57. ___	58. ___	59. ___	60. ___
Endoplasmic reticulum	61. ___	62. ___	63. ___	64. ___	65. ___
Cytoskeleton	66. ___	67. ___	68. ___	69. ___	70. ___
Complex cilia or flagella	71. ___	72. ___	73. ___	74. ___	75. ___
DNA molecules	76. ___	77. ___	78. ___	79. ___	80. ___
Chromosomes condense	81. ___	82. ___	83. ___	84. ___	85. ___

Chapter Objectives/Review Questions

This section lists general and detailed chapter objectives that can be used as review questions. You can make maximum use of these items by writing answers on a separate sheet of paper. Fill in answers where blanks are provided. To check for accuracy, compare your answers with information given in the chapter or glossary.

Page		Objectives/Questions
(51–52)	1.	Be able to cite the contributions of the following investigators to the cell theory: Galileo Galilei, Robert Hooke, Antony van Leeuwenhoek, Robert Brown, Theodor Schwann, Matthias Schleiden, and Rudolf Virchow.
(53)	2.	Describe these basic cellular features and their functions: plasma membrane, cytoplasm, and nucleus (nucleoid).
(53)	3.	A _____ is one-millionth of a meter long.
(56)	4.	Be able to explain the reason there is a limit to the size that a cell can attain by growth.
(53)	5.	A _____ is one-billionth of a meter long.
(54–55)	6.	Briefly describe the operation of the following: light microscopes, phase-contrast microscopes, transmission electron microscopes, and scanning electron microscopes.
(57)	7.	Describe the basic structure of prokaryotic cells; cite an example of these cells.
(58)	8.	Give the function and cellular location of the following basic eukaryotic organelles and structures: nucleus, endoplasmic reticulum, Golgi bodies, lysosomes, transport vesicles, mitochondria, ribosomes, cytoskeleton, chloroplasts, central vacuoles, and cell walls.
(61)	9.	Describe the nature of the nuclear envelope and relate its function to its structure.
(62)	10.	_____ are sites where the protein and RNA subunits of ribosomes are assembled.
(62)	11.	Eukaryotic _____ is threadlike, with a great number of proteins attached to it like beads on a string; this material is really the _____ and may look different as the life of the cell progresses.
(62–65)	12.	Explain how the endoplasmic reticulum, Golgi bodies, lysosomes, and a variety of vesicles function together as the cytomembrane system.
(65)	13.	Relate the function of the mitochondrion to its structure.
(65)	14.	Give the function of the following plant organelles: chloroplasts, chromoplasts, and amyloplasts.
(66)	15.	Describe the details of the structure of the chloroplast, the site of photosynthesis.
(66)	16.	Mature, living plant cells often have a large, fluid-filled _____ that can store amino acids, sugars, ions, and toxic wastes.
(69)	17.	Microtubules, microfilaments, and intermediate filaments are all main components of the _____.
(69)	18.	_____ and _____ propel eukaryotic cells through their environment; the microtubule organization in these organelles is a _____ array.
(69–70)	19.	_____ are cytoplasmic masses that organize microtubules and sometimes include _____ in animal cells.
(70–71)	20.	Be able to completely define the following concepts: cell walls, extracellular matrix, middle lamella, tight junctions, adhering junctions, gap junctions, and plasmodesmata.

Integrating and Applying Key Concepts

1. Which parts of a cell constitute the minimum necessary for keeping the simplest of living cells alive?
2. How did the existence of a nucleus, compartments, and extensive internal membranes confer selective advantages on cells that developed these features?

Critical Thinking Exercises

1. A section of animal tissue was treated with a chemical that stained nucleic acid. Upon examination in the microscope, the nuclei were seen to be heavily stained. Which of the following would be the best conclusion you could draw from this observation?

   a. Nucleic acid is a polymer.
   b. Polymers are found in all parts of the cell.
   c. The nuclei contain nucleic acid.
   d. These cells contain more nucleic acid than most other cells.
   e. Nucleic acid is composed of nucleotide monomers.

ANALYSIS

a. This is a true statement, but it is not supported by the observation. In order to conclude that nucleic acid is a polymer, you would have to demonstrate that it could be hydrolyzed into a number of small, similar subunit molecules.

b. Again, this is true but cannot be concluded from the observation in this experiment. Had all parts of the cell been stained, this conclusion would have been valid.

c. This statement is supported by the observation. The chemical stains DNA. The chemical stains the nuclei. Therefore, the nuclei contain DNA. DNA is a polymer. Therefore, the nuclei contain at least one polymer. Notice that in order to make this conclusion, you must assume that the chemical does *not* stain any other, nonpolymeric, molecules.

d. This statement might be true, but it is a comparison to other cells that were not observed. Other cells might stain just as heavily as these.

e. Once again, this is a true statement, but it depends on other observations. Unless you know more about the stain and its chemistry, you cannot infer anything about the structure of the substances with which it reacts.

2. The text says that nuclei have "two or more" nucleoli, but all the figures in the text show nuclei with only one nucleolus. Which of the following statements best reconciles this apparent contradiction?

a. The cells in the figures do have more than one nucleolus, but only one appeared in the section that was photographed. The others were cut off.

b. The extra nucleoli failed to react with the stain that formed the image.

c. Some cells have only one nucleolus.

d. The extra nucleoli fell out of the section as it was handled during preparation.

e. Cells have more than one nucleolus, but they develop nucleoli one at a time and go through a phase in which they have only one nucleolus.

ANALYSIS

a. This is likely to have happened. In the microscope, especially an electron microscope, we are always looking at a very small part of the specimen—a small area of a thin slice. *Most* of the specimen has been cut off, and most of the rest is outside the field of view.

b. Stains are often very specific and react only with certain parts of the cell. However, we have no evidence that nucleoli vary in their chemical composition. If they are chemically alike, they should all stain alike.

c. This might be true, but it does not resolve the contradiction. It *is* the contradiction.

d. Sometimes structures do fall out of sections, are dissolved by the chemicals used to prepare the specimens, or are otherwise destroyed. However, if something falls out, a visible hole is left in the section. These figures do not show any holes.

e. This might be true, but it is just a more elaborate version of choice (c) and does not resolve the contradiction.

3. Imagine five cells, each of a different shape and all requiring oxygen. Which of the following would have the most satisfactory oxygen supply?

a. A cube 1 mm on a side.
b. A cylinder 1 mm long and 0.1 mm in diameter.
c. A cylinder 10 mm long and 0.1 mm in diameter.
d. A flat disk 0.1 mm thick and 1 mm in diameter.
e. A sphere 1 mm in diameter.

ANALYSIS

First you must assume that all the cells consume oxygen at the same rate for each unit of cell volume. Then you must calculate the surface-to-volume ratio for each cell type. Whichever has the greatest ratio has the most satisfactory oxygen supply.

From geometry:

	Surface Area	Volume
Cube	$6x^2$	x^3
Cylinder	$2\pi r(r+h)$	$\pi r^2 h$
Sphere	πd^2	$(\pi d^3)/6$

Therefore:

	Surface Area (mm^2)	Volume (mm^3)	Ratio
a.	6	1	6
b.	0.105π	0.0025π	42
c.	1.005π	0.025π	40
d.	0.6π	0.025π	24
e.	3.1416	0.524	6

You might argue that cells (b), (c), and (d) all have equivalent oxygen supplies because the maximum diffusion distance to any part of the cell is 0.5 mm in all three cases.

Now evaluate a spherical cell with a diameter of 1.24 mm and compare its values to those of the 1-mm cube.

Answers

Answers to Interactive Exercises

GENERALIZED PICTURE OF THE CELL (4-I)
1. a. Galileo Galilei: the first to record biological observations through a microscope (insect eye);
b. Robert Hooke: introduced the term "cell"; c. Antony van Leeuwenhoek: constructed lenses; the first to observe a bacterium and many other microscopic organisms; d. Robert Brown: first observed and named the nucleus of a cell; e. Matthias Schleiden: reported that animal tissues are composed of cells; distilled the meaning of the first two parts of the cell theory; f. Theodor Schwann: concluded that all plant tissues are composed of cells and that the nucleus is very important; g. Rudolf Virchow: concluded that all cells arise from preexisting cells.
2. plasma membrane (C); 3. cytoplasm (A); 4. nucleus (B).
5. Bacterial cells lack an organized nucleus; bacterial DNA is in an organized part of the cytoplasm called the nucleoid.
6. T; 7. F; 8. T; 9. T; 10. F; 11. micrograph; 12. light; 13. Resolution; 14. nanometer; 15. transmission; 16. scanning; 17. nucleus.

PROKARYOTIC CELLS—THE BACTERIA (4-II)
1. Bacteria; 2. plasma; 3. cytoplasm; 4. Ribosomes; 5. nucleoid; 6. nucleus.

EUKARYOTIC CELLS (4-III)
1. I; 2. D; 3. E; 4. J; 5. G; 6. A; 7. B; 8. C; 9. H; 10. F.

THE NUCLEUS (4-IV)
1. F; 2. T; 3. F; 4. T; 5. F.

THE CYTOMEMBRANE SYSTEM (4-V)
1. exocytic vesicles (C); 2. Golgi bodies (F); 3. transport vesicles (E); 4. rough endoplasmic reticulum (H); 5. endocytic vesicles (A); 6. lysosomes (G); 7. peroxisomes (B); 8. smooth endoplasmic reticulum (I); 9. nucleus (D).

MITOCHONDRIA (4-VI)
1. F; 2. F; 3. T; 4. T; 5. F.

SPECIALIZED PLANT ORGANELLES (4-VII)
1. C; 2. A; 3. D; 4. F; 5. E; 6. B.

THE CYTOSKELETON (4-VIII)
1. cytoskeleton; 2. microtubules (microfilaments); 3. microfilaments (microtubules); 4. transient; 5. permanent; 6. tubulin; 7. actin; 8. environment; 9. 9 + 2; 10. MTOCs; 11. centrioles; 12. basal; 13. microtubules; 14. plane.

CELL SURFACE SPECIALIZATIONS (4-IX)
1. D; 2. E; 3. F; 4. B; 5. G; 6. C; 7. A.

Answers to Self-Quiz

1. ribosomes (H); 2. Golgi complex (body) (J); 3. rough endoplasmic reticulum (P); 4. nucleolus (K); 5. nuclear envelope (A); 6. mitochondrion (O); 7. chloroplast (M); 8. central vacuole (C); 9. cell (plasma) membrane (N); 10. cell wall (B); 11. flagellum (G); 12. ribosomes (H); 13. Golgi complex (body) (J); 14. nucleus (I); 15. nuclear envelope (A); 16. centrioles (F); 17. mitochondrion (O); 18. lysosome (or vacuole) (E); 19. microfilaments (L); 20. cytoplasm (D); 21. c; 22. c; 23. d; 24. b; 25. a; 26. c; 27. a; 28. d; 29. b; 30. d; 31. +; 32. +; 33. +; 34. √; 35. ; 36. √; 37. √; 38. √; 39. √; 40. √; 41. +; 42. +; 43. ; 44. √; 45. ; 46. ; 47. +; 48. ; 49. √; 50. ; 51. ; 52. √; 53. √; 54. √; 55. √; 56. √; 57. √; 58. √; 59. √; 60. √; 61. ; 62. √; 63. √; 64. √; 65. √; 66. ; 67. √; 68. √; 69. √; 70. √; 71. ; 72. +; 73. +; 74. +; 75. √; 76. √; 77. √; 78. √; 79. √; 80. √; 81. ; 82. √; 83. √; 84. √; 85. √.

5

MEMBRANE STRUCTURE
AND FUNCTION

Interactive Exercises

FLUID MEMBRANES IN A LARGELY FLUID WORLD (5-I, pp. 75–80)

1. Define *lipid bilayer;* explain the natural arrangement of phospholipid molecules in a watery environment that gives basic structure to cell membranes and lends relative impermeability to water-soluble molecules._____

_____

_____

_____

2. Three types of lipids in cell membranes are phospholipids, glycolipids, and sterols. What general structural feature do they share that is important to the structure of cell membranes?_____

_____

_____

3. The steroid _____ is abundant in animal cell membranes but is not found in plant cell membranes; plant cell membranes have _____ instead.

4. Generally describe the fluid mosaic model of membrane structure._____

_____

_____

_____

Label-Match

Identify each indicated part of the accompanying illustration by supplying proper labels in the blanks provided. Choose from these answers: receptor protein, carrier protein, transport system, channel protein, and recognition protein. Complete the exercise by matching the labels with their functions from the list below; enter the letter of the function in the parentheses following each label.

5. _____ _____ ( )

6. _____ _____ ( )

7. _____ _____ ( )

8. _____ _____ ( )

9. _____ _____ ( )

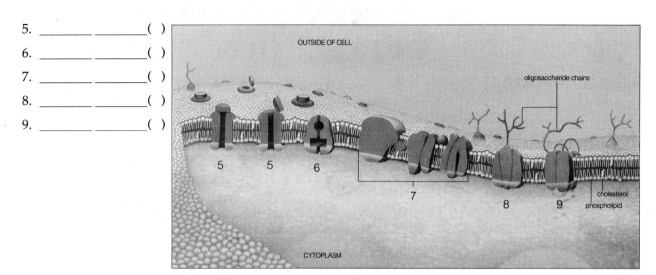

A. Serves as a molecular fingerprint at the cell surface; identifies the cell as being of a certain type. Some have oligosaccharide chains projecting from the cell surface.
B. Acts like a switch that turns cellular behavior or metabolism on or off when particular substances such as hormones bind to it.
C. Provides a pore through which ions or other water-soluble substances move from one side of the membrane to the other. Some pathways are perpetually open, but others have gates that open or close to permit passage of specific ions.
D. Binds specific substances and changes its shape in ways that shunt substances across the membrane. Some do this passively, while others use energy to pump substances in a specific direction.
E. A series of delivery proteins that act as a unit.

Fill-in-the-Blanks

(10)_____-_____ and (11)_____-_____ are two relatively new methods of preparing membranes for study with the electron microscope. A (12)_____ _____ serves as a sort of fluid sea matrix in which diverse (13)_____ are suspended like icebergs. Together, the two components form the (14) "_____." (15)_____ have fatty acid tails (which repel water) and heads with (16)_____ and alcohol groups (which dissolve in water). Membrane surface receptors are often (17)_____ groups attached to regions of membrane (18)_____ and membrane lipids. When activated, some receptors bring about changes in cell (19)_____ or behavior. In multicelled organisms, surface receptors also function in identifying cells of like type during the formation of (20)_____. The overall structure of the lipid bilayer of cell membranes allows it to function as a (21)_____ between the fluids inside and outside the cell.

True-False

If false, explain why.

___ 22. In a plasma membrane, the hydrophilic tails point inward, tail to tail, and form a region that excludes water.

___ 23. The fatty acid tails of phospholipids are hydrophobic.

___ 24. Most membrane proteins are lipoproteins with short-chain oligosaccharide sugars extending into the extracellular fluid.

___ 25. Molecular movements and packing variations of phospholipid molecules in membranes account for the membrane being more fluid than solid.

___ 26. Solid packing of phospholipid molecules is disrupted by their movements, the presence of short tails, and unsaturated tails that tend to kink at double-bond sites.

DIFFUSION (5-II, p. 81)

1. Describe what is meant by concentration gradient; relate this to the existence of gradients in molecule or ion concentration, pressure, temperature, or electric charge._____

   _____

   _____

2. Define *simple diffusion;* give an example._____

   _____

3. Define *bulk flow* and give two examples._____

   _____

Fill-in-the-Blanks

(4)_____ refers to the number of molecules (or ions) of a substance in a given volume of space. A (5)_____ can exist between two regions that differ in (6)_____ , pressure, temperature, or net electric charge. Diffusion is driven by the (7)_____ _____ inherent in all individual molecules as they move from a region of (8)_____ concentration to a region of (9)_____ concentration. In (10)_____ _____ , different ions and molecules present in a fluid move together in the same direction, often in response to a pressure gradient.

True-False

If false, explain why.

___ 11. Diffusion accounts for the greatest volume of substances moved into and out of cells.

___ 12. Bulk flow requires an expenditure of ATP to move large amounts of materials across membranes and may occur in response to a pressure gradient.

OSMOSIS (5-III, pp. 82–83)

1. Define *osmosis;* describe one example._____

   _____

   _____

   _____

2. Define *tonicity;* describe the conditions of relative solute concentrations with solutions that are isotonic, hypotonic, and hypertonic._____

_____

_____

_____

_____

3. Describe what causes turgor pressure._____

_____

4. List the two forces that, when summed, yield what is called water potential._____

_____

Fill-in-the-Blanks

The plasma membrane is (5)_____ _____ ; some molecules travel rapidly across the membrane, others cross it more slowly, and some are kept from crossing it at all. (6)_____ is one of the few molecules that can move freely into and out of the cell. Red blood cells shrivel and shrink when placed in a(n) (7) [choose one] ( ) hypotonic, ( ) isotonic, ( ) hypertonic solution. If plant cells were placed in a hypotonic solution, the cells would exhibit (8)_____ _____.

True-False

If false, explain why.

____ 9. Osmosis occurs in response to a concentration gradient that involves unequal concentrations of water molecules.

____ 10. If an animal cell were to be placed in a hypertonic solution, it would swell and perhaps burst.

____ 11. Placing plant cells in distilled water will ensure that their plasma membranes do not shrink away from the cell wall.

____ 12. Physiological saline is 0.9% NaCl; red blood cells placed in such a solution will not gain or lose water; therefore, one could state that the tonicity of red blood cells is hypertonic.

____ 13. A solution of 80% solvent, 20% solute is more concentrated than a solution of 70% solvent, 30% solute.

MOVEMENT OF WATER AND SOLUTES ACROSS CELL MEMBRANES (5-IV, pp. 84–87)

1. What types of molecules diffuse easily across the lipid bilayer of the plasma membrane? List several.

_____

_____

2. Define and contrast active transport and passive transport mechanisms._____

_____

_____

3. Explain how the membrane transport mechanism called "facilitated diffusion" operates in terms of the membrane proteins involved._____

_____

_____

4. Describe the mechanisms and salient features of all active transport systems; cite an actual example.

_____

_____

_____

5. Define *exocytosis* and *endocytosis* in terms of the vesicles involved; relate these terms to the concepts of phagocytosis and pinocytosis._____

_____

_____

6. Explain the concept of receptor-mediated endocytosis._____

_____

_____

True-False

If false, explain why.

___ 7. Active transport depends on proteins that serve either as carriers or as fixed channels across the plasma membrane.

___ 8. One passive transport system, the sodium-potassium pump, assists in maintenance of high potassium concentrations and low sodium concentrations inside the cell.

___ 9. The secretion of mucus is achieved by exocytosis.

___ 10. Facilitated diffusion is a type of diffusion that requires the cell to expend ATP molecules.

___ 11. Pinocytosis, sometimes referred to as cell drinking, is a form of exocytosis.

Fill-in-the-Blanks

12. _____ _____ systems cannot operate without direct energy outlays by the cell.

Matching

Choose all the appropriate answers for each.

13. ___ receptor-mediated endocytosis
14. ___ pinocytosis
15. ___ sodium-potassium pump
16. ___ phagocytosis
17. ___ simple diffusion
18. ___ active transport
19. ___ exocytosis
20. ___ facilitated diffusion
21. ___ passive transport
22. ___ endocytosis

A. Involves solute transport through membrane proteins following concentration gradients; involves no energy boost
B. Membrane proteins assisting a solute only in the direction that simple diffusion would take it
C. Cytoplasmic vesicles fusing with plasma membrane and releasing contents to the outside of the cell
D. An example of active transport
E. Plasma membrane region enclosing particles and pinching off a vesicle that moves into the cytoplasm
F. Process by which specific molecules cause coated pits to form in the plasma membrane and pit receptors bind specific molecules that sink into the cytoplasm to form an endocytic vesicle
G. "Cell drinking," a type of endocytosis
H. Small, electrically neutral molecules moving directly across the lipid bilayer along a concentration gradient
I. "Cell eating," a type of endocytosis
J. Membrane proteins actively moving specific solutes into and out of cells; involves an energy boost

Chapter Terms

The following page-referenced terms are important—those that were in boldface type in the chapter. Refer to the instructions given in Chapter 1, p. 4.

plasma membrane (74)   recognition protein (80)   osmosis (82)   passive transport (84)

internal cell membranes (75)   receptor protein (80)   tonicity (82)   facilitated diffusion (84)

phospholipid (76)   concentration gradient (81)   turgor pressure (83)   active transport (86)

lipid bilayer (76)   simple diffusion (81)   water potential (83)   exocytosis (86)

channel protein (80)   bulk flow (81)   active transport (84)   endocytosis (86)

carrier protein (80)

Self-Quiz

In the blank following each ion, molecule, or structure, enter the name(s) of one correct membrane transport mechanism.

1. H_2O _____ ; 2. CO_2 _____ ; 3. Na^+ _____ ;

4. protein particles _____ ; 5. glucose _____ ; 6. O_2 _____ ;

7. K^+ _____ ; 8. amino acids _____ ; 9. fluid droplets _____ ;

10. ingestion of an alien cell by a white blood cell after coating _____ .

___ 11. White blood cells use _____ to devour disease agents invading your body.
    a. diffusion
    b. bulk flow
    c. osmosis
    d. phagocytosis

___ 12. _____ cells depend on the calcium-pump mechanism.
    a. Intestine
    b. Nerve
    c. Muscle
    d. Amoeba

___ 13. Water is such an excellent solvent primarily because _____.
    a. it forms spheres of hydration around substances and can form hydrogen bonds with many polar substances
    b. it has a high heat of fusion
    c. of its cohesive properties
    d. it is a liquid at room temperature

___ 14. In a lipid bilayer, tails point inward and form a(n) _____ region that excludes water.
    a. acidic
    b. basic
    c. hydrophilic
    d. hydrophobic

___ 15. A protistan adapted to life in a freshwater pond is collected in a bottle and transferred to a saltwater bay. Which of the following is likely to happen?
    a. The cell bursts.
    b. Salts flow out of the protistan cell.
    c. The cell shrinks.
    d. Enzymes flow out of the protistan cell.

___ 16. Which of the following is *not* a form of active transport?
    a. sodium-potassium pump
    b. endocytosis
    c. exocytosis
    d. bulk flow

___ 17. Which of the following is *not* a form of passive transport?
    a. osmosis
    b. facilitated diffusion
    c. bulk flow
    d. exocytosis

___ 18. O_2, CO_2, H_2O, and other small, electrically neutral molecules move across the cell membrane by _____.
    a. facilitated diffusion
    b. receptor-mediated endocytosis
    c. simple diffusion
    d. active transport

___ 19. Ions such as H⁺, Na⁺, K⁺, and Ca⁺⁺ move across cell membranes by

    _____.

    a. facilitated diffusion
    b. receptor-mediated endocytosis
    c. bulk flow
    d. active transport

___ 20. Coated pits, receptors, and transport vesicle formation participate in

    _____.

    a. facilitated diffusion
    b. receptor-mediated endocytosis
    c. simple diffusion
    d. active transport

Chapter Objectives/Review Questions

This section lists general and detailed chapter objectives that can be used as review questions. You can make maximum use of these items by writing answers on a separate sheet of paper. Fill in answers where blanks are provided. To check for accuracy, compare your answers with information given in the chapter or glossary.

Page   *Objectives/Questions*

(74–75)   1. Materials are exchanged between cytoplasm and external cell environment across the _____.

(76)   2. Describe the general structure of a phospholipid molecule.

(76–77)   3. Explain the behavior of many phospholipid molecules in water.

(76–80)   4. Fully detail the most recent version of the fluid mosaic model of membrane structure.

(78–79)   5. Two methods of studying cell membrane structure are the freeze _____ and the freeze _____ methods.

(80)   6. Membrane functions are carried out by membrane _____.

(80)   7. _____ proteins serve as pores through which ions or other water-soluble substances move through membranes; _____ proteins bind specific substances and change shape in ways that move substances across membranes; _____ proteins are like molecular fingerprints at the cell surface; _____ proteins are like switches that turn on or off when specific substances bind to them.

(81)   8. Molecules moving to regions where they are less concentrated are moving down their _____ gradient.

(81)   9. Random movement of like molecules or ions down a concentration gradient is called simple _____.

(81)   10. Define *bulk flow* and understand how it enhances diffusion rates.

(82)   11. When salt is dissolved in water, which is the solute and which is the solvent?

(82)   12. Explain osmosis in terms of a differentially permeable membrane.

(82–83)   13. Define *tonicity* and understand the meanings of *isotonic, hypertonic,* and *hypotonic.*

(83)   14. When water moves into a plant cell by osmosis, the internal _____ pressure developed pushes on the wall.

(84)   15. List substances that can move directly through the plasma membrane.

(84)   16. Different types of _____ move substances like glucose and ions across the membrane by active and passive transport mechanisms.

(84–86)   17. Be able to distinguish between facilitated diffusion and active transport in terms of movement mechanisms and types of materials moved.

(86–87)   18. Define *exocytosis* and *endocytosis;* explain the details of receptor-mediated endocytosis.

Integrating and Applying Key Concepts

1. If there were no such thing as active transport, how would the lives of organisms be affected?

Critical Thinking Exercises

1. A certain enzyme (E) can catalyze the formation of a product (P) from the amino acid (A). A solution of E is placed at position 2 on a strip of gel, as shown in the diagram. A solution of A is placed at position 7. At which position is P most likely to appear first? Assume that both A and E can diffuse through the water in the gel.

ANALYSIS

P will form first at the point where the diffusing E and A molecules first meet. This will be between 2 and 7. Because E is a protein, a polymer of amino acids, it is larger than A. Larger molecules diffuse more slowly than smaller molecules. Therefore, A will diffuse farther in a given time than E, and the meeting point will be closer to 2 than to 7. Unless you know *how much* faster A moves than E, you cannot choose between positions 3 and 4.

2. An apparatus consists of two chambers separated by a membrane. The membrane is permeable to water but not to starch. On the left side, you place nonradioactive distilled water. On the right side, you place an equal volume of a starch solution in which some of the water molecules are radioactive. What do you predict will be the distribution of starch, total water, and radioactive water between the two compartments several hours later? Answer separately for each substance.
   a. All on the left
   b. More on the left than on the right
   c. Equal amounts on the left and the right
   d. More on the right than on the left
   e. All on the right

ANALYSIS

Starch — The membrane is not permeable to starch, so it will remain in the starting position, all on the right.

Total water — The membrane is permeable to water, so water molecules will continually cross it in both directions. Because the starch makes the water concentration lower on the right than on the left, however, water molecules cross from left to right faster than they go back, and the amount of water on the right will increase. Eventually, the osmotic pressure due to the increased volume on the right will increase the rate of movement from right to left and make it equal to the opposite movement, and no further change will occur in the distribution of total water.

Radioactive water — Even though the *net* movement of total water is to the right, many water molecules are moving to the left. Some of them will be radioactive. No radioactive molecules are on the left at first, so the net movement is to the left, and from then on there will be radioactivity on the left. What the concentrations of radioactive water become depends on the duration of the experiment; they will steadily approach equality. However, because the total water will always be greater on the right, even if the concentrations become equal on the two sides, there will always be a greater amount of radioactive water on the right.

3. Both ethanol (CH_3CH_2OH) and acetate (CH_3COO^-) can enter certain bacterial cells and be used as a nutrient. You place samples of these cells in various concentrations of the two substances and measure the

rate at which the substance enters the cells in each sample. You plot the data as shown below, but the experimental notes are destroyed before you have labeled the graph. Can you infer which line represents which substance?

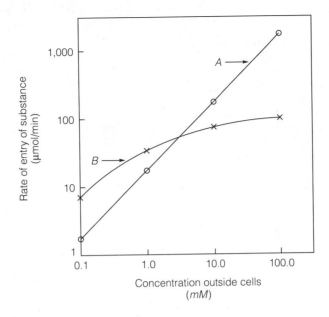

ANALYSIS

The two molecules are almost the same size, so there is no help there. But ethanol has no charge, while acetate is ionized. That means that ethanol could be permeable to cell membranes, but acetate would likely be impermeable. In this case, acetate would have to enter the cell by active transport or facilitated diffusion, both processes that require a carrier protein. Now a substance crossing a membrane by diffusion will cross faster and faster as the concentration rises and as molecules arrive at the membrane more and more often. That is happening in curve A. The only limit will be due to the solubility of the substance. On the other hand, crossing a membrane on a carrier takes a certain amount of time, during which that carrier is not available to transport another molecule that arrives. If the concentration of the substance is high enough, the available carriers will be refilled immediately after they release a transported molecule and thus will be transporting at their maximum capacity. Even if the concentration rises further and molecules arrive more often, the carriers will not transport the substance any faster. This situation is called saturation of the carriers and seems to be happening in curve B. Furthermore, at low concentrations curve B has higher entry rates than curve A, which is also characteristic of active transport.

Answers

Answers to Interactive Exercises

FLUID MEMBRANES IN A LARGELY FLUID WORLD (5-I)

1. The "bilayer" is composed of two layers of phospholipid molecules with all fatty acid tails sandwiched between the hydrophilic heads; this arrangement is the structural basis for all cell membranes. In water, lipid molecules cluster spontaneously as hydrophobic interactions with surrounding water molecules force them into two layers.

2. These three lipids are mostly hydrophobic with hydrophilic heads.
3. cholesterol; phytosterols.
4. The fluid mosaic membrane is a mosaic of embedded or surface lipids and proteins (mostly glycoproteins extending outward). The two layers of phospholipid molecules are prevented from solid packing by their fluid movement behavior and structure.
5. channel protein (C); 6. carrier protein (D); 7. transport system (E); 8. recognition protein (A); 9. receptor protein (B); 10. Freeze-fracturing (Freeze-etching); 11. freeze- etching (freeze-fracturing); 12. lipid bilayer;

13. proteins; 14. mosaic; 15. Phospholipids;
16. phosphate; 17. glycoprotein; 18. proteins;
19. metabolism; 20. tissues; 21. barrier; 22. F; 23. T;
24. F; 25. T; 26. T.

DIFFUSION (5-II)

1. "Gradient" refers to differences in concentrations of ions or molecules when one compares different regions; gradients or differences can also exist in pressure, temperature, or electric charge found in different regions.
2. "Simple diffusion" refers to the random movement of similar molecules or ions from regions of high concentration to regions of low concentration (down a concentration gradient). An example is drops of dye placed in water.
3. Bulk flow is the tendency for all the different substances in a fluid to move together in the same direction in response to a pressure gradient; sometimes bulk flow enhances diffusion rates. Bulk flow occurs in the circulatory systems of animals and the vascular systems of plants.
4. Density; 5. gradient; 6. concentration; 7. random collisions; 8. greater (higher); 9. lesser (lower); 10. bulk flow; 11. T; 12. F.

OSMOSIS (5-III)

1. Osmosis is the movement of water across a differentially permeable membrane in response to solute concentration gradients, pressure gradients, or both. An example: a sealed bag (with differentially permeable walls) containing table sugar or molasses is placed in distilled water; water moves by osmosis into the bag and increases turgor pressure in the bag.
2. *Tonicity* refers to the relative concentrations of solutes in two fluids. *Isotonic* refers to equal solute concentrations in both fluids; unequal solute concentrations result in one fluid being hypotonic (less solutes) and the other fluid being hypertonic (more solutes). Water moves from a hypotonic fluid to a hypertonic fluid.
3. Turgor pressure is caused by water moving into a plant cell by osmosis; it refers to the developing internal pressure on the cell wall.
4. turgor pressure and wall pressure.
5. differentially permeable; 6. Water; 7. hypertonic;
8. turgor pressure; 9. T; 10. F; 11. T; 12. F; 13. F.

MOVEMENT OF WATER AND SOLUTES ACROSS CELL MEMBRANES (5-IV)

1. O_2, CO_2, H_2O, ethanol, and other small, electrically neutral molecules.
2. Glucose and other large, water-soluble, electrically neutral molecules are moved across the membrane by proteins of active and passive transport mechanisms. In active transport, cellular energy activates the proteins to move a solute against its concentration gradient; in passive transport, the proteins do not require an energy boost.
3. Membrane proteins with facilitated diffusion (passive transport) functions assist particular solutes across the membrane in the direction simple diffusion would take them. Membrane proteins of facilitated diffusion open a space to only one side of the membrane at a time; binding of hydrophilic groups (in the protein space) to water-soluble molecules changes the shape of the protein that allows solute molecule passage.
4. Small ions, small charged molecules, and large molecules are usually pumped across a cell membrane against their concentration gradient. Proteins of the active transport system spanning the lipid bilayer selectively bind and move solutes by active transport when they receive an energy boost. Examples are the sodium-potassium pump and the calcium pump.
5. In exocytosis, cytoplasmic vesicles move to the plasma membrane, fuse with it, and then release their contents to the cell's exterior. In endocytosis, plasma membrane regions enclose particles at or near the cell's surface and then pinch off to form cytoplasmic vesicles.
6. In receptor-mediated endocytosis, pits coated with protein lattices form on the cytoplasmic side through special plasma membrane regions. The pits are coated on the outside with molecule-specific surface receptors. When the specific molecules bind to the receptors, the pit sinks into the cytoplasm as an endocytic vesicle forms.
7. T; 8. T; 9. T; 10. F; 11. F; 12. Active transport; 13. F (D); 14. G (D); 15. D (D); 16. I (D); 17. H; 18. J (D); 19. C (D); 20. B; 21. A; 22. E (D).

Answers to Self-Quiz

1. osmosis; 2. simple diffusion; 3. active transport;
4. endocytosis (phagocytosis); 5. facilitated diffusion;
6. simple diffusion; 7. active transport; 8. facilitated diffusion; 9. endocytosis (pinocytosis);
10. receptor-mediated endocytosis; 11. d; 12. c; 13. a;
14. d; 15. c; 16. d; 17. d; 18. c; 19. d; 20. b.

6

GROUND RULES OF METABOLISM

Interactive Exercises

In most cases, answers for specific molecules use abbreviations.

ENERGY AND LIFE (6-I, pp. 92–93)

1. Define metabolism._____

_____

_____

2. _____ is a capacity to make things happen, to cause change, to do work.

3. State the first law of thermodynamics._____

_____

_____

4. State the second law of thermodynamics._____

_____

_____

5. _____ is a measure of the degree of randomness or disorder in systems such as a living thing, an

automobile, or a house.

6. What is the explanation for the maintenance of a high degree of organization in the world of life?

_____

_____

True-False

If false, explain why.

___ 7. The first law of thermodynamics states that entropy is constantly increasing in the universe.

___ 8. Your body steadily gives off heat equal to that from a 100-watt light bulb.

___ 9. When you eat a potato, some of the stored chemical energy of the food is converted into mechanical energy that moves your muscles.

___ 10. Energy is the capacity to accomplish work.

___ 11. The amount of low-quality energy in the universe is decreasing.

Labeling

In the blank preceding each item, indicate whether the first law of thermodynamics (I) or the second law of thermodynamics (II) is best described.

12. ___ Cooling of a cup of coffee

13. ___ Evaporation of gasoline into the atmosphere

14. ___ A hydroelectric plant at a waterfall producing electricity

15. ___ The creation of a snowman by children

16. ___ The glow of an incandescent bulb following the flow of electrons through a wire

17. ___ A typesetter arranging the type for a page of the biology text

18. ___ The movement of a gas-powered automobile

19. ___ The glow of a firefly

20. ___ Humans running the 100-meter dash following usual food intake

21. ___ The death and decay of an organism

THE NATURE OF METABOLISM (6-II, pp. 93–96)

1. Define exergonic reactions; give an example._____

_____

2. Define endergonic reactions; give an example._____

_____

Fill-in-the-Blanks

The substances present at the end of a reaction, the (3)_____ , may have less or more energy than did the starting substances, the (4)_____ . Most reactions are (5)_____ in that they can proceed in forward and reverse directions. Such reactions tend to approach (6)_____ , a state in which the reactions are proceeding at about the same (7)_____ in both directions.

Sequence

Rank the chemical bonds in terms of their energies, with "A" being the strongest and "C" the weakest.

8. ___ Ionic bond

9. ___ Covalent bond

10. ___ Hydrogen bond

Labeling

Classify each of the following reactions as *endergonic* or *exergonic*.

11. _____ Burning wood at a campfire.

12. _____ The product of a chemical reaction has more energy than the reactants.

13. _____ Glucose + oxygen $\longrightarrow$ carbon dioxide + water + energy

14. _____ The reactants of a chemical reaction have more energy than the product.

15. _____ The reaction releases energy.

Matching

Choose the most appropriate answer for each.

16. ___ intermediates

17. ___ degradative pathways

18. ___ dynamic equilibrium

19. ___ cofactors

20. ___ end products

21. ___ metabolic pathway

22. ___ reactants

23. ___ energy carriers

24. ___ biosynthetic pathways

25. ___ enzymes

A. An orderly series of reactions catalyzed by enzymes
B. Sequence of reactions in which small organic molecules are assembled into larger organic molecules
C. Mainly ATP; donates energy to reactions
D. Small molecules and metal ions that assist enzymes or serve as carriers
E. Substances (substrates or precursors) able to enter into a reaction
F. Compounds formed between the beginning and end of a metabolic pathway
G. Sequence of stepwise reactions in which organic compounds are broken down
H. Proteins that catalyze reactions
I. Condition in which rate of forward reaction equals rate of reverse reaction
J. Substances present at the end of a metabolic pathway

Matching

Study the sequence of reactions below. Identify the components of the reactions by selecting items from the following list and entering the correct letter in the appropriate blank.

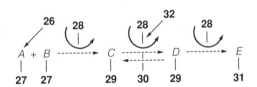

26. ___
27. ___
28. ___
29. ___
30. ___
31. ___
32. ___

A. energy carrier
B. cofactor
C. intermediates
D. reactants
E. end product
F. reversible reaction
G. enzymes

ENZYMES (6-III, pp. 96–99)

1. Define enzymes in terms of chemical structure and function._____

   _____

2. Define substrate; cite an example._____

   _____

   _____

3. List four characteristics that enzymes have in common._____

   _____

   _____

   _____

Matching

Match the items on the sketch below with the list of descriptions. Some answers may require more than one letter.

4. ____

5. ____

6. ____

7. ____

8. ____

9. ____

A. Transition state, the time of the most precise fit between enzyme and substrate
B. Complementary active site of the enzyme
C. Enzyme, a protein with catalytic power
D. Product or reactant molecules that an enzyme can specifically recognize
E. Product or reactant molecule
F. Bound enzyme-substrate complex

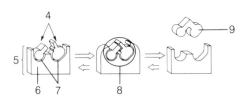

Fill-in-the-Blanks

(10)_____ are highly selective proteins that act as catalysts, which means that they greatly enhance the rate at which specific reactions approach (11)_____. The specific substance on which a particular enzyme acts is called its (12)_____ ; this substance fits into the enzyme's crevice, which is called its (13)_____ _____. The (14)_____-_____ model describes how a substrate contacts the substrate without a perfect fit. Enzymes increase reaction rates by lowering the required (15)_____ _____. (16)_____ and (17)_____ are two important factors that influence the rates of enzyme activity. Extremely high fevers can destroy the three-dimensional shape of metabolism, which may adversely affect (18)_____ and cause death. When (19)_____ occurs, weak bonds holding the enzyme in its three-dimensional shape break. Molecules that can bind with enzymes and interfere with their function as catalysts are called (20)_____. (21)_____ enzymes have control sites where specific substances can bind and alter enzyme activity. The situation in which the end product binds to the first enzyme in a metabolic pathway and prevents product formation is known as (22)_____ _____.

True-False

If false, explain why.

___ 23. Enzyme shape may change during the interaction between enzyme and substrate.

___ 24. The active site is a crevice shape on the reactant molecule.

___ 25. For two reactant molecules to become product molecules, the reactant molecules must first collide with a certain minimum energy.

___ 26. Enzymes enhance reaction rates by increasing the activation energy required.

___ 27. High temperatures can denature enzymes and affect reaction rates; the pH level does not seem to affect enzymes and their action.

COFACTORS (6-IV, p. 99)

Fill-in-the-Blanks

Nonprotein substances that aid enzymes in their catalytic task are called (1)_____ ; they include some large organic molecules that function as (2)_____. (3)_____ and (4)_____ are coenzymes that have roles in the breakdown of glucose and other carbohydrates. (5)_____ is a coenzyme with a central role in photosynthesis; its abbreviation is (6)_____ when it is loaded with protons and electrons. Some metal (7)_____ such as Fe^{++} also serve as cofactors as components of cytochrome molecules serving as carrier proteins in cell membranes.

ELECTRON TRANSFERS IN METABOLIC PATHWAYS (6-V, pp. 99–100)

Fill-in-the-Blanks

The release of energy from glucose in cells proceeds in controlled steps, so that (1)_____ molecules form along the route from glucose to carbon dioxide and water. At each step in a metabolic pathway, a specific (2)_____ lowers the activation energy for the formation of an intermediate compound. At each step in the pathway, only some energy is released. In chloroplasts and mitochondria, the liberated electrons released from the breaking of chemical bonds are sent through (3)_____ _____ systems; these systems consist of enzymes and (4)_____ , bound in a cell membrane, that transfer electrons in a highly organized sequence. A molecule that donates electrons in the sequence is being (5)_____ , while molecules accepting electrons are being (6)_____. Oxidation-reduction means an (7)_____ transfer. Electron transport systems "intercept" excited electrons and make use of the (8)_____ they release. If we think of the electron transport system as a staircase, electrons at the top of the staircase have the (9) [choose one] ( ) most, ( ) least energy. As the electrons are transferred from one electron carrier to another, some (10)_____ can be harnessed to do biological (11)_____. One type of biological work occurs when energy released during electron transfers is used to bond a (12)_____ group to ADP.

Matching

Match the lettered statements to the numbered items on the sketch below.

13. ___
14. ___
15. ___
16. ___
17. ___

A. Represent the cytochrome molecules in an electron transport system
B. Electrons at their highest energy level
C. Released energy harnessed and used to produce ATP
D. Electrons at their lowest level
E. The separation of hydrogen atoms into protons and electrons

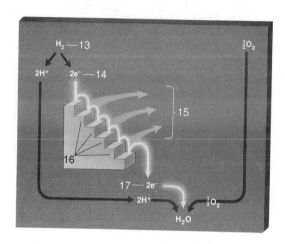

ATP: THE UNIVERSAL ENERGY CARRIER (6-VI, pp. 100–102)

Fill-in-the-Blanks

ATP is constructed of the nitrogenous base (1)_____ , (2) the sugar _____ , and three

(3)_____ groups. When ATP is hydrolyzed, a molecule of (4)_____ in the presence of an

appropriate (5)_____ is used to split ATP into (6)_____ , a (7)_____ group, and (most

important) usable (8)_____ , which is easily transferred to other molecules in the cell. The hydrolysis

of ATP provides (9)_____ for biosynthesis, active transport across cell membranes and molecular

displacements, such as those required for muscle contraction. ATP directly or indirectly delivers energy to

almost all (10)_____ pathways. In the (11)_____ / _____ cycle, a phosphate group is linked

to adenosine diphosphate, and adenosine triphosphate donates a phosphate group elsewhere and reverts

back to adenosine diphosphate. Adding a phosphate to a molecule is called (12)_____. When this

occurs, the molecule increases its store of (13)_____ and becomes primed to enter a specific

(14)_____. The (15)_____ / _____ cycle provides a renewable means of conserving and

transferring energy to specific reactions.

Identifying

Identify the molecule below and label its parts.

16. _____

17. _____

18. _____

19. The name of this molecule is

_____ _____.

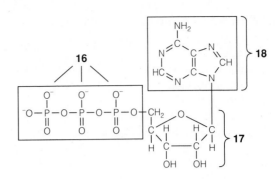

Chapter Terms

The following page-referenced terms are important; most were in boldface type in the chapter. Refer to the instructions given in Chapter 1, p. 4.

metabolism (91)
energy (92)
first law of
    thermodynamics (92)
second law of
    thermodynamics (92)
entropy (92)
dynamic equilibrium (94)
metabolic pathway (96)

degradative pathways (96)
biosynthetic pathways (96)
reactants (96)
intermediates (96)
enzymes (96)
cofactors (96)
energy carriers (96)
end products (96)
substrates (96)

active site (96)
induced-fit model (96)
activation energy (97)
inhibitors (98)
allosteric control (99)
feedback inhibition (99)
cofactors (99)
NAD+ (99)

FAD (99)
NADP+ (99)
electron transport
    systems (100)
ATP (100)
ATP/ADP cycle (102)
ADP (102)
phosphorylation (102)

Self-Quiz

___ 1. An important principle of the second law of thermodynamics states that _____.
   a. energy can be transformed into matter, and because of this we *can* get something for nothing
   b. energy can be destroyed only during nuclear reactions, such as those that occur inside the sun
   c. if energy is gained by one region of the universe, another place in the universe also must gain energy in order to maintain the balance of nature
   d. matter tends to become increasingly more disorganized

___ 2. Essentially, the first law of thermodynamics states that _____ .
   a. one form of energy cannot be converted into another
   b. entropy is increasing in the universe
   c. energy cannot be created or destroyed
   d. energy cannot be converted into matter or matter into energy

___ 3. An enzyme is best described as _____.
   a. an acid
   b. protein
   c. a catalyst
   d. a fat
   e. both (b) and (c)

___ 4. Which is *not* true of enzyme behavior?
   a. Enzyme shape may change during catalysis.
   b. The active site of an enzyme orients its substrate molecules, thereby promoting interaction of their reactive parts.
   c. All enzymes have an active site where substrates are temporarily bound.
   d. An individual enzyme can catalyze a wide variety of different reactions.

___ 5. When NAD⁺ combines with hydrogen, the NAD⁺ is _____.
   a. reduced
   b. oxidized
   c. phosphorylated
   d. denatured

___ 6. A substance that gains electrons is _____.
   a. oxidized
   b. a catalyst
   c. reduced
   d. a substrate

___ 7. In _____ pathways, carbohydrates, lipids, and proteins are broken down in stepwise reactions that lead to products of lower energy.
   a. intermediate
   b. biosynthetic
   c. induced
   d. degradative

___ 8. With regard to major function, NAD⁺, FAD, and NADP⁺ are classified as _____.
   a. enzymes
   b. phosphate carriers
   c. cofactors that function as coenzymes
   d. end products of metabolic pathways

___ 9. When a phosphate bond is linked to ADP, the bond formed _____.
   a. absorbs a large amount of free energy when the phosphate group is attached during hydrolysis
   b. is formed when ATP is hydrolyzed to ADP and one phosphate group
   c. is usually found in each glucose molecule; this is why glucose is chosen as the starting point for glycolysis
   d. releases a large amount of usable energy when the phosphate group is split off during hydrolysis.

___ 10. An allosteric enzyme _____.
   a. has an active site where substrate molecules bind and another site that binds with intermediate or end-product molecules
   b. is an important energy-carrying nucleotide
   c. carries out either oxidation reactions or reduction reactions but not both
   d. raises the activation energy of the chemical reaction it catalyzes

Chapter Objectives/Review Questions

This section lists general and detailed chapter objectives that can be used as review questions. You can make maximum use of these items by writing answers on a separate sheet of paper. Fill in answers where blanks are provided. To check for accuracy, compare your answers with information given in the chapter or glossary.

Page	Objectives/Questions
(91)	1. _____ is the controlled capacity to acquire and use energy for stockpiling, breaking apart, building, and eliminating substances in ways that contribute to survival and reproduction.
(92)	2. Define *energy*; be able to state the first and second laws of thermodynamics.
(92)	3. Explain what is meant by a "system" as related to the laws of thermodynamics.
(92)	4. _____ is a measure of the degree of randomness or disorder of systems.
(92–93)	5. Explain how the world of life maintains a high degree of organization.
(93–94)	6. Explain what is meant by bond energy; rank hydrogen bonds, covalent bonds, and ionic bonds in terms of their relative bond strength.
(94)	7. Reactions that show a net loss in energy are said to be _____ ; reactions that show a net gain in energy are said to be _____.
(94)	8. What is meant by a "reversible" reaction?
(94–95)	9. Describe the condition known as dynamic equilibrium.
(95–96)	10. What is the function of metabolic pathways in cellular chemistry?
(96)	11. Give the function of each of the following participants in metabolic pathways: reactants, intermediates, enzymes, cofactors, energy carriers, and end products.

Integrating and Applying Key Concepts

A piece of dry ice left sitting on a table at room temperature vaporizes. As the dry ice vaporizes into CO_2 gas, does its entropy increase or decrease? Tell why you answered as you did.

Critical Thinking Exercises

1. A major step in the discovery of enzymes came in the nineteenth century when Hans and Edward Büchner attempted to preserve nonliving extracts made from ground and pressed yeast cells by adding sugar in order to "pickle" the extracts and prevent the growth of spoilage microorganisms. Which of the following assumptions were they most likely making?

   a. Yeast extracts could not metabolize sugar.
   b. No microorganisms could enter the extracts.
   c. The temperature was appropriate for yeast growth.
   d. Yeast extracts contain proteins and nucleic acids.
   e. Yeast extracts destroy sugar.

ANALYSIS

   a. The hypothesis was that the sugar would preserve the extracts, which could not happen if the extracts destroyed the sugar. The Buchners must have assumed that the extracts could not metabolize sugar.
   b. Microorganisms are the agents that cause spoilage of organic materials. If no microorganisms could enter the extracts, they would not spoil, and there would be no need to try to preserve them. The Buchners must not have made assumption (b).
   c. The extracts were nonliving and could not have grown yeast cells at any temperature. This assumption is unnecessary. However, the Buchners must have assumed that the temperature was adequate for growth of spoilage microorganisms. If they had had a refrigerator, they might never have learned about enzymes.
   d. This is a true statement, but it need not be assumed in order for the hypothesis to be made. Prevention of spoilage depends on the effect of sugar on the spoilage bacteria, not on the kinds of organic molecules in the extract.
   e. If yeast extracts could destroy sugar, the sugar could not be used to preserve the extracts. If the Buchners had made this assumption, they would not have been able to make the hypothesis.

In fact, the yeast extracts did metabolize the sugar in the same way that intact living yeast cells did. From this it was inferred that metabolism did not require living cells, as had been believed, but that inanimate molecules were enough to make the reactions go. These molecules are called *enzymes*. Now, consider that blackberry cells contain the same enzymes. Given that fact, how can sugar be used to preserve crushed blackberry cells—blackberry jam? Why don't the enzymes in the blackberry extract metabolize the sugar just as the yeast enzymes did?

2. A flask contains a solution of a polymer. A solution of an enzyme is added to the flask. A student makes the hypothesis that the polymer is protein and that the enzyme catalyzes the hydrolysis of protein. What would the student expect to find in the flask several hours later?

ANALYSIS

The key to this question is to remember that an enzyme is also a protein. Thus, a protein-digesting enzyme is likely to catalyze its own hydrolysis. This leads to the prediction that some of the original polymer will be hydrolyzed *and* some of the enzyme will be hydrolyzed. The product will be amino acids or small fragments of protein, and the flask will also contain some leftover protein and some undestroyed enzyme. The amount of each substance predicted to be present depends on what is assumed about the rate of the reaction, the original concentrations, and the exact time of incubation.

3. Metabolic pathways branch when two enzymes have the same substrate but produce different products. For example, the substance phosphoenolpyruvate (PEP) is the substrate for two different enzymes—phosphoenolpyruvate carboxykinase (PEPCK) and pyruvate kinase (PK). Each enzyme catalyzes the formation of a different product, and each product is metabolized by the enzymes of a different pathway. These two enzymes are both dependent on pH but in very different ways.

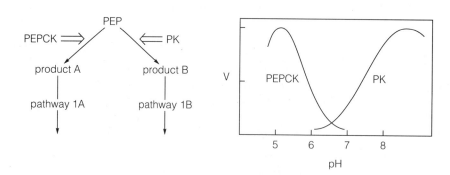

Suppose an oyster, at high tide, has a tissue pH of about 7.3. When the tide goes out, the oyster closes its shells, acidic wastes can no longer escape to the sea, and the internal pH drops to about 6. What happens to PEP metabolism under these conditions?

ANALYSIS

At pH 7.3, PK is functioning at about 50 percent of maximal activity, and PEPCK is almost inactive. Thus, almost all the PEP goes on pathway 1B. When the pH drops to 6, the situation is reversed; PEPCK functions at about 50 percent of maximum, and PK is inactivated. Under these conditions, PEP follows pathway 1A. The change in pH resulting ultimately from the movement of the tide causes a switch in the metabolism of the oyster. Notice that neither enzyme normally encounters its optimum pH in the living animal.

Reference: Hochachka, P. W., and G. N. Somero (1973). *Strategies of Biochemical Adaptation* (Philadelphia: Saunders), p. 49.

Answers

Answers to Interactive Exercises

ENERGY AND LIFE (6-I)
1. Metabolism is the controlled capacity to acquire and use energy for stockpiling, breaking apart, building, and eliminating substances in ways that contribute to survival and reproduction. Metabolism is sometimes defined as the sum total of chemical function in an organism.
2. Energy.
3. The first law of thermodynamics states that the total amount of energy in the universe remains constant. More energy cannot be created; existing energy cannot be destroyed. Energy can only be converted from one form to another.
4. The second law of thermodynamics states that the spontaneous direction of energy flow is from high-quality to low-quality forms. With each conversion, some energy is randomly dispersed in a form (usually heat) that is not readily available to do work.
5. Entropy.
6. The world of life maintains a high degree of organization only because it is being resupplied with energy lost from someplace else.
7. F; 8. T; 9. T; 10. T; 11. F; 12. II; 13. II; 14. I; 15. I; 16. I; 17. II; 18. I; 19. I; 20. I; 21. II.

THE NATURE OF METABOLISM (6-II)
1. Exergonic reactions show a net loss of energy (energy out); an example is the breakdown of food molecules in the human body as the reactants become products.
2. Endergonic reactions show a net gain in energy (energy in); an example is the construction of starch and other large molecules from smaller, energy-poor molecules.
3. products; 4. reactants; 5. reversible; 6. equilibrium; 7. rate; 8. B; 9. A; 10. C.
11. exergonic; 12. endergonic; 13. exergonic; 14. exergonic; 15. exergonic; 16. F; 17. G (A); 18. I; 19. D; 20. J; 21. A; 22. E; 23. C; 24. B (A); 25. H; 26. A; 27. D; 28. G; 29. C; 30. F; 31. E; 32. B.

ENZYMES (6-III)
1. Enzymes are usually protein molecules with enormous catalytic power (a few RNA forms have been found to act as enzymes).

2. Substrates are specific molecules that an enzyme can chemically recognize, bind briefly to itself, and bring about a specific structural change; an example is the enzyme thrombin, which helps break a specific bond in a specific protein.
3. Enzymes do not cause reactions that would not happen on their own; enzymes are not changed in reactions—they can be used over and over; each enzyme is highly selective about its substrates; an enzyme can recognize both the reactants and the products of a given reaction as its substrate.
4. D; 5. F; 6. C; 7. B; 8. A; 9. E; 10. Enzymes; 11. equilibrium; 12. substrate; 13. active site; 14. induced-fit; 15. activation energy; 16. Temperature (pH); 17. pH (temperature); 18. metabolism; 19. denaturation; 20. inhibitors; 21. Allosteric; 22. feedback inhibition; 23. T; 24. F; 25. T; 26. F; 27. F.

COFACTORS (6-IV)
1. cofactors; 2. coenzymes; 3. NAD^+ (FAD); 4. FAD (NAD^+); 5. $NADP^+$; 6. NADPH; 7. ions.

ELECTRON TRANSFERS IN METABOLIC PATHWAYS (6-V)
1. intermediate; 2. enzyme; 3. electron transport; 4. cofactors; 5. oxidized; 6. reduced; 7. electron; 8. energy; 9. most; 10. energy; 11. work; 12. phosphate; 13. E; 14. B; 15. C; 16. A; 17. D.

ATP: THE UNIVERSAL ENERGY CARRIER (6-VI)
1. adenine; 2. ribose; 3. phosphate; 4. water; 5. enzyme; 6. ADP; 7. phosphate; 8. energy; 9. energy; 10. metabolic; 11. ATP/ADP; 12. phosphorylation; 13. energy; 14. reaction; 15. ATP/ADP; 16. three phosphate groups; 17. ribose sugar; 18. adenine, a nitrogenous base; 19. adenosine triphosphate.

Answers to Self-Quiz

1. d; 2. c; 3. e; 4. d; 5. a; 6. c; 7. d; 8. c; 9. d; 10. a.

7

ENERGY-ACQUIRING PATHWAYS

Interactive Exercises

In most cases, answers for specific molecules use abbreviations.

SUNLIGHT, RAIN, AND CELLULAR WORK (Vignette, pp. 104–105)

1. Define autotroph; distinguish between photosynthetic autotrophs and chemosynthetic autotrophs.

   _____

   _____

   _____

2. Explain how heterotrophs obtain nourishment; give examples of heterotrophic organisms._____

   _____

   _____

   _____

3. Briefly state the relationship between photosynthesis and respiration._____

   _____

   _____

Fill-in-the-Blanks

(4)_____ obtain carbon and energy from the physical environment; their carbon source is

(5)_____ _____. (6) _____ autotrophs obtain energy from sunlight. (7)_____

autotrophs are represented by a few kinds of bacteria; they obtain energy by stripping (8)_____ from sulfur or other inorganic substances. (9)_____ feed on autotrophs, each other, and organic wastes; representatives include (10)_____ , fungi, many protistans, and most bacteria. Although energy stored in organic compounds such as glucose may be released by several pathways, the pathway known as (11)_____ _____ releases the most energy.

PHOTOSYNTHESIS (7-I, pp. 105–107)

1. In the space below, supply the missing information to complete the summary equation for photosynthesis:

$$12 \underline{\quad} + \underline{\quad} CO_2 \longrightarrow \underline{\quad} O_2 + C_6H_{12}O_6 + 6 \underline{\quad}$$

2. Supply the appropriate information to state the equation (above) for photosynthesis in words:

_____ molecules of water plus six molecules of _____ (in the presence of pigments, enzymes, and ultraviolet light) yield six molecules of _____ plus one molecule of _____ plus _____ molecules of water.

Fill-in-the-Blanks

The two major sets of reactions of photosynthesis are the (3)_____-_____ reactions and the (4)_____-_____ reactions. (5)_____ _____ and (6)_____ are the reactants of photosynthesis, and the end product is usually given as (7)_____. The internal membranes and channels of the chloroplast are the (8) _____ membrane system and are organized into stacks, called (9) _____. Spaces inside the thylakoid disks and channels form a continuous compartment where (10)_____ ions accumulate to be used to produce ATP. The semifluid interior area surrounding the grana is known as the (11)_____ and is the area where the products of photosynthesis are produced.

LIGHT-DEPENDENT REACTIONS (7-II, pp. 107–112)

Fill-in-the-Blanks

The light-capturing phase of photosynthesis takes place on (1)_____ membranes, which are arranged in stacks called (2) _____. This system forms a single compartment separate from the (3)_____ portion of the chloroplast. A (4)_____ is a packet of light energy. Thylakoid membranes contain (5)_____ , which absorb photons of light. The principal pigments are the (6)_____ , which reflect green wavelengths but absorb (7)_____ and (8) _____ wavelengths. (9)_____ are pigments that absorb violet and blue wavelengths but reflect yellow, orange, and red. A cluster of 200 to 300 of these pigment proteins is a (10)_____. When pigments absorb (11)_____ energy, an (12)_____ is transferred from a photosystem to an (13)_____ molecule. (14)_____ refers to the attachment of phosphate to ADP or other organic molecules. Due to the input of light energy, electrons flow through a transport system where ATP is produced in a process specifically called (15) _____. Electrons then end up in (16)_____ chlorophyll at the end of this transport chain.

17. In the table below, identify and state the role of each item given in cyclic photophosphorylation of the light-dependent reactions.

a. Photosystem I _____

b. Electron acceptor _____

c. P700 _____

d. Electron transport system _____

e. ADP _____

f. Electrons _____

Identifying

The diagram below illustrates noncyclic photophosphorylation. Identify each numbered part of the illustration.

18. _____ _____

19. _____ _____

 _____

20. _____ _____

21. _____ _____

22. _____

23. _____

24. _____

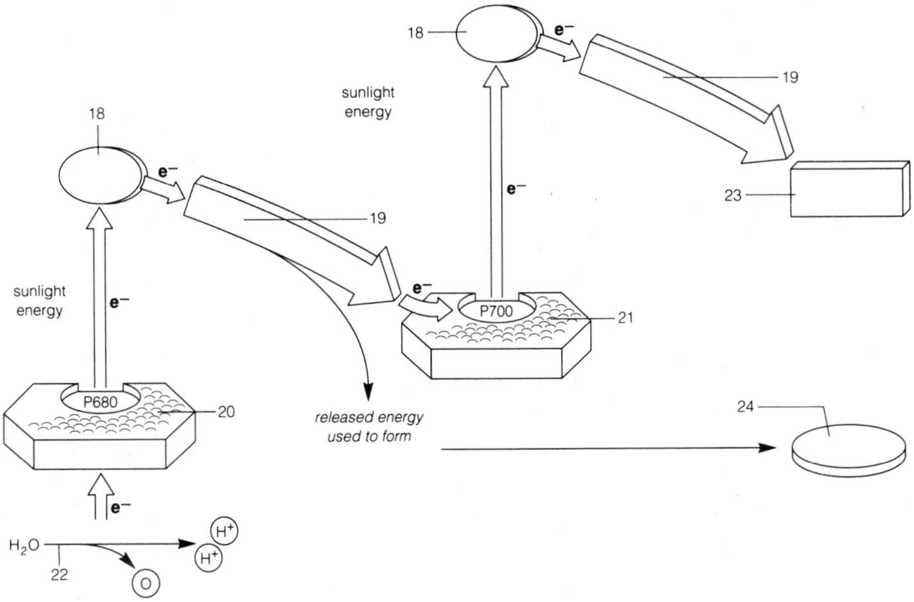

Fill-in-the-Blanks

Electrons ejected from chlorophyll molecules in photosystems on thylakoid membranes pass through one or two (25)_____ _____ systems. When light energy powers the formation of ATP molecules by electrons passing through electron transport systems, it is called (26)_____. Pathways of this type are either (27)_____ or (28)_____. The special chlorophyll found in photosystem I is referred to as (29)_____. The (30)_____ pathway is the simplest and operates with electrons traveling in a circle and producing only ATP. Today, land plants rely mostly on the (31)_____ pathway, which creates ATP and NADPH as energy carriers. This more complex pathway begins at photosystem II, which has a special chlorophyll molecule, (32)_____. This molecule absorbs light energy and ejects an electron that passes over an electron transport system to be accepted by chlorophyll (33)_____ of photosystem I. (34)_____ absorbs light energy, excites electrons, and ejects them to a second (35)_____ _____. Two electrons and one hydrogen (36)_____ are attached to (37)_____ , forming NADPH. (38)_____ splits water and releases oxygen, protons, and electrons. Electrons flow from split water to replace those given up by chlorophyll (39)_____ in photosystem II. (40)_____ from split water accumulates in the atmosphere and makes aerobic respiration possible. Hydrogen ions from split water accumulate inside (41)_____ compartments and build up concentration and electric gradients. The force of these gradients moves hydrogen ions through (42)_____ proteins, the force used to produce ATP. The concept that concentration and electric gradients across a membrane drive ATP formation is known as the (43)_____ theory.

LIGHT-INDEPENDENT REACTIONS (7-III, pp. 112–115)

Label-Match

Identify each part of the illustration below. Complete the exercise by matching and entering the letter of the proper function description in the parentheses following each label.

1. _____ _____ ( )

2. _____ _____ _____ ( )

3. _____ ( )

4. _____ _____ ( )

5. _____ _____ _____ _____ ( )

6. _____ ( )

7. _____ _____ ( )

8. _____-_____ _____ ( )

9. _____ _____ ( )

A. A three-carbon sugar, the first sugar produced; goes on to form sugar phosphate and RuBP

B. Typically used at once to form carbohydrate end products of photosynthesis

C. A five-carbon compound produced from PGALs; attaches to incoming CO_2

D. A compound that diffuses into leaves; attached to RuBP by enzymes in photosynthetic cells

E. Includes all the chemistry that "fixes" carbon into an organic compound

F. Three-carbon compounds formed from the splitting of the six-carbon intermediate compound

G. A molecule that was reduced in noncyclic photophosphorylation; furnishes hydrogen atoms to construct sugar molecules

H. A product of the light-dependent reactions; necessary in the light-independent chemistry to energize molecules for reactions

I. Includes all the chemistry that converts PGA to PGAL and PGAL to RuBP and sugar phosphates

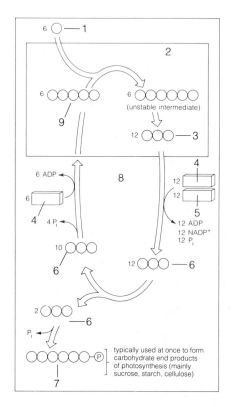

Fill-in-the-Blanks

The light-independent reactions can proceed without sunlight as long as (10)_____ and (11)_____ are available. The reactions begin when an enzyme links (12)_____ _____ to (13)_____ _____ , a five-carbon compound. The resulting six-carbon compound is highly unstable and breaks apart at once into two molecules of a three-carbon compound, (14)_____. This entire reaction sequence is called carbon dioxide (15)_____. ATP gives a phosphate group to each (16)_____. This intermediate compound takes on H^+ and electrons from NADPH to form (17)_____. It takes (18)_____ carbon dioxide molecules to produce twelve PGAL. Most of the PGAL becomes rearranged into new (19)_____ molecules—which can be used to fix more (20)_____ _____ . Two (21) _____ are joined together to form a (22)_____ _____ , primed for further reactions. The Calvin-Benson cycle yields enough RuBP to replace those used in carbon dioxide (23)_____. ADP, $NADP^+$, and phosphate leftovers are sent back to the (24)_____-_____ reaction sites, where they are again converted to (25)_____ and (26)_____. (27)_____ _____ formed in the cycle serves as a building block for the plant's main carbohydrates. When RuBP attaches to oxygen instead of carbon dioxide, (28)_____ results; this is typical of (29)_____ plants in hot, dry conditions. If less PGA is available, leaves produce a reduced amount of (30) _____ . C4 plants can still construct carbohydrates when the ratio of carbon dioxide to (31)_____ is unfavorable because of the attachment of carbon dioxide to (32)_____ in certain leaf cells.

CHEMOSYNTHESIS (7-IV, p. 115)

Fill-in-the-Blanks

Organisms that obtain energy from oxidation of inorganic substances such as ammonium compounds, and iron or sulfur compounds, are known as (1)_____ autotrophs. Such organisms use this energy to build (2)_____ compounds. As an example, some soil bacteria use ammonia molecules as an energy source, stripping them of (3)_____ and (4)_____ ; this leaves (5)_____ and (6)_____ ions that are readily washed out of the soil, thus lowering its (7) _____.

Chapter Terms

The following page-referenced terms are important; they were in boldface type in the chapter. Refer to the instructions given in Chapter 1, p. 4.

autotrophs (105)
heterotrophs (105)
photosynthesis (105)
aerobic respiration (105)
thylakoid membrane (107)

light-dependent reactions (107)
photosystems (110)
electron transport systems (110)
cyclic photophosphorylation (110)
noncyclic photophosphorylation (110)

photolysis (111)
chemiosmotic theory (112)
light-independent reactions (112)
Calvin-Benson cycle (113)
carbon dioxide fixation (113)

Self-Quiz

___ 1. The electrons that are passed to NADPH during noncyclic photophosphorylation were obtained from _____.
   a. water
   b. CO_2
   c. glucose
   d. sunlight

___ 2. Cyclic photophosphorylation functions mainly to _____.
   a. fix CO_2
   b. make ATP
   c. produce PGAL
   d. regenerate ribulose biphosphate

___ 3. Chemosynthetic autotrophs obtain energy by oxidizing such inorganic substances as _____.
   a. PGA
   b. PGAL
   c. sulfur
   d. water

___ 4. The ultimate electron and hydrogen acceptor in noncyclic photophosphorylation is _____.
   a. $NADP^+$
   b. ADP
   c. O_2
   d. H_2O

___ 5. C4 plants have an advantage in hot, dry conditions because _____.
   a. their leaves are covered with thicker wax layers than those of C3 plants
   b. their stomates open wider than those of C3 plants, thus cooling their surfaces
   c. special leaf cells possess a means of capturing CO_2 even in stress conditions
   d. they are also capable of carrying on photorespiration

___ 6. Chlorophyll is_____.
   a. on the outer chloroplast membrane
   b. inside the mitochondria
   c. in the stroma
   d. in the thylakoids

___ 7. Thylakoid disks are stacked in groups called_____.
   a. grana
   b. stroma
   c. lamellae
   d. cristae

___ 8. Plant cells produce O_2 during photosynthesis by_____.
   a. splitting CO_2
   b. splitting water
   c. degradation of the stroma
   d. breaking up sugar molecules

___ 9. Plants need _____ and _____ to carry on photosynthesis.
   a. oxygen; water
   b. oxygen; CO_2
   c. CO_2; H_2O
   d. sugar; water

___ 10. The two products of the light-dependent reactions that are required for the light-independent chemistry are _____ and _____.
   a. CO_2; H_2O
   b. O_2; NADPH
   c. O_2; ATP
   d. ATP; NADPH

Chapter Objectives/Review Questions

This section lists general and detailed chapter objectives that can be used as review questions. You can make maximum use of these items by writing answers on a separate sheet of paper. Fill in answers where blanks are provided. To check for accuracy, compare your answers with information given in the chapter or glossary.

Page   *Objectives/Questions*

(105)   1. Distinguish between organisms known as autotrophs and those known as heterotrophs.
(105)   2. _____ autotrophs can obtain their energy from sunlight; _____ autotrophs get energy by stripping electrons from sulfur or some other inorganic substance.

Page	Objectives/Questions
(105)	3. _____ is the metabolic pathway that is most efficient in releasing the energy stored in organic compounds.
(105–106)	4. List the major stages of photosynthesis and state what occurs in those sets of reactions.
(114)	5. Study the general equation for photosynthesis as shown on p. 114 of the main text until you can remember the reactants and products. Reproduce the equation from memory on another piece of paper.
(114)	6. The energy-poor molecules in the photosynthetic equations are _____ and _____.
(106)	7. Describe the details of a familiar site of photosynthesis, the green leaf. Begin with the layers of a leaf cross-section and complete your description with the minute structural sites within the chloroplast where the major sets of photosynthetic reactions occur.
(107)	8. The flattened channels and disklike compartments inside the chloroplast are organized into stacks, called _____ , which are surrounded by a semifluid interior, the _____ ; this is the _____ membrane system.
(108, 110)	9. Describe how the pigments found on thylakoid membranes are organized into photosystems and how they relate to photon light energy.
(108, 110)	10. Describe the role that chlorophylls and the other pigments found in chloroplasts play to initiate the light-dependent reactions. After consulting Figure 7.3 of the main text, state which colors of the visible spectrum are absorbed by (a) chlorophyll a, (b) chlorophyll b, and (c) carotenoids.
(109)	11. State what T. Englemann's 1882 experiment with *Cladophora* revealed.
(110)	12. Describe the function of electron transport systems in the thylakoid membrane.
(110–111)	13. Contrast cyclic pathways and noncyclic pathways (photophosphorylation) in terms of the substances produced and the number of transport chains used.
(111)	14. Explain what the water split during photolysis contributes to both cyclic and noncyclic pathways of the light-dependent reactions.
(110–111)	15. Two energy-carrying molecules produced in the noncyclic pathways are _____ and _____ ; explain why these molecules are necessary for the light-independent reactions.
(112–113)	16. Explain how the chemiosmotic theory is related to thylakoid compartments and the production of ATP.
(111)	17. After evolution of the noncyclic pathway, _____ accumulated in the atmosphere and made _____ respiration possible.
(112)	18. Explain why the light-independent reactions are called by that name.
(113)	19. Describe the process of carbon dioxide fixation by stating which reactants are necessary to initiate the process and what stable products result from the process.
(113–114)	20. Describe the Calvin-Benson cycle in terms of its reactants and products.
(114)	21. State the fate of all the sugar phosphates produced by photosynthetic autotrophs.
(114–115)	22. Describe the mechanism by which C4 plants thrive under hot, dry conditions; distinguish this CO_2-capturing mechanism from that of C3 plants.

Integrating and Applying Key Concepts

Suppose that humans acquired all the enzymes needed to carry out photosynthesis. Speculate about the attendant changes in human anatomy, physiology, and behavior that would be necessary for those enzymes to actually carry out photosynthetic reactions.

Critical Thinking Exercises

1. An important early experiment on photosynthesis showed that the rate of carbon dioxide fixation was greater when the light was delivered in a series of short flashes instead of by constant illumination. In both cases, the intensity and *total* duration of the light were the same; only the distribution of the light was varied. Which of the following is the best interpretation of this observation?

a. At least two different pigments trap light energy during photosynthesis.
b. ATP and NADPH are intermediates formed in the light-dependent reactions and consumed in the light-independent reactions.
c. Flashing light is a more efficient energy source for photosynthesis than continuous light.
d. High concentrations of carbon dioxide were continually present during this experiment.
e. Photosynthesis takes place in two steps, one light-dependent, the other independent of light.

ANALYSIS

a. This conclusion does not follow from the observations. There is no evidence here about the wavelength-dependence of photosynthesis or light absorption by the chloroplasts.
b. This conclusion is too detailed. No measurements were made of the amount of ATP or NADPH at any time in the experiment, nor was either of these compounds varied. It is not even indicated that either was present at all.
c. This is not an interpretation of the observations; it is a *statement* of the observations. The rate of photosynthesis was greater when the same amount of light energy was delivered in flashes than when it was continuous.
d. This is an assumption that is necessary in order to interpret the experiment. If carbon dioxide concentration were reduced during one of the experiments, photosynthesis would be expected to slow, because carbon dioxide is a substrate for a photosynthetic enzyme.
e. This interpretation is valid. If photosynthesis were a single-phase process, it would proceed at the same rate whenever the light was on. On the other hand, if the light energy is trapped by one process and then transferred to a second process that can continue in the dark, *and* if the second process is slower than the first, the energy transfer between the phases will be a bottleneck. The energy-trapping molecules of the first process will soon be "filled" when the light is on, and no more light energy can be trapped until the second process has had time to "empty" the energized molecules. Any light that arrives when no energy acceptors are available will be wasted. But if the light is on only briefly, the acceptors will be regenerated during the brief dark period, and more of the incident light will be absorbed when the light comes on again.

2. Observation: Plants release oxygen gas. Assumption: Chlorophyll is the pigment that absorbs the light energy for photosynthesis. Hypothesis: Oxygen is a product of photosynthesis. Which of the following is predicted by this hypothesis?

a. Water is a reactant as well as a product in photosynthesis.
b. At least some of the oxygen atoms in carbon dioxide appear in the starch formed by photosynthesis.
c. Oxygen is released at a greater rate when plants are illuminated with red light than when they receive green light.
d. The leaf surface contains pores through which carbon dioxide and oxygen can move.
e. The oxygen atoms in water do not appear in the starch formed by photosynthesis.

ANALYSIS

a. The hypothesis does not predict the source of the oxygen, only that it will be released when photosynthesis occurs. Furthermore, it does not predict that oxygen atoms will be found in products of photosynthesis. If anything, the hypothesis makes it less likely that oxygen will be released in any form other than oxygen gas.
b. Once again, the hypothesis does not predict the source or other destinations of the oxygen, only that oxygen gas will be released in coordination with the rate of photosynthesis.
c. Chlorophyll absorbs red light but not green light. If we assume that chlorophyll absorbs the light energy that drives photosynthesis, we expect that photosynthesis will go much faster in red light than in green light. The hypothesis then predicts that oxygen evolution will be faster when photosynthesis is faster, that is, in red light.
d. Because plants are observed to release oxygen and absorb carbon dioxide, we expect that there will be a pathway for those gases to leave and enter the plant. This would be true whatever hypothesis was made about how the plant produces the oxygen.

e. The hypothesis can predict only that oxygen atoms from some substrate of photosynthesis will be released as gas when a plant is illuminated. It cannot predict which molecule will be the source or where else oxygen atoms might go.

3. Before 1930, it was thought that photosynthesis split carbon dioxide. The carbon atoms were thought to combine with water to form starch, and the oxygen atoms were thought to be released as free oxygen gas. Which of the following operations could provide evidence contradicting this view?

a. Expose a plant in the light to carbon dioxide that contains radioactive carbon.
b. Increase the light intensity on a plant.
c. Increase the concentration of water vapor in the atmosphere around a plant.
d. Compare the total amounts of starch and oxygen produced by a plant over a certain length of time.
e. Demonstrate that a plant cell can release oxygen gas when no carbon dioxide is available.

ANALYSIS

a. This would allow the carbon atom to be traced but would give no information about the source of the light-dependent oxygen.
b. This would be expected to increase the rate of oxygen evolution but would give no information about the source of the oxygen atoms.
c. Water is consumed in photosynthesis and contains oxygen. However, increasing the availability of a substrate would be expected to increase the rate of the process regardless of the specific pathway taken by the substance. Both the older view and possible alternatives predict the same outcome from this operation. Furthermore, the water used in photosynthesis is brought to the leaves via the roots and stem, not from atmospheric vapor.
d. The relative amounts of starch and oxygen produced would be the same regardless of the identity of the source materials.
e. The older hypothesis predicts that no oxygen can be evolved in the absence of carbon dioxide. If a plant can produce oxygen when no carbon dioxide is present, the oxygen source must be some other substance.

Answers

Answers to Interactive Exercises

VIGNETTE: SUNLIGHT, RAIN, AND CELLULAR WORK

1. Autotrophs obtain carbon and energy from the physical environment (self-nourishing). Photosynthetic autotrophs, such as plants, some protistans, and some bacteria, obtain energy from sunlight; chemosynthetic autotrophs (a few kinds of bacteria) get energy by stripping electrons from sulfur or other inorganic substances.
2. Heterotrophs feed on autotrophs, each other, and organic wastes. Animals, fungi, many protistans, and most bacteria are heterotrophs.
3. Carbon and energy enter the web of life primarily by photosynthesis. This energy is stored in organic compounds and released by several pathways. The pathway called aerobic respiration releases the most energy.
4. Autotrophs; 5. carbon dioxide; 6. Photosynthetic; 7. Chemosynthetic; 8. electrons; 9. Heterotrophs; 10. animals; 11. aerobic respiration.

PHOTOSYNTHESIS (7-I)

1. $12 H_2O + 6CO_2 \longrightarrow 6O_2 + C_6H_{12}O_6 + 6H_2O$
2. Twelve molecules of water plus six molecules of carbon dioxide (in the presence of pigments, enzymes, and visible light) yield six molecules of oxygen plus one molecule of glucose plus six molecules of water.
3. light-dependent (light-independent);
4. light-independent (light-dependent); 5. Carbon dioxide; 6. water; 7. glucose; 8. thylakoid; 9. grana; 10. hydrogen; 11. stroma.

LIGHT-DEPENDENT REACTIONS (7-II)

1. thylakoid; 2. grana; 3. stroma; 4. photon; 5. pigments; 6. chlorophylls; 7. red (blue); 8. blue (red); 9. Carotenoids; 10. photosystem; 11. light (photon); 12. electron; 13. acceptor; 14. Phosphorylation; 15. photophosphorylation; 16. P700; 17. a. Photosystem I: a pigment cluster dominated by P700; b. Electron acceptor: a molecule that accepts electrons ejected from chlorophyll P700 and then passes electrons down the electron transport system;

c. P700: a special chlorophyll molecule that absorbs wavelengths of 700 nanometers and then ejects electrons; d. Electron transport system: electrons flow through this system, which is composed of a series of molecules bound in the thylakoid membrane that drive photophosphorylation; e. ADP: ADP undergoes photophosphorylation in cyclic photophosphorylation to become ATP; f. Electrons: electrons representing energy are ejected from P700 to an electron acceptor but move over the electron transport system, where some of the energy is used to produce ATP.

18. electron acceptor; 19. electron transport system; 20. photosystem II; 21. photosystem I; 22. photolysis; 23. NADPH; 24. ATP; 25. electron transport; 26. photophosphorylation; 27. cyclic (noncyclic); 28. noncyclic (cyclic); 29. P700; 30. cyclic; 31. noncyclic; 32. P680; 33. P700; 34. P700; 35. transport system; 36. ion; 37. NADP$^+$; 38. Photolysis; 39. P680; 40. Oxygen; 41. thylakoid; 42. channel; 43. chemiosmotic.

LIGHT-INDEPENDENT REACTIONS (7-III)

1. carbon dioxide (D); 2. carbon dioxide fixation (E); 3. phosphoglycerate (F); 4. adenosine triphosphate (H); 5. nicotinamide adenine dinucleotide phosphate (G); 6. phosphoglyceraldehyde (A); 7. sugar phosphates (B); 8. Calvin-Benson cycle (I); 9. ribulose bisphosphate (C); 10. ATP (NADPH); 11. NADPH (ATP); 12. carbon dioxide; 13. ribulose bisphosphate; 14. PGA; 15. fixation; 16. PGA; 17. PGAL; 18. six; 19. RuBP; 20. carbon dioxide; 21. PGALs; 22. sugar phosphate; 23. fixation; 24. light-dependent; 25. ATP (NADPH); 26. NADPH (ATP); 27. Sugar phosphate; 28. photorespiration; 29. C4; 30. food; 31. oxygen; 32. oxaloacetate.

CHEMOSYNTHESIS (7-IV)

1. chemosynthetic; 2. organic; 3. protons (electrons); 4. electrons (protons); 5. nitrate (nitrite); 6. nitrite (nitrate); 7. fertility.

Answers to Self-Quiz

1. a; 2. b; 3. c; 4. a; 5. c; 6. d; 7. a; 8. b; 9. c; 10. d.

8

ENERGY-RELEASING PATHWAYS

ATP-PRODUCING PATHWAYS
AEROBIC RESPIRATION
     Overview of the Reactions
     Glycolysis
     Krebs Cycle
     Electron Transport Phosphorylation
     Glucose Energy Yield

ANAEROBIC ROUTES
     Alcoholic Fermentation
     Lactate Fermentation
     Anaerobic Electron Transport
ALTERNATIVE ENERGY SOURCES
IN THE HUMAN BODY
     Commentary: Perspective on Life

Interactive Exercises

In most cases, answers for specific molecules use abbreviations.

ATP-PRODUCING PATHWAYS (8-I, pp. 119–120)

1. Although various organisms utilize different energy sources, what is the usual form of chemical energy that will drive metabolic reactions?_____

2. Describe the function of oxygen in the main degradative pathway, aerobic respiration._____
   _____
   _____

3. List the most common anaerobic pathways and describe the conditions in which they function._____
   _____
   _____

Fill-in-the-Blanks

Virtually all forms of life depend on a molecule known as (4)_____ as their primary energy carrier.
Plants produce adenosine triphosphate during (5)_____ , but plants and all other organisms also can
produce ATP through chemical pathways that degrade food molecules. The main degradative pathway
requires free oxygen and is called (6) _____ _____. Other degradative pathways are
(7)_____ , in that something other than oxygen serves as the final electron acceptor in
energy-releasing reactions. (8)_____ and anaerobic (9)_____ _____ are the most common
anaerobic pathways.

AEROBIC RESPIRATION (8-II, pp. 120–128)

1. Complete the equation below, which summarizes the degradative pathway known as aerobic respiration:

$$\_\_\_\_ + \_\_\_\_ H_2O \longrightarrow 6\_\_\_\_ + 6\_\_\_\_$$

2. Supply the appropriate information to state the equation (above) for aerobic respiration in words:

One molecule of glucose plus six molecules of _____ (in the presence of appropriate enzymes)

yield _____ molecules of carbon dioxide plus _____ molecules of water.

Fill-in-the-Blanks

There are three stages of aerobic respiration. In the first stage, (3)_____ , glucose is partially degraded to (4) _____. By the end of the second stage, which includes the (5) _____ cycle, glucose has been completely degraded to carbon dioxide and (6)_____. Neither of the first two stages produces much (7)_____. During both stages, protons and (8)_____ are stripped from intermediate compounds and delivered to a (9)_____ system. That system is used in the third stage of reactions, electron transport (10)_____ , which produces a high yield of (11)_____. (12)_____ accepts the "spent" electrons from the transport system and keeps it cleared for repeated operation.

Labeling

In exercises 13–17, identify the structure or location; in exercises 18–21, identify the chemical substance involved.

13. _____ _____ of mitochondrion
14. _____ _____ of mitochondrion
15. _____ _____ of mitochondrion
16. _____ _____ of mitochondrion

18. _____
19. _____
20. _____
21. _____

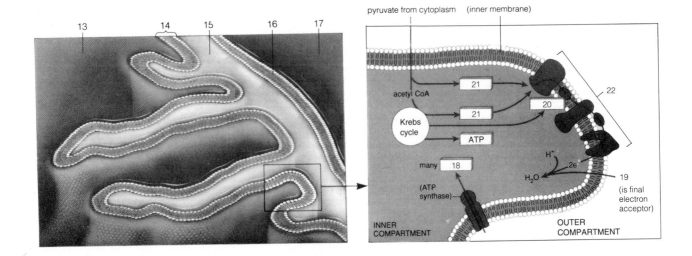

Fill-in-the-Blanks

If sufficient oxygen is present, the end product of glycolysis enters (23)_____ pathways (acetyl-CoA formation), the (24)_____ cycle, plus (25)_____ _____ phosphorylation, during which processes (26)_____ (number) additional (27)_____ molecules are generated. In the preparatory conversions prior to the Krebs cycle and within the Krebs cycle, the food molecule fragments are further broken down into (28)_____ _____. During these reactions, hydrogen atoms (with their (29)_____) are stripped from the fragments and transferred to the energy carriers (30)_____ and (31)_____. NADH delivers its electrons to the highest possible point of entry into a transport system; from each NADH enough H^+ is pumped to produce (32)_____ (number) ATP molecules. $FADH_2$ delivers its electrons at a lower point of entry into the transport system; fewer H^+ are pumped, and (33)_____ (number) ATPs are produced. The electrons are then sent down highly organized (34)_____ systems located in the inner membrane of the mitochondrion; hydrogen ions are pumped into the outer mitochondrial compartment. According to (35)_____ theory, the hydrogen ions accumulate and then follow a gradient to flow through ATP (36)_____: channel proteins that lead into the inner compartment. The energy of the hydrogen ion flow across the membrane is used in forming (37)_____. Electrons leaving the electron transport system combine with hydrogen ions and (38)_____ to form water. These reactions occur only in (39)_____. From glycolysis (in the cytoplasm) to the final reactions occurring in the mitochondria, the aerobic pathway commonly yields (40)_____ (number) ATP or (41)_____ (number) ATP for every glucose molecule degraded.

ANAEROBIC ROUTES (8-III, pp. 128–130)

Fill-in-the-Blanks

(1)_____ organisms can synthesize and stockpile energy-rich carbohydrates and other food molecules from inorganic raw materials. (2)_____ is partially dismantled by the glycolytic pathway, during which process some of its stored energy remains in two (3)_____ molecules. Some of the energy of glucose is released during the breakdown reactions and used in forming the energy carriers (4)_____ and (5)_____. These reactions take place in the cytoplasm. If (6)_____ is not present in sufficient amounts, the end product of glycolysis enters (7)_____ pathways; in some bacteria and muscle cells, it is converted into such products as (8) _____ , or in yeast cells it is converted into (9)_____ and (10)_____ _____.

Multiple-Choice

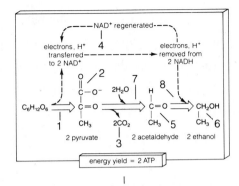

___ 11. Which of the above figures represents alcoholic fermentation?
   a. I
   b. II

___ 12. In the figures above, where does glycolysis occur?
   a. 7
   b. 1
   c. 4
   d. 8

___ 13. In the figures above, what represents the end product of glycolysis?
   a. 1
   b. 2
   c. 3
   d. 4

___ 14. Which of the above figures represents chemistry occurring in animal cells?
   a. I
   b. II

___ 15. In the figures above, select the site where ATP molecules are produced.
   a. 7
   b. 1
   c. 4
   d. 6

___ 16. In the figures above, to what site is the coenzyme NAD carrying electrons and hydrogens?
   a. 1
   b. 5
   c. 6
   d. 7

___ 17. Which of the figures above represents chemistry occurring in yeast cells?
   a. I
   b. II

Fill-in-the-Blanks

Anaerobic electron transport is an energy-releasing pathway occurring among the (18)_____.

Sulfate-reducing bacteria produce (19)_____ by stripping electrons from a variety of compounds and sending them through membrane transport systems. The inorganic compound (20)_____ (SO_4^{--}) serves as the final electron acceptor and is converted into sulfide (H_2S). Other kinds of bacteria produce ATP by stripping electrons from nitrate (NO_3^{--}), leaving (21)_____ (NO_2^-) as the end product. These bacteria are important in the global cycling of (22)_____.

ALTERNATIVE ENERGY SOURCES IN THE HUMAN BODY (8-IV, pp. 130–131)

True-False

If false, explain why.

___ 1. Glucose is the only carbon-containing molecule that can be fed into the glycolytic pathway.

___ 2. Fats are efficient long-term energy-storage molecules.

___ 3. Simple sugars, fatty acids, and glycerol that remain after a cell's biosynthetic needs have been met are generally sent to the cell's respiratory pathways for energy extraction.

___ 4. Carbon dioxide and water, the products of aerobic respiration, generally get into the blood and are carried to gills or lungs, kidneys, and skin, where they are expelled from the animal's body.

___ 5. Energy is recycled along with materials.

___ 6. The first forms of life on Earth were most probably photosynthetic eukaryotes.

___ 7. Photosynthesis produces molecular oxygen as a by-product.

___ 8. Energy flows only from forms rich in potential energy to forms with fewer usable stores of energy.

Labeling

Identify the process or substance indicated in the illustration below.

9. _____ _____

10. _____

11. _____

12. _____ _____

13. _____ _____

14. _____ _____

15. _____

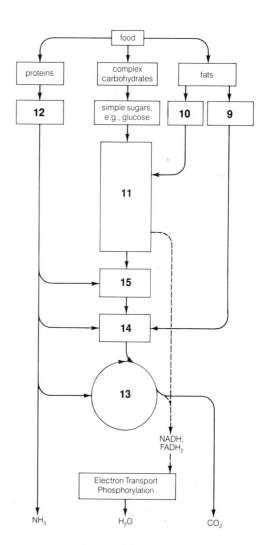

Chapter Terms

The following page-referenced terms are important; they were in boldface type in the chapter. Refer to the instructions given in Chapter 1, p. 4.

ATP (120)
aerobic respiration (120)
fermentation (120)
anaerobic electron transport (120)

glycolysis (122)
PGAL (122)
substrate-level phosphorylation (122)
Krebs cycle (124)

chemiosmotic theory (126)
alcoholic fermentation (128)
lactate fermentation (129)
anerobic electron transport (130)

Self-Quiz

___ 1. Glycolysis would quickly halt if the process ran out of _____ , which serves as the hydrogen and electron acceptor.
  a. $NADP^+$
  b. ADP
  c. NAD^+
  d. H_2O

___ 2. The ultimate electron acceptor in aerobic respiration is _____.
  a. NADH
  b. carbon dioxide (CO_2)
  c. oxygen ($1/2 O_2$)
  d. ATP

___ 3. When glucose is used as an energy source, the largest amount of ATP is generated by the _____ portion of the entire respiratory process.
  a. glycolytic pathway
  b. acetyl-CoA formation
  c. Krebs cycle
  d. electron transport phosphorylation

___ 4. The process by which about 10 percent of the energy stored in a sugar molecule is released as it is converted into two small organic-acid molecules is
  _____.
  a. photolysis
  b. glycolysis
  c. fermentation
  d. the dark reactions

___ 5. During which of the following phases of aerobic respiration is ATP produced directly by substrate-level phosphorylation?
  a. glucose formation
  b. ethyl-alcohol production
  c. acetyl-CoA formation
  d. the Krebs cycle

___ 6. What is the name of the process by which reduced NADH transfers electrons along a chain of acceptors to oxygen so as to form water and in which the energy released along the way is used to generate ATP?
  a. glycolysis
  b. acetyl-CoA formation
  c. the Krebs cycle
  d. electron transport phosphorylation

___ 7. Pyruvic acid can be regarded as the end product of _____ .
  a. glycolysis
  b. acetyl-CoA formation
  c. fermentation
  d. the Krebs cycle

___ 8. Which of the following is *not* ordinarily capable of being reduced at any time?
  a. NAD
  b. FAD
  c. oxygen, O_2
  d. water

___ 9. ATP production by chemiosmosis involves _____.
  a. H^+ concentration and electric gradients across a membrane
  b. ATP synthases
  c. formation of ATP in the inner mitochondrial compartment
  d. all of the above

___ 10. During the fermentation pathways, a net yield of two ATP is produced from _____ ; the NAD^+ necessary for _____ is regenerated during the reactions.
  a. the Krebs cycle; glycolysis
  b. glycolysis; electron transport phosphorylation
  c. the Krebs cycle; electron transport phosphorylation
  d. glycolysis; glycolysis

Matching

Match the following components of respiration to the list of words below. Some components may have more than one answer.

11. ___ lactic acid

12. ___ NAD$^+$ → NADH

13. ___ carbon dioxide

14. ___ NADH$^+$ → NAD$^+$

15. ___ pyruvate

16. ___ ATP produced by substrate-level

     phosphorylation

17. ___ glucose

18. ___ citrate

19. ___ oxygen

20. ___ water

A. Glycolysis
B. Preparatory conversions prior to the Krebs cycle
C. Fermentation
D. Krebs cycle
E. Electron transport

Chapter Objectives/Review Questions

This section lists general and detailed chapter objectives that can be used as review questions. You can make maximum use of these items by writing answers on a separate sheet of paper. Fill in answers where blanks are provided. To check for accuracy, compare your answers with information given in the chapter or glossary.

Page   *Objectives/Questions*

(119–120)   1. No matter what the source of energy might be, organisms must convert it to _____ , a form of chemical energy that can drive metabolic reactions.

(120)   2. The main degradative pathway is _____ respiration.

(120)   3. List the main anaerobic degradative pathways and the types of organisms that use them.

(120)   4. Give the overall equation for the aerobic respiratory route; indicate where energy occurs in the equation.

(120)   5. In the first of the three stages of aerobic respiration, _____ is partially degraded to pyruvate.

(120)   6. By the end of the second stage of aerobic respiration, which includes the _____ cycle, _____ has been completely degraded to carbon dioxide and water.

(120)   7. Do the first two stages of aerobic respiration yield a high or low quantity of ATP?

(120)   8. The third stage of aerobic respiration is called electron transport _____ ; it yields many ATP molecules.

(120)   9. Explain, in general terms, the role of oxygen in aerobic respiration.

(122)   10. Glycolysis occurs in the _____ of the cell.

(122)   11. Explain the purpose served by molecules of ATP reacting first with glucose and then with fructose-6-phosphate in the early part of glycolysis (see Figure 8.4 in the text).

(122)   12. Four ATP molecules are produced by _____-_____ phosphorylation for every two used during glycolysis. Consult Figure 8.5 in the text.

(122)   13. Glycolysis produces _____ (number) NADH, _____ (number) ATP (net) and _____ (number) pyruvate molecules for each glucose molecule entering the reactions.

(125)   14. Consult Figure 8.6 in the text. State the events that happen during acetyl-CoA formation and explain how the process of acetyl-CoA formation relates glycolysis to the Krebs cycle.

(124)   15. State the factors that cause pyruvic acid to enter the acetyl-CoA formation pathway.

(124)   16. What happens to the CO_2 produced during acetyl-CoA formation and the Krebs cycle?

Page	Objectives/Questions

(127) 17. Consult Figure 8.8 in the text and predict what will happen to the NADH produced during acetyl-CoA formation and the Krebs cycle.

(125) 18. Calculate the number of ATP molecules produced during the Krebs cycle for each glucose molecule that enters glycolysis.

(125–126) 19. Explain how chemiosmotic theory operates in the mitochondrion to account for the production of ATP molecules.

(126) 20. Briefly describe the process of electron transport phosphorylation by stating what reactants are needed and what the products are. State how many ATP molecules are produced through operation of the transport system.

(127–128) 21. Be able to account for the total *net yield* of thirty-six ATP molecules produced through aerobic respiration; that is, state how many ATPs are produced in glycolysis, the Krebs cycle, and the electron transport system.

(128) 22. List some places where there is very little oxygen present and where anaerobic organisms might be found.

(128) 23. Describe what happens to pyruvate in anaerobic organisms. Then explain the necessity for pyruvate to be converted to a fermentative product.

(128–130) 24. State which factors determine whether the pyruvate (pyruvic acid) produced at the end of glycolysis will enter into the alcoholic fermentation pathway, the lactate fermentation pathway, or the acetyl-CoA formation pathway.

(128–130) 25. In fermentation chemistry, _____ molecules from glycolysis are accepted to construct either lactate or ethyl alcohol; thus, a low yield of _____ molecules continues in the absence of oxygen.

(130–131) 26. List some sources of energy (other than glucose) that can be fed into the respiratory pathways.

(130) 27. Explain what cells do with simple sugars, amino acids, fatty acids, and glycerol that exceed what the cells need for synthesizing their own assortments of more complex molecules.

(131) 28. Predict what your body would do to synthesize its needed carbohydrates and fats if you switched to a diet of 100-percent protein.

(132–133) 29. After reading "Perspective on Life" in the main text, outline the supposed evolutionary sequence of energy-extraction processes.

(132) 30. Closely scrutinize the diagram of the carbon cycle in the Commentary; be able to reproduce the cycle from memory.

Integrating and Applying Key Concepts

How is the "oxygen debt" experienced by runners and sprinters related to aerobic and anaerobic respiration in humans?

Critical Thinking Exercises

1. Some of the mitochondrial electron transport carriers (the cytochromes) absorb light at characteristic wavelengths when they are carrying the extra electrons from NADH but not when they give up the electrons to the next carrier. Thus, the light absorption bands are all present when mitochondria are incubated in the absence of oxygen or when cyanide is present. Cyanide blocks the transfer of electrons from the last cytochrome to oxygen. When cyanide is absent and oxygen is introduced, the cytochromes stop absorbing light. There are three cytochromes in animal mitochondria, called *a*, *b*, and *c*. Another inhibitor of mitochondrial electron transfer is urethane. When urethane is added to aerobic mitochondria, cytochrome *b* begins to absorb light, but *a* and *c* do not. Finally, purified cytochrome *c*, even in the presence of oxygen, continues to absorb light. Use these observations to decide the order in which the cytochromes occur in the mitochondrial electron transport system.

ANALYSIS

The last observation demonstrates that cytochrome *c* cannot release its electrons directly to oxygen molecules. In other words, it is not last in the chain. Urethane blocks electron transfer. In the presence of urethane, cytochromes *a* and *c* give up their electrons to oxygen, but cytochrome *b* is prevented from transferring its electrons and continues to absorb light. This means that cytochrome *b* must occur first and transfer its electrons to one of the other cytochromes. Since *b* is first and *c* is not last, the order must be *b, c, a*.

2. Suppose you inoculate some yeast cells into a nutrient medium with glucose as an energy source. You seal the culture vessel, and at intervals you measure the growth of the cells and express the results as relative cell number. Your data are as shown:

Time (hrs)	Relative Cell Number
0	0.085
1	0.104
2	0.151
3	0.213
4	0.322
6	0.354
8	0.381
10	0.408

Make a graph of the data. What can you conclude about conditions in the culture vessel?

ANALYSIS

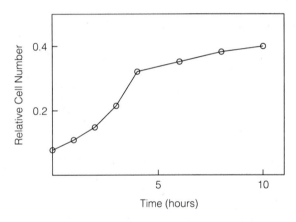

The graph shows that the yeast grew much faster for the first four hours than from four to ten hours. It is possible that some toxic metabolic product accumulated in the flask and inhibited growth, but in that case a steady decrease in growth would be expected, not a sudden change. Another hypothesis is that some essential nutrient was exhausted from the medium, but that would lead to the prediction of no growth at all after the nutrient was finished. Now, remember that yeast can use two ways to metabolize glucose, aerobic and anaerobic, and that aerobic metabolism produces many more ATP molecules per glucose molecule than does anaerobic metabolism. If we assume that growth rate depends on the supply of ATP and that only glucose metabolism produces ATP in this culture, we can make the hypothesis that the culture ran out of oxygen after four hours of aerobic metabolism and rapid growth. After that time, the cells were forced to ferment the glucose and grew more slowly on the limited ATP production.

3. Plant cells have both chloroplasts and mitochondria. They release oxygen when they are illuminated, and they consume oxygen when they are in the dark. Some investigators calculate the overall rate of photosynthesis of a sample of plant cells by adding the cells' rate of oxygen production in the light to their

rate of oxygen consumption in the dark. Which of the following assumptions are the investigators most likely making?

a. Respiration stops in the light.
b. Respiration slows in the light.
c. Respiration is unchanged in the light.
d. Respiration accelerates in the light.
e. Photosynthesis continues at a low rate in the dark.

ANALYSIS

The measured rate of oxygen production in the light is a *net* rate. It is the rate of photosynthetic oxygen production *minus* the rate of respiratory oxygen consumption, if any. If you know the rate of respiration, you can add it back to the net oxygen production and get the gross rate of production, the overall rate of photosynthesis. These investigators assume that the rate of respiration in the dark, when photosynthetic oxygen production is zero, does not change when the light comes on.

Answers

Answers to Interactive Exercises

ATP-PRODUCING PATHWAYS (8-I)
1. Adenosine triphosphate (ATP).
2. Oxygen accepts the "spent" electrons from the electron transport system and keeps that system clear for repeated operation.
3. Fermentation and anaerobic electron transport. Some organisms (including humans) use fermentation pathways when oxygen supplies are low; many microbes rely exclusively on anaerobic pathways that do not require an "outside" election acceptor such as oxygen.
4. ATP; 5. photosynthesis; 6. aerobic respiration; 7. anaerobic; 8. Fermentation; 9. electron transport.

AEROBIC RESPIRATION (8-II)
1. $C_6H_{12}O_6 + 6H_2O \longrightarrow 6CO_2 + 6H_2O$
2. One molecule of glucose plus six molecules of water (in the presence of appropriate enzymes) yield six molecules of carbon dioxide plus six molecules of water.
3. glycolysis; 4. pyruvate; 5. Krebs; 6. water; 7. ATP; 8. electrons; 9. transport; 10. phosphorylation; 11. ATP; 12. Oxygen; 13. inner compartment; 14. inner membrane; 15. outer compartment; 16. outer membrane; 17. cytoplasm; 18. ATP; 19. oxygen (O_2); 20. $FADH_2$; 21. NADH; 22. electron transport system;

23. aerobic; 24. Krebs; 25. electron transport; 26. thirty-four; 27. ATP; 28. carbon dioxide; 29. electrons; 30. NAD^+ (FAD); 31. FAD (NAD^+); 32. three; 33. two; 34. transport; 35. chemiosmotic; 36. synthases; 37. ATP; 38. oxygen; 39. mitochondria; 40. thirty-six (thirty-eight); 41. thirty-eight (thirty-six).

ANAEROBIC ROUTES (8-III)
1. Autotrophic; 2. Glucose; 3. pyruvate; 4. ATP (NADH); 5. NADH (ATP); 6. oxygen (O_2); 7. fermentation; 8. lactate; 9. ethanol; 10. carbon dioxide; 11. a; 12. b; 13. b; 14. b; 15. b; 16. c; 17. a; 18. bacteria; 19. ATP; 20. sulfate; 21. nitrite; 22. nitrogen.

ALTERNATIVE ENERGY SOURCES IN THE HUMAN BODY (8-IV)
1. F; 2. T; 3. T; 4. T; 5. F; 6. F; 7. T; 8. T; 9. fatty acids; 10. glycerol; 11. glycolysis; 12. amino acids; 13. Krebs cycle; 14. acetyl-CoA; 15. pyruvate.

Answers to Self-Quiz

1. c; 2. c; 3. d; 4. b; 5. d; 6. d; 7. a; 8. d; 9. d; 10. d; 11. C; 12. A, B, D; 13. B, (C), D; 14. C, E; 15. A, B; 16. A, D; 17. A; 18. D; 19. E; 20. A, C, D, E.

9

CELL DIVISION AND MITOSIS

Interactive Exercises

In most cases, reference to specific molecules is by abbreviation.

DIVIDING CELLS: THE BRIDGE BETWEEN GENERATIONS (9-I, pp. 137–139)

1. Define reproduction.

2. Distinguish the *general* functions of mitosis and meiosis from the function of cytokinesis.

3. Define and contrast the functions of somatic cells and germ cells.

4. Describe the general structure of a eukaryotic chromosome; include the following terms: sister chromatids and centromere.

5. State the essence of the division processes called mitosis and meiosis in terms of their effect on the chromosome number of a species during its life cycle; include the following terms: homologous chromosomes, haploid, and diploid._____

_____

_____

_____

_____

Fill-in-the-Blanks

Making a copy of oneself is called (6)_____ , and the process begins at the molecular level. Instructions for producing each new cell and its important chemicals reside in (7)_____. (8)_____ and (9)_____ are nuclear division mechanisms. (10)_____ is the process of dividing parental cell cytoplasm between two daughter cells. In animal cells, mitosis occurs in (11)_____ cells, and meiosis occurs in (12)_____ cells and leads to the production of (13)_____. Sperm nucleus and egg nucleus fuse in fertilization to form a (14)_____. The combination of DNA and proteins forms an elongated structure, the eukaryotic (15) _____. The microtubule attachment site on a chromosome is the (16)_____. A pair of chromosomes that are physically similar and whose genetic information deals with the same traits is said to be (17)_____. Chromosome numbers of (18)_____ cells do not change during the process of (19)_____. Chromosome numbers of (20)_____ cells are reduced by one-half during the process of (21)_____. A cell possessing half the chromosomes of a diploid cell of its species is said to be (22)_____.

True-False

If false, explain why.

_____ 23. Mitosis and meiosis are cytoplasmic division mechanisms.

_____ 24. In animals, mitosis occurs in germ cells.

_____ 25. In animals, meiosis results in the formation of haploid gametes.

_____ 26. The two elongated portions of a duplicated chromosome are also known as chromosomes.

_____ 27. Homologous chromosomes look alike but contain genetic information that controls different traits.

28. Complete the missing portions of the following table of cell division mechanisms.

Mechanisms	Used By
a. _____ and	Single-celled eukaryotes for asexual reproduction
b. _____	Multicelled eukaryotes for bodily growth and asexual reproduction in some species
c. Meiosis and	All eukaryotes for ?
d. cytokinesis	
e. _____ _____	Bacterial cells; the mechanism whereby one bacterial cell divides into two

MITOSIS AND THE CELL CYCLE (9-II, pp. 140)

Labeling

Identify the stage in the cell cycle indicated by each number.

1. _____
2. _____
3. _____
4. _____
5. _____
6. _____
7. _____
8. _____
9. _____
10. _____

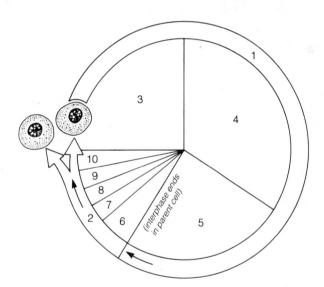

Matching

Link each time span identified below with the most appropriate number in the preceding labeling section.

11. ___ Period after replication of DNA during which the cell prepares for division by further growth and protein synthesis

12. ___ Period of asexual cell division

13. ___ Time when DNA replication occurs

14. ___ Period of cell growth before DNA replication

15. ___ Period when chromosomes are not visible

16. ___ Period of cytoplasmic division

STAGES OF MITOSIS (9-III, pp. 140–145)

Label-Match

Identify each of the mitotic stages shown below by entering the correct stage in the blank beneath the sketch. Select from late prophase, metaphase, early prophase, telophase, and anaphase. Complete the exercise by matching and entering the letter of the correct phase description in the parentheses following each label.

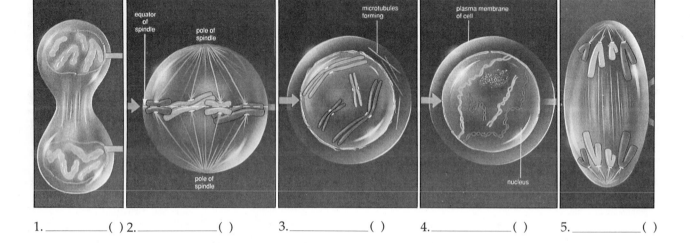

1. _____ ( ) 2. _____ ( ) 3. _____ ( ) 4. _____ ( ) 5. _____ ( )

A. The DNA and associated proteins of duplicated chromosomes begin condensing.
B. Sister chromatids of each chromosome will now be separated from each other and moved to opposite poles.
C. Sister chromatids of each chromosome are attached to the spindle; all chromosomes are now lined up at the spindle equator.
D. Chromosomes decondense; new nuclear membranes begin forming. Cytokinesis often occurs before the end of this phase.
E. Chromosomes continue to condense; microtubules that will form the spindle start to assemble outside of the nucleus. Centrioles are moved to opposite poles by microtubules; the nuclear envelope begins breaking up with the transition to metaphase.

True-False

If false, explain why.

____ 6. Chromosome movements occurring during mitosis are the responsibility of the spindle apparatus.

____ 7. In the early stages of mitosis, the spindle apparatus establishes four poles toward which chromosomes will move.

____ 8. In many cells, the MTOC includes a pair of centrioles that determines where the poles of division are established.

____ 9. Chromosomes move to the poles under their own power during anaphase.

____ 10. The stages of mitosis, in order, are prophase, anaphase, metaphase, and telophase.

____ 11. During anaphase, the microtubules that connect the kinetochore of the chromosome to its pole shorten.

____ 12. When the transition from metaphase to anaphase occurs, the nuclear envelope breaks up.

CYTOKINESIS (9-IV, pp. 145–146)

True-False

If false, explain why.

___ 1. Cytokinesis usually coincides with the mitotic period from metaphase through telophase.

___ 2. In the cytokinesis of animal cells, a cleavage furrow forms around each cell's midsection.

___ 3. Cell plate formation occurs as part of mitosis in most land plants.

___ 4. The formation of a disklike structure, the cell plate, precedes formation of the cleavage furrow in the cytokinesis of most land plants.

___ 5. At the cleavage furrow of animal cells in cytokinesis, contractile microfilaments pull the plasma membrane inward and eventually separate the cytoplasm into two masses.

Chapter Terms

The following page-referenced terms are important; they were in boldface type in the chapter. Refer to the instructions given in Chapter 1, p. 4.

reproduction (137)	sister chromatids (138)	cell cycle (140)	kinetochore (144)
mitosis (138)	centromere (138)	interphase (140)	anaphase (144)
meiosis (138)	homologous chromosomes (139)	spindle apparatus (141)	telophase (145)
cytokinesis (138)	haploid (139)	prophase (142)	cleavage furrow (145)
chromosome (138)	diploid (139)	metaphase (143)	cell plate formation (146)

Self-Quiz

___ 1. The replication of DNA occurs _____.
   a. between the growth phases of interphase
   b. immediately before prophase of mitosis
   c. during prophase of mitosis
   d. during prophase of meiosis

___ 2. In the cell life cycle of a particular cell, _____.
   a. mitosis directly precedes S
   b. mitosis directly precedes G_1
   c. G_2 directly precedes S
   d. G_1 directly precedes S and G_2
   e. mitosis and S directly precede G_1

___ 3. In eukaryotic cells, which of the following can occur during mitosis?
   a. two mitotic divisions to maintain the parental chromosome number
   b. the replication of DNA
   c. a long growth period
   d. the disappearance of the nuclear envelope and nucleolus

___ 4. *Haploid* refers to _____ , and *diploid* refers to _____.
   a. half the parental chromosome number; having two chromosomes of each type in somatic cells
   b. having two chromosomes of each type in somatic cells; twice the parental chromosome number
   c. having two chromosomes of each type in somatic cells; half the parental chromosome number
   d. twice the parental chromosome number; having two chromosomes of each type in somatic cells

___ 5. If a parent cell has sixteen chromosomes and undergoes meiosis, the resulting cells will have _____ chromosomes.
   a. sixty-four
   b. thirty-two
   c. sixteen
   d. eight
   e. four

___ 6. If a parent cell has sixteen chromosomes and undergoes mitosis, the resulting cells will have _____ chromosomes.
   a. sixty-four
   b. thirty-two
   c. sixteen
   d. eight
   e. four

___ 7. Chromosomes are duplicated during the _____ phase.
   a. M
   b. D
   c. G_1
   d. G_2
   e. S

___ 8. The correct order of the stages of mitosis is _____.
   a. prophase, metaphase, telophase, anaphase
   b. telophase, anaphase, metaphase, prophase
   c. telophase, prophase, metaphase, anaphase
   d. anaphase, prophase, telophase, metaphase
   e. prophase, metaphase, anaphase, telophase

___ 9. During _____ , sister chromatids of each chromosome are separated from each other, and those former partners are then chromosomes moved toward opposite poles.
   a. prophase
   b. metaphase
   c. anaphase
   d. telophase

___ 10. In the process of cytokinesis, cleavage furrows are associated with _____ cell division, and cell plate formation is associated with _____ cell division.
   a. animal; animal
   b. plant; animal
   c. plant; plant
   d. animal; plant

Chapter Objectives/Review Questions

This section lists general and detailed chapter objectives that can be used as review questions. You can make maximum use of these items by writing answers on a separate sheet of paper. Fill in answers where blanks are provided. To check for accuracy, compare your answers with information given in the chapter or glossary.

Page   *Objectives/Questions*

(137)   1. _____ means producing a new generation of cells or multicelled individuals.
(138)   2. Name the substance that contains the instructions for making proteins.
(138)   3. Mitosis and meiosis refer to the division of the cell's _____ ; division of the cell's cytoplasm is known as _____ .
(138)   4. Distinguish between somatic cells and germ cells as to their location and function.
(138)   5. Define asexual and sexual reproduction.
(138)   6. Name the prokaryotic cell division mechanism.
(138)   7. The eukaryotic chromosome is composed of _____ and _____ .
(138)   8. The two attached threads of a duplicated chromosome are known as sister _____ .
(138)   9. Describe the function of the portion of a chromosome known as a centromere.
(139) 10. Describe the effects of mitosis and meiosis on the chromosome number of a species.
(139) 11. Contrast the meaning of "haploid" and "diploid" chromosome number.
(139) 12. How is the diploid chromosome number of a species restored?
(140) 13. Be able to list, in order, the various activities occurring in the eukaryotic cell life cycle.
(140) 14. Interphase of the cell cycle consists of G_1, _____ , and G_2.
(140) 15. *S* is the time in the cell cycle when _____ replication occurs.

Page	Objectives/Questions
(141)	16. Describe the mechanism of chromosome movement through the different stages of mitosis.
(141)	17. What is the MTOC responsible for the microtubular spindle in many cells?
(142–145)	18. Be able to describe the cellular events occurring in the prophase, metaphase, anaphase, and telophase of mitosis.
(144)	19. Each chromatid has a _____ , a specialized grouping of proteins to which several spindle microtubules will become attached.
(143)	20. The "_____" is the time when the nuclear envelope breaks up prior to metaphase.
(145–146)	21. Compare and contrast cytokinesis as it occurs in plant mitosis and animal mitosis; use the following terms: cleavage furrow and cell plate formation.

Integrating and Applying Key Concepts

Runaway cell division is characteristic of cancer. Imagine the various points of the mitotic process that might be sabotaged in cancerous cells in order to halt their multiplication. Then try to imagine how one might discriminate between cancerous and normal cells in order to guide those methods of sabotage most effective in combating cancer.

Critical Thinking Exercises

1. DNA content was estimated in several individual cells from a culture of diploid cells.

Cell	DNA Content
a	7.4
b	5.5
c	9.2
d	11.1
e	10.9
f	8.7
g	5.4
h	5.7

In which part of interphase would you say each cell was measured? How much DNA would you expect to find in a cell in anaphase of mitosis?

ANALYSIS

These cultured cells are growing by mitosis, which distributes half the parent cell's DNA to each daughter cell. Each daughter then replicates its DNA and divides again. Cells in G_2 have completed DNA replication and are about to enter mitosis. After mitosis, they pass through G_1 with half as much DNA as they had in G_2. Cells in S phase have intermediate amounts of DNA. Cells d and e have approximately twice as much DNA as cells b and g, so we can conclude that d and e are in G_2 and b and g are in G_1. Cells a, c, and f have intermediate values; they are probably in S phase and have replicated only part of their DNA. Cell h has a little more DNA than cells b and g. It might have begun replication and should be assigned to S phase. On the other hand, the accuracy of all measuring methods is limited. The difference between 5.7 units and 5.5 units might be less than the measuring method can accurately resolve, in which case we should regard the two cells as having the same amount of DNA and place both in G_1.

A cell in anaphase should have about 11 units of DNA. It has completed DNA replication, separated the sister chromatids, and distributed them into two groups, but all the chromatin is still in a single cell.

2. Your text says that separation of the kinetochores of sister chromatids is begun in response to a "specific chemical signal." Which of the following statements would provide the strongest evidence to support the text's hypothesis?

   a. When a prophase nucleus is transplanted into an interphase cell, separation of chromatids does not occur at the expected time.
   b. The force exerted by the elongating spindle gradually increases until it becomes great enough to break the attachment between the kinetochores.
   c. When ATP production is blocked by treating a prophase cell with a chemical inhibitor, the chromatids fail to separate.
   d. When a prophase nucleus is transplanted into an anaphase cell, the chromatids of the transplanted nucleus immediately separate.
   e. When a prophase nucleus is isolated and held in fresh culture medium, its chromatids separate on schedule.

ANALYSIS

   a. This is negative evidence. If this happened, we could conclude that the environment in an interphase cell does not promote separation of chromatids, but it would not necessarily support the hypothesis that a chemical signal is required. It might be that a chemical signal prevents separation and is enzymatically removed at a certain time, releasing the mitotic process.
   b. This is an alternative hypothesis. To state this is to argue *against* the text's hypothesis.
   c. Again, this is negative evidence. It shows that continued supply of ATP is necessary for mitosis, as it is for all life processes, but it does not indicate that ATP or any other substance acts as a chemical signal to begin chromatid separation.
   d. This would be an observation of chromatid separation earlier than normal. If it occurred, we could eliminate all hypotheses that involve development of the chromosomes or the spindle. It would show that the chemical environment of the anaphase cell triggers chromatid separation. However, this could still be due either to the presence of a chemical signal to start separation or to the removal of a chemical inhibitor that prevents separation.
   e. If this were observed, it would be evidence against the text's hypothesis. It would be an observation of chromatid separation in the absence of any cellular signal molecules.

3. Many hypotheses have been proposed to explain cytokinesis in animal cells. Three popular ideas follow:

   a. The chromatids produce a chemical signal at anaphase that causes the cell membrane to contract near the equator of the spindle.
   b. Spindle poles cause the cell membrane to relax near the ends of the spindle. This in turn allows normal surface tension to pull the membrane inward near the equator of the spindle.
   c. When spindle fibers overlap, as at the middle of the spindle, they stimulate the nearby cell membrane to contract.

A sea urchin egg was deformed into a ring by pressing a glass bead into the middle of the cell. The cell proceeded normally through the cell cycle and underwent mitosis with normal chromosome movements, but the pattern of cytokinesis in two successive cell divisions was altered.

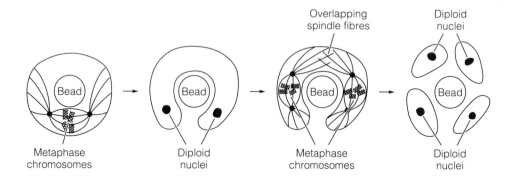

Using these observations, evaluate the three hypotheses. State what pattern of cytokinesis would be predicted by each hypothesis. Is each hypothesis consistent with the observed outcome, or must it be rejected?

ANALYSIS

a. This hypothesis predicts that cleavage furrows will form only near the arrays of chromosomes. This occurred in the first division, but in the second division an extra cleavage furrow formed where there were no chromosomes. This hypothesis is rejected.

b. This hypothesis predicts that a cleavage furrow will form between any two spindle poles. This accounts for the extra furrow in the second division, but it predicts another furrow in the first division. Because that other furrow did not form, this hypothesis must be rejected.

c. This hypothesis predicts formation of a cleavage furrow wherever spindle fibers overlap, whether chromosomes are present or not. In the first division, only one furrow is predicted, because the spindle poles are too far apart for fibers to meet in the upper part of the cell. When there are two spindles in the second division, they are close enough together that fibers overlap where there are no chromosomes, and a third furrow forms there. This hypothesis accounts for the observed pattern of division.

Reference: Rappaport, R. 1986. *Int. Rev. Cytology* 105: 245–281.

Answers

Answers to Interactive Exercises

DIVIDING CELLS: THE BRIDGE
BETWEEN GENERATIONS (9-I)

1. Reproduction is the production of a new generation of cells or multicelled individuals.
2. Mitosis and meiosis are division mechanisms of the nucleus; cytokinesis is the division of the cytoplasm.
3. Somatic cells are body cells of multicelled organisms; germ cells are the cells in a multicelled organism that undergo meiosis to produce gametes.
4. A eukaryotic chromosome is composed of DNA with various proteins; when duplicated prior to cell division, the chromosome is composed of two identical DNA threads known as sister chromatids; they are attached at a centromere region.
5. With mitosis, the chromosome number remains constant, division after division; during meiosis, the chromosome number is reduced by half during gamete production. Chromosomes occur in homologous pairs with hereditary instructions on each that deal with the same traits. During mitosis, each daughter cell receives both members of each homologous pair (diploid number); during meiosis, each gamete receives one member of each homologous pair of chromosomes (haploid number).
6. reproduction; 7. DNA; 8. Mitosis (Meiosis); 9. meiosis (mitosis); 10. Cytokinesis; 11. somatic; 12. germ; 13. gametes; 14. zygote; 15. chromosome;

16. centromere; 17. homologous; 18. somatic; 19. mitosis; 20. germ; 21. meiosis; 22. haploid; 23. F; 24. F; 25. T; 26. F; 27. F; 28. a. Mitosis and b. cytokinesis are used by single-celled eukaryotes for asexual reproduction and by multicelled eukaryotes for bodily growth and asexual reproduction in some species; c. Meiosis and d. cytokinesis are used by all eukaryotes for gamete formation in sexual reproduction; e. Prokaryotic fission is used by bacterial cells, the mechanism whereby one bacterial cell divides into two.

MITOSIS AND THE CELL CYCLE (9-II)

1. interphase; 2. mitosis; 3. G_1; 4. S; 5. G_2; 6. prophase; 7. metaphase; 8. anaphase; 9. telophase; 10. cytokinesis; 11. 5; 12. 2; 13. 4; 14. 3; 15. 1; 16. 10.

STAGES OF MITOSIS (9-III)

1. telophase (D); 2. metaphase (C); 3. late prophase (E); 4. early prophase (A); 5. anaphase (B); 6. T; 7. F; 8. T; 9. F; 10. F; 11. T; 12. F.

CYTOKINESIS (9-IV)

1. F; 2. T; 3. F; 4. F; 5. T.

Answers to Self-Quiz

1. a; 2. d; 3. d; 4. a; 5. d; 6. c; 7. e; 8. e; 9. c; 10. d.

10

A CLOSER LOOK AT MEIOSIS

Interactive Exercises

ON ASEXUAL AND SEXUAL REPRODUCTION (10-I, pp. 151–152)

1. List the basic distinctions between asexual reproduction and sexual reproduction. _____

_____

_____

_____

_____

Labeling

For each item listed below, enter A if it applies to asexual reproduction and S if it applies to sexual reproduction.

2. ___ brings together new combinations of alleles in offspring

3. ___ involves only one parent

4. ___ the first cell of a new individual inheriting two genes for every trait

5. ___ new plants sprouting up along runners of the strawberry plant

6. ___ involves two parents

7. ___ the production of "clones"

8. ___ a male octopus inserting a sperm packet into a cavity under the female's mantle

9. ___ involves meiosis and fertilization

10. ___ the splitting of a flatworm into parts, with each part dividing by mitosis to produce a new flatworm

11. ___ produces some of the variation in traits that forms the basis of evolutionary change

12. ___ the instructions in every pair of genes identical in all individuals of a species

OVERVIEW OF MEIOSIS (10-II, p. 153)

1. To consider the major aspects of meiosis, complete the following table by sketching the chromosome conditions described (left column) in the center column and indicating in the right column whether the cells shown are haploid (*n*) or diploid (2*n*). The cell initially has only one pair of homologous chromosomes. The use of shaded and open chromosomes indicates that one member of the homologous pair came from a maternal source, the other from a paternal source. Thus, this model germ cell has a diploid number of 2. Assume that crossing over does not occur.

Meiosis—Overview Descriptions	*Stages*	*Cell(s)* 2*n* or *n*
a. One pair of homologous chromosomes prior to S of interphase in a germ cell.	◯	
b. While the germ cell is in S of interphase, chromosomes are duplicated in DNA replication; the two sister chromatids are attached at the centromere.	◯	___
c. During meiosis I, each duplicated chromosome lines up with its partner, homologue to homologue.	◯	___
d. Also during meiosis I, the chromosome partners separate from each other in anaphase I; cytokinesis occurs, and each chromosome goes to a different cell.	◯ ◯	___
e. During meiosis II, the two cells formed during meiosis I divide; the sister chromatids of each chromosome are separated from each other in anaphase II; four haploid cells (potential gametes) are formed.	◯ ◯ ◯ ◯	___

True-False

If false, explain why.

___ 2. Meiosis is a nuclear division process that, in some organisms, occurs in somatic cells.

___ 3. Germ cells that undergo meiosis necessarily have a haploid number of chromosomes.

___ 4. A pair of homologous chromosomes generally have the same length, the same shape, and genes that affect different traits.

_____ 5. The sex chromosomes present exceptions to the concept of homologous chromosomes, but they function as homologues during meiosis.

_____ 6. As meiosis reduces the diploid (2*n*) chromosome number to the haploid (*n*) number of gametes, the essence is that one chromosome of each homologous pair is sent to a different gamete.

STAGES OF MEIOSIS (10-III, pp. 154–157)

1. When meiosis I occurs in a germ cell, the actual events occurring during prophase I result in some major recombinations of genes between the chromosomes inherited from father (paternal) and the chromosomes inherited from mother (maternal). Compose a paragraph that relates the special activities happening at this time of cell division. Include the following concepts: homologous chromosomes, synapsis, crossing over, genetic recombination, and chiasma.

_____

_____

_____

_____

_____

_____

_____

Label-Match

Identify each of the meiotic stages shown below by entering the correct stage of either meiosis I or meiosis II in the blank beneath the sketch. Select from prophase I, metaphase I, anaphase I, telophase I, prophase II, metaphase II, anaphase II, and telophase II. Complete the exercise by matching and entering the letter of the correct phase description in the parentheses following each label.

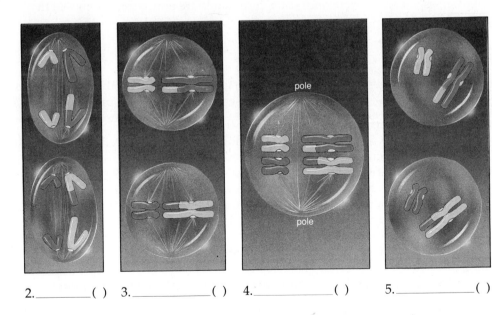

2._____ ( )   3._____ ( )   4._____ ( )   5._____ ( )

  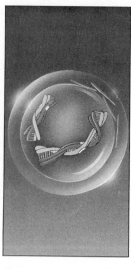 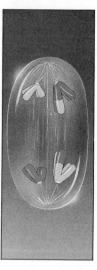

6. _____ ( )   7. _____ ( )   8. _____ ( )   9. _____ ( )

A. Spindle apparatus forms in one diploid cell; nuclear envelope breaks down and homologues align randomly at spindle equator.
B. Sister chromatids of each chromosome are still attached at the centromere; spindles have not yet formed in two haploid cells.
C. Four daughter nuclei form; following cytokinesis, each cell (gamete) has a haploid number of chromosomes, all in the unduplicated state.
D. Each homologue (in the duplicated state) is separated from its partner, and the two are moving to opposite poles.
E. Each chromosome condenses, then pairs with its homologue; crossing over and recombination occur.
F. Each chromosome is aligned at the spindle equator; this is occurring in two haploid cells.
G. A haploid number of chromosomes (still duplicated) ends up at each pole.
H. Each chromosome splits; what were once sister chromatids are now chromosomes in their own right and are moved to opposite poles.

Matching

Assume for the following linkages that the cell starts with two pairs of already duplicated chromosomes and that each chromosome consists of two chromatids. Choose the one most appropriate answer for each term.

10. ___ anaphase I
11. ___ anaphase II
12. ___ metaphase I
13. ___ metaphase II
14. ___ prophase I
15. ___ prophase II
16. ___ telophase I
17. ___ telophase II

A. Meiosis I is completed by this stage.
B. One member of each homologous pair is separated from its mate; both are guided to opposite poles.
C. Two pairs of chromosomes are lined up at the spindle apparatus equator.
D. Chromosomes become clearly visible; nuclear region has two pairs of already duplicated chromosomes.
E. Each chromosome splits; sister chromatids are separated and moved to opposite poles.
F. Each daughter cell ends up with one chromosome of each type and may function as a gamete.
G. Spindle apparatus re-forms; each cell has two chromosomes, each consisting of two chromatids.
H. Two chromosomes are lined up at spindle apparatus equator.

MEIOSIS AND THE LIFE CYCLES (10-IV, pp. 158–160)

Sequence

Arrange the following entities in correct order of development, entering a 1 by the stage that appears first and a 5 by the stage that completes the process of spermatogenesis. Refer to Figures 10.11 and 10.12 in the text.

1. ___ primary spermatocyte

2. ___ sperm

3. ___ spermatid

4. ___ spermatogonium

5. ___ secondary spermatocyte

Matching

Choose the most appropriate answer to match with each oogenesis concept.

6. ___ primary oocyte

7. ___ oogonium

8. ___ secondary oocyte

9. ___ ovum and three polar bodies

10. ___ first polar body

A. The cell in which synapsis, crossing over, and recombination occur

B. A cell that is equivalent to a diploid germ cell

C. A haploid cell formed after division of the primary oocyte that does not form an ovum at second division

D. Haploid cells, only one of which functions as an egg

E. A haploid cell formed after division of the primary oocyte, the division of which forms a functional ovum

True-False

If false, explain why.

___ 11. The stage that completes meiosis II in male animals is the spermatid stage.

___ 12. The spore in a plant life cycle is the haploid product of meiosis.

___ 13. Oogenesis in female, multicellular animals results in four equal-size eggs.

___ 14. All cells composing the human body are diploid except for any gametes that may be produced.

___ 15. The exact gene combinations you now possess could have arisen from the meeting of any sperm from your father with any egg from your mother.

16. Compose a paragraph that summarizes the various methods utilized in sexual reproduction that bring about new combinations of alleles. _____

_____

_____

_____

_____

MEIOSIS COMPARED WITH MITOSIS (10-V, p. 160)

Matching

The cell model used in this exercise has two pairs of homologous chromosomes, one long pair and one short pair. Match the descriptions to the sketches below.

1. ___ beginning of meiosis II following interkinesis

2. ___ a daughter cell at the end of meiosis II

3. ___ metaphase I of meiosis

4. ___ metaphase of mitosis

5. ___ G₁ in a daughter cell following mitosis

6. ___ prophase of mitosis

A   B   C   D   E   F

The following questions refer to the sketches above; enter answers in the blanks following each question.

7. How many chromosomes are present in cell E? _____

8. How many chromatids are present in cell E? _____

9. How many chromatids are present in cell C? _____

10. How many chromatids are present in cell D? _____

11. How many chromosomes are present in cell F? _____

Chapter Terms

The following page-referenced terms are important; they were in boldface type in the chapter. Refer to the instructions given in Chapter 1, p. 4.

asexual reproduction (151)	sister chromatids (153)	metaphase I (155)	ovum (158)
sexual reproduction (151)	prophase I (154)	anaphase I (155)	spores (158)
allele (152)	crossing over (154)	interkinesis (156)	fertilization (158)
homologous chromosomes (153)	genetic recombination (154)	sperm (158)	

Self-Quiz

___ 1. Which of the following does *not* occur in prophase I of meiosis?
   a. a cytoplasmic division
   b. a cluster of four chromatids
   c. homologues pairing tightly
   d. crossing over

___ 2. Crossing over is one of the most important events in meiosis because _____.
   a. it produces new combinations of alleles on chromosomes
   b. homologous chromosomes must be separated into different daughter cells
   c. the number of chromosomes allotted to each daughter cell must be halved
   d. homologous chromatids must be separated into different daughter cells

___ 3. Crossing over _____.
   a. generally results in pairing of homologues and binary fission
   b. is accompanied by gene-copying events
   c. involves breakages and exchanges being made between sister chromatids
   d. alters the composition of chromosomes and results in new combinations of alleles being channeled into the daughter cells

___ 4. The appearance of chromosome ends lapped over each other in meiotic prophase I provides evidence of _____.
   a. meiosis
   b. crossing over
   c. chromosomal aberration
   d. fertilization
   e. spindle fiber formation

___ 5. Which of the following does *not* increase genetic variation?
   a. crossing over
   b. random fertilization
   c. prophase of mitosis
   d. random homologue alignments at metaphase I

___ 6. Which of the following is the most correct sequence of events in animal life cycles?
   a. meiosis ⟶ fertilization ⟶ gametes ⟶ diploid organism
   b. diploid organism ⟶ meiosis ⟶ gametes ⟶ fertilization
   c. fertilization ⟶ gametes ⟶ diploid organism ⟶ meiosis
   d. diploid organism ⟶ fertilization ⟶ meiosis ⟶ gametes

___ 7. In sexually reproducing organisms, the zygote is _____.
   a. an exact genetic copy of the female parent
   b. an exact genetic copy of the male parent
   c. unlike either parent genetically
   d. a genetic mixture of male parent and female parent

___ 8. Which of the following is the most correct sequence of events in plant life cycles?
   a. fertilization ⟶ zygote ⟶ sporophyte ⟶ meiosis ⟶ spores ⟶ gametophytes ⟶ gametes
   b. fertilization ⟶ sporophyte ⟶ zygote ⟶ meiosis ⟶ spores ⟶ gametophytes ⟶ gametes
   c. fertilization ⟶ zygote ⟶ sporophyte ⟶ meiosis ⟶ gametes ⟶ gametophyte ⟶ spores
   d. fertilization ⟶ zygote ⟶ gametophyte ⟶ meiosis ⟶ gametes ⟶ sporophyte ⟶ spores

___ 9. The cell in the diagram is a diploid that has three pairs of chromosomes. From the number and pattern of chromosomes, the cell _____.

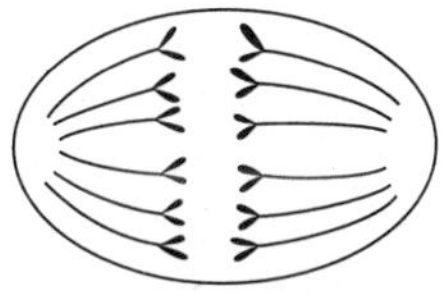

   a. could be in the first division of meiosis
   b. could be in the second division of meiosis
   c. could be in mitosis
   d. could be in neither mitosis nor meiosis, because this stage is not possible in a cell with three pairs of chromosomes

_____10. You are looking at a cell from the same organism as in the previous question. Now the cell _____.

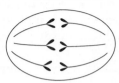

a. could be in the first division of meiosis
b. could be in the second division of meiosis
c. could be in mitosis
d. could be in neither mitosis nor meiosis, because this stage is not possible in this organism

---

Chapter Objectives/Review Questions

This section lists general and detailed chapter objectives that can be used as review questions. You can make maximum use of these items by writing answers on a separate sheet of paper. Fill in answers where blanks are provided. To check for accuracy, compare your answers with information given in the chapter or glossary.

Page   Objectives/Questions

(151)   1. "One parent always passes on a duplicate of all its genes to offspring" describes _____ reproduction.
(151–152) 2. Describe the basic aspects of sexual reproduction.
(152)   3. Sexual reproduction puts together new combinations of _____ in offspring.
(152)   4. Genetic variation is acted upon by agents of natural selection and so is a basis of _____ change.
(153)   5. _____ chromosomes have the same length, shape, and genes that affect the same characteristics; they also line up with each other during meiosis.
(153)   6. Describe the relationship between the following terms: homologous chromosomes, diploid, and haploid.
(153)   7. State the exact difference between the haploid (*n*) number of chromosomes and the diploid (2*n*) number of chromosomes.
(153)   8. In any sexually reproducing species, the haploid number always has _____ chromosome from each of the homologous pairs of chromosomes.
(153)   9. While germ cells are in interphase, each chromosome is duplicated by a process called DNA _____ ; each chromosome is then composed of two sister _____ .
(153)   10. What are the two different divisions of meiosis called?
(153)   11. In meiosis I, homologous chromosomes pair; each homologue consists of _____ chromatids.
(153)   12. During meiosis II, the sister _____ of each _____ are separated from each other.
(153)   13. If the diploid chromosome number for a particular plant species is 18, the haploid number is _____ .
(154–157) 14. Be able to relate cellular events occurring during the stages of meiosis I and meiosis II; consider carefully the definition of a chromosome, chromatids, and the change in chromosome number.
(154)   15. During prophase I, each threadlike chromosome and its homologue are drawn together by a process called _____ .
(154)   16. Following synapsis, _____ _____ occurs; then nonsister chromatids break at one or more sites along their length and exchange gene segments. This results in gene _____ .
(155–156) 17. Explain what is meant by the statement "each pair of homologues is assorted into gametes independently of the other pairs present in the cell."
(158)   18. Describe spermatogenesis in male animals.
(158)   19. Describe oogenesis in female animals.
(158)   20. Compare and contrast the generalized life cycles for plants and animals.
(158)   21. Meiosis in the animal life cycle results in haploid _____ ; meiosis in the plant life cycle results in haploid _____ .
(160)   22. List the three events responsible for new combinations of alleles in offspring of the sexual reproduction process.

Integrating and Applying Key Concepts

A few years ago, it was claimed that the actual cloning of a human being had been accomplished. Later, this claim was admitted to be fraudulent. If sometime in the future cloning of humans becomes possible, speculate about the effects on human populations of reproduction without sex.

Critical Thinking Exercises

1. The graph below represents the amount of DNA per cell in a small population of plant cells.

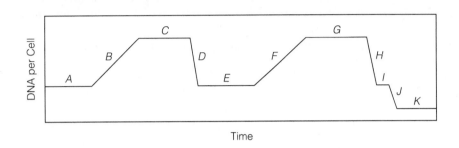

a. What phase of the cell cycle occurs at *B*?
b. What event occurs at *D*?
c. If there are *x* cells at *A*, how many are there at *G*?
d. What phase of the cell cycle occurs at *C* and *G*?
e. If the total DNA in the population at *A* is *y*, how much is there at *G*? at *I*?
f. If there are *n* chromosomes per cell at *E*, how many are there at *G*? at *I*? at *K*?
g. If there are *n* chromatids per cell at *E*, how many are there at *G*? at *I*?
h. What happens next to a cell at *K*?

ANALYSIS

a. At *B*, the amount of DNA per cell is doubling; this could only be S phase.
b. At *D*, the amount of DNA per cell is reduced by half; this could happen in either mitosis, meiosis I, or meiosis II. If the division at *D* were meiosis I, it would be followed by a second reduction without an intervening S phase. Conversely, if it were meiosis II, it would have been preceded by a reduction. Thus, the division at *D* is mitotic. Meiosis, two successive divisions without DNA replication between, occurs at *H–I–J*.
c. Cell number increases when cells divide. This occurs only once between *A* and *G*—specifically, at *D*. Thus, there are twice as many cells, 2*x*, at G.
d. *C* and *G* both occur immediately after S phases, in which DNA content doubles. Thus, *C* and *G* are both G_2.
e. Two S phases, *B* and *F*, occur between *A* and *G*; therefore, there is four times as much total DNA, 4*y*, in the cell population at *G* as there is at *A*. After *G*, only cell divisions occur, one at *H*. Although this distributes the DNA among more cells, it does not change the total amount. The amount at *I* is also 4*y*.
f. *F* is an S phase; during this phase, DNA is replicated, but the number of chromosomes is constant. Thus, at *G* there are still *n* chromosomes per cell. *H* is meiosis I, during which homologous chromosomes are distributed to the daughter cells and the number per cell is reduced by half. At *I*, there are 0.5*n* chromosomes per cell. *J* is meiosis II, during which chromatids separate and chromosome number per daughter cell is conserved. At *K*, there are still 0.5*n* chromosomes per cell.
g. At *G*, there are *n* chromosomes per cell; because this follows an S phase, each chromosome consists of two chromatids. At *G*, there are 2*n* chromatids per cell. *I* follows meiosis I, so each chromosome still consists of two chromatids, and there is a total of *n* chromosomes per cell.
h. These are plant cells. In plants, meiosis forms spores; thus, the next thing that happens, if conditions are adequate, is an S phase.

2. A student says, "Because two homologous chromosomes in a cell do not contain the same information, loss of a piece of only one homologue will result in an altered cell." Which of the following assumptions is the student most likely making?

    a. DNA carries the genetic instructions.
    b. The chromosome was broken during meiosis.
    c. All the information in a given cell is not expressed.
    d. Information from both homologues is expressed in a cell.
    e. Information on one homologue blocks expression of information on the other.

ANALYSIS

    a. This is a valid statement, according to current evidence, but it is not necessary for the student's prediction. The student is only assuming that *some* component of the chromosome carries the genetic instructions.
    b. The student is predicting the consequences of chromosome breakage; no assumption about the mechanism or cause of breakage is needed.
    c. Making this assumption would lead to the prediction that some chromosome breakage would *not* result in changes in the cell, because the information in the lost part of the chromosome would not be expressed even if it was still there.
    d. This assumption is necessary. If one homologue contained unexpressed information, that information could be lost without causing any changes in the characteristics of the cell.
    e. This is a variant of the assumption that not all the information in a cell is expressed; it just proposes a mechanism for the blockage. Making this assumption would lead to a different prediction.

3. The letters in the drawing below represent alleles on the chromosomes of a cell.

If the cell above underwent meiosis, which of the following combinations of alleles would most likely be present in the same gamete?

    a. *A*, *B*, and *C*   b. *A*, *B*, and *D*   c. *A* and *E*   d. *E* and *F*   e. *F* and *D*

ANALYSIS

    a. *A*, *B*, and *C* are present on the same chromatid. Remember that sister chromatids are identical unless crossing over occurs. Thus, these alleles will be in the same gamete, unless *both* chromatids undergo crossing over that separates the alleles. That event has a relatively low probability in a single cell.
    b. *A* and *G* are on the same position of homologous chromosomes, and *D* occurs only on the chromosome that carries *G*. *A* will only occur on the same chromosome as *D* if a crossover occurs. Since *B* is present on both chromosomes, it will not be affected by any crossover.

c. *A* and *E* are on nonhomologous chromosomes. The probability that they will separate together in meiosis I is one in two.

d. *E* and *F* are at the same position on homologues. As long as meiosis I proceeds normally, they will not move into the same gamete.

e. *F* and *D* are on nonhomologous chromosomes. The probability that they will separate together in meiosis I is the same as for *A* and *E*, namely, one in two.

Answers

Answers to Interactive Exercises

ON ASEXUAL AND SEXUAL REPRODUCTION (10-I)

1. In asexual reproduction, one parent always passes on a duplicate of all its genes to offspring; there is only genetic variation due to mutation. Sexual reproduction involves two parents, each with two genes for every trait. Both parents pass on one of each gene to offspring by meiosis, gamete formation, and fertilization. Sexual reproduction brings together new allele combinations in offspring.
2. S; 3. A; 4. S; 5. A; 6. S; 7. A; 8. S; 9. S; 10. A; 11. S; 12. A.

OVERVIEW OF MEIOSIS (10-II)

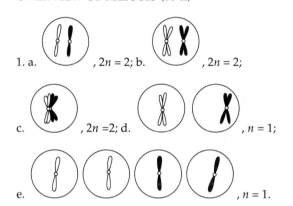

1. a.   , 2n = 2; b.   , 2n = 2;

c.   , 2n =2; d.   , n = 1;

e.   , n = 1.

2. F; 3. F; 4. F; 5. T; 6. T.

STAGES OF MEIOSIS (10-III)

1. During prophase I, homologous chromosomes (each duplicated) synapse (X and Y pair only at one end).

While synapsing, crossing over occurs, and nonsister chromatids break at one or more points and exchange chromosome segments. Crossing over results in genetic recombination, which increases variation in offspring traits. After gene segments are exchanged, the four chromatids repel each other between crossover points but remain temporarily attached. The crosslike appearance due to temporary attachment between two nonsister chromatids is a chiasma.

2. anaphase II (H); 3. metaphase II (F); 4. metaphase I (A); 5. prophase II (B); 6. telophase II (C); 7. telophase I (G); 8. prophase I (E); 9. anaphase I (D); 10. B; 11. E; 12. C; 13. H; 14. D; 15. G; 16. A; 17. F.

MEIOSIS AND THE LIFE CYCLES (10-IV)

1. 2; 2. 5; 3. 4; 4. 1; 5. 3; 6. A; 7. B; 8. E; 9. D; 10. C; 11. F; 12. T; 13. F; 14. T; 15. F.

16. During prophase I of meiosis, crossing over and genetic recombination occur. During metaphase I of meiosis, the two members of each homologous chromosome assort independently of the other pairs. Fertilization is a chance mix of different combinations of alleles from two different gametes.

MEIOSIS COMPARED WITH MITOSIS (10-V)

1. C; 2. F; 3. D; 4. A; 5. B; 6. E; 7. 4; 8. 8; 9. 4; 10. 8; 11. 2.

Answers to Self-Quiz

1. a; 2. a; 3. d; 4. b; 5. c; 6. b; 7. d; 8. a; 9. c; 10. b.

11

OBSERVABLE PATTERNS
OF INHERITANCE

MENDEL'S INSIGHTS INTO THE
PATTERNS OF INHERITANCE
 Mendel's Experimental Approach
 Some Terms Used in Genetics
 The Concept of Segregation
 Testcrosses
 The Concept of Independent Assortment
VARIATIONS ON MENDEL'S THEMES
 Dominance Relations

Interactions Between Different Gene Pairs
 Comb Shape in Poultry
 Hair Color in Mammals
Multiple Effects of Single Genes
Environmental Effects on Phenotype
Variable Gene Expression in a Population
 Penetrance and Expressivity
 Continuous Variation

Interactive Exercises

MENDEL'S INSIGHTS INTO THE PATTERNS OF INHERITANCE (11-I, pp. 165–172)

1. Describe how pea plants are fertilized in nature. _stemen (pollen) flow into the carpel_

2. Explain what is meant by a "true-breeding" pea plant; describe how Mendel planned to manipulate
the reproduction of true-breeding garden pea plants to learn about heredity. _a certain phenotype produces only that certain phenotype; he took stem from an "T.B." plant + put it into
the carpel of another type of "T.B." plant._

Matching

Choose the one most appropriate answer for each.

3. _F_ genotype
4. _B_ alleles
5. _C_ heterozygous
6. _G_ dominant allele
7. _E_ phenotype
8. _I_ genes
9. _D_ recessive allele
10. _H_ homozygous
11. _J_ diploid cell
12. _A/B_ locus

A. All the different molecular forms of a gene that exist
B. Particular location of a gene on a chromosome
C. Describes an individual for which two alleles of a pair are different
D. Gene whose effect is masked by its partner
E. Refers to an individual's observable traits
F. Refers to the genes present in an individual organism
G. Gene whose effect "masks" the effect of its partner
H. Describes an individual for which two alleles of a pair are the same
I. Units of instructions for producing a specific trait, each of which has a particular location on a chromosome
J. Has a pair of genes for each trait, one on each of two homologous chromosomes

Fill-in-the-Blanks

A (13)_____ plant is one that is heterozygous and is produced from two parents that have bred true for contrasting forms of a single trait. (14)_____ is the symbol used for the parental generation; (15)_____ is the symbol used for first-generation offspring; (16)_____ is the symbol used for second-generation offspring. The separation of A and a as homologues move to different gametes during meiosis is known as Mendel's concept of (17)_____. Crossing a first-generation hybrid (possibly of unknown genotype) back to a plant known to be a true-breeding recessive plant is known as Mendel's (18)_____. From this cross, a ratio of (19)_____ is expected. When F_1 offspring inherit two gene pairs, neither of which consists of identical alleles, the cross is known as a (20)_____ cross. "Gene pairs assorting into gametes independently of other gene pairs located on nonhomologous chromosomes" describes (21)_____ _____.

22. In garden pea plants, Tall (T) is dominant over dwarf (t). In the cross $Tt \times tt$, the Tt parent would produce a gamete carrying T (tall) and a gamete carrying t (short) through segregation; the tt parent could only produce gametes carrying the t (short) gene. By use of the Punnett-square method (refer to Figure 11.6 in the text), determine the genotype and phenotype probabilities of offspring from the above cross, $Tt \times tt$:_____ Tt _____

	T	t
t	Tt	tt
t	Tt	tt

2:2
50/50

Although the Punnett-square (checkerboard) method is a favored method for solving genetics problems, there is a quicker way. Six different outcomes are possible from monohybrid crosses. Studying the following *relationships* allows us to obtain the result of any monohybrid cross by *inspection*.

1. $AA \times AA$ = all AA
   (Each of the four blocks of the Punnett square would be AA.)
2. $aa \times aa$ = all aa
3. $AA \times aa$ = all Aa
4. $AA \times Aa$
   or   = $1/2\ AA, 1/2\ Aa$
   $Aa \times AA$
   (Two blocks of the Punnett square are AA, and two blocks are Aa.)
5. $aa \times Aa$
   or   = $1/2\ aa, 1/2\ Aa$
   $Aa \times aa$
6. $Aa \times Aa$ = $1/4\ AA, 1/2\ Aa, 1/4\ aa$
   (One block in the Punnett square is AA, two blocks are Aa, and one block is aa.)

Using the gene symbols in exercise 22, apply the six Mendelian ratios listed above to solve the following monohybrid crosses by *inspection*. State results as genotype ratios.

23. $TT \times TT$ = _____

24. $Tt \times Tt$ = _____

25. $Tt \times tt$ = _____

26. $tt \times tt$ = _____

When working genetics problems dealing with two gene pairs (dihybrid crosses), we can visualize the independent assortment of gene pairs located on nonhomologous chromosomes into gametes by use of a fork-line device. Assume that in man, pigmented eyes (B) are dominant (an eye color other than blue) over blue (b), and right-handedness (R) is dominant over left-handedness (r). To learn to solve a problem, cross the parents BbRr × BbRr. A sixteen-block Punnett square is required, with gametes from each parent arrayed on two sides of the Punnett square (refer to Figure 11.10 in the text). The gametes receive genes through independent assortment using a fork-line method:

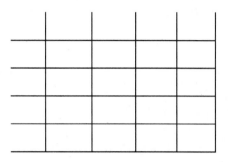

Array the gametes above on two sides of the Punnett square; combine these haploid gametes to form diploid zygotes within the squares. In the blank spaces in exercises 27–30, enter the probability ratios derived within the Punnett square for the phenotypes listed:

27. ___ pigmented eyes, right-handed

28. ___ pigmented eyes, left-handed

29. ___ blue eyes, right-handed

30. ___ blue eyes, left-handed

The inspection method for solving monohybrid crosses illustrated as a part of exercises 23–26 above can also be applied to an algebraic method for solving dihybrid crosses, although some writing is required. We must keep in mind the six possible outcomes from monohybrid crosses. The ratios are combined, as in the following example, where all the factors of one ratio are multiplied by all the factors of the other.

The dihybrid cross, Aabb × AaBb, can be solved as follows:

$$1AA \quad \begin{cases} 1bb & = & 1AAbb \\ 1Bb & = & 1AABb \end{cases}$$

$$2Aa \quad \begin{cases} 1bb & = & 2Aabb \\ 1Bb & = & 2AaBb \end{cases}$$

$$1aa \quad \begin{cases} 1bb & = & 1aabb \\ 1Bb & = & 1aaBb \end{cases}$$

The answer (right vertical column) can be read as 1/8 AAbb (or 2/16 blocks in a sixteen-block Punnett square); 1/8 AABb; 2/8 Aabb; 2/8 AaBb; 1/8 aabb; 1/8 aaBb.

31. Albinos cannot form the pigments that normally produce skin, hair, and eye color, so albinos have white hair and pink eyes and skin (because the blood shows through). To be an albino, one must be homozygous recessive for the pair of genes that codes for the key enzyme in pigment production. Suppose a woman of normal pigmentation with an albino mother marries an albino man. State the possible kinds of pigmentation possible for this couple's children, and specify the ratio of each kind of child the couple is likely to have. Show the genotype(s) and state the phenotype(s).

_____

_____

_____

_____

_____

32. In horses, black coat color is influenced by the dominant allele (B), and chestnut coat color is influenced by the recessive allele (b). Trotting gait is due to a dominant gene (T), pacing gait to the recessive allele (t). A homozygous black trotter is crossed to a chestnut pacer.

    a. What will be the appearance of the F_1 and F_2 generations?_____

    b. Which phenotype will be most common?_____

    c. Which genotype will be most common?_____

    d. Which of the potential offspring will be certain to breed true?_____

VARIATIONS ON MENDEL'S THEMES (11-II, pp. 172–178)

Genes that are not always dominant or recessive may blend to produce a phenotype of a different appearance. This is termed incomplete dominance. In four o'clock plants, red flower color is determined by gene R, and white flower color by R' while the heterozygous condition, RR', is pink. Determine the phenotypes and genotypes of the offspring from the following crosses:

    1. $RR \times R'R'$ _____

    2. $R'R' \times R'R'$ _____

    3. $RR \times RR'$ _____

    4. $RR \times RR$ _____

    5. Describe what is meant by a multiple allele system._____

_____

The three genes, I^A, I^B, and i produce proteins found on the surfaces of red blood cells that determine the four blood types in the ABO system, A, B, AB, and O. Genes I^A and I^B are both dominant over i but not over each other. They are codominant. Recognize that blood types A and B may be heterozygous or homozygous (I^AI^A, I^Ai or I^BI^B, I^Bi), while blood type O is symbolized by ii. Indicate the genotypes and phenotypes of the offspring and their probabilities from the parental combinations in exercises 6–10.

    6. $I^Ai \times I^AI^B$ _____

    7. $I^Bi \times I^Ai$ _____

    8. $I^AI^A \times ii$ _____

    9. $ii \times ii$ _____

   10. $I^AI^B \times I^AI^B$ _____

11. How would one explain a child of blood type O (*ii*) being born to parents of genotypes $I^A I^B \times I^A i$?

_____

_____

Interactions exist between different gene pairs that can produce some effect on the phenotype. Interactions between genes can produce changes in the normally expected dihybrid ratios. Sometimes two genes *cooperate* to produce a phenotype that neither gene can produce alone. Work the following exercises to understand this type of interaction.

In poultry, the genes for rose comb (*R*) and pea comb (*P*) produce walnut comb whenever they occur together (*R_P_*); single-combed individuals have the homozygous condition for both genes (*rrpp*).

12. Give the F_1 and F_2 phenotypic results of a cross of a pure-breeding rose comb (*RRpp*) with a

pure-breeding pea comb (*rrPP*)._____

13. Give the phenotypic results of a cross of *Rrpp* with *rrPp*._____

In other types of gene interactions, two alleles of a gene mask the expression of alleles of another gene, and some expected phenotypes never appear. Epistasis is the term used to describe such interactions; they are common among gene pairs affecting fur or skin color in mammals. Work the following exercises to understand epistasis interactions.

In sweet peas, genes *C* and *P* are necessary for colored flowers. In the absence of either (_ _ *pp* or *cc* _ _), or both (*ccpp*), the flowers are white. What will be the color of the offspring of the following crosses and in what proportions?

14. $CcPp \times ccpp$ = _____

15. $CcPP \times Ccpp$ = _____

16. $Ccpp \times ccPp$ = _____

True-False

If false, explain why.

____ 17. If two F_1 red-flowered snapdragons are crossed, three-fourths of their offspring (the F_2) will be red-flowered, and one-fourth will be white-flowered.

____ 18. Interaction among alleles of two gene pairs affecting the coat color of Labrador retrievers is an example of recessive epistasis.

____ 19. A child of blood type AB has a mother with type A blood. The child's father could not have type O.

____ 20. Human eye color, hair color, and height depend on the additive effects of genes.

____ 21. In Himalayan rabbits homozygous for genes $c^h c^h$, development of fur color depends on what the light intensity is at the time the hair is growing.

____ 22. Environmental effects on the phenotype of the water buttercup are expressed by the development of two types of flowers, one type underwater and the other type above water.

____ 23. A person's becoming thirty pounds overweight is one way in which the external environment can influence the phenotypic expression of human genotype.

____ 24. In fruit flies, a gene called "polymorph" is known to affect several different characteristics; this is an example of pleiotropy.

____ 25. When only seven of nine people who carry a particular dominant gene express that gene, the expressivity is said to be 7/9, or 77 percent.

____ 26. The dominant gene in humans that is responsible for immobile, bent fingers is not expressed in some individuals and shows up on one or both hands of others. This illustrates expressivity.

Chapter Terms

The following page-referenced terms are important; they were in boldface type in the chapter. Refer to the instructions given in Chapter 1, p. 4.

true-breeding (166) homozygous recessive (167) Punnett-square method (169) multiple allele system (173)
cross-fertilization (167) heterozygous (167) testcross (170) epistasis (174)
genes (167) genotype (167) incomplete dominance (172) pleiotropy (175)
alleles (167) phenotype (167) codominance (172) continuous variation (178)
homozygous dominant (167) probability (169) ABO blood typing (172)

Self-Quiz

____ 1. Mendel's principle of independent assortment states that _____.
   a. one allele is always dominant to another
   b. hereditary units from the male and female parents are blended in the offspring
   c. the two hereditary units that influence a certain trait separate during gamete formation
   d. each hereditary unit is inherited separately from other hereditary units

____ 2. One of two or more alternative forms of a gene for a single trait is a(n) _____.
   a. chiasma
   b. allele
   c. autosome
   d. locus

____ 3. In the F_2 generation of a monohybrid cross involving complete dominance, the expected *phenotypic* ratio is _____.
   a. 3 : 1
   b. 1 : 1 : 1 : 1
   c. 1 : 2 : 1
   d. 1 : 1

____ 4. In the F_2 generation of a cross between a red snapdragon (homozygous) and a white snapdragon, the expected phenotypic ratio of the offspring is _____.
   a. 3/4 red, 1/4 white
   b. 100 percent red
   c. 1/4 red, 1/2 pink, 1/4 white
   d. 100 percent pink

____ 5. The results of a testcross reveal that all of the offspring resemble the parent being tested. That parent is necessarily _____.

   a. heterozygous
   b. polygenic
   c. homozygous
   d. recessive

____ 6. When gametes from one individual undergo sexual fusion with gametes from another, it is called _____.
   a. blending
   b. cross-fertilization
   c. true-breeding
   d. independent assortment

____ 7. A single gene that affects several seemingly unrelated aspects of an individual's phenotype is said to be

   _____.
   a. pleiotropic
   b. epistatic
   c. mosaic
   d. continuous

____ 8. Suppose two individuals, each heterozygous for the same characteristic, are crossed. The characteristic involves complete dominance. The expected genotypic ratio of their progeny is _____.
   a. 1 : 2 : 1
   b. 1 : 1
   c. 100 percent of one genotype
   d. 3 : 1

____ 9. If the two homozygous classes in the F_1 generation of the cross in exercise 8 are allowed to mate, the observed genotypic ratio of the offspring will be _____.
   a. 1 : 1
   b. 1 : 2 : 1
   c. 100 percent of one genotype
   d. 3 : 1

___ 10. Assume that genes *A*, *B*, and *C* are located on different chromosomes. How many different types of gametes could an individual of genotype *AaBbCc* produce during meiosis if crossing over did not occur?

    a. two
    b. eight
    c. six
    d. twelve

Chapter Objectives/Review Questions

This section lists general and detailed chapter objectives that can be used as review questions. You can make maximum use of these items by writing answers on a separate sheet of paper. Fill in answers where blanks are provided. To check for accuracy, compare your answers with information given in the chapter or glossary.

Page   Objectives/Questions

(166–167)   1. What was the prevailing method of explaining the inheritance of traits before Mendel's work with pea plants?

(166–167)   2. Garden pea plants are naturally _____-fertilizing, but Mendel took steps to _____-fertilize them for his experiments.

(167)   3. Define allele.

(167)   4. Explain the difference between an individual that is heterozygous and one that is homozygous.

(167)   5. How many alleles are present in the genotype *Tt? tt? TT?*

(167)   6. What symbols are used to distinguish the dominant allele from the recessive allele?

(167)   7. Define genotype and phenotype.

(168–169)   8. State Mendel's concept of segregation.

(169)   9. The results of genetic crosses are interpreted according to the rules of _____.

(169)   10. Be able to use the Punnett-square method of solving genetics problems.

(170)   11. Define testcross and give an example.

(169–170)   12. Define dihybrid cross and distinguish it from monohybrid cross.

(170–172)   13. State the principle of independent assortment as formulated by Mendel.

(170–172)   14. Solve the following problems—both of which deal with two different traits on different chromosomes—by using the Punnett-square method or by multiplying the separate probabilities.
    a. The abilities to bark and to erect their ears are dominant traits in some dogs; keeping silent while trailing prey and droopy ears are recessive traits. If a barking dog with erect ears (heterozygous for both traits) is mated with a droopy-eared dog that keeps silent while trailing prey, what kinds of puppies can result?
    b. A spinach plant with straight leaves and white flowers was crossed with another spinach plant with curly leaves and yellow flowers. All of the F_1 generation have curly leaves and white flowers. If any two individuals of the F_1 generation are crossed, what phenotypic ratio will result? Show all genotypes.

(172–173)   15. Distinguish between complete dominance, incomplete dominance, and codominance.

(173)   16. Give an example of a multiple allele system.

(173)   17. Cite an example of a gene interaction in which two gene pairs cooperate to produce a phenotype that neither could produce alone.

(174)   18. Define epistasis and give an example.

(175)   19. Explain why sickle-cell anemia is a good example of pleiotropy.

(176–177)   20. Himalayan rabbits and water buttercups are good examples of environmental effects on _____.

(177)   21. Define penetrance and expressivity.

(177–179)   22. List reasons to explain the variation in gene expression in populations.

Integrating and Applying Key Concepts

Solve the following genetics problem: In garden peas, one pair of alleles controls the height of the plant, and a second pair of alleles controls flower color. The allele for tall (D) is dominant to the allele for dwarf (d), and the allele for purple (P) is dominant to the allele for white (p). A tall plant with purple flowers crossed with a tall plant with white flowers produces 3/8 tall purple, 1/8 tall white, 3/8 dwarf purple, and 1/8 dwarf white. What is the genotype of the parents?

Critical Thinking Exercises

1. Two black female mice were bred with the same brown male. In three litters, female A produced nine black offspring and eight brown offspring, while female B produced nineteen black offspring and no brown offspring. Assuming that coat color is determined by a single pair of genes and exhibits simple dominance in mice, which of the following is the most likely set of parental genotypes?

   a. The male is homozygous for brown, female A is heterozygous, and female B is homozygous for black.
   b. The male is heterozygous, female A is heterozygous, and female B is homozygous for black.
   c. The male is homozygous for brown, and female A and female B are both heterozygous for black.
   d. The male is heterozygous, and both females are homozygous for black.
   e. All three parents are heterozygous.

   ANALYSIS

   a. In this choice, female A is heterozygous and black; therefore, black is the dominant allele. About half the gametes of female A will carry the dominant allele and lead to black offspring, and about half will carry the recessive allele for brown fur and produce brown offspring in this mating with a brown homozygous male. Gametes of homozygous-black female B, on the other hand, will all carry the dominant black allele and produce black offspring in any mating. This set of parental genotypes predicts the observed distribution of offspring phenotypes.
   b. This choice can be rejected, because the male and female A, brown and black, respectively, cannot both be heterozygous. With simple dominance, one genotype can't produce two different phenotypes.
   c. In this choice, the black females are heterozygous; therefore, black is the dominant allele. About half of female B's gametes would be expected to carry the recessive allele for brown and produce brown offspring in a mating with a brown homozygous male. This choice is not impossible, but the observed outcome is very unlikely if female B is heterozygous.
   d. In this choice, brown must be the dominant allele, because the male is heterozygous and brown. About half his gametes are predicted to carry the dominant brown allele and to produce brown offspring in a mating with homozygous black females. Female B produced no brown offspring. While not impossible, this deviation from the predicted outcome is very unlikely.
   e. This choice, like choice (b), is impossible, because the parents have different phenotypes, and therefore can't have the same genotypes.

2. Imagine that the doughnut is a sexually reproducing animal and that a single enzyme forms the hole. All doughnuts observed in the past have had holes. Suppose a mutation occurs in one gene of one doughnut such that it now codes for an inactive protein instead of the hole-forming enzyme. Which of the following predictions would be most likely in descendants of this mutant doughnut?

   a. All members of the next generation will have holes.
   b. Heterozygous doughnuts will not have holes.
   c. If the cell in which the mutation occurs undergoes mitosis, subsequent generations of doughnuts will all have holes.
   d. The next generation will all be holeless.
   e. Past generations of doughnuts have all had holes.

ANALYSIS

Because the native allele codes for an active enzyme that catalyzes hole formation, any doughnut that carries at least one copy of that allele will synthesize the enzyme and have a hole. This means that holed is dominant, holeless recessive.

a. If we assume that doughnuts breed randomly, a population of doughnuts that included some heterozygous individuals would be expected to produce at least a few homozygous recessive, therefore holeless, individuals over a period of many generations. All the doughnuts observed in the past have had holes, so our most likely conclusion is that they have all been homozygous for the hole-forming enzyme. In that case, all the potential mates for the mutant heterozygote are homozygous dominant, and all the offspring must be homozygous or heterozygous dominant and have holes.

b. Because we have concluded that the holed trait is dominant, heterozygotes will have holes.

c. The characteristics of future generations depend on whether the cell in which the mutation occurred produces gametes. If the mutation occurred in a somatic cell, the new allele will not be transmitted to future generations. If it occurred in a cell that produces gametes, it will be transmitted, and some future doughnuts will be holeless. If the mutant cell undergoes mitosis and the cells it produces then generate gametes, the number of mutant alleles in the next generation will be even greater.

d. For choice (a), we have argued that the next generation's members will all have holes.

e. This is an observation, not a prediction.

3. A continuously variable trait was measured in two groups of organisms. The data are given below.

Group 1

3.2	3.7	2.6	3.8	3.4	2.8
1.7	4.9	3.3	2.4	2.7	3.9
3.6	2.5	2.3	3.5	3.0	3.5
3.3	3.3	2.8	3.2	2.8	4.9
3.7	2.6	3.6	3.5	3.8	3.2
3.7	3.2	3.4	2.9	3.3	3.3
3.9	3.0	3.2	4.7	2.7	3.9
3.9	2.7	3.8	3.4	3.1	3.0
3.6	3.1	2.5	2.9	2.2	2.9
3.5	3.9	3.1	4.2	3.6	3.6
2.9	4.2	3.9	3.6	3.5	3.3
1.9	2.6	3.1	2.7	2.2	2.9
3.0	2.3	3.7	4.4	3.0	4.1
3.1	3.0	3.5	2.2	3.6	4.0
4.2	2.7	3.6	3.4	2.9	3.9
3.7	5.4				

Group 2

4.3	3.3	3.9	3.5	4.0	3.5
5.0	4.8	3.0	3.9	5.1	3.8
5.2	6.1	3.5	2.8	5.3	5.0
5.2	5.0	4.0	5.1	4.9	3.9
4.9	4.1	4.4	3.0	4.8	3.3
4.0	4.5	4.3	4.3	3.9	3.0
4.1	3.6	4.9	3.5	3.2	3.2
4.8	6.0	5.5	4.9	4.2	5.3
5.5	4.0	3.3	4.6	3.9	4.7
3.9	5.6	4.3	3.7	3.6	4.6
3.6	4.0	4.2	3.9	3.4	4.1
2.9	5.2	5.2	3.3	5.4	4.0
3.5	4.4	1.6	4.8	4.4	2.6
4.8	4.7	3.8			

Make frequency distributions of these data. Can you infer from these data whether the two groups were the same or different?

ANALYSIS

To facilitate comparison between the two groups, it is best to draw the two distributions on the same figure.

The first step is to decide how many categories there will be in the distribution. There are about eighty or ninety individuals in each group, and the range of values is about 4 units from the smallest measurement to the largest. If each category is 0.1 unit wide, there will be about forty categories in each group and therefore an average of two individuals in each category. That is likely to show insufficient variation in number of individuals in the categories. Wider categories will be more likely to show the shape of the distribution. The example uses categories 0.3 units wide and thus has about fourteen categories for each group, with an average of six individuals per category.

The second step is to decide how big the figure will be and to divide the available space into the appropriate number of units. The final step is to label the dimensions of the graph and draw the columns.

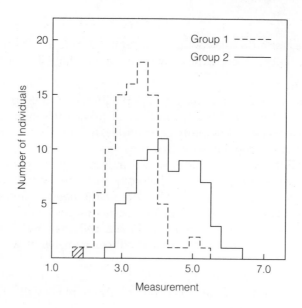

It is clear that the average measurement is different between the two groups. From this, you might conclude that the two groups were biologically different in some way. However, you should also notice that the ranges of the two groups overlap. This might lead you to conclude that the two groups were actually identical and that the difference was due to sampling error. Application of appropriate statistical techniques would allow you to estimate the probability that the two samples were drawn from a single, variable population.

In fact, these data are measurements of vital capacity, an indicator of lung size, in human physiology students. Group 1 consists of all females, group 2 of all males.

Answers

Answers to Interactive Exercises

MENDEL'S INSIGHTS INTO THE PATTERNS OF INHERITANCE (11-I)

1. Garden pea plants are self-fertilizing in nature.
2. True-breeding plants are those whose successive generations are exactly like the parents in one or more traits. Mendel stopped pea plants from self-fertilizing by opening their flower buds and removing the stamens. He promoted cross-fertilization by brushing pollen from another plant on the "castrated" flower bud.
3. F; 4. A; 5. C; 6. G; 7. E; 8. I; 9. D; 10. H; 11. J; 12. B; 13. monohybrid; 14. *P*; 15. *F₁*; 16. *F₂*; 17. segregation; 18. testcross; 19. 1 : 1; 20. dihybrid; 21. independent assortment; 22. genotype: 1/2 *Tt*: 1/2 *tt*; phenotype: 1/2 tall, 1/2 short; 23. all *TT*; 24. 1/4 *TT*: 1/2 *Tt*: 1/4 *tt* or 1 *TT*: 2 *Tt*: 1 *tt*; 25. 1/2 *Tt*: 1/2 *tt* or 1 *Tt*: 1 *tt*; 26. all *tt*. Summary: 27. 9/16 pigmented eyes, right-handed; 28. 3/16 pigmented eyes, left-handed; 29. 3/16 blue-eyed, right-handed; 30. 1/16 blue-eyed, left-handed (note Punnett square below).

	BR	*Br*	*bR*	*br*
BR	*BBRR*	*BBRr*	*BbRR*	*BbRr*
Br	*BBRr*	*BBrr*	*BbRr*	*Bbrr*
bR	*BbRR*	*BbRr*	*bbRR*	*bbRr*
br	*BbRr*	*Bbrr*	*bbRr*	*bbrr*

31. albino = *pp*

normal pigmentation = *PP* or *Pp*

Woman of normal pigmentation with an albino mother ⟶ *Pp*; received her recessive gene from her mother and her dominant gene (*P*) from her father. It is likely that half of the couple's children will be albinos (*pp*) and half will have normal pigmentation but be heterozygous (*Pp*).

Pp × pp
↓   ↓

kinds of ⓅP   ⓅP   only kind
gametes   of gamete
from Pp ⓟp   ⓟp   from pp

	ⓟp
ⓅP	Pp
ⓟp	pp

32. a. F_1: black trotter; F_2: nine black trotters, three black pacers, three chestnut trotters, one chestnut pacer; b. black trotter; c. *BbTt*; d. *bbtt*, chestnut pacers.

VARIATIONS ON MENDEL'S THEMES (11-II)

1. phenotype: all pink, genotype: all *RR'*; 2. phenotype: all white, genotype: all *R'R'*; 3. phenotype: 1/2 red, 1/2 pink; genotype: 1/2 *RR*, 1/2 *RR'*; 4. phenotype: all red, genotype: all *RR*.
5. The term *multiple alleles* is used when more than two forms of an allele exist for a given gene locus.
6. genotypes: 1/4 I^AI^A, 1/4 I^AI^B, 1/4 I^Ai, 1/4 I^Bi; phenotypes: 1/2 A, 1/4 AB, 1/4 B; 7. genotypes: 1/4 I^AI^B, 1/4 I^Bi, 1/4 I^Ai, 1/4 *ii*; phenotypes: 1/4 AB, 1/4 B, 1/4 A, 1/4 O; 8. genotypes: all I^Ai, phenotypes: all A; 9. genotypes: all *ii*, phenotypes: all O; 10. genotypes: 1/4 I^AI^A, 1/2 I^AI^B, 1/4 I^BI^B; phenotypes: 1/4 A, 1/2 AB, 1/4 B.
11. This could only be explained as a case of disputed paternity or, less likely, as a case of baby mix-up in a hospital.
12. F_1: *RrPp*, walnut comb; F_2: 9 R_P_ walnut, 3 R_pp rose, 3 rrP_ pea, 1 *rrpp* single; 13. 1/4 walnut, 1/4 rose, 1/4 pea, 1/4 single; 14. 1/4 color, 3/4 white; 15. 3/4 color, 1/4 white; 16. 1/4 color, 3/4 white; 17. F; 18. T; 19. T; 20. T; 21. F; 22. F; 23. T; 24. T; 25. F; 26. T.

Answers to Self-Quiz

1. d; 2. b; 3. a; 4. c; 5. c; 6. b; 7. a; 8. a; 9. c; 10. b.

Answer to Integrating and Applying Key Concepts

DdPp × Ddpp

12

CHROMOSOME VARIATIONS AND HUMAN GENETICS

Interactive Exercises

CHROMOSOMAL THEORY OF INHERITANCE (12-I, pp. 183–190)

Fill-in-the-Blanks

(1) _Flemming_ first described the behavior of threadlike bodies (chromosomes) during mitosis.
(2) _weismann_ first proposed that chromosomal number must be reduced by half during gamete formation;
the process became known as (3) _Meiosis_ . (4) _Homologous_ chromosomes resemble each other in length,
shape, and gene sequence.

5. Complete the Punnett square below, which brings Y-bearing and X-bearing sperm together randomly
    in fertilization, and then complete exercises 6–8.

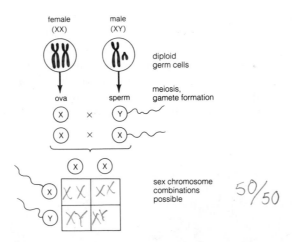

50/50

6. Males transmit their Y chromosome only to their ~~Males~~ sons.

7. Males receive their X chromosome only from their Mothers.

8. From what source(s) do females receive their two X chromosomes? one from their mom + one from their pop.

Matching

Choose the one most appropriate answer for each.

9. _F_ crossing over
10. _D_ linkage
11. _C_ karyotype
12. _H_ independent assortment
13. _A_ autosomes
14. _B_ segregation
15. _G_ Y-linked
16. _E_ SRY gene

A. All the chromosomes in a cell except the sex chromosomes
B. Channeling of homologous chromosomes into different gametes
C. Visual representation of the metaphase chromosomes of a cell arranged in order, from largest to smallest
D. The tendency of genes located on the same chromosome to travel together in inheritance
E. Apparently the master gene for sex determination
F. Exchange of parts of chromatids between nonsister chromatids during prophase I of meiosis; disrupts linkage
G. Genes carried on the smaller human sex chromosome
H. The way in which one pair of chromosomes segregates into different gametes does not influence how a different pair of chromosomes segregates into the same gamete

Early investigators realized that the frequency of crossing over (data from actual genetic crosses) could be used as a tool to map genes on chromosomes. Genetic maps (see Figure 12.7 in the text) are graphic representations of the relative positions and distances between genes on each chromosome (linkage group). Geneticists arbitrarily equate 1 percent of crossover (recombination) with one genetic map unit. In the following exercises, a line is used to represent a chromosome with its linked genes.

17. Which of the following represents a chromosome that has undergone crossover and recombination? It is assumed that the organism involved is heterozygous with the following genotype: $\begin{array}{c}A\\B\end{array}\Big|\Big|\begin{array}{c}a\\b\end{array}$. _____

    a. $\begin{array}{|c}A\\B\end{array}$ b. $\begin{array}{c|}a\\b\end{array}$ c. $\begin{array}{|c}B\\A\end{array}$ d.⃝ $\begin{array}{|c}a\\B\end{array}$

18. Following breeding experiments and a study of crossovers, the following map distances were recorded: C to B equals 35 genetic map units, gene B to A equals 10 map units, and gene A to C equals 45 map units. Which of the following maps is correct? _____

    a.⃝ A B   C   b. B   A C   c. B   C A   d. B A   C

19. Individuals with dominant gene N have the nail-patella syndrome and exhibit abnormally developed fingernails and absence of kneecaps; the recessive gene n is normal. Geneticists have learned that the genes determining the ABO blood groups and the nail-patella gene are linked, with both found on chromosome 9. A woman with blood type A and nail-patella syndrome marries a man with blood type O and normal fingernails and kneecaps. This couple has three children. One daughter has blood type A like her mother but has normal fingernails and kneecaps. A second daughter has blood type O with nail-patella syndrome. A son has blood type O and normal fingernails and kneecaps. The

chromosomes of the parents are diagramed below. Complete the exercise by writing the gene symbols of each child on his or her chromosomes.

$$\frac{An}{ON} \times \frac{On}{On} \qquad \text{(or } An/ON \times On/On\text{)}$$

$$\underline{On\!/\!An} \quad \underline{ON} \quad \underline{On}$$
$$\underline{An} \quad \underline{On} \quad \underline{On}$$

Explain how the son can have the genotype On/On when only one parent has a chromosome with genes O and n linked. _____

_____

CHROMOSOME VARIATIONS IN HUMANS (12-II, pp. 191–201)

1. List the characteristics that identify autosomal recessive inheritance._____

_____

_____

2. The autosomal allele that causes albinism (c) is recessive to the allele for normal pigmentation (C). A normally pigmented woman whose father is an albino marries an albino man whose parents are normal. They have three children, two normal and one albino. Give the genotypes for each person listed.

_____

3. List the characteristics that identify autosomal dominant inheritance._____

_____

4. Huntington's disorder is a rare form of autosomal dominant inheritance (H); the normal gene is h. The disease causes progressive degeneration of the nervous system with onset exhibited near middle age. An apparently normal man in his early twenties learns that his father has recently been diagnosed as having Huntington's disorder. What are the chances that the son will develop Huntington's disorder?

_____

5. List the characteristics that identify X-linked recessive inheritance._____

_____

_____

6. A color-blind man and a woman with normal vision whose father was color-blind have a son. Color blindness, in this case, is caused by an X-linked recessive gene. If only male offspring are considered,

what is the probability that their son is color-blind?_____

_____

7. Hemophilia A is caused by an X-linked recessive gene. A woman who is seemingly normal but whose father was a hemophiliac marries a normal man.

   a. What proportion of their sons will have hemophilia?_____

   b. What proportion of their daughters will have hemophilia?_____

   c. What proportion of their daughters will be carriers?_____

8. The following pedigree shows the pattern of inheritance of color blindness in a family (persons with the trait are indicated by black circles).

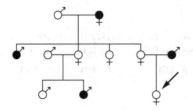

What is the chance that the third-generation female indicated by the arrow will have a color-blind son if she marries a normal male? A color-blind male?_____

_____

Label-Match

On rare occasions, chromosome structure becomes abnormally rearranged. Such changes may have profound effects on the phenotype of an organism. Label the following diagrams of abnormal chromosome structure as a deletion, a duplication, an inversion, or a translocation. Complete the exercise by matching and entering the letter of the proper description in the parentheses following each label.

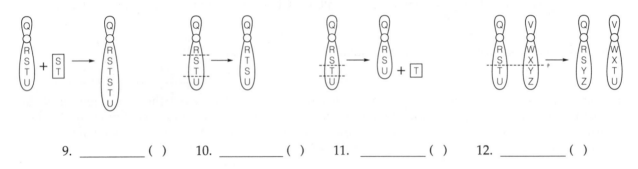

9. _____ ( )   10. _____ ( )   11. _____ ( )   12. _____ ( )

A. The loss of a chromosome segment
B. A gene sequence in excess of its normal amount in a chromosome
C. A chromosome segment that separated from the chromosome and then was inserted at the same place, but in reverse
D. The transfer of part of one chromosome to a nonhomologous chromosome

13. Define the following terms.

a. aneuploidy_____

_____

b. polyploidy_____

_____

14. Aneuploidy results from meiotic mistakes in the process of chromosome separation to daughter cells. The mistake may occur at first or second meiotic division. The name for these mistakes is nondisjunction, which means the chromosomes fail to separate (see Figure 12.19 in the text).

a. If the nondisjunction occurs at anaphase I of the first meiotic division, what may the number of abnormal gametes (for the chromosome involved in the nondisjunction) be?_____

b. If the nondisjunction occurs at anaphase II of the second meiotic division, what may the number of abnormal gametes (for the chromosome involved in the nondisjunction) be?_____

Labeling

Spermatogenesis of the male sex chromosomes is partly illustrated below. Show nondisjunctions of the X and the Y chromosomes occurring at anaphase II by completing the sketches in the spermatids and sperm. Then fertilize the normal haploid X-bearing eggs shown with the sperm (obtained following nondisjunction). Enter the chromosome constitutions in the blanks in the zygote column. Imagine in the primary spermatocyte, shown below, a duplicated X and a duplicated Y chromosome as meiosis I begins.

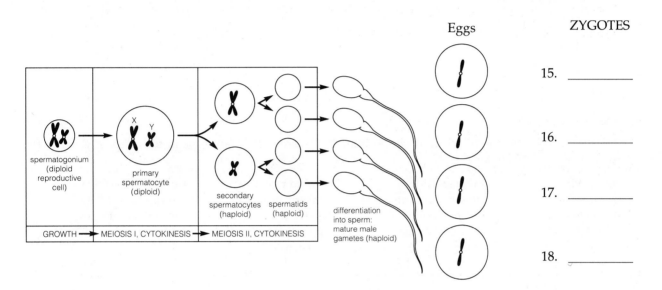

Eggs   ZYGOTES

15. _____

16. _____

17. _____

18. _____

19. Complete the table below by filling in the missing information.

Aneuploidy	Chromosome Constitution	Brief Description
a. Down syndrome	trisomy 21	
b. _____	XO	distorted female phenotype, ovaries nonfunctional, sterile
c. Klinefelter syndrome	_____	testes smaller than normal, some breast enlargement, some mental retardation
d. _____	_____	taller than average, some showing mild mental retardation, most phenotypically normal

Matching

Choose the one most appropriate answer for each. A letter may be used more than once or not at all.

20. ___ albinism

21. ___ amniocentesis

22. ___ chorionic villi analysis

23. ___ cleft lip

24. ___ galactosemia

25. ___ phenylketonuria

26. ___ karyotype analysis

27. ___ sickle-cell anemia

28. ___ Wilson's disorder

29. ___ pedigree chart

A. A phenotypic defect that can be helped by diet modification
B. A phenotypic defect that can be helped by environmental adjustments
C. A phenotypic defect that can be helped by surgical correction
D. A phenotypic defect that can be helped by chemotherapy
E. Used to arrive at a diagnosis
F. Used to determine the location of a specific gene on a specific chromosome
G. Used to depict genetic relationships among the members of families

---

Chapter Terms

The following page-referenced terms are important; they were in boldface type in the chapter. Refer to the instructions given in Chapter 1, p. 4.

karyotype (185)	pedigrees (191)	X-linked recessive inheritance (193)	translocation (195)
X-linked genes (189)	autosomal recessive inheritance (192)	deletion (195)	aneuploidy (197)
Y-linked genes (189)			polyploidy (197)
linkage (189)	autosomal dominant inheritance (192)	duplication (195)	nondisjunction (197)
crossing over (189)		inversion (195)	

Self-Quiz

___ 1. All the genes located on a given chromosome compose a _____ .
   a. karyotype
   b. genome
   c. wild-type allele
   d. linkage group

___ 2. Chromosomes other than those involved in sex determination are known as_____.
   a. nucleosomes
   b. heterosomes
   c. alleles
   d. autosomes

___ 3. If two genes are very close to each other on the same chromatid, _____.
   a. crossing over and recombination occur between them quite often
   b. they act as if they are assorted independently into gametes
   c. they will most probably be in the same linkage group
   d. they will be segregated into different gametes when meiosis occurs

___ 4. Karyotype analysis is _____ .
   a. a means of detecting and reducing mutagenic agents
   b. a surgical technique that separates chromosomes that have failed to segregate properly during meiosis II
   c. used in prenatal diagnosis to detect chromosomal mutations and metabolic disorders in embryos
   d. a process that substitutes normal alleles for defective ones

___ 5. Which of the following did Morgan and his research group *not* do?
   a. They isolated and kept under culture fruit flies with the sex-linked recessive white-eyed trait.
   b. They developed the technique of amniocentesis.
   c. They discovered X-linked genes.
   d. Their work reinforced the concept that each gene is located on a specific chromosome.

___ 6. Red-green color blindness is a sex-linked recessive trait in humans. A color-blind woman and a man with normal vision have a son. What are the chances that the son is color-blind? If the parents ever have a daughter, what is the chance for each birth that the daughter will be color-blind? (Consider only the female offspring.)
   a. 100 percent, 0 percent
   b. 50 percent, 0 percent
   c. 100 percent, 100 percent
   d. 50 percent, 100 percent
   e. none of the above

___ 7. Suppose a hemophilic male (sex-linked recessive allele) and a female carrier for the hemophilic trait have a nonhemophilic daughter with Turner syndrome. Nondisjunction could have occurred in

   _____ .
   a. both parents
   b. neither parent
   c. the father only
   d. the mother only

___ 8. Nondisjunction involving the X chromosome occurs during oogenesis and produces two kinds of eggs, XX and O (no X chromosome). If normal Y sperm fertilize two types, which genotypes are possible?
   a. XX and XY
   b. XXY and YO
   c. XYY and XO
   d. XYY and YO

___ 9. Of all phenotypically normal males in prisons, the type once thought to be genetically predisposed to becoming criminals was the _____ .
   a. XXY disorder
   b. XYY disorder
   c. Turner syndrome
   d. Down syndrome

___ 10. Amniocentesis is _____ .
   a. a surgical means of repairing deformities
   b. a form of chemotherapy that modifies or inhibits gene expression or the function of gene products
   c. used in prenatal diagnosis to detect chromosomal mutations and metabolic disorders in embryos
   d. a form of gene-replacement therapy

Chapter Objectives/Review Questions

This section lists general and detailed chapter objectives that can be used as review questions. You can make maximum use of these items by writing answers on a separate sheet of paper. Fill in answers where blanks are provided. To check for accuracy, compare your answers with information given in the chapter or glossary.

Page	Objectives/Questions
(183)	1. Identify the researcher who first described threadlike bodies in the nuclei of dividing cells; state what was so useful about this contribution.
(184)	2. In 1887, _____ proposed that a special division process must reduce the chromosome number by half before gametes form.
(184)	3. Relate the significance of the year 1900 to genetics knowledge.
(185–186)	4. Define sex chromosomes and autosomes ; generally distinguish the types of alleles that must be found on each.
(185)	5. Define karyotype and state why it is useful; describe how karyotyping is done.
(186–189)	6. Distinguish between sex-linked genes and sex-determining genes.
(187)	7. A newly identified region of the Y chromosome called _____ appears to be the master gene for sex determination.
(189)	8. Describe two significant contributions the Morgan research team made to our early understanding of genetics.
(186)	9. Explain how the meiotic segregation of sex chromosomes to gametes and subsequent random fertilization determine sex in many organisms.
(189)	10. The tendency of genes located on the same chromosome to end up together in the same gamete is called _____ .

Page   Objectives/Questions

(189–190) 11. State the relationship between the probability of crossing over (and subsequent recombination) and the distance between two genes located on the same chromosome.

(191)   12. List reasons why it is difficult to study inheritance patterns in humans. Researchers can identify inheritance patterns and track genetic abnormalities through several generations by constructing _____ charts.

(191)   13. A genetic _____ is a rare or less common occurrence, whereas a _____ causes mild to severe medical problems.

(192)   14. Describe the characteristics of autosomal recessive inheritance ; summarize the characteristics of galactosemia as an example.

(192–193) 15. Describe the characteristics of autosomal dominant inheritance ; summarize the characteristics of achondroplasia and Huntington's disorder as examples.

(193–194) 16. Describe the characteristics of X-linked inheritance; summarize the characteristics of hemophilia A as an example.

(195)   17. A _____ is a loss of a chromosome segment; a _____ is a gene sequence in excess of its normal amount in a chromosome; a chromosome segment that separated from the chromosome and then was inserted at the same place—but in reverse, is a _____ ; a _____ is the transfer of part of one chromosome to a nonhomologous chromosome.

(196–197) 18. When gametes or cells of an affected individual end up with one extra or one less than the parental number of chromosomes, it is known as _____ ; relate this to monosomy and trisomy.

(201)   19. Practice sketching a diagram of nondisjunction similar to Figure 12.19 in the text.

(198, 201) 20. Trisomy 21 is known as _____ syndrome; Turner syndrome has the chromosome constitution _____ ; XXY chromosome constitution is _____ syndrome; taller than average males with sometimes slightly depressed IQs have the _____ condition.

(197)   21. Distinguish between autosomal nondisjunction and sex chromosome nondisjunction.

(199)   22. Define phenotypic treatment and describe one example.

(200)   23. Explain the procedures used in two types of prenatal diagnosis, amniocentesis and chorionic villi analysis; compare the risks.

(199–200) 24. List some benefits of genetic screening and genetic counseling to society.

(200)   25. Discuss some of the ethical considerations that might be associated with a decision of induced abortion.

Integrating and Applying Key Concepts

The parents of a young boy bring him to their doctor. They explain that the boy does not seem to be going through the same vocal developmental stages as his older brother. The doctor orders a common cytogenetics test to be done, and it reveals that the young boy's cells contain two X chromosomes and one Y chromosome. Describe the test that the doctor ordered and explain how and when such a genetic result—XXY—most logically occurred.

Solve the following genetics problem (show all setups, genotypes, and phenotypes): A husband sues his wife for divorce, arguing that she has been unfaithful. His wife gave birth to a girl with a fissure in the iris of her eye, a sex-linked recessive trait. Both parents have normal eye structure. Can the genetic facts be used to argue for the husband's suit? Explain your answer.

Critical Thinking Exercises

1. The offspring of one mated pair of mammals included three males, all of which showed an X-linked recessive trait. The fourth zygote was female. Which of the following statements could be made about her with the most confidence?

   a. Even if the father shows the trait, she will not.
   b. If the father does not show the trait, she will not.

c. She will show the trait.
d. If the mother shows the recessive trait, the daughter will also show the recessive trait.
e. If the father shows the recessive trait, she will show the recessive trait.

ANALYSIS

A male mammal has one X chromosome from his mother and one Y chromosome from his father. These male offspring carry the recessive X-linked allele; thus, their mother is either heterozygous or homozygous for the allele, and no inference can be made about the father. The daughter will receive the father's only X chromosome and one of her mother's X chromosomes.

    a. If the father carries the allele on his only X chromosome, the daughter will inherit it. She will not show the recessive trait, however, if her mother is heterozygous and she happens to receive the mother's dominant X chromosome. If she receives another copy of the recessive allele from her mother, she will show the recessive trait.

    b. If the father does not show the trait, he will transmit to the daughter his only X chromosome bearing the dominant allele. His daughter will not show the trait regardless of the mother's contribution to the daughter's genotype.

    c. The daughter might or might not show the recessive trait, depending on the genotypes of the parents, which can be only partially inferred from the information about the sons.

    d. If the mother shows the recessive trait, she must be homozygous and will contribute a recessive allele to the daughter's genotype. However, a father who does not show the trait will necessarily contribute the dominant allele and produce only daughters who do not show the trait.

    e. A father who shows the trait will contribute the allele to all daughters, but the mother in this case could be heterozygous and contribute a dominant allele to her daughter. The result would be a dominant phenotype.

2. Study the pedigree diagram below. Is the trait being tracked dominant or recessive?

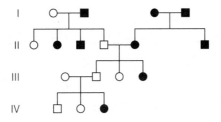

ANALYSIS

Lines I and II are inconclusive. In the first mating, one of the parents must be heterozygous, because the phenotypes are different, but the results do not indicate which is which. The second mating, between two like phenotypes, is also inconclusive. Both parents could be homozygous, but the same results would be predicted whether the trait is dominant or recessive. Both parents could be heterozygous, and we would expect variation in the offspring and could then infer which form is dominant. However, the sample size is too small to make any inference.

    Line III is inconclusive for the same reason as for the first mating on line I.

    Line IV leads to the conclusion that the trait is recessive, because the two parents who do not show the trait produce an offspring who shows the trait. Because the trait appears in the phenotype of one of the offspring, the allele must have been present in the parents. Yet their phenotypes do not carry the dominant allele. The only way this could happen with a dominant allele would be if the allele was formed by a new mutation in one of these parents, an unlikely event.

3. In an attempt to map the locations of five genes on a single chromosome, crossover frequencies between pairs of genes are measured. Some of the data are as follows:

A–B	35.5%
A–E	41.5%
B–E	6.0%
C–B	18.5%
D–C	36.0%
D–E	23.5%

What is the sequence of the five genes along the chromosome, and what are the distances between adjacent genes?

ANALYSIS

The basic assumption in this kind of study is that crossover frequency between two genes is proportional to the linear distance between them along the chromosome. One-percent crossover frequency defines one map unit of distance.

Many series of steps will lead to the solution to this problem. One of these follows:

1. Given:

    A _____41.5_____ E

2. Assume D is not between A and E.

    A _____41.5_____ E _____23.5_____ D

3. Assume B is to the left of A.

    B _____35.5_____ A _____41.5_____ E _____23.5_____ D

4. This makes B more than 6.0 units from E. Try placing B between A and E.

    A _____35.5_____ B _____6.0_____ E _____23.5_____ D

5. Assume C is to the right of D.

    A _____35.5_____ B _____6.0_____ E _____23.5_____ D _____36.0_____ C

6. This makes C more than 18.5 units from B. Try placing C to the left of D.

    A _____30.0_____ C _____5.5_____ B _____6.0_____ E _____23.5_____ D

7. Now C is less than 18.5 units from B. Try reversing the first assumption in step 2.

    A _____18.0_____ D _____23.5_____ E

8. Insert B according to steps 3 and 4.

    A _____18.0_____ D _____17.5_____ B _____6.0_____ E

9. Now, if C is to the left of D, it's obviously too far from B. Try it to the right of D.

    A _____18.0_____ D _____17.5_____ B _____6.0_____ E _____12.5_____ C

10. It works.

Answers

Answers to Interactive Exercises

CHROMOSOMAL THEORY OF INHERITANCE (12-I)
1. Flemming; 2. Weismann; 3. meiosis; 4. Homologous; 5. Two blocks of the Punnett square should be XX, and two blocks should be XY; 6. sons; 7. mother; 8. Females receive one X chromosome from their mother and one from their father; 9. F; 10. D; 11. C; 12. H; 13. A; 14. B; 15. G; 16. E; 17. d; 18. a; 19. Linked genes of the children: daughter, On/An; daughter, ON/On; son, On/On. The son has genotype On/On because a crossover must have occurred during meiosis in his mother to show the recombination On.

CHROMOSOME VARIATIONS IN HUMANS (12-II)

1. Males or females can carry the recessive allele on an autosome. Heterozygotes are generally symptom-free; homozygotes are affected. When both parents are heterozygous, there is a 50 percent chance that each child born to them will be heterozygous, also, and a 25 percent chance a child will be homozygous recessive; when both parents are homozygous, all their children will be affected.

2. The woman's mother is heterozygous normal, *Cc;* the woman is also heterozygous normal, *Cc.* The albino man, *cc,* has two heterozygous normal parents, *Cc.* The two normal children are heterozygous normal, *Cc;* the albino child is *cc.*

3. A dominant allele is always expressed to some extent. An affected individual will always have at least one affected parent.

4. Assuming the father is heterozygous with Huntington's disorder and the mother is normal, the chances are 1/2 that the son will develop the disease.

5. The mutated gene occurs on the X (not the Y) chromosome; heterozygous females may carry the trait but not express it; males, having only one X chromosome, are afflicted with the disorder.

6. If only male offspring are considered, the probability is 1/2 that the couple will have a color-blind son.

7. a. The probability is that half of the sons will have hemophilia; b. the probability is zero that a daughter will express hemophilia; c. the probability is that half of the daughters will be carriers.

8. If the woman marries a normal male, the chance that her son would be color-blind is 1/2. If she marries a color-blind male, the chance that her son would be color-blind is also 1/2.

9. duplication (B); 10. inversion (C); 11. deletion (A); 12. translocation (D).

13. a. Aneuploidy is the abnormal condition existing when gametes or cells of an affected individual may end up with one extra or one less than the parental number of chromosomes; b. Polyploidy is an abnormal condition existing when gametes or cells of an affected

individual may end up with three or more of each type of chromosome characteristic of the parental stock.

14. a. All gametes will be abnormal; b. Half the gametes will be abnormal.

15–18. From nondisjunction in the second meiotic division, resulting sperm contain either two X chromosomes, two Y chromosomes, or no sex chromosomes. The zygotes are as follows: 15. XXX; 16. XO; 17. XYY; 18. XO (the order may vary).

19. a. Down syndrome; trisomy 21; shorter than normal, mental retardation, skin fold over inner corner of eyelid; b. Turner syndrome; XO; distorted female phenotype, ovaries nonfunctional, sterile; c. Klinefelter syndrome; XXY; testes smaller than normal, some breast enlargement, some mental retardation; d. XYY condition; XYY; taller than average, some showing mild mental retardation, most phenotypically normal; 20. B; 21. E; 22. E; 23. C; 24. A; 25. A; 26. E; 27. B; 28. D; 29. G.

Answers to Self-Quiz

1. d; 2. d; 3. c; 4. c; 5. b; 6. a; 7. d; 8. b; 9. b; 10. c.

Answers to Integrating and Applying Key Concepts

Answer to genetics problem: Yes; the husband could not have supplied either of his daughter's recessive genes because his only X chromosome bears the *N* for normal iris.

$X^N Y$	$X^N X^n$	*N* = normal iris
father	mother	*n* = fissured iris
	$X^n X^n$ daughter's genotype	The mother must also carry the recessive gene in order to be her daughter's mother.

13

DNA STRUCTURE AND FUNCTION

DISCOVERY OF DNA FUNCTION
    A Puzzling Transformation
    Bacteriophage Studies
DNA STRUCTURE
    Components of DNA
    Patterns of Base Pairing

DNA REPLICATION
    Assembly of Nucleotide Strands
    A Closer Look at Replication
        Origin and Direction of Replication
        Energy and Enzymes for Replication
ORGANIZATION OF DNA IN CHROMOSOMES

Interactive Exercises

In most cases, answers for specific molecules use abbreviations.

DISCOVERY OF DNA FUNCTION (13-I, pp. 205–208)

1. Complete the table below which traces the discovery of DNA function.

Investigators	Year(s)	Contribution
a. Miescher	1868	
b.	1928	discovered the transforming principle in *Streptococcus pneumoniae;* live, harmless R cells were mixed with dead S cells, R cells became S cells
c. Avery (also MacLeod and McCarty)	1944	
d.	mid-1940s	studied the infectious cycle of bacteriophages
e. Hershey and Chase	1952	

DNA STRUCTURE (13-II, pp. 208–210)

1. List the three parts of a nucleotide. _Sugar, nitrogen, phosphate_

Labeling

Four nucleotides are illustrated below. In the blank, label each nitrogen-containing base correctly as guanine, thymine, cytosine, or adenine. In the parentheses following each blank, indicate whether that nucleotide base is a purine (pu) or a pyrimidine (py).

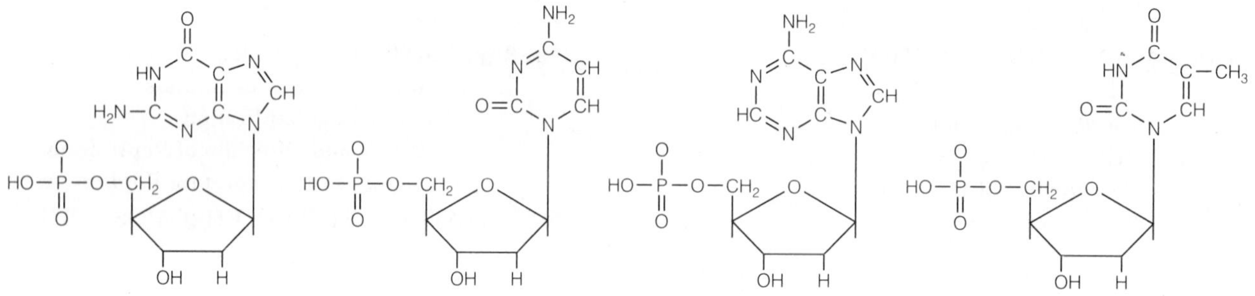

2. _____ ( )   3. _____ ( )   4. _____ ( )   5. _____ ( )

Label-Match

Identify each indicated part of the DNA illustration on page 141. Choose from these answers: phosphate group, purine, pyrimidine, nucleotide, and deoxyribose. Complete the exercise by matching and entering the letter of the proper structure description in the parentheses following each label.

The following DNA memory devices may be helpful: Use *pyrCUT* to remember that the single-ring pyrimidines are cytosine, uracil, and thymine; use *purAG* to remember that the double-ring purines are adenine and guanine; pyrimidine is a *long* name for a *narrow* molecule; purine is a *short* name for a *wide* molecule; to recall the number of hydrogen bonds between DNA bases, remember that AT = 2 and CG = 3.

6. _____ ( )

7. _____ _____ ( )

8. _____ ( )

9. _____ ( )

10. _____ ( )

11. _____ ( )

12. _____ ( )

A. The pyrimidine is thymine, because it has two hydrogen bonds.
B. A five-carbon sugar joined to two phosphate groups in the upright portion of the DNA ladder.
C. The purine is guanine because it has three hydrogen bonds.
D. The pyrimidine is cytosine because it has three hydrogen bonds.
E. The purine is adenine because it has two hydrogen bonds.
F. Composed of three smaller molecules: a phosphate group, five-carbon deoxyribose sugar, and a nitrogenous base (in this case, a pyrimidine).
G. A chemical group that joins two sugars in the upright portion of the DNA ladder.

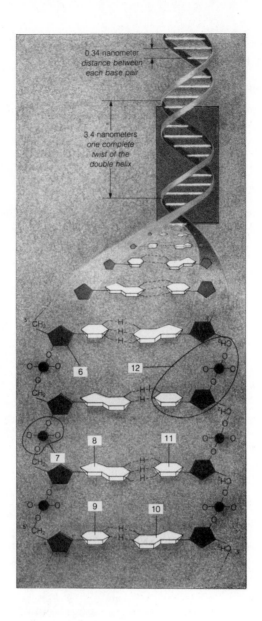

True-False

If false, explain why.

_I_ 13. DNA is composed of four different types of nucleotides.

_T_ 14. In the DNA of every species, the amount of adenine present always equals the amount of thymine, and the amount of cytosine always equals the amount of guanine (A = T and C = G).

_I_ 15. In a nucleotide, the phosphate group is attached to the nitrogen-containing base, which is attached to the five-carbon sugar.

_I_ 16. Watson and Crick built their model of DNA in the early 1950s.

_T_ 17. Guanine pairs with cytosine and adenine pairs with thymine by forming hydrogen bonds between them.

Fill-in-the-Blanks

Base (18)_____ between the two nucleotide strands in DNA is (19)_____ for all species (A-T; G-C). The base (20)_____ (determining which base follows the next in a nucleotide strand) is (21)_____ from species to species.

DNA REPLICATION (13-III, pp. 210–213)

Labeling

1. The term semiconservative replication refers to the fact that each new DNA molecule resulting from the replication process is "half-old, half-new." In the illustration below, complete the replication required in the middle of the molecule by adding the required letters representing the missing nucleotide bases. Recall that ATP energy and the appropriate enzymes are actually required in order to complete this process.

T - A   T - A
G - U   G - C
A - T   A - T
C - G   C - G
C - G   C - G
C - G   C - G
old   new   old   new

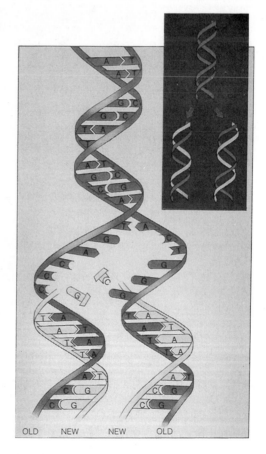

True-False

If false, explain why.

___ 2. The hydrogen bonding of adenine to guanine is an example of complementary base pairing.

___ 3. The replication of DNA is considered a conserving process because the same four nucleotides are used again and again during replication.

___ 4. Each parent strand remains intact during replication, and a new companion strand is assembled on each of those parent strands.

___ 5. Some of the enzymes associated with DNA assembly repair errors during the replication process.

Matching

Match the numbers on the illustration below to the letters representing structures and functions.

6. C
7. E
8. D
9. A
10. F
11. B

A. The enzyme (ligase) links short DNA segments together.
B. One parent DNA strand.
C. Enzymes unwind DNA double helix.
D. Enzymes add short "primer" segments to begin chain assembly.
E. Enzymes (DNA polymerase) assemble new DNA strands.
F. Newly forming DNA strand.

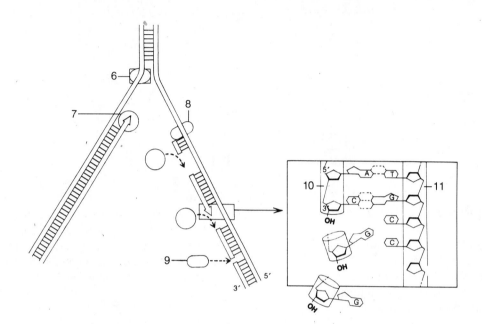

ORGANIZATION OF DNA IN CHROMOSOMES (13-IV, p. 213)

Fill-in-the-Blanks

Each chromosome has one (1) DNA molecule coursing through it. Eukaryotic DNA is complexed tightly with many (2) histones. Some (3) histone proteins act as spools to wind up small pieces of (4) DNA. A (5) nucleosome is a histone-DNA spool. The way the chromosome is packed is known to influence the activity of different (6) gene.

Chapter Terms

The following page-referenced terms are important; they were in boldface type in the chapter. Refer to the instructions given in Chapter 1, p. 4.

DNA (205)
bacteriophages (207)
nucleotide (208)

bases (208)
adenine (A) (208)
guanine (G) (208)

thymine (T) (208)
cytosine (C) (208)
semiconservative replication (210)

DNA polymerases (211)
nucleosome (213)

Crossword Puzzle: DNA Structure and Function

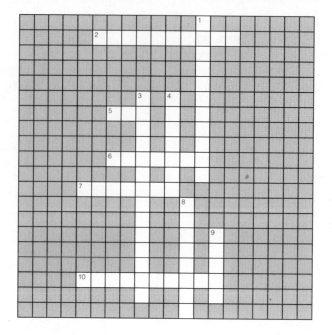

Across
2. DNA segment looped twice around a histone core
5. double-stranded helical molecule; carries genetic instructions
6. a purine; pairs with thymine in DNA
7. a pyrimidine; pairs with adenine in DNA
10. structural unit of nucleic acids

Down
1. major DNA replication enzymes
3. viruses that infect bacterial cells
4. a purine; pairs with cytosine in DNA
8. a pyrimidine; pairs with guanine in DNA
9. refers to the nucleotides in DNA; A, T, G, C

Self-Quiz

___ 1. Each DNA strand has a backbone that consists of alternating _____ .
   a. purines and pyrimidines
   b. nitrogen-containing bases
   c. hydrogen bonds
   d. sugar and phosphate molecules

___ 2. In DNA, complementary base pairing occurs between _____.
   a. cytosine and uracil
   b. adenine and guanine
   c. adenine and uracil
   d. adenine and thymine

___ 3. Adenine and guanine are _____ .
   a. double-ringed purines
   b. single-ringed purines
   c. double-ringed pyrimidines
   d. single-ringed pyrimidines

___ 4. Franklin used the technique known as _____ to determine many of the physical characteristics of DNA.
   a. transformation
   b. transmission electron microscopy
   c. density-gradient centrifugation
   d. X-ray diffraction

___ 5. The significance of Griffith's experiment that used two strains of pneumonia-causing bacteria is that _____ .

    a. the conserving nature of DNA replication was finally demonstrated

    b. it demonstrated that harmless cells had become permanently transformed through a change in the bacterial hereditary system

    c. it established that pure DNA extracted from disease-causing bacteria transformed harmless strains into "pathogenic strains"

    d. it demonstrated that radioactively labeled bacteriophages transfer their DNA but not their protein coats to their host bacteria

___ 6. The significance of the experiments in which ^{32}P and ^{35}S were used is that _____ .

    a. the semiconservative nature of DNA replication was finally demonstrated

    b. it demonstrated that harmless cells had become permanently transformed through a change in the bacterial hereditary system

    c. it established that pure DNA extracted from disease-causing bacteria transformed harmless strains into "killer strains"

    d. it demonstrated that radioactively labeled bacteriophages transfer their DNA but not their protein coats to their host bacteria

___ 7. Franklin's research contribution was essential in _____ .

    a. establishing the double-stranded nature of DNA

    b. establishing the principle of base pairing

    c. establishing most of the principal structural features of DNA

    d. all of the above

___ 8. When Griffith injected mice with a mixture of dead pathogenic cells—encapsulated S cells and living, unencapsulated R cells of pneumonia bacteria—he discovered that _____ .

    a. the previously harmless strain had permanently inherited the capacity to build protective capsules

    b. the dead mice teemed with living pathogenic (R) cells

    c. the killer strain R was encased in a protective capsule

    d. all of the above

___ 9. A single strand of DNA with the base-pairing sequence C-G-A-T-T-G is compatible only with the sequence _____ .

    a. C-G-A-T-T-G

    b. G-C-T-A-A-G

    c. T-A-G-C-C-T

    d. G-C-T-A-A-C

___ 10. The nucleosome is a _____ .

    a. subunit of a nucleolus

    b. coiled bead of histone-DNA

    c. DNA packing arrangement within a chromosome

    d. term synonymous with *gene*

    e. both (b) and (c)

Chapter Objectives/Review Questions

This section lists general and detailed chapter objectives that can be used as review questions. You can make maximum use of these items by writing answers on a separate sheet of paper. Fill in answers where blanks are provided. To check for accuracy, compare your answers with information given in the chapter or glossary.

Page   *Objectives/Questions*

(204)   1. Before 1952, _____ molecules and _____ molecules were suspected of housing the genetic code.

(205–208)   2. Summarize the research carried out by Miescher, Griffith, Avery and colleagues, and Hershey and Chase; state the specific advances made by each in the understanding of genetics.

(207)   3. Viruses called _____ were used in early research efforts to discover the genetic material.

(207–208)   4. Summarize the specific research that demonstrated that DNA, not protein, governed inheritance.

(208–209)   5. DNA is composed of double-ring nucleotides known as _____ and single-ring nucleotides known as _____ ; the two purines are _____ and _____ , while the two pyrimidines are _____ and _____.

(209)   6. Draw the basic shape of a deoxyribose molecule and show how a phosphate group is joined to it when forming a nucleotide.

(209)   7. Show how each nucleotide base would be joined to the sugar-phosphate combination drawn in objective 6.

(209)   8. List the pieces of information about DNA structure that Rosalind Franklin discovered through her X-ray diffraction research.

(209–210)   9. The two scientists who assembled the clues to DNA structure and produced the first model were _____ and _____.

(209–210)   10. Explain what is meant by the pairing of nitrogen-containing bases (base pairing), and explain the mechanism that causes bases of one DNA strand to join with bases of the other strand.

(210)   11. Describe how the Watson-Crick model reflects the constancy in DNA observed from species to species yet allows for variations from species to species.

(210)   12. Assume that the two parent strands of DNA have been separated and that the base sequence on one parent strand is A-T-T-C-G-C; the base sequence that will complement that parent strand is _____.

(210)   13. Describe how double-stranded DNA replicates from stockpiles of nucleotides.

(210)   14. Explain what is meant by "each parent strand is conserved in each new DNA molecule."

(210)   15. DNA replication is specifically referred to as _____ replication.

(210–213)   16. List four functions that enzymes perform during the process of replication.

(211)   17. During DNA replication, enzymes called DNA _____ assemble new DNA strands.

(213)   18. The basic histone-DNA packing unit of the chromosome is the _____.

(213)   19. List possible reasons for the highly organized packing of nucleoprotein into chromosomes.

Integrating and Applying Key Concepts

Review the stages of mitosis and meiosis, as well as the process of fertilization. Include what has now been learned about DNA replication and the relationship of DNA to a chromosome. As you cover the stages, be sure each cell receives the proper number of DNA threads.

Critical Thinking Exercises

1. If each nucleotide in DNA paired only with itself (A with A, C with C, and so on), which of the following statements would have the greatest validity?

   a. A pairs with T and C pairs with G.
   b. DNA could not replicate.
   c. DNA could not carry genetic information.
   d. The two strands of each DNA molecule would be identical.
   e. The diameter of the DNA molecule would be variable.

ANALYSIS

   a. This statement simply denies the conditions of the question. While this is the actual observed pairing pattern, the question concerns the consequences of a different pattern.
   b. The essence of the replication process is that base pairing is specific. It does not matter which base pairs with which. As long as each kind of base pairs with only one other kind, each strand will direct the synthesis of a complementary strand, and the double-stranded molecule will be replicated.
   c. The genetic information is encoded in the sequence of nucleotides along the strand. This is not affected by the pattern of pairing between the two strands.
   d. This statement is true to an extent. If you started at one end of the double-stranded molecule and followed both strands, they would have identical sequences of nucleotides. However, remember that

the two ends of a single strand of DNA are different and that the two strands in a double helix run in opposite directions. If you separated the two strands, placed them with their similar ends together, and read their sequences, you would find that the sequences were opposite each other.

e. A and G are larger than C and T. Thus, A-A and G-G pairs would create wide spots in the double-stranded molecule, and C-C and T-T pairs would be narrow.

2. Suppose you had a supply of nucleotides made with atoms that were twice as heavy as normal and you allowed some cells to replicate their DNA *once*, using only the heavy nucleotides. What would be the weight of the double-stranded DNA molecules in the resulting G_2 cells if replication is semiconservative? What would be their weight if replication is conservative?

a. All will have normal weight.
b. All will weigh twice normal.
c. Half will have normal weight, and half will weigh twice normal.
d. All will weigh 1.5 times normal.
e. The weight cannot be predicted, because we do not know the total number of base pairs in the genome.

ANALYSIS

Semiconservative replication produces two double-stranded molecules; each consists of one old strand and one new strand. If the old strands weigh x units, the new strands weigh $2x$. The double-stranded molecules thus normally weigh $2x$, and the newly produced ones weigh $3x$, or 1.5 times as much (d).

Conservative replication yields one molecule consisting of the two original strands and one consisting of two new strands. Thus, the conservative hypothesis predicts that half the molecules will weigh the normal $2x$ and half will weigh $4x$ (c).

This analysis depends on the assumption that the entire double-stranded molecule is replicated. If only part of it is replicated or if a mixture of normal and heavy nucleotides is available, then we still predict that semiconservative replication will produce one group of heavier than normal molecules and that conservative replication will produce one group of original, light molecules and one group of even heavier molecules.

In fact, this thinking led to the experiment that established that replication is semiconservative. The heavy nucleotides contained a heavy form of nitrogen atoms. While the nucleotides did not weigh twice normal, they were heavier enough that DNA with them could be separated from DNA without them. The two hypotheses generated two different predictions, and the observations allowed a clear-cut evaluation.

3. a. The DNA of *E. coli* contains 4×10^6 base pairs in a single double-stranded molecule. It is replicated in about 20 minutes. How many bases per second does each polymerase molecule install?
   b. The DNA of each human cell contains 3×10^9 base pairs. Assuming that the human enzymes function at the same rate as the bacterial enzymes, how long would it take for S phase in a human cell?
   c. At this rate, how many cells could be formed from a single fertilized human ovum in 9 months?
   d. A full-term human fetus has about 75 times as many cells as the correct answer to part (c). State two assumptions that were made in the calculation that could be changed to resolve the discrepancy.

ANALYSIS

a. To replicate the bacterial DNA, 8×10^6 bases must be polymerized. This is accomplished by four enzyme molecules, one on each strand in each direction from the initiation site. This means 2×10^6 bases per enzyme per 1,200 seconds, or 1,667 bases per enzyme per second.
b. The human DNA is 750 times as long as the bacterial DNA. If the enzymes work at the same rate, replication will take 750 times as long, or 15,000 minutes (250 hours; 10.4 days).
c. Gestation of 270 days allows twenty-six replication cycles of 10.4 days. This produces 2^{26}, or 67 million, cells from the original fertilized egg.
d. We assumed that the human enzymes work at the same rate as the bacterial *and* that the same number of enzyme molecules carry out the replication from a single initiation site. However, the human DNA exists in at least forty-six separate molecules (one per chromosome), and there are probably many initiation sites on each molecule. Thus, many more enzymes are working, the total time for replication is much shorter, and many more cells are produced in a given time. Furthermore, the eukaryotic enzymes might catalyze quicker polymerization.

Answers

Answers to Interactive Exercises

DISCOVERY OF DNA FUNCTION (13-I)
1. a. Miescher: identified "nuclein" from nuclei of pus cells and fish sperm; discovered DNA; b. Griffith: discovered the transforming principle in *Streptococcus pneumoniae*; live, harmless R cells were mixed with dead S cells; R cells became S cells; c. Avery: reported that the transforming substance in Griffith's bacteria experiments was probably DNA, the substance of heredity; d. Delbrück, Hershey, and Luria: studied the infectious cycle of bacteriophages; e. Hershey and Chase: worked with radioactive sulfur (protein) and phosphorus (DNA) labels; T4 bacteriophage and *E. coli* demonstrated that labeled phosphorus was in bacteriophage DNA and contained hereditary instructions for new bacteriophages.

DNA STRUCTURE (13-II)
1. A five-carbon sugar called deoxyribose, a phosphate group, and one of the four nitrogen-containing bases; 2. guanine (purine); 3. cytosine (pyrimidine); 4. adenine (purine); 5. thymine (pyrimidine); 6. deoxyribose (B); 7. phosphate group (G); 8. purine (C); 9. pyrimidine (A); 10. purine (E); 11. pyrimidine (D); 12. nucleotide (F); 13. T; 14. T; 15. F; 16. T; 17. T; 18. pairing; 19. constant; 20. sequence; 21. different.

DNA REPLICATION (13-III)
1.
T - A	T - A
G - C	G - C
A - T	A - T
C - G	C - G
C - G	C - G
C - G	C - G
old new	old new

2. F; 3. F; 4. T; 5. T; 6. C; 7. E; 8. D; 9. A; 10. F; 11. B.

ORGANIZATION OF DNA IN CHROMOSOMES (13-IV)
1. DNA; 2. histones; 3. histone; 4. DNA; 5. nucleosome; 6. genes.

Answers to Crossword Puzzle: DNA Structure and Function

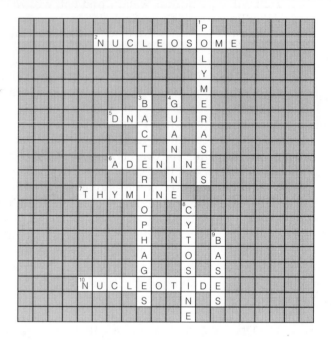

Answers to Self-Quiz

1. d; 2. d; 3. a; 4. d; 5. b; 6. d; 7. c; 8. d; 9. d; 10. e.

14

FROM DNA TO PROTEINS

PROTEIN SYNTHESIS
     The Central Dogma
     Overview of the RNAs
TRANSCRIPTION OF
DNA INTO RNA
     How RNA Is Assembled
     Messenger-RNA Transcripts

TRANSLATION
     The Genetic Code
     Codon-Anticodon Interactions
     Ribosome Structure
     Stages of Translation
MUTATION AND PROTEIN SYNTHESIS
     Commentary: Gene Mutation and Evolution

Interactive Exercises

PROTEIN SYNTHESIS (14-I, pp. 217–219)

Fill-in-the-Blanks

Which base follows the next in a strand of DNA is referred to as the base (1)_____. The region of

DNA that calls for the assembly of specific amino acids into a polypeptide chain is the (2)_____. The

two steps from genes to proteins are called (3)_____ and (4)_____. In (5)_____,

single-stranded molecules of RNA are assembled on DNA templates in the nucleus. In (6)_____ , the

RNA molecules are shipped from the nucleus into the cytoplasm, where they are used as templates for

assembling (7)_____ chains. Following translation, one or more chains become (8)_____ into the

three-dimensional shape of protein molecules. To summarize, a circular relationship exists between DNA,

RNA, and proteins. Proteins have (9)_____ and (10)_____ roles in cells, including control of

DNA. This flow of information in cells is the (11)_____ _____ of biology.

12. Three types of RNA are transcribed from DNA in the nucleus (from genes that code only for RNA).
Complete the following table, which summarizes information about these molecules.

RNA Molecule	Abbreviation	Description/Function
a. Ribosomal RNA	_____	_____
b. Messenger RNA	_____	_____
c. Transfer RNA	_____	_____

TRANSCRIPTION OF DNA INTO RNA (14-II, pp. 220–221)

1. List three ways in which a molecule of RNA is structurally different from a molecule of DNA. _____

   _____

   _____

2. Cite two similarities in DNA replication and transcription. _____

   _____

   _____

3. What are the three key ways in which transcription differs from DNA replication? _____

   _____

   _____

Sequence

Arrange the steps of transcription in hierarchical order with the earliest step first and the latest step last.

4. ___
5. ___
6. ___
7. ___
8. ___

A. The RNA strand grows along exposed bases until RNA polymerase meets a DNA base sequence that signals "stop."

B. RNA polymerase binds with the DNA promoter region to open up a local region of the DNA double helix.

C. An RNA polymerase enzyme locates the DNA bases of the promoter region of one DNA strand by recognizing DNA-associated proteins near a promoter.

D. RNA is released from the DNA template as a free, single-stranded transcript.

E. RNA polymerase moves stepwise along exposed nucleotides of one DNA strand; as it moves, the DNA double helix keeps unwinding.

9. Suppose the line below represents the DNA strand that will act as a template for the production of mRNA through the process of transcription. Fill in the blanks below the DNA strand with the sequence of complementary bases that will represent the message carried from DNA to the ribosome in the cytoplasm.

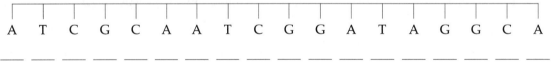

A   T   C   G   C   A   A   T   C   G   G   A   T   A   G   G   C   A

___ ___ ___ ___ ___ ___ ___ ___ ___ ___ ___ ___ ___ ___ ___ ___ ___ ___

(transcribed single-strand of mRNA)

Label-Match

Newly transcribed mRNA contains more genetic information than is necessary to code for a chain of amino acids. Before the mRNA leaves the nucleus for its ribosome destination, an editing process occurs as certain portions of nonessential information are snipped out. Identify each indicated part of the illustration below; use abbreviations for the nucleic acids. Complete the exercise by matching and entering the letter of the description in the parentheses following each label.

10. _____ ( )

11. _____ ( )

12. _____ ( )

13. _____ ( )

14. _____ ( )

15. _____ _____ _____ ( )

A. The actual coding portions of mRNA
B. Noncoding portions of the newly transcribed mRNA
C. Presence of cap and tail, introns snipped out and exons spliced together
D. Acquiring of a poly-A tail by the maturing mRNA transcript
E. The region of the DNA template strand to be copied
F. Reception of a nucleotide cap by the 5′ end of mRNA (the first synthesized)

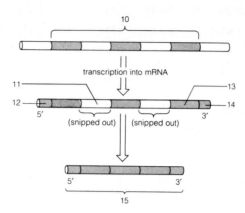

TRANSLATION (14-III, pp. 222–226)

Matching

1. ___ name for each base triplet in mRNA

2. ___ Khorana, Nirenberg, Ochoa, and Holley

3. ___ sixty-one

4. ___ the genetic code

5. ___ "wobble effect"

6. ___ ribosome

7. ___ Crick, Brenner, and others

8. ___ anticodon

9. ___ the "stop" codons

10. ___ three at a time

A. Composed of two subunits, the small subunit with P and A amino acid binding sites as well as a binding site for mRNA
B. Reading frame of the nucleotide bases in mRNA
C. Deduced the nature of the genetic code
D. UAA, UAG, UGA
E. A sequence of three nucleotide bases that can pair with a specific mRNA codon
F. Codon
G. The number of codons that actually specify amino acids
H. Freedom in codon-anticodon pairing at the third base
I. Term for how the nucleotide sequences of DNA and then mRNA correspond to the amino acid sequence of a polypeptide chain
J. Deciphered the genetic code

11. Complete the following table, which distinguishes the stages of translation.

Translation Stage	Description
a. _____	Special initiator tRNA loads onto small ribosomal subunit and recognizes AUG; small subunit binds with mRNA, and large ribosomal subunit joins small one.

Translation Stage	Description
b. _____ _____	Amino acids are strung together in sequence dictated by mRNA codons as the mRNA strand passes through the two ribosomal subunits; two tRNAs interact at P and A sites.
c. _____ _____	mRNA "stop" codon signals the end of the polypeptide chain; release factors detach the ribosome and polypeptide chain from the mRNA.

12. Given the following DNA sequence, deduce the composition of the mRNA transcript:

TAC AAG ATA ACA TTA TTT CCT ACC GTC ATC

‒‒‒ ‒‒‒ ‒‒‒ ‒‒‒ ‒‒‒ ‒‒‒ ‒‒‒ ‒‒‒ ‒‒‒ ‒‒‒
(mRNA transcript)

13. Deduce the composition of the tRNA anticodons that would pair with the above specific mRNA codons as these tRNAs deliver the amino acids (identified below) to the P and A binding sites of the small ribosomal subunit.

‒‒‒ ‒‒‒ ‒‒‒ ‒‒‒ ‒‒‒ ‒‒‒ ‒‒‒ ‒‒‒ ‒‒‒ ‒‒‒
(tRNA anticodons)

14. From the mRNA transcript in exercise 12, use Figure 14.6 of the text to deduce the composition of the amino acids of the polypeptide sequence.

‒‒‒ ‒‒‒ ‒‒‒ ‒‒‒ ‒‒‒ ‒‒‒ ‒‒‒ ‒‒‒ ‒‒‒ ‒‒‒
(amino acids)

Label-Match

A summary of the flow of genetic information in protein synthesis is useful as an overview. Identify the indicated parts of the illustration on the next page by filling in the blanks with the names of the appropriate *structures* or *functions*. Choose from the following: DNA, mRNA, tRNA, polypeptide, rRNA subunits, intron, exon, mature mRNA transcript, new mRNA transcript, anticodon, amino acids, ribosome-mRNA complex. Complete the exercise by matching and entering the letter of the description in the parentheses following each label.

15. _____ ( )

16. _____ _____ _____ ( )

17. _____ ( )

18. _____ ( )

19. _____ _____ _____ ( )

20. _____ ( )

21. _____ _____ ( )

22. _____ ( )

23. _____ ( )

24. _____ _____ ( )

25. _____ ( )

26. _____-_____ _____ ( )

27. _____ ( )

A. Coding portion of mRNA that will translate into proteins
B. Carries the genetic code from DNA in the nucleus to the cytoplasm
C. Transports amino acids to the ribosome and mRNA
D. The building blocks of polypeptides
E. Noncoding portions of newly transcribed mRNA
F. tRNA after delivering its amino acid to the ribosome-mRNA complex
G. Join at translation initiation
H. Holds the genetic code for protein production
I. Place where translation occurs
J. Includes introns and exons
K. A sequence of three bases that can pair with a specific mRNA codon
L. Snipping out of introns, only exons remaining
M. May serve as a functional protein (enzyme) or a structural protein

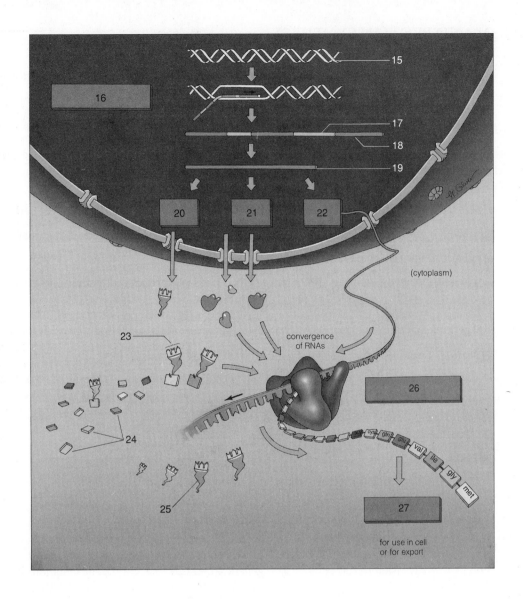

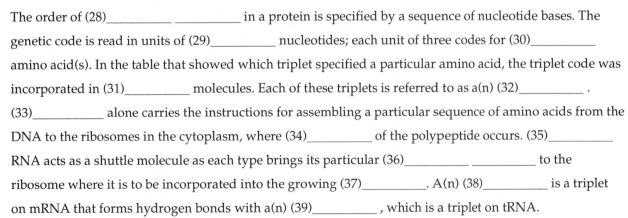

Fill-in-the-Blanks

The order of (28)_____ _____ in a protein is specified by a sequence of nucleotide bases. The

genetic code is read in units of (29)_____ nucleotides; each unit of three codes for (30)_____

amino acid(s). In the table that showed which triplet specified a particular amino acid, the triplet code was

incorporated in (31)_____ molecules. Each of these triplets is referred to as a(n) (32)_____ .

(33)_____ alone carries the instructions for assembling a particular sequence of amino acids from the

DNA to the ribosomes in the cytoplasm, where (34)_____ of the polypeptide occurs. (35)_____

RNA acts as a shuttle molecule as each type brings its particular (36)_____ _____ to the

ribosome where it is to be incorporated into the growing (37)_____. A(n) (38)_____ is a triplet

on mRNA that forms hydrogen bonds with a(n) (39)_____ , which is a triplet on tRNA.

MUTATION AND PROTEIN SYNTHESIS (14-IV, pp. 226–228)

1. Define the term *mutagen* and cite several examples._____

_____

Fill-in-the-Blanks

In addition to changes in chromosomes (crossing over, recombination, deletion, addition, translocation, and inversion), changes can also occur in the structure of DNA; these modifications are referred to as gene mutations. Complete the following exercise on types of spontaneous gene mutations.

Viruses, ultraviolet radiation, and certain chemicals are examples of environmental agents called

(2)_____ that may enter cells and damage strands of DNA. If A becomes paired with C instead of T

during DNA replication, this spontaneous mutation is a base-pair (3)_____. Sickle-cell anemia is a

genetic disease whose cause has been traced to a single DNA base pair; the result is that one (4)_____

_____ is substituted for another in the beta chain of (5)_____. A (6)_____ mutation is

defined as the insertion or deletion of one to several DNA base pairs; this puts the nucleotide sequence out

of phase, and abnormal proteins are produced. Some DNA regions "jump" to new DNA locations and

often inactivate the genes in their new environment; such (7)_____ elements may give rise to

observable changes in the phenotype of an organism.

Chapter Terms

The following page-referenced terms are important; they were in boldface type in the chapter. Refer to the instructions given in Chapter 1, p. 4.

base sequence (219)	central dogma (219)	promoter (221)	anticodon (223)
template (219)	ribosomal RNA (219)	exons (221)	ribosome (223)
gene (219)	messenger RNA (219)	introns (221)	polysome (226)
transcription (219)	transfer RNA (219)	genetic code (222)	gene mutation (226)
RNA (219)	uracil (220)	codon (223)	mutagens (226)
translation (219)			

Self-Quiz

___ 1. Transcription _____.
   a. occurs on the surface of the ribosome
   b. is the final process in the assembly of a protein
   c. occurs during the synthesis of any type of RNA by use of a DNA template
   d. is catalyzed by DNA polymerase

___ 2. _____ carry(ies) amino acids to ribosomes, where amino acids are linked into the primary structure of a polypeptide.
   a. mRNA
   b. tRNA
   c. Introns
   d. rRNA

___ 3. Transfer RNA differs from other types of RNA because it _____ .
   a. transfers genetic instructions from cell nucleus to cytoplasm
   b. specifies the amino acid sequence of a particular protein
   c. carries an amino acid at one end
   d. contains codons

___ 4. _____ dominates the process of transcription.
   a. RNA polymerase
   b. DNA polymerase
   c. Phenylketonuria
   d. Transfer RNA

___ 5. _____ and _____ are found in RNA but not in DNA.
   a. Deoxyribose; thymine
   b. Deoxyribose; uracil
   c. Uracil; ribose
   d. Thymine; ribose

___ 6. Each "word" in the mRNA language consists of _____ letters.
   a. three
   b. four
   c. five
   d. more than five

___ 7. If each nucleotide is coded for only one amino acid, how many different types of amino acids could be selected?
   a. four
   b. sixteen
   c. twenty
   d. sixty-four

___ 8. The genetic code is composed of _____ codons .

   a. three
   b. twenty
   c. sixteen
   d. sixty-four

___ 9. The cause of sickle-cell anemia has been traced to _____ .
   a. a mosquito-transmitted virus
   b. two DNA mutations that result in two incorrect amino acids in a hemoglobin chain
   c. three DNA mutations that result in three incorrect amino acids in a hemoglobin chain
   d. one DNA mutation that results in one incorrect amino acid in a hemoglobin chain

___ 10. An example of a mutagen is _____.
   a. a virus
   b. ultraviolet radiation
   c. certain chemicals
   d. all of the above

Chapter Objectives/Review Questions

This section lists general and detailed chapter objectives that can be used as review questions. You can make maximum use of these items by writing answers on a separate sheet of paper. Fill in answers where blanks are provided. To check for accuracy, compare your answers with information given in the chapter or glossary.

Page	Objectives/Questions
(219)	1. The flow of genetic information from DNA to the RNAs through transcription and from RNAs to protein in translation is known as the _____ _____ of molecular biology.
(218–219)	2. State how RNA differs from DNA in structure and function, and indicate what features RNA has in common with DNA.
(219)	3. _____ RNA combines with certain proteins to form the ribosome; _____ RNA carries genetic information for protein construction from the nucleus to the cytoplasm; _____ RNA picks up specific amino acids and moves them to the area of mRNA and the ribosome.
(221)	4. Describe the process of transcription and indicate three ways in which it differs from replication.
(220–221)	5. What RNA code would be formed from the following DNA code: TAC-CTC-GTT-CCC-GAA?
(221)	6. Transcription starts at a _____ , a specific sequence of bases on one of the two DNA strands that signals the start of a gene.
(221)	7. The first end of the mRNA to be synthesized is the _____ end; at the opposite end, the most mature transcripts acquire a _____ tail.
(221)	8. Actual coding portions of a newly transcribed mRNA are called _____ ; _____ are the noncoding portions.
(222)	9. Crick, Brenner, and others deduced the nature of the genetic _____.
(222–223)	10. State the relationship between the DNA genetic code and the order of amino acids in a protein chain.
(223)	11. Each base triplet in mRNA is called a _____.

(222–223) 12. Explain the nature of the genetic code and describe the kinds of nucleotide sequences that code for particular amino acids.

(222)   13. Scrutinize Figure 14.6 in the text and decide whether the genetic code in this instance applies to DNA, mRNA, or tRNA.

(222–223) 14. Describe how the three types of RNA participate in the process of translation.

(220–221) 15. Determine which mRNA codons would be formed from the following DNA code: TAC-CTC-GTT-CCC-GAA.

(222)   16. Explain how the DNA message TAC-CTC-GTT-CCC-GAA would be used to code for a segment of protein, and state what its amino acid sequence would be.

(227)   17. Using a diagram, summarize the steps involved in the transformation of genetic messages into proteins (see Figure 14.12 in the text).

(226)   18. List some of the environmental agents, or _____ , that can cause mutations.

(228)   19. Briefly describe the spontaneous DNA mutations known as base-pair substitution, frameshift mutation, and transposable element.

(228)   20. Cite an example of a change in one DNA base pair that has profound effects on the human phenotype.

Integrating and Applying Key Concepts

Genes code for specific polypeptide sequences. Not every substance in living cells is a polypeptide. Explain how genes might be involved in the production of a storage starch (such as glycogen) that is constructed from simple sugars.

Critical Thinking Exercises

1. During translation, tRNA molecules carry amino acids to the mRNA and position them in a specific sequence. Suppose a molecule of the amino acid cysteine, bonded to an appropriate tRNA molecule, was changed to the amino acid alanine. Where would the tRNA bearing the altered amino acid most likely interact with the mRNA?

   a. At the specific alanine codon.
   b. At the specific cysteine codon.
   c. At a site midway between the cysteine and alanine codons.
   d. The "defective" tRNA would be rejected by the ribosome.
   e. Such accidents cannot occur.

ANALYSIS

The positioning of the tRNA on the mRNA is accomplished by specific base pairing between the codon and the anticodon. In this case, the tRNA and its anticodon are not changed; only the amino acid that was already attached to a tRNA appropriate for cysteine is altered. The amino acid–tRNA complex will interact with the mRNA at the usual position—the cysteine codon. A site "midway between" any two codons is some other codon and will be occupied in turn by a tRNA, bearing its own amino acid, that has the complementary anticodon. The tRNA is not defective in any sense; it has not been changed. Finally, such an accident certainly could occur, because it is simply a chemical reaction. However, this was not an accident. It was deliberately caused as part of the experiment that demonstrated that the tRNA rather than the amino acid determined the interaction of the complex with the mRNA.

2. Because DNA is restricted to the nucleus and protein synthesis takes place in the cytoplasm, it was clear that some substance had to transfer the genetic information out of the nucleus. Identifying the substance was more difficult. In one experiment, radioactive nucleotide U was used to label newly synthesized RNA. The radioactivity was located in the nucleus after very short labeling periods and in the nucleus and the cytoplasm after longer labeling periods. This was interpreted to mean that RNA was synthesized in the nucleus and subsequently moved out to the cytoplasm, behavior expected of the hypothetical information transfer molecule. Which of the following is the best alternative explanation of these observations?

   a. RNA is synthesized in both the nucleus and the cytoplasm, but the cytoplasmic synthesis is slower and is only detectable after a longer labeling period.
   b. Labeled atoms move by chemical rearrangements from U to other molecules that are found in both nucleus and cytoplasm.
   c. The distribution is an artifact. The techniques of specimen preparation cause the redistribution of labeled RNA molecules.
   d. Radioactivity in the labeled molecules decayed before detection.
   e. RNA is synthesized in the nucleus and normally stays there. Radiation from the labeled molecules destroyed the nuclear membrane after sufficient exposure time and allowed the RNA to diffuse out to the cytoplasm.

ANALYSIS

   a. This is a valid interpretation of these observations. Stated as a hypothesis, this would predict the observed results. It took another experimental design, in which this hypothesis and the one given in the question predicted different results, to eliminate this hypothesis.
   b. This is always a concern in radioactive tracer experiments but is not an objection to this experiment, because this problem would put the label simultaneously into the nucleus and the cytoplasm. It actually appeared sequentially in the two compartments.
   c. This is another general concern in this type of experiment, but again it is not a problem here. Artifacts of preparation would not depend on the duration of the labeling period. Even the first labeled molecules would be redistributed, not limited to the nucleus.
   d. With some short-lived isotopes, this is a real problem, but with any isotope investigators design the experiment to be finished before radioactive decay is significant. In this experiment, decay would have proceeded at the same rate in all compartments and could not account for the distribution of radioactivity. Simply because the radioactivity was detectable, we can infer that it did not decay to a significant extent.
   e. An assumption in every radioactive tracer experiment is that the radioactivity does not damage the biological system being studied. There is always at least potential concern that it does cause some damage. In this case, we can conclude that it is not significant, because RNA is normally found in both the nucleus and the cytoplasm.

3. The existence of introns was first inferred from a technique called hybridization. DNA and mRNA were isolated from cells, mixed and heated, then cooled to allow base pairing between the two kinds of molecules. When the resulting DNA-RNA "hybrid" double-stranded molecules were examined with electron-microscopy, the DNA strands were looped away from the mRNA strands.

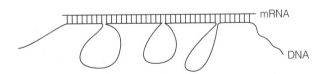

The DNA loops were thought to be nucleotide sequences that were not complementary to any part of the mRNA. Thus, the mRNA was made of sequences that occurred discontinuously along the DNA. Which of the following assumptions were these investigators most likely making?

a. Radioactivity causes mutations.

b. Heating DNA changes the sequence of its monomers.

c. The DNA loops were nucleotide sequences that were not complementary to any part of the RNA.

d. Only nucleic acids with complementary sequences of monomers can stick together by base pairing.

e. Only nucleic acids with the same sequence of monomers can stick together by base pairing.

ANALYSIS

a. Radioactivity does cause mutations, but in order to make inferences about the sequences of nucleotides, it must be assumed that any incident radioactivity does *not* change the sequences during the course of the experiment.

b. As in (a), it is necessary to assume the opposite. If heating changed the sequences of nucleotides, it would be impossible to infer the sequence of native DNA from the behavior of heated DNA.

c. This is the interpretation, not an assumption that was necessary in order to make the interpretation.

d. The conclusion is that the single-stranded loops are noncomplementary regions and the double-stranded regions are complementary. If noncomplementary regions could stick together, by base pairing between occasional complementary single bases, for example, the stated conclusion could not be reached.

e. If this assumption were made, the conclusion would have to be that the hybridized segments were identical sequences. This would be contradictory to the conclusion that the hybridization occurred between the mRNA and the portions of the DNA on which the RNA was transcribed.

Answers

Answers to Interactive Exercises

PROTEIN SYNTHESIS (14-I)
1. sequence; 2. gene; 3. transcription (translation);
4. translation (transcription); 5. transcription; 6. translation;
7. protein; 8. folded; 9. structural (functional);
10. functional (structural); 11. central dogma.
12. a. ribosomal RNA; rRNA; RNA molecule that associates with certain proteins to form the ribosome, the "workbench" on which polypeptide chains are assembled; b. messenger RNA; mRNA; RNA molecule that moves to the cytoplasm, complexes with the ribosome where translation will result in polypeptide chains; c. transfer RNA; tRNA; RNA molecule that moves into the cytoplasm, picks up a specific amino acid, and moves it to the ribosome where tRNA pairs with a specific mRNA code word for that amino acid.

TRANSCRIPTION OF DNA INTO RNA (14-II)
1. RNA molecules are single-stranded, while DNA has two strands; uracil substitutes in RNA molecules for thymine in DNA molecules; ribose sugar is found in RNA, while DNA has deoxyribose sugar.
2. Both DNA replication and transcription follow base-pairing rules; nucleotides are added to a growing RNA strand one at a time as in DNA replication.
3. Only one region of a DNA strand serves as a template for transcription; transcription requires different enzymes (three types of RNA polymerase); the results of transcription are single-stranded RNA molecules, but replication results in DNA, a double-stranded molecule.
4. C; 5. B; 6. E; 7. A; 8. D;
9. U-A-G-C-G-U-U-A-G-C-C-U-A-U-C-C-G-U;

10. DNA (E); 11. introns (B); 12. cap (F); 13. exons (A);
14. tail (D); 15. mature mRNA transcript (C).

TRANSLATION (14-III)
1. F; 2. J; 3. G; 4. I; 5. H; 6. A; 7. C; 8. E; 9. D; 10. B;
11. a. initiation; b. chain elongation; c. chain termination.
12. mRNA transcript: AUG UUC UAU UGU AAU AAA GGA UGG CAG UAG
13. tRNA anticodons: UAC AAG AUA ACA UUA UUU CCU ACC GUC AUC
14. amino acids: met phe tyr cys asn lys gly try gln stop (start)
15. DNA (H); 16. new mRNA transcript (J); 17. intron (E); 18. exon (A); 19. mature mRNA transcript (L);
20. tRNAs (C); 21. rRNA subunits (G); 22. mRNA (B);
23. anticodon (K); 24. amino acids (D); 25. tRNA (F);
26. ribosome-mRNA complex (I); 27. polypeptide (M);
28. amino acids; 29. three; 30. one; 31. mRNA; 32. codon;
33. mRNA; 34. assembly (synthesis); 35. Transfer;
36. amino acid; 37. protein (polypeptide); 38. codon;
39. anticodon.

MUTATION AND PROTEIN SYNTHESIS (14-IV)
1. Mutagens are environmental agents that attack a DNA molecule and modify its structure. Viruses, ultraviolet radiation, and certain chemicals are examples; 2. mutagens; 3. substitution; 4. amino acid;
5. hemoglobin; 6. frameshift; 7. transposable.

Answers to Self-Quiz

1. c; 2. b; 3. c; 4. a; 5. c; 6. a; 7. a; 8. d; 9. d; 10. d.

15

CONTROL OF GENE EXPRESSION

Interactive Exercises

THE NATURE OF GENE CONTROL (15-I, pp. 233–234)

1. All the cells in an organism possess the same genes, and every cell utilizes most of the same genes; yet specialized cells must activate only certain genes. Some agents of gene control have been discovered. Transcriptional controls are the most common. Complete the following table to summarize the agents of gene control.

Agents of Gene Control	Method of Gene Control
a. Repressor protein	
b. _____	encourages binding of RNA polymerases to DNA; this is positive control of transcription
c. Hormones	major agents of vertebrate gene control; signaling molecules that move through the bloodstream to affect gene expression in target cells
d. Promoter	
e. _____	short DNA base sequences between promoter and the start of a gene; a binding site for control agents

2. When do gene controls come into play? _____

_____

GENE CONTROL IN PROKARYOTES (15-II, pp. 234–236)

Label-Match

Escherichia coli, a bacterial cell living in mammalian digestive tracts, is able to exert a negative type of gene control over lactose metabolism. Use the numbered blanks to identify each part of the illustration below. Use abbreviations for nucleic acids. Choose from the following: lactose, regulator gene, repressor-operator complex, promoter, mRNA transcript, enzyme genes, lactose operon, lactose enzymes, repressor-lactose complex, repressor protein, RNA polymerase, and operator. Complete the exercise by matching and entering the letter of the proper function description in the parentheses following each label.

1. _____ _____ ( )
2. _____ _____ ( )
3. _____ _____ ( )
4. _____ ( )
5. _____ ( )
6. _____ _____ ( )
7. _____ - _____ _____ ( )
8. _____ _____ ( )
9. _____ - _____ _____ ( )
10. _____ ( )
11. _____ _____ ( )
12. _____ _____ ( )

A. Includes promoter, operator, and the lactose-metabolizing enzymes
B. Short DNA base sequence between promoter and the beginning of a gene
C. The nutrient molecule in the lactose operon
D. Major enzyme that catalyzes transcription
E. Capable of preventing RNA polymerases from binding with DNA
F. Prevents repressor from binding to the operator
G. Genes that produce lactose-metabolizing enzymes
H. Catalyze the digestion of lactose
I. Binds to operator and overlaps promoter; this prevents RNA polymerase from binding to DNA and initiating transcription
J. Specific base sequence that signals the beginning of a gene
K. Gene that contains coding for production of repressor protein
L. Carries genetic instructions to ribosomes for production of lactose enzymes

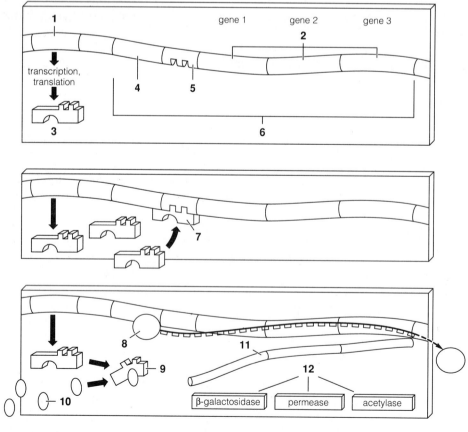

Fill-in-the-Blanks

(13)_____ _____ is a species of bacterium that lives in mammalian digestive tracts and provided some of the first clues about gene control. A(n) (14)_____ is any group of genes together with its promoter and operator sequence. Promoter and operator provide (15)_____ _____. The (16)_____ codes for the formation of mRNA, which assembles a repressor protein. The affinity of the (17)_____ for RNA polymerase dictates the rate at which a particular operon will be transcribed. Repressor protein allows (18)_____ _____ over the lactose operon. Repressor binds with operator and overlaps promoter when lactose concentrations are (19)_____. This blocks (20)_____ _____ from the genes that will process lactose. This (21) [choose one] ( ) blocks, ( ) promotes production of lactose-processing enzymes. When lactose is present, lactose molecules bind with the (22)_____ _____. Thus, repressor cannot bind to (23)_____ , and RNA polymerase has access to the lactose-processing genes. This gene control works well because lactose-degrading enzymes are not produced unless they are (24)_____.

Sequence

Arrange the following events that occur in the positive control of the nitrogen-related operon in the correct order of occurrence if *Escherichia coli* is first in a low-protein environment and then in a protein-rich environment.

25. ___
26. ___
27. ___
28. ___
29. ___
30. ___
31. ___
32. ___

A. RNA polymerase to promoter binding is enhanced.
B. A cascade of molecule-activating reactions brings a large number of activator molecules into service.
C. Enzymes remove phosphates from activator proteins to reverse the earlier response to low nitrogen.
D. Nitrogen again becomes plentiful in the environment of *Escherichia coli*.
E. Enzymes phosphorylate activator molecules to prepare them for binding with promoter.
F. Activator molecules bind with promoter.
G. Chances for *Escherichia coli* to assimilate any available nitrogen increase.
H. Synthesis of nitrogen-obtaining enzymes increases.

GENE CONTROL IN EUKARYOTES (15-III, pp. 236–241)

1. Although a complex organism such as a human being arises from a single cell, the zygote, differentiation occurs in development. Define differentiation and relate it to a definition of selective gene expression.

_____

_____

Label-Match

Identify each numbered part of the illustration below showing eukaryotic gene control. Choose from translational control, transport controls, transcriptional controls, transcript processing controls, and post-translational controls. Complete the exercise by correctly matching and entering the letter of the corresponding gene control description in the parentheses following each label.

2. _____ _____ ( )

3. _____ _____ _____ ( )

4. _____ _____ ( )

5. _____ _____ ( )

6. _____-_____ _____ ( )

A. Govern the rates at which mRNA transcripts that reach the cytoplasm will be translated into polypeptide chains at ribosomes

B. Govern modification of the initial mRNA transcripts in the nucleus

C. Govern how the polypeptide chains become modified into functional proteins

D. Dictate which mature mRNA transcripts will be shipped out of the nucleus and into the cytoplasm for translation

E. Influence when and to what degree a particular gene will be transcribed (if at all)

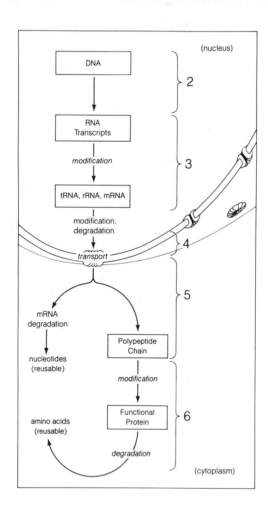

Fill-in-the-Blanks

All the cells in our bodies contain copies of the same (7)_____. These genetically identical cells become structurally and functionally distinct from one another through a process called (8)_____, which arises through (9)_____ gene expression in different cells. Cells depend on (10)_____, which govern transcription, translation, and enzyme activity. Controls that operate during transcription and transcript processing utilize (11)_____ proteins, especially (12)_____ that are turned on and off by the addition and removal of phosphate. (13)_____ is also controlled by the way eukaryotic DNA is packed with proteins in chromosomes. The (14)_____ is packing at the most basic level of chromosome structure and consists of (15)_____ looped around a core of histone proteins. Packing variations such as this may affect the accessibility and (16)_____ of different genes. A (17)_____ body is a condensed X chromosome; the process by which an XX randomly inactivates an X chromosome is called (18)_____. X chromosome inactivation produces adult human females who are (19)_____ for X-linked traits. This effect is shown in human females affected by (20)_____ _____ _____ and provides evidence for (21)_____ gene expression.

True-False

If false, explain why.

___ 22. "Lampbrush chromosomes" of meiotic prophase I seen in electron micrographs of amphibian eggs provide evidence of translational control.

___ 23. Polytene chromosomes of many insect larvae are unusual in that their DNA has been repeatedly replicated; thus, these chromosomes have multiple copies of the same genes. When these genes are undergoing transcription, they "puff."

___ 24. When the same primary mRNA transcript is processed in alternative ways, the result may be different mRNAs, each coding for a slightly different protein.

___ 25. When cells become cancerous, cell populations decrease to very low densities and stop dividing.

___ 26. All abnormal growths and massings of new tissue in any region of the body are called tumors.

___ 27. Malignant tumors have cells that migrate and divide in other organs.

___ 28. The term *metastasis* means the acquiring of the shape or form typical of that species.

___ 29. Oncogenes are genes that combat cancerous transformations.

___ 30. Proto-oncogenes rarely trigger cancer.

___ 31. The normal expression of proto-oncogenes is vital, even though their normal expression may be lethal.

Chapter Terms

The following page-referenced terms are important; they were in boldface type in the chapter. Refer to the instructions given in Chapter 1, p. 4.

repressor protein (234)	hormones (234)	differentiated (236)	oncogene (241)
activator protein (234)	operon (234)	metastasis (241)	proto-oncogenes (241)
promoter (234)	nitrogen-related operon (236)	cancer (241)	carcinogens (241)
operators (234)			

Self-Quiz

___ 1. _____ refers to the processes by which cells with identical genotypes become structurally and functionally distinct from one another according to the genetically controlled developmental program of the species.
   a. Metamorphosis
   b. Metastasis
   c. Cleavage
   d. Differentiation

___ 2. _____ binds to operator whenever lactose concentrations are low.
   a. Operon
   b. Repressor
   c. Promoter
   d. Operator

___ 3. Any gene or group of genes together with its promoter and operator sequence is a(n) _____.
   a. repressor
   b. operator
   c. promoter
   d. operon

___ 4. The operon model explains the regulation of _____ in prokaryotes.
   a. replication
   b. transcription
   c. induction
   d. Lyonization

___ 5. In multicelled eukaryotes, cell differentiation occurs as a result of _____.
   a. growth
   b. selective gene expression
   c. repressor molecules
   d. the death of certain cells

___ 6. One type of gene control discovered in female mammals is _____.
   a. a conflict in maternal and paternal alleles
   b. slow embryo development
   c. X chromosome inactivation
   d. operon

___ 7. Due to X inactivation of either the paternal or maternal X chromosome, human females with anhidrotic ectodermal dysplasia _____.
   a. completely lack sweat glands
   b. develop benign growths
   c. have mosaic patches of skin that lack sweat glands
   d. develop malignant growths

___ 8. Which of the following characteristics seems to be most uniquely correlated with metastasis?
   a. loss of nuclear-cytoplasmic controls governing cell growth and division
   b. changes in recognition proteins on membrane surfaces
   c. "puffing" in the chromosomes
   d. the massive production of benign tumors

___ 9. Genes with the potential to induce cancerous formations are known as _____.
   a. proto-oncogenes
   b. oncogenes
   c. carcinogens
   d. malignant genes

___ 10. _____ controls govern the rates at which mRNA transcripts that reach the cytoplasm will be translated into polypeptide chains at the ribosomes.
   a. Transport
   b. Transcript processing
   c. Translational
   d. Transcriptional

Chapter Objectives/Review Questions

This section lists general and detailed chapter objectives that can be used as review questions. You can make maximum use of these items by writing answers on a separate sheet of paper. Fill in answers where blanks are provided. To check for accuracy, compare your answers with information given in the chapter or glossary.

Page	Objectives/Questions

(234)   1. The elements of gene control operate in response to _____ conditions within a cell or its surroundings.

(234)   2. The negative control of _____ protein prevents the enzymes of transcription from binding to DNA; the positive control of _____ protein enhances the binding of RNA polymerases to DNA.

(234)   3. Explain how hormones act as a major agent of gene control.

(234)   4. A _____ is a specific base sequence that signals the beginning of a gene in a DNA strand.

(234)   5. Some control agents bind to _____ , which are short base sequences between a promoter and the beginning of a gene.

(234)   6. Gene expression is controlled through regulatory _____ , hormones, and other molecules that interact with DNA, _____ , and the protein products of _____ .

(234)   7. Give examples of when gene controls involving your cells might come into play.

(234)   8. A gene control arrangement in which the same promoter-operator sequence services more than one gene is called an _____ .

(234–235)   9. Describe the sequence of events that occurs on the chromosome of E. coli after you drink a glass of milk.

(234–235)   10. The cells of E. coli manage to produce enzymes to degrade lactose when those molecules are _____ and to stop production of lactose-degrading enzymes when lactose is _____ .

(235–236)   11. Nitrogen metabolism in E. coli is an example of _____ gene control; the genes coding for these enzymes are called the nitrogen-related _____ .

(236)   12. Define selective gene expression and explain how this concept relates to cell differentiation in multicelled eukaryotes.

(236)   13. Cell _____ occurs in multicelled eukaryotes as a result of _____ gene expression.

(236–237)   14. Be able to list and define the levels of gene control in eukaryotes.

(238–239)   15. DNA packing within the chromosome affords _____ control.

(237)   16. The most basic level of chromosome structure is the _____ , which consists of DNA looped around a histone core.

(238–239)   17. The "Lampbrush" chromosomes seen in amphibian eggs and larvae represent a change in chromosome structure that has been correlated directly with _____ of the genes necessary for growth.

(239)   18. Describe the gene control mechanism afforded many insect larvae by polytene chromosomes.

(239)   19. The condensed X chromosome seen on the edge of the nuclei of female mammals is known as the _____ body.

(239–240)   20. Explain how X chromosome inactivation provides evidence for selective gene expression; use the example of anhidrotic ectodermal dysplasia.

(239)   21. Describe the tissue effect known as Lyonization.

(241)   22. Define tumor and distinguish between benign and malignant tumors.

(241)   23. _____ is a process in which a cancer cell leaves its proper place and invades other tissues to form new growths.

(241)   24. Describe the relationship of proto-oncogenes, environmental irritants, and oncogenes.

(241)   25. Cigarette smoke, X-rays, gamma rays, and ultraviolet radiation are examples of _____ .

Integrating and Applying Key Concepts

Suppose you have been restricting yourself to a completely vegetarian diet for the past 6 months. Quite unexpectedly, you find yourself in a social situation that requires you to eat a half-pound sirloin steak. Would you expect to digest the steak as easily as you digest soybean burgers? Explain your yes or no answer in terms of transcriptional controls or feedback inhibition.

Critical Thinking Exercises

1. A statement in the text repeats a well-known hypothesis, "Differentiation arises through selective gene expression in different cells." Techniques used in experiments on differentiation include tissue culture and nuclear transplantation. Tissue culture is the isolation of cells from organisms and the maintenance and growth of these cells on artificial media. In a nuclear transplantation experiment, the nucleus is destroyed in one cell and replaced by a nucleus removed from another cell. In most experiments, the cells are from different individuals of the same species. Which of the following experimental results would be the strongest criticism of this hypothesis?

   a. A nucleus is transplanted from a differentiated cell into an enucleated egg cell. The egg develops normally into an adult of the species.
   b. Differentiated cells are placed in tissue culture. They lose their differentiated appearance and become like generalized embryonic cells (dedifferentiation).
   c. A nucleus is transplanted from a differentiated cell into an enucleated egg cell. The egg fails to develop normally. It divides a few times, then dies.
   d. A nucleus is transplanted from a differentiated cell into an enucleated egg cell. The egg divides several times, then the proteins are studied in the resulting cluster of cells. A protein is found that was translated from an allele present in the original egg cell nucleus but not in the transplanted nucleus.
   e. A single differentiated cell is placed in tissue culture. It dedifferentiates, then begins to divide, and develops into a normal adult of the species.

ANALYSIS

The alternative hypothesis is that genetic information is selectively lost from cells as they differentiate. According to this idea, a liver cell, for instance, contains only the "housekeeping" genes necessary for all cells and the specific genes for liver enzymes. It has no information for brain cell–specific proteins. If this hypothesis is valid, we should be able to adjust the environment around a differentiated cell nucleus (the cytoplasm) and make the nucleus express genes for proteins the cell did not have.

   a. This observation would support the hypothesis and, in fact, was one of the major pieces of evidence that has led to common acceptance of the hypothesis. The development had to be governed by genes in the transplanted (differentiated) nucleus, and the successful development indicates that a full set of genetic information is present.
   b. This observation could be taken as criticism of the hypothesis. It could be interpreted to mean that the differentiated cells did not have information to become any other type of cell. But it could also be interpreted to mean that the information was still present but irreversibly shut off, or that the environment was not able to turn it back on. In any case, the cells do have at least some information other than that for the specific type they were. Some cells do this; others remain differentiated in culture.
   c. This kind of negative information is inconclusive. It could mean that the differentiated nuclei were missing essential genetic information for development, or it could mean that the nuclei contained a full set of genes for the species but the cells were damaged by the transplantation process.
   d. This could be construed as a criticism of the hypothesis. This observation would indicate that at least part of the development was controlled by genetic information from the original nucleus, probably in the form of mRNA transcribed before enucleation. Why didn't the transplanted nucleus control the production of this protein? Possibly because it didn't have the genetic information for it. You would want to check for the product of the allele from the individual that donated the nucleus.
   e. Like (a), this observation would be support, not criticism, of the hypothesis. If it could direct development of a normal adult, the differentiated nucleus clearly had *not* lost any genes during differentiation. Some cells from plants are able to do this.

2. When it was discovered that the lactose-metabolizing enzymes were inducible, the first question to be answered was whether control was exerted on transcription or translation. One hypothesis was that lactose promoted transcription of appropriate mRNA to produce the enzymes. The other hypothesis was that lactose acted on ribosomes to promote the translation of mRNA that had already been transcribed but

was not being translated. One kind of study examined the time course of changes in amount of lactose-metabolizing enzymes and the mRNA for their synthesis in cells as lactose was added and then removed from the medium. If the lactose operon functions as you have studied it, which of the following patterns is the best prediction?

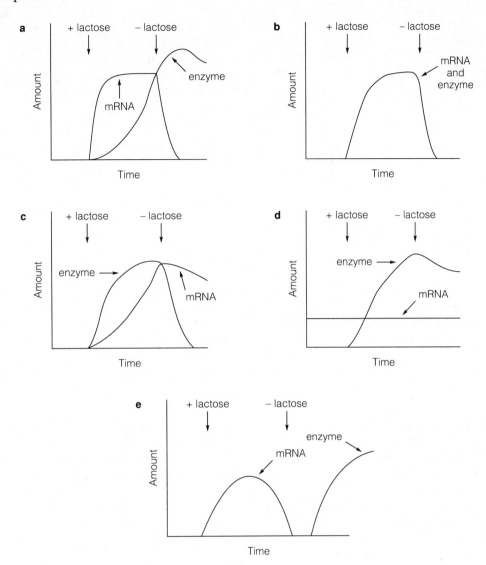

ANALYSIS

Only patterns (d) and (e) are inconsistent with the hypothesis. Pattern (d) is predicted by the alternative hypothesis, translational control. Transcriptional control predicts that more mRNA will be synthesized when lactose is present. Pattern (e) is not predicted by either hypothesis. It would be difficult to explain the synthesis of enzymes in the absence of the mRNA coding for them.

The hypothesis of transcriptional control predicts an increase in mRNA. Whether it would be simultaneous with or precede the increase in enzyme depends on assumptions about the relative speed of transcription and translation. Predictions about the time course of events after lactose is removed from the medium likewise depend on assumptions about the degradation rates of mRNA and enzyme. If you assume that transcription and translation proceed at equal rates and that proteins and mRNAs have similar life spans in the cell, you predict pattern (b). If you assume that translation is rapid relative to transcription and that mRNAs last longer in cytoplasm than do enzymes, you predict pattern (a). If you assume that transcription is the fast process and that mRNA is rapidly degraded in cytoplasm, you predict pattern (c). Pattern (a) was observed in *E. coli*.

3. Much of the development of the operon concept came from study of constitutive mutants for the lactose-metabolizing enzymes. These bacteria had lost control of production of the enzymes because of mutations that inactivated parts of the control system in the DNA. They produced the enzymes whether lactose was present or not. When a technique was developed to give the bacteria an extra copy of the genes for these enzymes and their control, it became apparent that there were two types of constitutive mutants. One type behaved like a dominant allele over the wild type, while the other type behaved like a recessive to the wild type. Which type of behavior, dominant or recessive, would be predicted for a mutation that coded for an inactive repressor protein? Which type would be predicted for a mutation that coded for an inactive operator sequence? What would be the consequences of a mutation that produced an inactive promoter? Would the allele for inactive promoter behave as a dominant or as a recessive?

ANALYSIS

A cell heterozygous for active and inactive repressor would be inducible. Assuming both alleles are transcribed, the cell would have active repressor and would shut off transcription of the genes for the enzymes unless lactose was present. Because the constitutive mutation is not expressed in the phenotype of the heterozygote, it is recessive.

In contrast, a cell heterozygous for active and inactive operator would be constitutive. Its normal operator would be repressed in the absence of lactose; it can bind repressor. But the inactive operator cannot bind repressor under any circumstances; it is a sequence of nucleotides with which the repressor does not react. The structural genes it governs will be transcribed even in the absence of lactose and the presence of active repressor. Because it is expressed in the phenotype of the heterozygote, it is dominant.

A mutation for an inactive promoter would not allow transcription of the structural genes whether the operator was occupied or not. Such a mutation would be uninducible. However, a heterozygote would be inducible because it would possess one normal promoter that could initiate transcription when lactose was present. Thus, an allele for inactive promoter would be recessive.

Answers

Answers to Interactive Exercises

THE NATURE OF GENE CONTROL (15-I)
1. a. Repressor protein; prevent transcription enzymes (RNA polymerases) from binding to DNA; this is negative transcription control; b. Activator protein; encourages binding of RNA polymerases to DNA; this is positive control of transcription; c. Hormones; major agents of vertebrate gene control; signaling molecules that move through the bloodstream to affect gene expression in target cells; d. Promoter; specific base sequences on DNA that serve as binding sites for control agents; before RNA assembly can occur on DNA, the enzymes must bind with the promoter site; e. Operators; short DNA base sequences between promoter and the start of a gene; a binding site for control agents. 2. In any organism, gene controls operate in response to chemical changes within the cell or its surroundings.

GENE CONTROL IN PROKARYOTES (15-II)
1. regulator gene (K); 2. enyzme genes (G); 3. repressor protein (E); 4. promoter (J); 5. operator (B); 6. lactose operon (A); 7. repressor-operator complex (I); 8. RNA polymerase (D); 9. repressor-lactose complex (F); 10. lactose (C); 11. mRNA transcript (L); 12. lactose enzymes (H); 13. *Escherichia coli;* 14. operon; 15. transcription controls; 16. regulator; 17. promoter; 18. negative control; 19. low; 20. RNA polymerase (mRNA transcription); 21. blocks; 22. repressor protein; 23. operator; 24. needed (required); 25. B; 26. E; 27. F; 28. A; 29. H; 30. G; 31. D; 32. C.

GENE CONTROL IN EUKARYOTES (15-III)
1. All cells in the body descend from the same zygote; as cells divide to form the body, they become specialized in composition, structure, and function—they differentiate through selective gene expression. 2. transcriptional control (E); 3. transcript processing control (B); 4. transport control (D); 5. translational control (A); 6. post-translational control (C); 7. DNA (genes); 8. differentiation; 9. selective; 10. controls; 11. regulatory; 12. activators; 13. Transcription; 14. nucleosome; 15. DNA; 16. activity; 17. Barr; 18. Lyonization; 19. mosaic; 20. anhidrotic ectodermal dysplasia; 21. selective; 22. F; 23. T; 24. T; 25. F; 26. T; 27. T; 28. F; 29. F; 30. T; 31. T.

Answers to Self Quiz

1. d; 2. b; 3. d; 4. b; 5. b; 6. c; 7. c; 8. a; 9. b; 10. c.

16

RECOMBINANT DNA AND GENETIC ENGINEERING

Interactive Exercises

RECOMBINATION IN NATURE: SOME EXAMPLES (16-I, pp. 246–247)

1. Define recombinant DNA technology._____

_____

_____

_____

2. Describe and distinguish between the bacterial chromosome and plasmids present in a bacterial cell.

_____

_____

_____

3. Explain the process of bacterial conjugation; relate the importance of F plasmids and the F pilus to the

process._____

_____

_____

_____

Label-Match

Bacterial conjugation proceeds through several steps. Identify the numbered items on the illustration below. Choose from the following: displaced DNA strand, bacterial chromosome, new donor cell—post conjugation, donor cell, nick, transferred DNA strand, recipient cell, donor cell—post conjugation, pilus, and donor cell plasmid. Complete the exercise by matching and entering the letter of the proper description in the parentheses following each label.

4. _____ _____ ( )

5. _____ _____ ( )

6. _____ _____ ( )

7. _____ ( )

8. _____ ( )

9. _____ _____ _____ ( )

10. _____ _____ _____ ( )

11. _____ _____ _____ ( )

12. _____ _____ —_____ _____ ( )

13. _____ _____ _____ —

_____ _____ ( )

A. Replication begins in donor cell
B. A conjugation tube; an appendage that binds the recipient cell and draws it tightly against the donor cell
C. Recipient plasmid circularizes
D. A circular DNA molecule; contains all the genes for growth and development
E. Replication begins in recipient cell
F. Begins entry of recipient cell
G. Contains bacterial chromosome and plasmid
H. A cut across one strand of DNA by an enzyme
I. Contains only a bacterial chromosome
J. Donor cell plasmid circularizes

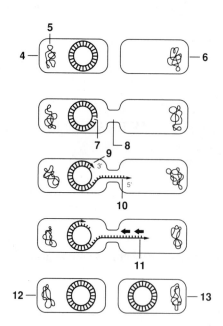

True-False

If false, explain why.

____ 14. Plasmids are organelles on the surfaces of which amino acids are assembled into polypeptides.

____ 15. Bacterial conjugation is a process by which a donor cell transfers DNA to a recipient cell.

____ 16. Although bacteriophages attack bacteria, the transfer of their genes to those bacteria is unknown.

___ 17. Gene transfer and recombination are common in nature.

___ 18. Bacterial conjugation is itself probably controlled by genes carried on a plasmid.

RECOMBINANT DNA TECHNOLOGY (16-II, pp. 248–250)

1. Complete the table below, which summarizes some of the basic tools and procedures used in recombinant DNA technology.

Tool/Procedure	Definition and Role in Recombinant DNA Technology
a. Restriction enzymes	
b. DNA ligase	
c. DNA library	
d. Cloned DNA	
e. Reverse transcriptase	
f. cDNA	
g. PCR	
h. cDNA probe	

Matching

Match the steps in the formation of a DNA library with the parts of the illustration below.

2. ___
3. ___
4. ___
5. ___
6. ___
7. ___

A. Joining of chromosomal and plasmid DNA using DNA ligase
B. Restriction enzyme cuts chromosomal DNA at specific recognition sites
C. Cut plasmid DNA
D. Recombinant plasmids containing cloned library
E. Fragments of chromosomal DNA
F. Same restriction enzyme is used to cut plasmids

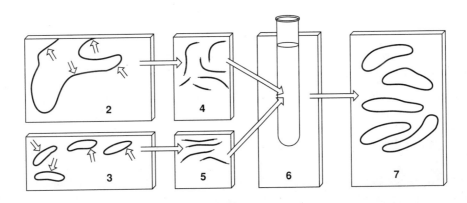

True-False

A genetic engineer used restriction enzymes to prepare fragments of DNA from two different species that were then mixed. Four of these fragments are illustrated below. Fragments (a) and (c) are from one species, (b) and (d) from the other species. Answer exercises 8–12. If false, explain why.

_____ TACA	_____ TTCA	_____ CGTA	_____ ATGT
a.	b.	c.	d.

___ 8. Some of the fragments represent sticky ends.

___ 9. The same restriction enzyme was used to cut fragments (b), (c), and (d).

___ 10. Different restriction enzymes were used to cut fragments (a) and (d).

___ 11. Fragment (a) will base-pair with fragment (d) but not with fragment (c).

___ 12. The same restriction enzyme was used to cut the different locations in the DNA of the two species shown.

Matching

Match the most appropriate letter with each numbered partner.

13. ___ polymerase chain reaction (PCR)
14. ___ DNA ligase
15. ___ cDNA
16. ___ DNA library
17. ___ cloned DNA
18. ___ bacterial conjugation
19. ___ recombinant DNA technology
20. ___ plasmids
21. ___ restriction enzymes
22. ___ genome

A. All the DNA in a haploid set of chromosomes
B. Process by which a gene is split into two strands and then copied over and over (most common type of gene amplification) by enzymes
C. A process by which a donor cell transfers DNA to a recipient cell
D. Connects DNA fragments
E. Small circular DNA molecules that carry only a few genes
F. DNA assembled through use of reverse transcriptase and coding on mRNA
G. Cuts DNA molecules
H. A collection of DNA fragments produced by restriction enzymes and incorporated into plasmids
I. Multiple, identical copies of DNA fragments from an original chromosome
J. Method of genetic engineering

RISKS AND PROSPECTS OF THE NEW TECHNOLOGY (16-III, pp. 251–256)

1. Explain the goal of the human genome project._____

_____

In exercises 2–8, summarize the results of the given experimentation dealing with genetic modifications of plants and animals.

2. Plants such as carrots that are regenerated from cultured cells:_____

_____

3. The bacterium, _Agrobacterium tumefaciens:_____

_____

_____

4. Cotton plants:_____

5. DNA delivery into plant protoplasts by chemicals, electric shock, or pistol blanks:_____

_____

_____

_____

6. Introduction of the rat and human somatotropin gene into fertilized mouse eggs:_____

_____

7. Insertion of the somatotropin gene into pigs: _____

_____

8. Insertion of genes into a sperm nucleus just after egg penetration:_____

_____

_____

Fill-in-the-Blanks

(9)_____ _____ _____ _____ refers to the use of restriction enzymes to cut human
DNA into fragments of specific lengths. (10)_____ refers to the fact that each person has a slightly
unique pattern of sites on DNA where enzymes make their cuts. Each person has a (11)_____
fingerprint or a unique pattern of RFLPs that can be used to map the human genome, apprehend criminals,
and resolve cases of disputed paternity and maternity. Bacterial host cells lack the proper (12)_____
_____ to translate cloned genes unless the (13)_____ have been cut out. Identification of the
order and identity of nucleotides in DNA is called (14)_____. Bacterial strains used in (15)_____
_____ may be initially harmless, but there is concern about danger to humans and the environment.
A strain of bacteria, *Pseudomonas syringae,* lives on stems and leaves of crop plants and makes them
susceptible to (16)_____ _____ . The harmful gene is called the "ice-forming" gene, and the
bacteria are known as (17)_____ - _____ bacteria; genetic engineers were able to remove the
harmful gene and test the modified bacterium on strawberry plants with no adverse effects. Inserting one
or more genes into the (18)_____ _____ of an organism for the purpose of correcting genetic
defects is known as (19)_____ _____. Attempting to modify a human trait by inserting genes
into sperm or eggs is called (20)_____ _____.

Chapter Terms

The following page-referenced terms are important; they were in boldface type in the chapter. Refer to the
instructions given in Chapter 1, p. 4.

recombinant DNA	DNA ligase (248)	nucleic acid	RFLPs (253)
technology (245)	DNA library (248)	hybridization (250)	ice-minus bacteria (254)
plasmid (246)	cloned DNA (248)	DNA probe (250)	gene therapy (257)
F plasmids (246)	PCR (249)	cDNA (250)	eugenic engineering (257)
restriction enzymes (248)		human genome project (251)	

Self-Quiz

___ 1. Small circular molecules of DNA in bacteria are called _____.
   a. plasmids
   b. desmids
   c. pili
   d. F particles
   e. transferins

___ 2. Bacteria reproduce sexually by _____.
   a. fission
   b. gametic fusion
   c. conjugation
   d. lysis
   e. none of the above, because bacteria only reproduce asexually

___ 3. Enzymes used to cut genes in recombinant DNA research are _____.
   a. ligases
   b. restriction enzymes
   c. transcriptases
   d. DNA polymerases
   e. replicases

___ 4. The total DNA in a haploid set of chromosomes of a species is its _____.
   a. plasmid
   b. enzyme potential
   c. genome
   d. DNA library
   e. none of the above

___ 5. An enzyme that heals random base-pairing of chromosomal fragments and plasmids is _____.
   a. reverse transcriptase
   b. DNA polymerase
   c. cDNA
   d. DNA ligase

___ 6. A DNA library is _____.
   a. a collection of DNA fragments produced by restriction enzymes and incorporated into plasmids
   b. cDNA plus the required restriction enzymes
   c. mRNA-cDNA
   d. composed of mature mRNA transcripts

___ 7. Amplification results in _____.
   a. plasmid integration
   b. bacterial conjugation
   c. cloned DNA
   d. production of DNA ligase

___ 8. Any DNA molecule that is copied from mRNA is known as _____.
   a. cloned DNA
   b. cDNA
   c. DNA ligase
   d. hybrid DNA

___ 9. The most commonly used method of DNA amplification is _____.
   a. polymerase chain reaction
   b. gene expression
   c. genome mapping
   d. RFLPs

___ 10. Restriction fragment length polymorphisms are valuable because _____.
   a. they reduce the risks of genetic engineering
   b. they provide an easy way to sequence the human genome
   c. they allow fragmenting DNA without enzymes
   d. they provide DNA fragment sizes unique to each person

Chapter Objectives/Review Questions

This section lists general and detailed chapter objectives that can be used as review questions. You can make maximum use of these items by writing answers on a separate sheet of paper. Fill in answers where blanks are provided. To check for accuracy, compare your answers with information given in the chapter or glossary.

Page   *Objectives/Questions*

(245)   1. List the means by which natural genetic experiments occur.
(245)   2. Define recombinant DNA technology.
(246)   3. _____ are small, circular, self-replicating molecules of DNA or RNA within a bacterial cell.

Page	Objectives/Questions
(246)	4. Bacterial _____ is a process whereby one bacterial cell transfers DNA to another cell.
(247)	5. Lambda bacteriophage is a _____ that can integrate its DNA into the chromosome of bacterial host cells; the modified bacterial chromosome is replicated and passed on in latent form to succeeding generations of bacteria when it may leave the chromosome and begin an infectious cycle.
(248)	6. Some bacteria produce _____ enzymes that cut apart DNA molecules injected into the cell by viruses; such DNA fragments or "_____ ends" often have staggered cuts capable of base-pairing with other DNA molecules cut by the same _____ enzymes.
(248)	7. Base-pairing between chromosomal fragments and cut plasmids is made permanent by DNA _____.
(248)	8. Be able to explain what a DNA library is; review the steps used in creating such a library.
(248–249)	9. List and define the two major methods of DNA amplification.
(248)	10. Multiple, identical copies of DNA fragments produced by restriction enzymes are known as _____ DNA.
(250)	11. A special viral enzyme, _____ _____ , presides over the process by which mRNA is transcribed into DNA.
(250)	12. Define cDNA.
(249)	13. Polymerase chain reaction is the most commonly used method of DNA _____.
(249)	14. Explain how gel electrophoresis is used to sequence DNA.
(250)	15. How is a cDNA probe used to identify a desired gene carried by a modified host cell?
(250)	16. Why do researchers prefer to work with cDNA when working with human genes?
(251)	17. Tell about the human genome project and its implications.
(251–253)	18. DNA fragments from different people that show slightly unique banding patterns are known as restriction _____ length _____.
(253)	19. List some practical genetic uses of RFLPs.
(254)	20. A strain of *Pseudomonas syringae* that was genetically engineered so as to not damage crop plants is known as the _____-_____ bacteria.
(254–256)	21. Be able to briefly relate the significance of the following genetic engineering experiments: "Ti" plasmid from *Agrobacterium tumefaciens*; delivery of a firefly gene into cultured tobacco plant cells; resistance by cotton plants to worm attacks; use of chemicals, electric shocks, and bullets to deliver genes into cultured plant cells; introduction of the somatotropin gene into fertilized mouse eggs; insertion of the somatotropin gene into pigs; insertion of genes into a sperm nucleus just after egg penetration.
(257)	22. Define gene therapy and eugenic engineering.

Integrating and Applying Key Concepts

How could scientists guarantee that *Escherichia coli*, the human intestinal bacterium, will not be transformed into a severely pathogenic form and released into the environment if researchers use the bacterium in recombinant DNA experiments?

Critical Thinking Exercises

1. You isolate an mRNA of interest from a eukaryotic cell and use it to prepare a cDNA. You incorporate the cDNA into the genome of a prokaryote. The cell grows well in culture but fails to synthesize the protein coded by the original mRNA. You have independent evidence that the cDNA was successfully inserted into the prokaryotic genome and that it replicates properly. Which of the following hypotheses best explains the failure?

   a. The prokaryotic cell had no enzymes to edit the introns out of its RNA transcripts.
   b. The cDNA was complementary, not identical, to the mRNA.
   c. There were no introns in the mRNA, and thus the transcripts from the cDNA were incomplete.
   d. The promoter was not transcribed into the original mRNA.
   e. The mRNA had a poly-A "tail" that produced a confused message.

ANALYSIS

a. This is a true statement; prokaryotes have neither introns nor the enzymes to remove them. However, this is irrelevant. The mRNA had already been edited when it was used to produce the cDNA. No introns are present to be edited.

b. The original cDNA strand was complementary—it must be in order to be transcribed as a new copy of the mRNA.

c. It is true that the original introns had been edited before the mRNA was used. Thus, the new transcripts would be incomplete relative to the original eukaryotic DNA, but they are complete in terms of the information for the sequence of amino acids in the protein.

d. This is probably true, and it would mean that the cDNA contained no promoter. Unless the cDNA happened to be inserted into the prokaryotic genome adjacent to a promoter, it could not be transcribed, and the protein would not be synthesized in the cell.

e. The mRNA probably did have a poly-A tail, and it would copy as a poly-T sequence in the cDNA, leading eventually to poly-A tails on the new transcripts. This would not be a confused message, however, since the original message also had the poly-A tail and was translated readily. Furthermore, the stop translation signal must occur before the tail is reached, making the tail irrelevant to the information content of the message.

2. You plan to use PCR to make multiple copies of a certain gene. You isolate the DNA regions flanking the gene and prepare short cDNA primers from them. You get some fresh DNA containing the gene, heat-denature it, add the primers, and proceed as in Figure 16.7. Which of the following assumptions are you most likely making?

   a. Other genes in the DNA do not have the same flanking sequences as the gene you want to copy.
   b. The gene you want to copy doesn't repeat either of the flanking sequences in its middle.
   c. The flanking sequences on both ends of the gene you want are identical to each other.
   d. The DNA you want to copy has no introns.
   e. Polymerization proceeds faster on this gene in one direction than in the other direction.

ANALYSIS

a. If any other genes in the DNA do have the same flanking sequences, your cDNA primers will bind to them, too, and you will produce a mixture of copies of all the genes.

b. If there is another copy of the flanking sequence in the middle of the gene, a cDNA primer will bind to it, and polymerization will start at two positions. As long as some ligase is present, this will cause no problems. In fact, it will make the process quicker.

c. As long as you use cDNA primers complementary to both of the flanking sequences, they don't have to be identical.

d. This depends on what you want to do with the copies of the DNA. If they are to be inserted into a prokaryotic cell in expectation of producing a protein, they can't have introns, because the prokaryotic cell will not be able to edit the transcripts. But if you want to study the sequence of the DNA, attach a fluorescent label to it and probe a genome, use it to hybridize RNA molecules from the original cell type, or use it for any other such purpose, the presence or absence of introns is immaterial.

e. You must make the opposite assumption—that polymerization proceeds at the same rate in both directions. If it does not, you will get DNA molecules of two different sizes, and your purification technique depends on all of them being the same size.

3. You have purified two RFLPs and want to join them end to end, making a single DNA molecule of special interest. You mix the two fragments in equal proportions with some DNA ligase that you know will attach the ends of the fragments. After incubation, you harvest the DNA fragments and separate them by electrophoresis. On the diagram below, track 1 shows the bands formed by the original two fragments. On tracks 2 and 3, draw the bands predicted when ligase joins the two fragments to form a new, larger sequence. The actual patterns are shown on tracks 4 and 5. Can you explain the results? What could you do to confirm your interpretation?

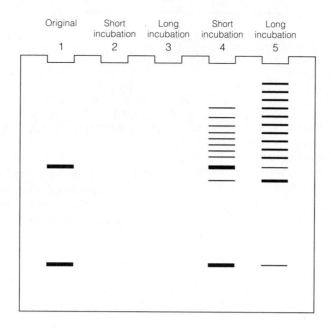

ANALYSIS

The expected results are shown in tracks 2 and 3 below. The newly formed larger molecules would all be of a single size and would form a single band with a lower mobility than the original fragments. As incubation time increases, the new band will become larger and the original bands smaller.

In fact, many new bands appear, formed by combinations of more than one of each of the original fragments. Remember that RFLPs are formed by digestion with an enzyme that recognizes specific nucleotide sequences and hydrolyzes the strands, leaving a single-stranded end protruding. These tabs are on both ends of the fragments and are complementary to each other. Thus, any two fragments can attach to each other by base-pairing and be linked by ligase. The new molecule still has a single-stranded tab at each end and can attach to another fragment, and another, and another. This accounts for the new, larger bands that appear on the gels. The band that is a little faster than the slower of the original bands contains molecules formed by two of the smaller, faster original fragments.

To confirm this interpretation, incubate the mixture of ligated fragments with the enzyme that was used to form them in the first place. The mixture of large pieces should be digested to the original two fragments.

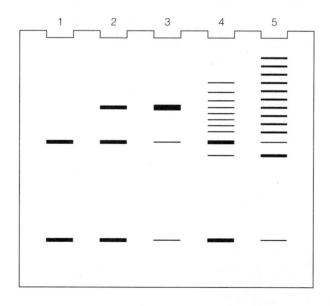

Answers

Answers to Interactive Exercises

RECOMBINATION IN NATURE:
SOME EXAMPLES (16-I)

1. DNA from different species can be cut, spliced together, and then inserted into bacteria or other types of rapidly dividing cells, which multiply the recombinant DNA molecules in quantity. Genes can be isolated, modified, and reinserted into the organism (or transplanted to a different one).
2. The bacterial chromosome, a circular DNA molecule, contains all the genes necessary for normal growth and development. Plasmids, small, circular molecules of "extra" DNA, carry only a few genes and are self-replicating.
3. Genes carried on an F plasmid permit conjugation, a process by which one bacterial cell transfers DNA to another. Some F plasmid genes code for the proteins necessary to construct an F pilus, a long appendage that can latch onto a recipient cell and draw it right up against the donor.
4. donor cell (G); 5. bacterial chromosome (D);
6. recipient cell (I); 7. nick (H); 8. pilus (B); 9. donor cell plasmid (A); 10. displaced DNA strand (F);
11. transferred DNA strand (E); 12. donor cell—post conjugation (J); 13. new donor cell—post conjugation (C); 14. F; 15. T; 16. F; 17. T; 18. T.

RECOMBINANT DNA TECHNOLOGY (16-II)

1. a. bacterial enzymes that cut apart DNA molecules injected into the cell by viruses; several hundred have been identified; b. a replication enzyme that joins the short fragments of DNA; c. a collection of DNA fragments produced by restriction enzymes and incorporated into plasmids; d. after a DNA library is inserted into a host cell's cloning vector (often a plasmid), repeated replications and divisions of the host cells produce multiple, identical copies of DNA fragments, or cloned DNA; e. a viral enzyme that allows mRNA to be transcribed into DNA; f. any DNA molecule "copied" from mRNA; g. most common method of DNA amplification; DNA containing a gene of interest is split into two single strands that enzymes convert back to double-stranded forms; h. a nucleic acid hybridization technique used to identify bacterial colonies harboring the DNA (gene) of interest; a short

nucleotide sequence is assembled from radioactively labeled subunits (part of the sequence must be complementary to that of the desired gene).
2. B; 3. F; 4. E; 5. C; 6. A; 7. D; 8. T; 9. F; 10. F; 11. T; 12. F; 13. B; 14. D; 15. F; 16. H; 17. I; 18. C; 19. J; 20. E; 21. G; 22. A.

RISKS AND PROSPECTS OF THE
NEW TECHNOLOGY (16-III)

1. Researchers are working to sequence the estimated 3 billion nucleotides present in human chromosomes.
2. Such culturing of cells apparently increases mutation rates and provides genetic novelties; researchers have successfully inserted genes into cultured cells.
3. The plasmid of *Agrobacterium* can be used as a vector to introduce desired genes into cultured plant cells; *Agrobacterium* was used to deliver a firefly gene into cultured tobacco plant cells.
4. Certain cotton plants have been genetically engineered for resistance to worm attacks.
5. Chemicals or electric shocks have been used to deliver DNA directly into plant protoplasts; a pistol has been used with blanks to drive DNA-coated, microscopic tungsten particles into plant protoplasts with some success.
6. In separate experiments the rat and human somatotropin genes became integrated into the mouse DNA. The mice grew much larger than their normal littermates.
7. The gene was expressed, but the pigs developed arthritis-like symptoms and other disorders.
8. The eggs are very vulnerable, and the procedure has a high failure rate; even when gene delivery is successful, there is no control as to where in the DNA the inserted gene will end up.
9. Restriction fragment length polymorphisms;
10. Polymorphisms; 11. genetic; 12. splicing enzymes;
13. introns; 14. sequencing; 15. genetic engineering;
16. frost damage; 17. ice-minus; 18. body cells; 19. gene therapy; 20. eugenic engineering.

Answers to Self-Quiz

1. a; 2. c; 3. b; 4. c; 5. d; 6. a; 7. c; 8. b; 9. a; 10. d.

17

EMERGENCE OF EVOLUTIONARY THOUGHT

GROWING AWARENESS OF CHANGE
    The Great Chain of Being
    Questions from Biogeography
    Questions from Comparative Anatomy
    Questions About Fossils
ATTEMPTS TO RECONCILE THE EVIDENCE
WITH PREVAILING BELIEFS
    Cuvier's Theory of Catastrophism
    Lamarck's Theory of Desired Evolution

DARWIN'S JOURNEY
    Voyage of the *Beagle*
    Evolution by Natural Selection:
    The Theory Takes Form
EVOLUTION—A STUDY OF
PROCESS AND PATTERN

Interactive Exercises

GROWING AWARENESS OF CHANGE (17-I, pp. 261–263)

1. The beginnings of biological science can be traced to the ancient Greeks. Most notable among several great Greek thinkers was Aristotle. Describe Aristotle's view of nature._____

_____

_____

2. Briefly state the prevailing view that most fourteenth- and fifteenth-century scholars held regarding the origins of the different kinds of living things._____

_____

_____

_____

_____

3. Global explorations of the sixteenth century "expanded" the natural world of European scholars. It soon became apparent that the diversity of organisms in the world was overwhelming. What was the crucial question that scholars then began to ask regarding the origin of species?_____

_____

4. During the eighteenth century, anatomists began to ask questions about organisms they were studying. What questions did they ask when they carefully compared human arms, whale flippers, and bat wings?

_____

_____

_____

5. Why would it be disturbing to people of the eighteenth century to learn that humans had body parts (the coccyx) that looked exactly like the bones of a tail? _____

_____

_____

6. Define the word *fossil*. _____

7. Summarize the work and ideas of the mid-eighteenth-century geologists; how did these ideas challenge the prevailing beliefs about the origin of species? _____

_____

_____

_____

8. How did George-Louis Leclerc de Buffon attempt to reconcile the new observations about nature with the traditional view of creation? _____

_____

_____

_____

Matching

Select the single best answer.

9. ___ comparative anatomy

10. ___ biogeography

11. ___ fossils

12. ___ school of Hippocrates

13. ___ evolution

14. ___ Great Chain of Being

15. ___ Buffon

16. ___ species

17. ___ Aristotle

18. ___ stratification

A. Horizontal layering of sedimentary rocks beneath the earth's surface
B. Came to view nature as gradual levels of organization
C. Each kind of being
D. Changes in lines of descent of organisms over time
E. Extended from the lowest forms of life to humans and on to spiritual beings
F. Suggested that perhaps species had originated in more than one place and perhaps had been modified over time
G. Studies of body structure comparisons and patterning
H. Studies of the world distribution of plants and animals
I. Suggested that the gods were not the cause of the sacred disease
J. Recognizable remains or body impressions of organisms living in the past

ATTEMPTS TO RECONCILE THE EVIDENCE WITH PREVAILING BELIEFS (17-II, pp. 264)

1. In the nineteenth century, there were notable attempts to reconcile evidence collected (which suggested that organisms had changed with time) with traditional concepts that did not allow for changes to occur. Complete the following table to summarize two well-known theories that account for change in organisms while acknowledging long-held beliefs.

Naturalist	Years	Theory Name	Theory Summary
a. Georges Cuvier	1769–1832	catastrophism	
b. Jean-Baptiste Lamarck	1744–1829	desired evolution	

DARWIN'S JOURNEY (17-III, pp. 264–270)

1. Several key players and events in the life of Charles Darwin led him to his conclusions about natural selection and evolution. Summarize these influences by completing the table below.

Event/Person	Importance to Synthesis of Evolutionary Theory
a. John Henslow	
b. _____	British ship that carried Darwin to South America as naturalist for that voyage
c. _____	English geologist who wrote *Principles of Geology,* a book read by Darwin that argued that very slow and gradual processes now molding the earth's surface had also been at work in the past
d. Thomas Malthus	
e. _____	English naturalist who arrived at the same conclusion as Darwin regarding evolution; papers released jointly by him and Darwin first announced their thinking to the world

True-False

If false, explain why.

____ 2. There is considerable evidence indicating that species have remained unchanged since the time of their creation.

____ 3. One expects to find the most recent fossil organisms in the upper layers of Earth's sedimentary rocks.

____ 4. Malthus believed that the food supply of a population increases faster than the population increases.

____ 5. Lyell's ideas on overpopulation and Malthus's ideas on slow geologic change provoked Darwin's thinking on how organisms change with time.

____ 6. When resources become scarce, competition for similar resources promotes greater specialization among the competitors.

___ 7. Darwin first learned about the ideas of evolution when he received a letter from Wallace.

___ 8. *Archaeopteryx* was a unique fossil that supplied a "missing link" for the idea that a group of reptiles evolved into birds.

___ 9. All natural populations possess reproductive capacity to exceed the resources required to sustain them.

___ 10. The members of a natural population show little variation in their traits, and much of this variation is passed on through generations.

___ 11. A population in nature might change because nature would select individuals with advantageous traits and eliminate others.

Chapter Terms

The following page-referenced terms are important; they were in boldface type in the chapter. Refer to the instructions given in Chapter 1, p. 4.

fossils (263)   evolution (263)   theory of natural selection (269) .   natural selection (269)

Self-Quiz

___ 1. An acceptable definition of evolution is
_____.
   a. changes in organisms that are extinct
   b. changes in organisms since the flood
   c. changes in organisms over time
   d. changes in organisms in only one place

___ 2. The two scientists most closely associated with the concept of evolution are
_____.
   a. Lyell and Malthus
   b. Henslow and Buffon
   c. Henslow and Malthus
   d. Darwin and Wallace

___ 3. Studies of the distribution of organisms on the earth is known as _____.
   a. the Great Chain of Being
   b. stratification
   c. geological evolution
   d. biogeography

___ 4. Buffon suggested that _____.
   a. tail bones in a human have no place in a perfectly designed body
   b. perhaps species originated in more than one place and have been modified over time
   c. the force for change in organisms was a built-in drive for perfection, up the Chain of Being
   d. gradual processes now molding the earth's surface had also been at work in the past

___ 5. In Lyell's book *Principles of Geology,* he suggested that _____.
   a. tail bones in a human have no place in a perfectly designed body
   b. perhaps species originated in more than one place and have been modified over time
   c. the force for change in organisms was a built-in drive for perfection, up the Chain of Being
   d. gradual processes now molding the earth's surface had also been at work in the past

___ 6. One of the central ideas of Lamarck's theory of desired evolution was that _____.
   a. tail bones in a human have no place in a perfectly designed body
   b. perhaps species originated in more than one place and have been modified over time
   c. the force for change in organisms was a built-in drive for perfection, up the Chain of Being
   d. gradual processes now molding the earth's surface had also been at work in the past

___ 7. The theory of catastrophism is associated with _____.
   a. Darwin
   b. Cuvier
   c. Buffon
   d. Lamarck

8. The idea that any population tends to outgrow its resources and that its members must compete for what is available belonged to _____.
   a. Malthus
   b. Darwin
   c. Lyell
   d. Henslow

9. The ideas that all natural populations have the reproductive capacity to exceed the resources required to sustain them and that members of a natural population show great variation in their traits are associated with _____.
   a. catastrophism
   b. desired evolution
   c. natural selection
   d. Malthus's idea of survival

10. *Archaeopteryx* is evidence for _____.
    a. the theory that birds descended from reptiles
    b. evolution
    c. an evolutionary "link" between two major groups of organisms
    d. the existence of fossils
    e. all of the above

Chapter Objectives/Review Questions

This section lists general and detailed chapter objectives that can be used as review questions. You can make maximum use of these items by writing answers on a separate sheet of paper. Fill in answers where blanks are provided. To check for accuracy, compare your answers with information given in the chapter or glossary.

Page	Objectives/Questions
(262)	1. State Aristotle's view of nature and describe what is meant by the Great Chain of Being.
(262)	2. _____ refers to the study of the world distribution of organisms.
(262)	3. It was the study of _____ that raised the following question: If all species were created at the same time in the same place, why were certain species found in only some parts of the world and not others?
(263)	4. Give examples of the type of evidence found by comparative anatomists that suggested living things may have changed with time.
(263)	5. A _____ is defined as the recognizable remains or body impressions of organisms that lived in the past.
(263)	6. The modification of species over time is called _____.
(264)	7. Be able to discuss Cuvier's theory of catastrophism.
(264)	8. What was the main force for change in organisms as described in Lamarck's theory of desired evolution?
(264–270)	9. Be able to state the significance of the following: Henslow, H.M.S. *Beagle*, Lyell, Darwin, Malthus, Wallace, and *Archaeopteryx*.
(266–268)	10. What conclusion was Darwin led to when he considered the various species of finches living on the separate islands of the Galapagos?
(269–270)	11. Outline the key correlations of the theory of natural selection.

Integrating and Applying Key Concepts

An Austrian biologist named Paul Kammerer (1880–1926) once reported that he had found solid evidence that the nuptial pads (used to grip the female during mating) on the forelimbs of the male midwife toad were *acquired* by the "energy and diligence" of the parent toad and were then transmitted to offspring for the benefit of future generations. When Kammerer presented his evidence, several observing scientists suspected that his specimens were forgeries. Scientists attempted to repeat Kammerer's experiments, but they were unrepeatable. Kammerer incurred the wrath of the international scientific community and eventually ended his life by his own hand. Considering what you have learned in this chapter, what is your assessment of Kammerer's claim?

Critical Thinking Exercises

1. Which of the following mechanisms would be necessary in order for Lamarck's theory of evolution to be valid?

    a. The nervous system causes specific mutations in DNA.
    b. Mutations occur at random locations in DNA.
    c. All parts of the body contribute to the genetic information in the gamete.
    d. Proteins direct their own replication.
    e. One parent's gametes are diploid.

ANALYSIS

    a. According to Lamarck's theory, the nervous system of one organism influences the characteristics of the organism's offspring. Since we now know that the DNA encodes the genetic information transmitted from one generation to the next, in order to accept Lamarck's ideas we would have to find that the nervous system could direct changes in the DNA.
    b. Random mutations would be the opposite of Lamarck's idea. Lamarck's theory required mutations directed by the feelings and activities of the previous generation.
    c. This concept of gamete formation was popular for many years. If bits of information collected into the gamete from all parts of the body, it is conceivable that changes in specific body parts brought about by activity could cause changes in the information they transmit to the gamete. However, we now know that the gamete's DNA is transmitted intact by replication, mitosis, and meiosis within the germ cell line only.
    d. If proteins directed their own replication, then the gamete would have to contain all the types of protein present in the whole organism. Even if the gamete did have all the proteins, the problem of how the nervous system could bring about specific changes in the molecules still remains.
    e. If one type of gamete was diploid, the characteristics of the next generation would be determined only by one parent. Presumably, the other gamete would merely activate or support development of the diploid one. However, this still leaves unanswered the problem of how the nervous system could cause directed changes in the molecules that transmit the genetic information.

2. Sequences of strata in sedimentary rocks are very useful in the working out of evolutionary sequences. Occasionally, however, anomalous patterns are observed that seem to be due to later inversion of many strata. Which of the following observations would be evidence of secondary inversion of strata?

    a. Unusual sequences of fossils in strata.
    b. Human and dinosaur fossils in the same strata.
    c. Strata tilted from the horizontal plane.
    d. Raindrop imprints and ripple marks on the bottoms of strata.
    e. Sudden changes from one type of rock to another in the horizontal plane within one layer.

ANALYSIS

    a. This is the observation that has to be explained. Often the sequence is consistent with an inversion and can be explained by inversion, but independent evidence is required.
    b. This would be evidence that the accepted evolutionary sequence and timetable must be revised. It would not be evidence of inversion, although it might be explained by breaking and mixing of strata. It has never been observed.
    c. This would be evidence that strata could be moved vertically by natural forces. It would indicate that inversions are at least possible, but it would not be evidence that any specific set of strata in a given place had been inverted.
    d. Raindrops can strike only the top of a layer, and ripple marks can only be formed by water on top of a layer. Such fossil marks found on the bottom of a layer of rock are clear evidence that the layer has turned over after it became rock.

e. This kind of discontinuity is evidence that the series of layers broke and that the rock on one side of the break became displaced vertically so that the broken edge of one layer now abuts the edge of a different layer. This is not an inversion, but it is one cause of earthquakes.

3. It is possible to determine the size and growth rate of the human population in the last couple of centuries and by calculation extend the curve backward in time. This leads some to the conclusion that about 10,000 years ago there were two humans on earth. Which of the following is the strongest criticism of this calculation and conclusion?

    a. Growth of the human population has been exponential.
    b. The calculation assumes a constant growth rate.
    c. Early humans lacked agriculture and industrial technology.
    d. The earth is more than 10,000 years old.
    e. The average life span of early humans was less than that of modern humans.

ANALYSIS

    a. This is not a criticism of the calculation; it is an assumption essential to the calculation. Without assuming some mathematical form for the curve, you cannot calculate beyond the measured portion.
    b. Unless there is constant growth rate, the curve is irregular and cannot be calculated. The calculation must assume constant growth rate, but independent evidence indicates that the growth rate of the human population was near zero for long periods and has increased rapidly in recent centuries.
    c. This fact explains the increases in population growth rate and is another criticism of the conclusion. Although the calculation itself does not require that the first humans practiced agriculture, those who make this calculation assume that the first humans did have agriculture and significant technology.
    d. This simply denies the conclusion without using any argument based on evidence to criticize it.
    e. This is another facet of the changing growth rate of the human population. As life span increases, death rate falls and the population grows at a faster rate. However, when the maximum possible life span is reached, the aging population generates an increasing death rate and population growth slows and stops, assuming constant birthrate.

Answers

Answers to Interactive Exercises

GROWING AWARENESS OF CHANGE (17-I)
1. Aristotle viewed nature as gradual levels of organization, from lifeless matter through complex forms of animal life.
2. A Great Chain of Being was visualized that extended from the lowest form to humans and on to spiritual beings. Each kind of being had a separate place in the divine order of things and had not changed since creation; each was a link in the chain. Once all the links were discovered, named, and described, the meaning of life would be revealed.
3. Where did the great diversity of organisms fit into the Great Chain?
4. Why are these different structures made of the same materials? Why do these organs have similar locations in the body? Why do these structures form and develop in similar ways in the animal embryo?
5. This was disturbing because the human body was thought to be perfectly designed and should not have parts of a tail.
6. Fossils are the recognizable remains or body impressions of organisms that lived in the past.

7. Geologists began mapping the horizontal layers of sedimentary rocks they felt had been slowly deposited. Different layers held different kinds of fossils that lived in the past. These findings challenged the traditional view of a recent creation.
8. Buffon suggested that if all species were created at the same time and place they would not be found throughout the world; mountains and oceans would have blocked their dispersal. Perhaps species originated in more than one place, and perhaps species became modified over time.
9. G; 10. H; 11. J; 12. I; 13. D; 14. E; 15. F; 16. C; 17. B; 18. A.

ATTEMPTS TO RECONCILE THE EVIDENCE WITH PREVAILING BELIEFS (17-II)
1. a. Creation occurred once, and this populated the world with species; many species were destroyed in a global catastrophe. The few survivors repopulated the world, but their fossils simply had not been found yet. Various catastrophes continued, with survivors left each time to repopulate. b. Life was created long ago in a simple state but changed to an improved state because of a built-in drive for perfection, up the Chain of Being.

Nerve fibers directed "Fluida" to body parts in need of change. Characteristics can be acquired during an individual's life through environmental pressure and internal "desires," and offspring can inherit the desired changes.

DARWIN'S JOURNEY (17-III)

1. a. John Henslow: Cambridge botanist who perceived and respected Darwin's real interests in nature; arranged that Darwin be offered the position of ship's naturalist aboard H.M.S. *Beagle.* b. H.M.S. *Beagle:* British ship that carried Darwin to South America as naturalist for that voyage. c. Charles Lyell: English geologist who wrote *Principles of Geology,* a book read by Darwin that argued that very slow and gradual processes now molding the earth's surface had also been at work in the past. d. Thomas Malthus: English clergyman and economist whose essay, read by Darwin, suggested that any population tends to outgrow its resources and that its members must compete for what is available. e. Alfred Wallace: English naturalist who arrived at the same conclusion as Darwin regarding evolution; papers released jointly by him and Darwin first announced their thinking to the world. 2. F; 3. T; 4. F; 5. F; 6. T; 7. F; 8. T; 9. T; 10. F; 11. T.

Answers to Self-Quiz

1. c; 2. d; 3. d; 4. b; 5. d; 6. c; 7. b; 8. a; 9. c; 10. e.

18

MICROEVOLUTION

Interactive Exercises

MICROEVOLUTIONARY PROCESSES (18-I, pp. 273–280)

1. A central idea forming the basis of evolutionary processes is the statement that "individual organisms do not evolve, populations do." Define the population referred to in the statement._____

_____

2. What is meant by the genetic variation that exists in populations?_____

_____

_____

3. How does the environment affect phenotypic variation?_____

_____

4. In populations, genetic variation arises in individuals through five events. List these five sources of variation._____

_____

_____

5. List the conditions that must be met before genetic equilibrium (or nonevolution) will occur (any order)._____

_____

_____

6. For the following situation, assume that the conditions listed in exercise 5 do exist; therefore, there should be no change in gene frequency, generation after generation. Consider a population of hamsters in which dominant gene B produces black coat color and recessive gene b produces gray coat color (two alleles are responsible for color). The dominant gene has a frequency of 80 percent (or 0.80). It would follow that the frequency of the recessive gene is 20 percent (or 0.20). From this, the assumption is made that 80 percent of all sperm have gene B and 80 percent of all eggs have gene B. Also, 20 percent of all sperm carry gene b and 20 percent of all eggs carry gene b. (See pages 276–277 in the text.)

   a. Calculate the probabilities of all possible matings in the Punnett square.

Sperm

		0.80 B	0.20 b
Eggs	0.80 B	BB	Bb
	0.20 b	Bb	bb

   b. Summarize the genotype frequencies of the F_1 generation:

Genotypes	Phenotypes
___BB	
___Bb	___% black
___bb	___% gray

   c. Further assume that the individuals of the F_1 generation produce another generation and the assumptions of the Hardy-Weinberg principle still hold. What are the frequencies of the sperm produced?

Parents (F_1)	B sperm	b sperm
___BB	___	___
___Bb	___	___
___bb	___	___
Totals =	___	___

The egg frequencies can be similarly calculated. Note that the gamete frequencies of the F_2 generation are the same as the gamete frequencies of the last generation. Phenotype percentage also remains the same. Thus, the gene frequencies did not change between the F_1 and the F_2 generations. Again, given the assumptions of the Hardy-Weinberg equilibrium, gene frequencies do not change generation after generation.

7. In a population, 81 percent of the organisms are homozygous dominant, and 1 percent are homozygous recessive. Find the following.

   a. the percentage of heterozygotes_____

   b. the frequency of the dominant allele_____

   c. the frequency of the recessive allele_____

8. In a population of 200 individuals, determine the following for a particular locus if $p = 0.8$.

   a. the number of homozygous dominant individuals_____

   b. the number of homozygous recessive individuals_____

   c. the number of heterozygous individuals if $p = 0.8$_____

9. If the percentage of gene D is 70 percent in a gene pool, find the percentage of gene d._____

10. If the frequency of gene R in a population is 0.60, what percentage of the individuals are heterozygous Rr?_____

11. Distinguish between the two extreme cases of genetic drift, the founder effect and bottlenecks._____

_____

_____

_____

Labeling

The traits of individuals in a population are often classified as morphological, physiological, or behavioral. In the list of traits below, enter M in the blank if the trait is morphological, P if the trait is physiological, and B if the trait is behavioral.

12. ____ Frogs have a three-chambered heart.

13. ____ The active transport of Na^+ and K^+ ions is unequal.

14. ____ Humans and orangutans possess an opposable thumb.

15. ____ An organism periodically seeks food.

16. ____ Some animals have a body temperature that fluctuates with the environmental temperature.

17. ____ The platypus is a strange mammal; it has a bill like a duck and lays eggs.

18. ____ Some vertebrates exhibit greater parental protection of offspring than others.

19. ____ During short photoperiods, the pituitary gland releases small quantities of gonadotropins.

20. ____ Red grouse defend large, multipurpose territories where they forage, mate, nest, and rear young.

21. ____ Lampreys have an elongated cylindrical body without scales.

Matching

Select the single best answer.

22. ___ gene flow
23. ___ allele frequencies
24. ___ neutral mutations
25. ___ microevolution
26. ___ lethal mutation
27. ___ alleles
28. ___ mutation
29. ___ genetic drift
30. ___ genetic equilibrium
31. ___ natural selection

A. A heritable change in DNA
B. Zero evolution
C. Different molecular forms of a gene
D. Change in allele frequencies as individuals leave or enter a population
E. Change or stabilization of allele frequencies due to differences in survival and reproduction among variant members of a population
F. Random fluctuation in allele frequencies over time due to chance
G. Change in which expression of mutated gene always leads to the death of the individual
H. The abundance of each kind of allele in the entire population
I. Neither harmful nor helpful to the individual
J. Changes in allele frequencies brought about by mutation, genetic drift, gene flow, and natural selection

Fill-in-the-Blanks

A(n) (32)_____ is a group of individuals of the same species that occupy a given area at a specific time. The (33)_____-_____ principle allows researchers to establish a theoretical reference point (baseline) against which changes in allele frequency can be measured. Variation can be expressed in terms of (34)_____ _____ , the relative abundance of different alleles carried by the individuals in that population. The stability of allele ratios that would occur if all individuals had equal probability of surviving and reproducing is called (35)_____ _____. Over time, allele frequencies tend to change through infrequent but inevitable (36)_____ , which are the original source of genetic variation. Random fluctuation in allele frequencies over time due to chance occurrence alone is called (37)_____ _____ ; it is more pronounced in small populations than in large ones. (38)_____ flow associated with immigration and/or emigration also changes allele frequencies. (39)_____ _____ is the differential survival and reproduction of individuals of a population that differ in one or more traits. (40)_____ _____ is the most important microevolutionary process.

EVIDENCE OF NATURAL SELECTION (18-II, pp. 280–284)

1. Complete the following table, which defines three types of natural selection effects.

Effect	Characteristics	Example
a. Stabilizing selection		
b. Directional selection		
c. Disruptive selection		

Identifying

2. Identify the three curves below as stabilizing selection, directional selection, or disruptive selection.

a._____ b._____ c._____

Fill-in-the-Blanks

When individuals of different phenotypes in a population differ in their ability to survive and reproduce, their alleles are subject to (3)_____ selection. (4)_____ selection favors the most common phenotypes in the population. (5)_____ selection occurs when a specific change in the environment causes a heritable trait to occur with increasing frequency and the whole population tends to shift in a parallel direction. (6)_____ selection favors the development of two or more distinct polymorphic varieties such that they become increasingly represented in a population and the population splits into different phenotypic variations. (7)_____ provide an excellent example of stabilizing selection, because they have existed essentially unchanged for hundreds of millions of years. Kettlewell used a (8) mark-_____-_____ method to test the hypothesis that in England more dark peppered moths survived as soot and other pollutants darkened the environment. The shift in color of peppered moths from light to dark provides a good example of (9)_____ selection. The increasing resistance of subsequent populations of disease-carrying flies to an insecticide known as DDT is an example of (10)_____ selection. Two different but successful adaptations of beak types in a small population of Galapagos finches represent (11)_____ selection.

SELECTION AND THE MAINTENANCE OF DIFFERENT PHENOTYPES (18-III, pp. 284–285)

Fill-in-the-Blanks

Differences in appearance between males and females of a species are known as (1)_____ _____. Among birds and mammals, females act as agents of (2)_____ when they choose their mates. (3)_____ selection is based on any trait that gives the individual a competitive edge in mating and producing offspring. The more colorful, showier, and larger appearance of male pheasants when compared with females is the result of (4)_____ selection. In Central Africa, persons with sickle-cell trait (HbS/HbA) have a greater chance of surviving (5)_____ infection than persons lacking the sickle-cell gene. This is a special case of (6)_____ selection known as balanced (7)_____.

SPECIATION (18-IV, pp. 286–290)

1. Define speciation._____

2. Write a complete definition of the biological concept of species._____

_____

_____

3. Describe the process of divergence._____

_____

_____

4. The illustration below depicts the divergence of one species into two as time passes. Answer the questions below the illustration.

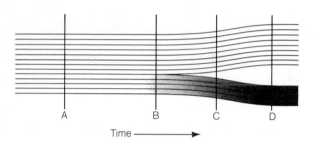

Time ———→

a. In what areas is there distinctly one species?_____

b. Between what letters does divergence begin?_____

c. Does this divergence appear to be slow or rapid?_____

d. What letter represents the time when divergence is probably complete?_____

5. Study the illustration below, which shows the possible course of wheat evolution; similar genomes are designated by the same letter. Answer the questions below the diagram.

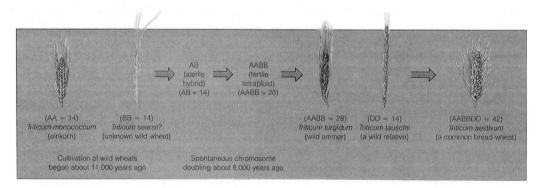

a. How many pairs of chromosomes are found in *Triticum monococcum?*_____

b. How are the *T. monococcum* chromosomes designated?_____

c. How many pairs of chromosomes are found in *T. searsii?*_____

d. How are the *T. searsii* chromosomes designated?_____

e. Why is the hybrid designated AB sterile?_____

f. What cellular event must have occurred to create the *T. turgidum* genome?_____

g. Describe the genome of the plant arising from the cross of *T. turgidum* and *T. tauschii* (not shown on the diagram)._____

h. What cellular event must have occurred to create the *T. aestivum* genome?_____

i. What is the source of the A genome in *T. aestivum*? The source of the B genome in *T. aestivum*? The source of the D genome in *T. aestivum*? _____

6. Complete the table below to summarize reproductive isolating mechanisms.

Reproductive Isolating Mechanism	Description/Example
a. Mechanical isolation	
b. _____	Incompatibilities between the sperm of one species and the egg (or female reproductive system) of another may prevent fertilization; example: two species of sea urchin.
c. Hybrid inviability and infertility	
d. _____	Behavior such as complex courtship rituals can be a strong isolating mechanism; example: courtship rituals of many bird species.
e. _____	Differences in reproductive timing may serve as an isolating mechanism; example: two closely related species of cicadas.

Matching

Select the single best answer.

7. ___ allopatric speciation

8. ___ sympatric speciation

9. ___ polyploidy

10. ___ lineage

11. ___ gradualism

12. ___ punctuation

A. A single line of descent (a twig or branch on the "family tree")
B. Speciation resulting from geographic isolation
C. Slow divergence over long time spans
D. Rapid changes during speciation; quick spurt of changes when first diverging from parental lineage
E. Offspring receiving three or more of each parental chromosome due to meiotic nondisjunction; a method of sympatric speciation
F. Formation of genetic barriers in a single population, followed by speciation

Fill-in-the-Blanks

(13)_____ is the process whereby species are formed. A (14)_____ is composed of one or more populations of individuals who can interbreed under natural conditions and produce fertile, reproductively isolated offspring. (15)_____ can be described as a process in which differences in alleles accumulate between populations. Any aspect of structure, function, or behavior that prevents interbreeding is a (16)_____ _____ mechanism. Physical barriers that prevent gene flow as in the case of large rivers changing course or forests giving way to grasslands are (17)_____ barriers. Wheat is an example of a species formed by (18)_____ and (19)_____.

Chapter Terms

The following page-referenced terms are important; they were in boldface type in the chapter. Refer to the instructions given in Chapter 1, p. 4.

population (273)
genetic variation (274)
allele frequencies (275)
genetic equilibrium (275)
microevolution (276)
mutation (276)

lethal mutation (277)
neutral mutations (277)
genetic drift (278)
gene flow (279)
natural selection (280)
stabilizing selection (280)

directional selection (283)
mark-release-recapture
  method (283)
disruptive selection (284)
sexual selection (285)
speciation (286)

divergence (286)
reproductive isolating
  mechanism (286)
allopatric speciation (288)
sympatric speciation (288)
polyploidy (288)

Self-Quiz

___ 1. A species is _____.
   a. a group of individuals that have identical characteristics
   b. a group of individuals that look alike and can interbreed
   c. a group of individuals that can interbreed under natural conditions and produce offspring
   d. a group of individuals that can interbreed under natural conditions, produce fertile offspring, and are reproductively isolated

___ 2. The unit of evolution is the _____.
   a. population
   b. individual
   c. fossil
   d. missing link

___ 3. Changes in allele frequencies brought about by mutation, genetic drift, gene flow, and natural selection are called _____ processes.
   a. genetic equilibrium
   b. microevolution
   c. founder effect
   d. independent assortment

For questions 4–6, choose from the following answers:
   a. disruptive selection
   b. genetic drift
   c. directional selection
   d. balanced polymorphism

___ 4. Increasing resistance of cockroaches to pesticides is an example of _____.

___ 5. The founder effect is a special case of _____.

___ 6. Different alleles that code for different forms of hemoglobin have led to _____ in the midst of the malarial belt that extends through West and Central Africa.

For questions 7–9, choose from the following answers:
   a. isolating mechanisms
   b. polyploidy
   c. genetic equilibrium
   d. sexual selection

___ 7. Hardy and Weinberg invented an ideal population that was in _____ and used it as a baseline against which to measure evolution in real populations.

___ 8. Plants that blossom at different times and chromosome sets that can no longer match up effectively during fertilization are examples of _____.

___ 9. Many domestic varieties of fruits, vegetables, and grains have evolved larger vegetative and reproductive structures through _____.

For questions 10–12, choose from the following answers:
   a. speciation
   b. divergence
   c. a reproductive isolating mechanism
   d. polyploidy

___ 10. Two species of sage plants with differently shaped floral parts that prevent pollination by the same pollinator serve as an example of _____.

___ 11. _____ occurs when reproductively isolated populations accumulate allele frequency differences between them over time.

___ 12. By the process of _____ , a population of individuals becomes reproductively isolated from other such populations.

Chapter Objectives/Review Questions

This section lists general and detailed chapter objectives that can be used as review questions. You can make maximum use of these items by writing answers on a separate sheet of paper. Fill in answers where blanks are provided. To check for accuracy, compare your answers with information given in the chapter or glossary.

Page		Objectives/Questions
(273)	1.	A _____ is a group of individuals occupying a given area and belonging to the same species.
(273–274)	2.	Distinguish between morphological, physiological, and behavioral traits.
(274)	3.	What does genetic variation mean in terms of genes?
(274)	4.	Genetic variation manifests itself as variation in _____.
(275)	5.	Review the five categories through which genetic variation occurs among individuals.
(275)	6.	The abundance of each kind of allele in the whole population is referred to as allele _____.
(275)	7.	A point at which allele frequencies for a trait remain stable through the generations is called genetic _____.
(276)	8.	Be able to list the five conditions that must exist before conditions for the Hardy-Weinberg principle are met.
(276)	9.	Changes in allele frequencies brought about by mutation, genetic drift, gene flow, and natural selection are called _____.
(276)	10.	A _____ is a random, heritable change in DNA.
(278)	11.	Random fluctuations in allele frequencies over time due to chance are called _____.
(279)	12.	Allele frequencies change as individuals leave or enter a population; this is gene _____.
(280)	13.	_____ _____ is defined as the change or stabilization of allele frequencies due to differences in survival and reproduction among variant members of a population.
(277)	14.	Define lethal mutation and neutral mutation.
(278)	15.	Distinguish the founder effect from a bottleneck.
(280)	16.	When the most common phenotypes in a population are favored, _____ selection is operating.
(283)	17.	_____ selection is operating when allele frequencies shift in a steady, consistent course in response to a new environment or a directional change in the old one.
(284)	18.	When both ends of the phenotypic range are favored and intermediate forms are selected against, _____ selection is occurring.
(285)	19.	Define sexual dimorphism.
(285)	20.	_____ selection is based on any trait that gives an individual a competitive edge in mating and producing offspring.
(284–285)	21.	Define balanced polymorphism.
(286)	22.	Be able to completely define what a species is according to biologists.
(286)	23.	What part does divergence play in speciation?
(286–287)	24.	Define each of the following reproductive isolating mechanisms: mechanical isolation, gamete isolation, hybrid inviability and infertility, behavioral isolation, and isolation in time.
(288)	25.	When new species form as a result of geographic isolation, the process is called _____ speciation; _____ speciation follows barriers that arise within the boundaries of a single population.
(288–290)	26.	Be able to define polyploidy and discuss its implications for speciation.

Integrating and Applying Key Concepts

Can you imagine any way in which directional selection may have occurred or may be occurring in humans? Which factors do you suppose are the driving forces that sustain the trend? Do you think the trend could be reversed? If so, by what factor(s)?

Critical Thinking Exercises

1. It is often stated that most new mutations are recessive. Which of the following assumptions would be most important in arguing for this statement?

   a. Most individuals are heterozygous for most of their genes.
   b. Most individuals are homozygous for most of their genes.
   c. Most existing alleles are recessive.
   d. Most existing alleles are dominant.
   e. Mutations are rare events.

   ANALYSIS

   a. Because the statement concerns the pattern of expression of *new* mutations, the frequencies and pattern of occurrence of existing alleles are largely irrelevant.
   b. This assumption is subject to the same argument as (a).
   c. If most existing alleles are recessive, they already code for a protein that is inactive in some way. Therefore, any changes in the amino acid sequence of the protein are likely to simply continue their recessive status and not change the phenotype.
   d. If most existing alleles are dominant, they code for active proteins or enzymes. Changes in such proteins would be most likely to inactivate them. These would represent new phenotypes and would be recessive to the active alleles.
   e. The statement concerns the qualitative effects of mutations, not their frequency.

2. A sample of butterflies contained 50 percent yellow-winged individuals and 50 percent black-winged individuals. In this species, wing color is determined by a single gene with two alleles, and the allele for black is dominant. Which of the following statements about the allele frequencies in the sample would most likely be true? Do not assume that this sample was obtained from a population in genetic equilibrium.

   a. The frequency of the yellow allele is greater than that of the black allele.
   b. The allele frequency of yellow is 3 times the allele frequency of black.
   c. The allele frequency of yellow is twice the allele frequency of black.
   d. The allele frequencies of black and yellow are equal.
   e. The allele frequency of black is greater than the allele frequency of yellow.

   ANALYSIS

   The frequencies cannot be calculated exactly in this sample, because a black-winged individual could be either homozygous or heterozygous. First assume that all black individuals are homozygous. In that case, there are equal numbers of black and yellow alleles in the sample, and the frequencies are both 50 percent. Next assume the opposite extreme, that all black individuals are heterozygous. In that case, half the individuals contribute two yellow alleles to the pool, and the other half contribute one yellow and one black. The frequency of yellow is then 3 times the frequency of black. The actual distribution is most likely somewhere between these two extremes. The only choice that is impossible is (e).

3. Now assume that the population of butterflies sampled in exercise 2 is in Hardy-Weinberg equilibrium. Also assume that the sample is random and large enough that the allele frequencies in the sample equal the allele frequencies in the population. What is the frequency of the allele for yellow wings in the population?

ANALYSIS

Yellow is the recessive allele; thus, the frequency of the yellow phenotype is the square of the frequency of the yellow allele. The sample is 50 percent yellow-winged. The frequency of the allele is the square root of 0.500, which equals 0.707. If you extend the solution and find that $p = 0.293$, in problem 2, (a) is clearly the only true answer if Hardy-Weinberg equilibrium exists.

Answers

Answers to Interactive Exercises

MICROEVOLUTIONARY PROCESSES (18-I)

1. A population is a group of individuals occupying a given area and belonging to the same species.
2. Some number of the genes in one individual have a slightly different molecular structure than their counterpart in another individual.
3. The environment can mediate phenotypic variation by influencing how the genes governing different traits are expressed.
4. Gene mutation, abnormal changes in chromosome structure or number, crossing over and genetic recombination at meiosis, independent assortment of chromosomes at meiosis, and fertilization between genetically different gametes
5. No mutation; very large population; isolation from other populations of the same species; all members survive, mate, and reproduce (no selection); and random mating
6. a. 0.64 *BB*, 0.16 *Bb*, 0.16 *Bb*, and 0.04 *bb*. b. genotypes: 0.64 *BB*, 0.32 *Bb*, and 0.04 *bb*; phenotypes: 96% black, 4% gray.

c. *Parents*	*B sperm*	*b sperm*
0.64 *BB*	0.64	0
0.32 *Bb*	0.16	0.16
0.04 *bb*	0	0.04
Totals =	0.80	0.20

7. Find (b) first, then (c), and finally (a). a. $2pq = 2 \times (0.9) \times (0.1) = 2 \times (0.09) = 0.18 = 18\%$, which is the percentage of heterozygotes; b. $p^2 = 0.81$, $p = \sqrt{0.81} = 0.9$ = the frequency of the dominant allele; c. $p + q = 1$, $q = 1 - 0.9 = 0.1$ = the frequency of the recessive allele.
8. a. homozygous dominant = $p^2 \times 200 = (0.8)^2 \times 200 = 0.64 \times 200 = 128$ individuals; b. homozygous recessive = $q^2 \times 200 = (0.2)^2 \times 200 = (0.04) \times (200) = 8$ individuals; c. heterozygotes = $2pq \times 200 = 2 \times 0.8 \times 0.2 \times 200 = 0.32 \times 200 = 64$ individuals. Check: $128 + 8 + 64 = 200$.
9. If $p = 0.70$, since $p + q = 1$, $0.70 + q = 1$; then $q = 0.30$, or 30 percent.
10. If $p = 0.60$, since $p + q = 1$, $0.60 + q = 1$; then $q = 0.40$; thus, $2pq = 0.48$, or 48 percent.
11. In the founder effect, a few individuals leave a population and establish a new one; by chance, allele frequencies will differ from the original population. In bottlenecks, disease, starvation, or some other stressful situation nearly eliminates a population; relative allele frequencies are randomly changed.

12. M; 13. P; 14. M; 15. B; 16. P; 17. M; 18. B; 19. P; 20. B; 21. M; 22. D; 23. H; 24. I; 25. J; 26. G; 27. C; 28. A; 29. F; 30. B; 31. E; 32. population; 33. Hardy-Weinberg; 34. gene frequency; 35. genetic equilibrium; 36. mutations; 37. genetic drift; 38. Gene; 39. Natural selection; 40. Natural selection.

EVIDENCE OF NATURAL SELECTION (18-II)

1. a. the most common phenotypes are favored; average human birth weight of 7 pounds; b. allele frequencies shift in a steady, consistent direction in response to a new environment or a directional change in an old one; light to dark forms of peppered moths; c. forms at both ends of the phenotypic range are favored, and intermediate forms are selected against; finches on the Galapagos Islands.
2. a. directional; b. disruptive; c. stabilizing; 3. natural; 4. Stabilizing; 5. Directional; 6. Disruptive; 7. Nautiloids; 8. release-recapture; 9. directional; 10. directional; 11. disruptive.

SELECTION AND THE MAINTENANCE OF DIFFERENT PHENOTYPES (18-III)

1. sexual dimorphism; 2. selection; 3. Sexual; 4. sexual; 5. malaria; 6. stabilizing; 7. polymorphism.

SPECIATION (18-IV)

1. Speciation is the process by which species originate.
2. A species is a taxonomic unit of one or more populations of individuals that can interbreed under natural conditions and produce fertile offspring and that are reproductively isolated from other such units.
3. Divergence is a process that blocks gene flow in some degree between local population units. When divergence becomes great enough, members of two populations will not be able to breed successfully; they become two separate species.
4. a. A–B, some of B–C; b. B and C; c. slow; d. D; 5. a. 7 pairs; b. AA; c. 7 pairs; d. BB; e. The chromosomes fail to pair in meiosis; f. Fertilization of nonreduced gametes produced a fertile tetraploid; g. a sterile hybrid, ABD; h. Fertilization of nonreduced gametes resulted in a fertile hexaploid; i. *Triticum monococcum, T. searsii, T. tauschii.*
6. a. differences in reproductive structures or other body parts may prevent members of two populations from interbreeding; example: two species of sage plants and their pollinators; b. isolation of gametes; c. even

when fertilization occurs between the gametes of different species, the resulting embryo usually dies due to physical or chemical incompatibilities; example: sometimes hybrids such as the mule and zebroid are vigorous but sterile; d. behavioral isolation; e. isolation in time.

7. B; 8. F; 9. E; 10. A; 11. C; 12. D; 13. Speciation; 14. species; 15. Divergence; 16. reproductive isolating; 17. geographic; 18. polyploidy (hybridization); 19. hybridization (polyploidy).

Answers to Self-Quiz

1. d; 2. a; 3. b; 4. c; 5. b; 6. d; 7. c; 8. a; 9. b; 10. c; 11. b; 12. a.

19

LIFE'S ORIGINS AND MACROEVOLUTIONARY TRENDS

Interactive Exercises

EVIDENCE OF MACROEVOLUTION (19-I, pp. 293–299)

1. Define macroevolution._____

_____

2. State the simple fact that we can use to interpret the evidence of large-scale patterns and trends._____

_____

3. Cite some examples of fossils, the most direct evidence available that organisms lived long ago._____

_____

4. What conditions favor fossil formation?_____

_____

Sequence

Earth history has been divided into four great eras that are based on four abrupt transitions in the fossil record. The oldest era has been subdivided. Arrange the eras in correct chronological sequence from the oldest to the youngest.

5. ___

6. ___

7. ___

8. ___

9. ___

A. Mesozoic
B. Cenozoic
C. Proterozoic
D. Archean
E. Paleozoic

Labeling

Evidence for macroevolution comes from comparative morphology and comparative biochemistry. For each of the following items, place an M in the blank if the evidence comes from comparative morphology and a B if the evidence comes from comparative biochemistry.

10. ___ Using neutral mutations to date the divergence of two species from a common ancestor

11. ___ Similar embryonic stages, such as aortic arches, that persist in different vertebrates

12. ___ Similarities in vertebrate forelimbs when comparing the wings of pterosaurs, birds, bats and porpoise flippers

13. ___ Comparison of the proportional development changes in a chimpanzee skull and a human skull

14. ___ The amino acid sequence in the cytochrome *c* of humans precisely matching the sequence in chimpanzees

15. ___ Distantly related vertebrates, such as sharks, penguins, and porpoises, showing a similarity to one another in their proportion, position, and function of body parts

16. ___ Establishing a rough measure of evolutionary distance between two organisms by use of DNA hybridization studies that demonstrate the extent to which the DNA from one species base-pairs with another

17. ___ The striking resemblances between certain cactuses of North America and euphorbs of Africa even though these two plant families are rather distant

18. ___ Storks and new-world vultures being more closely related to each other than flamingos and ibises are to each other according to DNA hybridization studies

19. ___ Regulatory genes controlling the rate of growth of different body parts and being able to produce large differences in the development of two very similar embryos

Fill-in-the-Blanks

The establishment of large-scale patterns, trends, and rates of change among groups of species is known as

(20)_____ . Most of the information about the history of life on Earth comes from (21)_____ .

Earth history can be divided into five great eras; beginning with the oldest, they are the (22)_____ ,

the (23)_____ , the (24)_____ , the (25)_____ , and the (26)_____ . These eras are

based on abrupt (27)_____ in the (28)_____ record. Evidence for macroevolution comes from

detailed comparisons of body form and structural patterns of major taxa; this is known as (29)_____

_____ . It is believed that variations among genetically similar adult vertebrates come about by

mutations of (30)_____ _____ . The wings of birds and bats show genetic relationships and are termed (31)_____ structures. Such departure from a common ancestral form is known as

morphological (32)_____. Penguin wings and shark fins perform similar functions in a similar

environment; these structures are said to be (33)_____ and demonstrate morphological

(34)_____. Divergence between two species can be dated using accumulation rates of neutral

(35)_____. DNA (36)_____ is a method of converting DNA of different species to

single-stranded forms and allowing them to recombine as a measure of similarity.

MACROEVOLUTION AND EARTH HISTORY (19-II, pp. 299–316)

1. Refer to Figure 19.16 in the text. To review some of the important events that occurred in the geologic
past, complete the table below by entering the geologic era (Archean, Proterozoic, Paleozoic, Mesozoic,
and Cenozoic) and the approximate time in millions of years since the time of the events.

Era	Time	Events
a. _____	_____	A few reptile lineages give rise to mammals and the dinosaurs.
b. _____	_____	Formation of Earth's crust, early atmosphere, oceans; chemical evolution leading to the origin of life.
c. _____	_____	Origin of amphibians.
d. _____	_____	Origin of animals with hard parts.
e. _____	_____	Rocks 3.5 billion years old contain fossils of well-developed prokaryotic cells that probably lived in tidal mud flats.
f. _____	_____	Flowering plants emerge, gymnosperms begin their decline.
g. _____	_____	In the Carboniferous, there were major radiations of insects and amphibians; gymnosperms present, origin of reptiles.
h. _____	_____	Oxygen accumulated in the atmosphere.
i. _____	_____	Before the close of this era, the first photosynthetic bacteria had evolved.
j. _____	_____	Insects, amphibians, and early reptiles flourished in the swamp forests of the Permian.
k. _____	_____	Most of the major animal phyla evolved in rather short order.
l. _____	_____	Dinosaurs ruled.
m. _____	_____	Origin of aerobic metabolism; origin of protistans, algae, fungi, animals.
n. _____	_____	Grasslands emerge and serve as new adaptive zones for plant-eating mammals and their predators.
o. _____	_____	Humans destroy habitats and many species.
p. _____	_____	The first ice age triggered the first global mass extinction; reef life everywhere collapsed.
q. _____	_____	The invasion of land begins; small stalked plants establish themselves along muddy margins, and the lobe-finned fishes ancestral to amphibians move onto land.

Matching

Several probable stages in the physical and chemical evolution of life have been identified. Match the numbered statements and conditions below to the four listed categories. Letters may be used more than once.

A. Early Earth and its atmosphere
B. Synthesis of biological molecules
C. Self-replicating systems
D. The first plasma membrane

2. ___ Most likely, the first living cells were little more than membrane-bound sacs holding nucleic acids that served as templates for protein synthesis.

3. ___ Now suppose a clay template also attracted nucleotides.

4. ___ Farther out from the center, Earth was forming along with other planets.

5. ___ We also know that lightning, hot volcanic ash, and even shock waves have enough energy to drive the construction of biological molecules under abiotic conditions.

6. ___ During the 300 million years after the first rains began, organic compounds accumulated in the shallow waters of Earth. The interacting molecules DNA, RNA, and proteins were formed.

7. ___ This first atmosphere had very little free oxygen—a condition that favored the origin of life.

8. ___ Sidney Fox heated amino acids under dry conditions to form protein chains, which he placed in hot water. The cooled chains self-assembled into small, stable spheres.

9. ___ All the components found in biological molecules were present on the early earth.

10. ___ Simple systems of enzymes, coenzymes, and RNA have been created in the laboratory.

11. ___ Early on, water vapor must have been released from the breakdown of rocks during volcanic eruptions, but it would have evaporated in the intense heat blanketing the crust.

12. ___ Clay crystals at the bottom of tidal flats and estuaries may have been the first template for protein synthesis.

13. ___ Stanley Miller mixed hydrogen, methane, ammonia, and water in a reaction mixture. He recirculated the mixture and bombarded it with a spark discharge to simulate lightning. Within one week, many amino acids and other organic compounds had formed.

Sequence

Refer to Figure 19.15 in the text. Study of the geologic record reveals that, as the major events in the evolution of the earth and its organisms occurred, there were periodic major *extinctions* of organisms followed by major *radiations* of organisms. Arrange the letters of the extinctions and radiations listed below in the approximate order in which they occurred, from youngest to oldest.

14. ____
15. ____
16. ____
17. ____
18. ____
19. ____
20. ____
21. ____
22. ____
23. ____
24. ____
25. ____
26. ____
27. ____

A. Pangea, worldwide ocean forms; shallow seas squeezed out. Major radiations of reptiles, gymnosperms.

B. Glaciations as Gondwana crosses South Pole. Mass extinction of many marine organisms.

C. Mass extinction of many marine invertebrates, most fishes.

D. Pangea breakup begins. Rich marine communities. Major radiations of dinosaurs.

E. Asteroid impact? Mass extinction of all dinosaurs and many marine organisms.

F. Recovery, radiations of marine invertebrates, fishes, dinosaurs. Gymnosperms the dominant land plants. Origin of mammals.

G. Major glaciations. Modern humans emerge and begin what may be the greatest mass extinction of all time on land, starting with Ice Age hunters.

H. Unprecedented mountain building as continents rupture, drift, collide. Major climatic shifts; vast grasslands emerge. Major radiations of flowering plants, insects, birds, mammals. Origin of earliest human forms.

I. Gondwana moves south. Major radiations of marine invertebrates, early fishes.

J. Laurasia forms. Gondwana moves north. Vast swamplands, early vascular plants. Radiation of fishes continues. Origin of amphibians.

K. Pangea breakup continues; broad inland seas form. Major radiations of marine invertebrates, fishes, insects, dinosaurs; origin of angiosperms (flowering plants).

L. Asteroid impact? Mass extinction of many organisms in seas, some on land; dinosaurs, mammals survive.

M. Mass extinction. Nearly all species in seas and on land perish.

N. Tethys sea forms. Recurring glaciations. Major radiations of insects, amphibians. Spore-bearing plants dominant; gymnosperms present; origin of reptiles.

Chronology of Events

Refer to Figure 19.16 in the text. From the list of evolutionary events above (A–N), select the events that occurred in a particular era of time by circling the appropriate letters.

28. Cenozoic: A - B - C - D - E - F - G - H - I - J - K - L - M - N

29. Cenozoic–Mesozoic border: A - B - C - D - E - F - G - H - I - J - K - L - M - N

30. Mesozoic: A - B - C - D - E - F - G - H - I - J - K - L - M - N

31. Mesozoic–Paleozoic border: A - B - C - D - E - F - G - H - I - J - K - L - M - N

32. Paleozoic: A - B - C - D - E - F - G - H - I - J - K - L - M - N

Fill-in-the-Blanks

About (33) [choose one] ( ) 10, ( ) 4.6, ( ) 3.6, ( ) 3.2 billion years ago, the cloud that formed our solar system had flattened out into a slowly rotating disk. By (34) [choose one] ( ) 4, ( ) 3.8, ( ) 3.2, ( ) 2 billion years ago, the earth was hurtling through space as a thin-crusted inferno. In time, the crust cooled and rain fell, stripping mineral salts from parched rocks, and the oceans were formed. (35)_____ _____ probably were the absorbing agents that served to assemble (36)_____ _____ into proteins. Stanley Miller constructed a reaction chamber containing circulating hydrogen, methane, ammonia, and

(37)_____. Within one week, many (38)_____ _____ and other organic compounds had

formed. Sidney Fox heated dry amino acids to form (39)_____ chains. These were placed in hot water,

and they self-assembled into small, stable, membranous (40)_____ that showed properties of

(41)_____ _____. (42)_____ _____ refers to the earth's slablike plates floating on a hot,

plastic (43)_____. Gondwana and other land masses crunched to form a single world continent called

(44)_____ that broke up; its movements continue today. These and other devastating changes in

Earth's environment provoked two trends that occurred in the evolution of life, (45)_____ _____

and (46)_____ _____. In (47)_____ _____ , a lineage fills the environment with new

species through bursts of microevolution. This occurs when there are unfilled (48)_____ _____.

At times, evolutionary access results when a (49)_____ _____ occurs in a species.

   Rocks 3.5 billion years old contain fossils of (50)_____ _____ that probably existed in tidal

mud flats. Little or no free (51)_____ was in the atmosphere, so (52)_____ , an anaerobic

pathway, must have been the first metabolic pathway to develop. Between (53) [choose one] ( ) 3.2, ( ) 2.5,

( ) 1 billion and 700 million years ago, (54)_____ _____ dominated the shallow seas.

Mound-shaped bacterial cell mats found in rock formations more than 2 billion years old are called

(55)_____. Oxygen accumulated in the (56)_____ atmosphere. This prevented further synthesis

of (57)_____ _____ and opened up new worldwide (58)_____ _____ in which

organisms such as multicelled plants, fungi, and animals evolved that carried on (59)_____

respiration. By 1.2 billion years ago, the green (60)_____ had evolved. The Paleozoic era is subdivided

into the Cambrian, Ordovician, Silurian, Devonian, Carboniferous, and Permian periods. Nearly all animal

phyla had evolved by the end of the (61)_____ period; (62)_____ were the dominant animals.

Gondwana drifted southward and seas flooded land to open new, shallow marine environments for

flourishing reef organisms during the (63)_____ period. In the area of the South Pole, glaciers formed

on Gondwana, and many forms of reef life became extinct. Gondwana drifted northward during the

Silurian and the Devonian. Reef organisms recovered as a major radiation of (64)_____ fishes (with

armor plates and massive jaws) occurred. (65)_____ - _____ fishes that were ancestral to

amphibians began to invade the land as small stalked plants began to evolve in muddy land–water

margins. Another global (66)_____ _____ occurred at the Devonian–Carboniferous boundary.

During the (67)_____ period, major adaptive radiations of insects and treelike coal-forming plants

occurred in swamp forests. When Pangea was formed by collisions of all land masses, nearly all land and

sea species perished in the greatest known mass extinction as the (68)_____ period drew to a close.

Great changes in geology and climate occurred, but the small reptilian ancestors of dinosaurs, birds,

snakes, and lizards survived. The Mesozoic era is subdivided into Triassic, Jurassic, and Cretaceous

periods and lasted about 175 million years. The major radiation of reptiles had ended abruptly before

the Mesozoic, but divergences in a few lineages now gave rise to (69)_____ and reptiles, including

dinosaurs. During a (70)_____ _____ at the Triassic–Jurassic boundary, many marine organisms were lost. (71)_____ ruled the earth during the Jurassic and early Cretaceous periods. There is physical evidence that Earth's collision with a(n) (72)_____ may have brought about dinosaur extinction. In the early Cretaceous period, (73)_____ _____ emerged and began a major radiation amid the decline of gymnosperms that continues today. When the (74)_____ era began, major mountain ranges were formed, the great land masses underwent major renovations, and climate changes took place that greatly altered the course of evolution. New environments promoted a great diversification of (75)_____.

Chapter Terms

The following page-referenced terms are important; they were in boldface type in the chapter. Refer to the instructions given in Chapter 1, p. 4.

macroevolution (293)
fossil (294)
lineages (295)
comparative morphology (295)

regulatory genes (296)
morphological
   divergence (296)

morphological
   convergence (297)
plate tectonic theory (303)
gradualistic model (303)

punctuational model (304)
mass extinction (304)
adaptive radiation (304)
key innovation (305)

Self-Quiz

For questions 1–8, choose from the following answers:
   a. Archean
   b. Cenozoic
   c. Mesozoic
   d. Paleozoic
   e. Proterozoic

____ 1. Dinosaurs and gymnosperms were the dominant forms of life during the _____ era.

____ 2. The Alps, Andes, Himalayas, and Cascade Range were born during major reorganization of land masses early in the _____ era.

____ 3. The composition of Earth's atmosphere changed during the _____ era from anaerobic to aerobic.

____ 4. Invertebrates, primitive plants, and primitive vertebrates were the principal groups of organisms on Earth during the _____ era.

____ 5. Before the close of the _____ era, the first photosynthetic bacteria had evolved.

____ 6. The _____ era ended with the greatest of all extinctions, the Permian extinction.

____ 7. Late in the _____ era, flowering plants arose and underwent a major radiation.

____ 8. The _____ era included adaptive zones into which plant-eating mammals and their predators radiated.

For questions 9–11, choose from the following answers.
   a. convergence
   b. divergence
   c. adaptive radiation

____ 9. Penguins and porpoises serve as examples of _____.

____ 10. Kangaroos, koala bears, and opossums serve as examples of _____.

____ 11. Evolution of the major groups of mammals following dinosaur extinction occurred by _____.

For questions 12–15, choose from the following answers:
    a. key innovation
    b. punctuation
    c. gradualism
    d. extinction

___ 12. An example of a _____ is the modification of forelimbs into wings; this opened new adaptive zones.

___ 13. The vertebrate invasion of the land at the end of the Devonian was followed by rapid evolution; this is an example of _____.

___ 14. The disappearance of dinosaurs from the earth is an example of _____.

___ 15. Fossil foraminiferans (tiny shelled protists) show a constant change of form (over lengthy time periods) within one lineage; this is an example of _____.

For questions 16–20, choose from the following answers:
    a. homologous
    b. analogous
    c. macroevolution
    d. plate tectonics
    e. regulatory genes

___ 16. _____ structures resemble one another due to common descent.

___ 17. _____ is the study of the slabs floating on the earth's underlying mantle.

___ 18. Examples of _____ structures are the shark fin and the penguin wing.

___ 19. _____ refers to changes in groups of species.

___ 20. _____ control the rate of growth in different body parts.

Chapter Objectives/Review Questions

This section lists general and detailed chapter objectives that can be used as review questions. You can make maximum use of these items by writing answers on a separate sheet of paper. Fill in answers where blanks are provided. To check for accuracy, compare your answers with information given in the chapter or glossary.

Page   *Objectives/Questions*

(293)   1. _____ refers to the large-scale patterns, trends, and rates of change among groups of species.
(293)   2. Evolution proceeds by _____ of organisms that already exist.
(294)   3. Define fossil in the broad sense.
(295)   4. Another term for a line of descent is a _____.
(295)   5. Arrange the great eras of Earth's history in the correct order from oldest to youngest; how is the age of a rock layer determined?
(295)   6. What type of macroevolutionary evidence is collected by workers in comparative morphology?
(296)   7. Make a general statement about embryos of different organisms within the same major group.
(296)   8. _____ genes control the rate of growth of different body parts.
(296)   9. Structures that are similar due to descent from a common ancestor are _____ structures.
(296–297) 10. Relate the term *morphological divergence* to homologous structures.
(297–298) 11. Similar body parts used for similar function in evolutionarily remote lineages are said to be _____ to one another; cite an example of morphological convergence.
(298)   12. Comparative _____ is a field that examines genes and gene products to determine evolutionary relationships.
(298–299) 13. Gel _____ is used to study protein and nucleic acid structures.
(299–301) 14. Be able to generally "tell the story" of the origin of life; include statements relating to the following: primitive Earth and its atmosphere, synthesis of biological molecules, self-replicating systems, the first plasma membranes.
(300–301) 15. Cite the contributions of Stanley Miller and Sidney Fox to studies of life's origin.

Integrating and Applying Key Concepts

Imagine that in the next decade three more Chernobyl-type disasters happen, the oceans acquire critical levels of carcinogenic pesticides that work their way up the food chains, and the ozone layer shrinks dramatically in the upper atmosphere. Describe the macroevolutionary events that you believe might happen.

As Earth becomes increasingly loaded with carbon dioxide and various industrial waste products, how do you think living forms on Earth will evolve to cope with these changes?

Critical Thinking Exercises

1. The text gives the impression that parts of the Proterozoic oceans were like a modern lake undergoing eutrophication, densely populated by photosynthetic prokaryotic organisms. However, eutrophic lakes become anaerobic, while the Proterozoic sea began as anaerobic and produced the oxygen now present in the atmosphere. Which of the following statements most likely accounts for the difference?

    a. There were no eukaryotic cells in the Proterozoic.
    b. There were no aerobic bacteria in the Proterozoic.
    c. There were no decomposers at all in the Proterozoic.
    d. Proterozoic photosynthesis did not use water as an electron donor.
    e. The modern atmosphere already has so much oxygen that extra oxygen produced by photosynthesis cannot diffuse into the air.

ANALYSIS

    a. Being eukaryotic or prokaryotic has no bearing on metabolic pathways. Both types of cells include species that consume and produce oxygen.
    b. Aerobic pathways could not come into existence until the atmosphere contained oxygen. In the Proterozoic, there were no organisms to consume the oxygen that the photosynthetic bacteria produced, so the oxygen concentration rose steadily. In modern lakes, however, overgrowth of photosynthetic organisms supports overgrowth of aerobic decomposers that consume all the available oxygen.
    c. Anaerobic decomposers could have been present. They would have recycled carbon but left the oxygen to accumulate.
    d. There is no evidence of this, and, more important, it would not account for the accumulation of oxygen. If photosynthesis used a different electron donor, a different product, such as sulfur, would be expected to accumulate.
    e. Whenever there is a concentration difference between two compartments, a substance can diffuse. But again, this statement, even if true, does not account for the removal of oxygen from modern eutrophic lakes. This statement predicts that the oxygen released by photosynthetic bacteria will be trapped in the lake.

2. One concept of evolution has been based on the idea that a new kind of organism would emerge by a sudden, large mutation in an individual. This new type of individual would then exploit a new way of life and proliferate as a new species. Which of the following statements makes the strongest argument against this hypothesis?

    a. A single, sudden mutation cannot produce that much difference in an individual.
    b. Such a large mutation would most likely be lethal.
    c. Natural selection takes longer than one generation to produce a new species.
    d. There would be no mate for the first individual of the new species.
    e. The new species would begin with too little genetic variation and would rapidly become extinct.

ANALYSIS

    a. Mutations in genes that control the developmental program, that is, genes that influence the expression of other genes, can have dramatic effects on the phenotype.
    b. Many, perhaps most, mutations with extensive effects on development would likely be lethal. However, the hypothesis does not require that all or even a majority of such mutations be viable, only some.
    c. The statement is true, but the hypothesis proposes an alternative mechanism to gradual natural selection to produce new species.
    d. If the mutant individual is a member of a new species, by definition it would be unable to reproduce with any member of the old species. Since the mutation would not be likely to occur in more than one individual, that individual would have no potential mate and would have no evolutionary future.
    e. Low genetic diversity is a disadvantage to a species but does not necessarily lead to extinction. This mechanism would be an extreme example of the founder effect, but mutation would soon generate more diversity in the new population.

3. One early objection to evolutionary theory calculated that the earth was too young for evolution to have happened. The calculation began with the estimated temperature of the original hot planet, applied principles of heat transfer from the planet to space, and computed the time required for the earth to cool to its present temperature. Which of the following assumptions did the calculation most likely require?

    a. The rate of heat loss from Earth was constant throughout the period.
    b. Earth is too young for evolution to have occurred.
    c. Heat moves spontaneously from warmer to cooler bodies.
    d. Earth is a closed system.
    e. No source replaced the heat that Earth lost.

ANALYSIS

    a. The calculation could take into account the decreasing rate of heat loss as the difference in temperature between the cooling earth and the heat sink of space diminished.
    b. This is the conclusion, not an assumption required by it.
    c. This is a necessary condition for the computation, but it is not an assumption. It is an established conclusion with great observational support.
    d. If this assumption were made, it would allow for no cooling of Earth. The initial heat of the planet would be closed into the planet.
    e. If heat were added to Earth as it cooled, its temperature would drop less rapidly. And if the addition of heat was constant, the earth would reach a steady state in which the heat loss is equal to the heat gain and the temperature is constant. The calculation was made before radioactivity was recognized. It now appears that the energy released from radioactive atoms within the earth generates enough heat to maintain Earth's temperature.

Answers

Answers to Interactive Exercises

EVIDENCE OF MACROEVOLUTION (19-I)

1. Macroevolution refers to the large-scale patterns, trends, and rates of change among groups of species.
2. Evolution proceeds by modifications of organisms that already exist.
3. Fossilized skeletons, shells, leaves, seeds, and tracks.
4. To be preserved as a fossil, body parts or impressions must be buried (burial medium varies) before they decompose; there must be no disturbance of this material.
5. D; 6. C; 7. E; 8. A; 9. B; 10. B; 11. M; 12. M; 13. M; 14. B; 15. M; 16. B; 17. M; 18. B; 19. M; 20. macroevolution; 21. fossils; 22. Archean; 23. Proterozoic; 24. Paleozoic; 25. Mesozoic; 26. Cenozoic; 27. changes; 28. fossil; 29. comparative morphology; 30. regulatory genes; 31. homologous; 32. divergence; 33. analogous; 34. convergence; 35. mutations; 36. hybridization.

MACROEVOLUTION AND EARTH HISTORY (19-II)

1. a. Mesozoic, 240–205; b. Archean, 4,600–3,800; c. Paleozoic, 435–360; d. Paleozoic, 550–500; e. Archean, 4,600–3,800; f. Mesozoic, 135–65; g. Paleozoic, 360–280; h. Proterozoic, 2,500–570; i. Archean, 3,800–2505; j. Paleozoic, 360–280; k. Paleozoic, 550–500; l. Mesozoic, 181–65; m. Proterozoic, 2,500–570; n. Cenozoic, 65–1.65; o. Cenozoic, 1.65–present; p. Paleozoic, 435; q. Paleozoic, 435–360; 2. D; 3. C; 4. A; 5. B; 6. C (B); 7. A; 8. D (B); 9. B; 10. C; 11. A; 12. B; 13. B; 14. G; 15. H; 16. E; 17. K; 18. D; 19. L; 20. F; 21. M; 22. A; 23. N; 24. C; 25. J; 26. B; 27. I; 28. G, H; 29. E; 30. D F, K, L; 31. M; 32. A, B, C, I, J, N; 33. 4.6; 34. 3.8; 35. Clay crystals; 36. amino acids; 37. water (steam); 38. amino acids; 39. protein; 40. spheres; 41. cell membranes; 42. Plate tectonics; 43. mantle; 44. Pangea; 45. mass extinctions (adaptive radiations); 46. adaptive radiations (mass extinctions); 47. adaptive radiation; 48. adaptive zones; 49. key innovation; 50. prokaryotic cells (anaerobic bacteria); 51. oxygen (O_2); 52. fermentation; 53. 2.5; 54. photosynthetic bacteria; 55. stromatolites; 56. Proterozoic; 57. organic compounds (chemicals); 58. adaptive zones; 59. aerobic; 60. algae; 61. Cambrian; 62. trilobites; 63. Ordovician; 64. predatory; 65. Lobe-finned; 66. mass extinction; 67. Carboniferous; 68. Permian; 69. mammals; 70. mass extinction; 71. Dinosaurs (Reptiles); 72. asteroid; 73. flowering plants; 74. Cenozoic; 75. species.

Answers to Self-Quiz

1. c; 2. b; 3. e; 4. d; 5. a; 6. d; 7. c; 8. b; 9. a; 10. b; 11. c; 12. a; 13. b; 14. d; 15. c; 16. a; 17. d; 18. b; 19. c; 20. e.

20

CLASSIFICATION OF LIFE'S DIVERSITY

SYSTEMATICS
TAXONOMY
    Recognizing Species
    The Linnean Scheme
PHYLOGENY
    Three Interpretive Approaches
    Portraying Relationships Among Organisms

Doing Science: Constructing a Family Tree
Called a Cladogram
CLASSIFICATION
    Classification as a Retrieval System
    Five-Kingdom Classification

Interactive Exercises

SYSTEMATICS (20-I, pp. 319–320)

1. Explain the objectives of a branch of biology called systematics._____

    _____

2. How do microevolution and macroevolution deal with the process and the patterns of evolution?

    _____

    _____

    _____

3. In the reconstruction of evolutionary history, only two sources of information are available; list them.

    _____

Fill-in-the-Blanks

(4)_____ is the process of categorizing phylogenetic information into a retrieval system consisting of

many hierarchical levels, or ranks. Identifying organisms and assigning names to them comprise the

activity called (5)_____. The goal of (6)_____ _____ is to identify evolutionary patterns that

unite organisms. These three fields of inquiry are the means by which (7)_____ evaluates patterns of

diversity.

TAXONOMY (20-II, pp. 320–322)

Matching

Choose the single most appropriate letter.

1. ___ black bear
2. ___ lineage
3. ___ binomial system
4. ___ *Pinus strobus* and *Pinus banksiana*
5. ___ species
6. ___ genus
7. ___ family
8. ___ taxa
9. ___ species of the same lineage
10. ___ Linnaeus

A. A group of similar species
B. The taxon most often studied
C. Groups of organisms evolving independently of other such organisms
D. Share evolutionary history
E. Classification level that includes similar genera
F. Originated the modern practice of providing two scientific names for organisms
G. An example of a common name
H. Species belonging to one genus
I. Species connected by genetic contact in the space of the environment and through time
J. System of assigning a two-part Latin name to species

PHYLOGENY (20-III, pp. 323–324)

1. Complete the following table to summarize three interpretive approaches to phylogeny.

Phylogeny Approach	Description
a. Evolutionary systematics	
b. Phenetics	
c. Cladistics	

2. What are monophyletic lineages?_____

_____

3. Define a phylogeny._____

4. What does phylogenetic reconstruction depend on?_____

5. Describe the construction of a cladogram._____

_____

6. What is meant by a derived trait?_____

_____

7. What is the essential information portrayed by a cladogram?_____

_____

_____

8. Study the cladogram of seven taxa below; the short numbered bars on the stem of the cladogram represent shared derived traits. Various taxa are indicated by letters. Answer the questions that follow the cladogram.

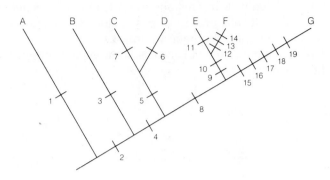

a. On the cladogram above, what types of information could be represented by the numbered traits?

_____

b. Which derived trait is shared by taxa EFG?_____

c. Which derived trait is shared by taxa CDEFG?_____

d. Which is the unique derived trait shared by taxa C and D?_____

e. What does it mean if some taxa are closer together on the cladogram than others?_____

_____

f. What is an outgroup?_____

g. Which taxon on the cladogram represents the outgroup condition?_____

h. What is a monophyletic taxon?_____

_____

i. For the cladogram above, list the monophyletic taxa shown as well as shared derived traits._____

_____

CLASSIFICATION (20-IV, pp. 324–328)

1. Give some reasons why the knowledge of phylogeny is very valuable to researchers._____

_____

_____

_____

2. Arrange the following jumbled taxonomic categories in proper order, with the most inclusive listed

first: family, class, species, kingdom, genus, phylum (or division), and order._____

_____

3. Arrange the following jumbled taxa and their categories in proper order, with the most inclusive listed first: genus: *Archibaccharis*; kingdom: Plantae; order: Asterales; species: *lineariloba*; class:

Dicotyledonae; division: Anthophyta; family: Asteraceae._____

_____

Matching

Choose the one best answer for each.

4. ___ Monera   A. Multicelled heterotrophs that feed by extracellular digestion and absorption

5. ___ Protista   B. Single-celled prokaryotes, some autotrophs, others heterotrophs
       C. Diverse multicelled heterotrophs, including predators and parasites

6. ___ Fungi   D. Multicelled photosynthetic autotrophs

7. ___ Plantae   E. Diverse single-celled eukaryotes, some photosynthetic autotrophs, many heterotrophs

8. ___ Animalia

Chapter Terms

The following page-referenced terms are important; they were in boldface type in the chapter. Refer to the instructions given in Chapter 1, p. 4.

phylogeny (318)   classification (320)   cladistics (323)   Monera (328)
systematics (320)   taxa (320)   monophyletic (323)   Protista (328)
taxonomy (320)   lineage (320)   homologous structures (324)   Fungi (328)
phylogenetic   binomial system (322)   cladogram (324)   Plantae (328)
    reconstruction (320)   genus (322)   classification (327)   Animalia (328)

Self-Quiz

The evolutionary history of a group that considers all ancestors and descendants is its (1)_____. When evolutionary relationships are viewed as a branching tree, a branch represents a single line of descent or (2)_____. Evolution can be viewed either by the (3)_____ model, which envisions speciation as accounting for only a small amount of large-scale change, or by the (4)_____ model, which sees higher taxa originating from the rapid crossing of adaptive thresholds. Linnaeus devised a two-name system for cataloging diverse life forms; the first name is the (5)_____ name, and the second name is the (6)_____ name. Systems of classification begin with these two names and are based on several categories, such as families and orders, that become increasingly more (7)_____ until all members are included in one of the five (8)_____ of life.

___ 9. Phylogeny is _____.
- a. identifying organisms and assigning names to them
- b. producing a "retrieval system" consisting of several levels
- c. an evolutionary history of organisms, both living and extinct
- d. a study of the adaptive responses of various organisms

___ 10. Taxonomy is defined as _____.
- a. identifying organisms and assigning names to them
- b. producing a "retrieval system" consisting of several levels
- c. an evolutionary history of organisms, both living and extinct
- d. a study of the adaptive responses of various organisms

___ 11. Activities associated with classification are best defined as _____.
- a. identifying organisms and assigning names to them
- b. producing a "retrieval system" consisting of several levels
- c. an evolutionary history of organisms, both living and extinct
- d. a study of the adaptive responses of various organisms

___ 12. All species that are connected by genetic contact in the space of the environment and through time represent a _____.
- a. population
- b. kingdom
- c. phylum
- d. lineage

___ 13. Correct identification of an organism often requires knowledge of _____.
   a. anatomy
   b. biochemistry
   c. physiology
   d. ecology and behavior
   e. all of the above

___ 14. The significant contribution of Linnaeus was to develop _____.
   a. the idea that organisms should be placed in two kingdoms
   b. the binomial system
   c. cladograms
   d. the idea that a species could have several common names

___ 15. The process of grouping organisms according to similarities derived from a common ancestor is known as _____.
   a. taxonomy
   b. classification
   c. phylogenetic systematics
   d. cladistics
   e. both c and d

___ 16. Independently evolving lineages that share a common evolutionary heritage are termed _____.
   a. phenetic
   b. polyphyletic
   c. monophyletic
   d. homologous

___ 17. Phylogenetic reconstruction depends on the identification of _____.
   a. direct information about ancestors and descendants
   b. analogous characters
   c. adaptive responses of organisms
   d. homologous characters

___ 18. All living organisms have eukaryotic cell structure except the _____.
   a. Animalia
   b. Plantae
   c. Fungi
   d. Monera
   e. Protista

Chapter Objectives/Review Questions

This section lists general and detailed chapter objectives that can be used as review questions. You can make maximum use of these items by writing answers on a separate sheet of paper. Fill in answers where blanks are provided. To check for accuracy, compare your answers with information given in the chapter or glossary.

Page	Objectives/Questions
(320)	1. Identifying organisms and assigning names to them is a field of systematic inquiry known as _____.
(320)	2. The goal of _____ reconstruction is to identify the evolutionary patterns that unite different organisms.
(320)	3. The task of _____ is to categorize phylogenetic information into a "retrieval system" consisting of many hierarchical levels, or ranks.
(320)	4. What is meant by "all species of the same lineage"?
(321–322)	5. List some of the problems encountered in attempting to identify a species.
(321–322)	6. Why doesn't outward appearance alone provide enough information to assign an organism to a particular species?
(322)	7. Linnaeus developed the _____ system, by which species are assigned a two-part Latin name.
(322)	8. Explain the purpose of applying a genus name and a species name to an organism.
(322)	9. Give reasons why using common names alone for organisms is often very unsatisfactory.
(323)	10. Compare and contrast the concepts of evolutionary systematics, phenetics, and cladistics with respect to their methods of grouping organisms.
(323)	11. A _____ lineage is one that evolves independently and whose organisms share a common ancestry.
(324)	12. Phylogenetic reconstruction depends on _____ structures.

(324–325) 13. Be able to explain the construction of a cladogram and the information it represents.
(324–325) 14. Be able to interpret a simple cladogram; define derived trait and outgroup.
(324–326) 15. Give reasons why a knowledge of phylogeny is useful to researchers.
(326–327) 16. Be able to arrange the series of ever more inclusive taxa of a classification system in proper order: kingdom, phylum (or division), class, order, family, genus, species.
(328)   17. Single-celled prokaryotes belong to kingdom _____ ; diverse, single-celled eukaryotes are placed in kingdom _____ ; kingdom _____ contains multicelled heterotrophs that feed by extracellular digestion; multicelled photosynthetic autotrophs are in kingdom _____ ; diverse multicelled heterotrophs that include parasites are classified in kingdom _____ .

Integrating and Applying Key Concepts

Water crowfoot plants (family Ranunculaceae) often have two or more distinct leaf types within the same species and on the same plant. One type is capillary (hairlike), and is found submerged in the water or growing well above the water. The other type is laminate (flat and expanded) and is found floating on the water or submerged. In some *Ranunculus* species, three different leaf shapes form on a plant in a sequence. Suppose two taxonomists describe the same crowfoot plant as being two different and distinct species due to these two different leaf shapes. Can you think of difficulties that might arise when other researchers are required to refer to these "two species" in the course of their work?

Critical Thinking Exercises

1. Which of the following would best test the hypothesis that two populations of very similar squirrels, isolated from contact with each other, are different species?

   a. Examine many individuals to determine whether there are any morphological differences between the two populations.
   b. Examine DNA from many individuals to assess genetic differences between the two populations.
   c. Examine the environments in the two locations to determine whether the same selection pressures would be operating.
   d. Capture live individuals from both populations and determine whether they can crossbreed and produce viable, fertile offspring.
   e. Look for hybrids between the two populations in the wild.

ANALYSIS

   a. Morphological differences are not a reliable guide to separating species. Most species have a variety of characteristics within the species, and many species are very similar.
   b. Even DNA sequences are not necessarily species-specific. Individual humans can be identified on the basis of their DNA sequences. There is a great deal of genetic diversity within species.
   c. This might lead to predictions that certain characteristics will be similar in the two populations, but it does not define species.
   d. Reproductive isolation is the factor that separates two species. If the two populations interbreed, they are classifiable as a single species. If they do not, it may be that they are two separate species, or it may be that they will not reproduce in captivity.
   e. Because the two populations are spatially isolated, there can be no hybrids.

2. An enzyme was purified from eight organisms, and the amino acid sequence was determined from the same region of each. The region that was studied was forty-five amino acids long. The organism labeled Mx162 had a certain amino acid sequence in its enzyme. Each of the other seven organisms was scored for the number of amino acids it had in common with Mx162. For example, Mx162 has a certain amino acid at position 31; an organism that has the same amino acid at position 31 is scored one point. The data follow.

Organism	Number of Amino Acids in Common with Mx162
Copia	6
EcA	26
EcB	26
Gypsy	12
HIV	15
MtP	14
Typ912	8

Using these data to indicate similarity of organisms, prepare a cladogram to express the evolutionary relationships of the organisms.

ANALYSIS

The simplest form of cladogram prepared from these data would look something like A below. Alternatively, you could make more elaborate assumptions and cluster the lines of descent as shown in B. To do this more rigorously, you would need information on the degrees of similarity between other pairs of organisms. Now evaluate your cladogram with the following additional information.

Symbol	Source of Enzyme
Copia	*Drosophila*
EcA	*E. coli* strain A
EcB	*E. coli* strain B
Gypsy	*Drosophila*
HIV	Human Immunodeficiency Virus—causes AIDS
MtP	Mammalian mitochondria
Mx162	*Myxococcus xanthus*—a primitive bacterial species
Typ912	Yeast

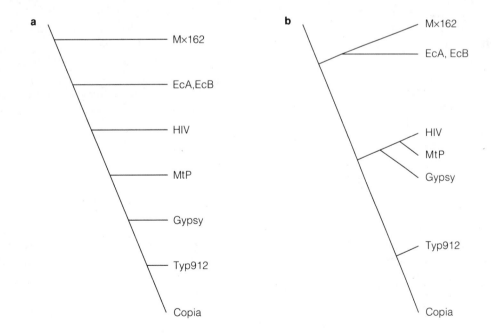

3. The table below presents some of the characteristics of twenty-six species of American coniferous trees. Classify the species into genera and families, using these observations. Do not be concerned about names; simply gather like species into groups representing genera, and like genera into groups representing families. Make as many groups as you think the observations justify.

No.	Cones				Leaves				Remarks
No.	*Texture*	*Shape*	*Grouping*	*Position*	*Type*	*Position*	*Number*	*Cross-section*	*Remarks*
1	wo	cy	so	pe	ne	sp	5	rd	Needles ridged and blunt
2	le	ov	cl	tr	sc	op			
3	le	ob	cl	pe	sc	op			Leaf tips curved inward, cones toothed
4	le	ov	cl	tr	sc	op			
5	pa	cy	so	er	ne	sp	1	fl	Blunt needles with two ribs
6	wo	ov	so	pe	ne	sp	2	rd	Needles ridged, sharp-tipped
7	wo	ov	so	tr	sc	sp			
8	pa	cy	so	pe	ne	sp	1	sq	Needles sharp-tipped, ridged
9	wo	cy	so	pe	ne	sp	5	rd	Needles stiff, sharp-tipped, twisted
10	wo	ob	so	pe	ne	sp	3	rd	Needles stiff
11	wo	cy	so	pe	ne	sp	2	ob	Needles stout and twisted
12	le	ov	so	tr	sc	op			Leaves with toothed edges
13	pa	cy	so	pe	ne	sp	1	fl	Needles flexible with rounded tip
14	wo	ob	so	pe	ne	sp	5	rd	Needles stiff
15	le	ov	cl	tr	sc	op			
16	wo	ob	so	tr	sc	sp			Scales sharp-pointed
17	wo	ob	cl	pe	sc	op			Scales pointed
18	pa	cy	so	pe	ne	sp	1	sq	Needles sharp-tipped and ridged
19	pa	cy	so	er	ne	sp	1	fl	Needles blunt and flexible
20	pa	cy	so	er	ne	sp	1	fl	Needles blunt and flexible
21	wo	ob	cl	pe	sc	op			Scales triangular and sharp
22	wo	ob	so	pe	ne	sp	3	rd	Needles stiff
23	wo	ob	so	pe	ne	sp	1	rd	Needles stiff, sharp-tipped, curved
24	le	ov	cl	tr	ne	wh			Needles awl-shaped
25	wo	ob	so	pe	ne	op	2	fl	Needles flexible, round tips, grooved
26	wo	ov	so	pe	ne	sp	5	rd	Needles ridged, twigs flexible

Abbreviations:

Cone texture: le = leathery, pa = papery, wo = woody
Cone shape: ob = oblong, ov = ovoid (almost spherical), cy = cylindrical
Cone grouping: cl = clustered, so = solitary
Cone position: er = erect (pointing upward from twig), pe = pendulous (pointing downward from twig),
        tr = transverse (pointing horizontally from twig)
Leaf type: ne = needlelike, sc = scalelike
Leaf position: (refers to arrangement of leaves along the twig) op = opposite, sp = spiral, wh = whorls
Leaf number: refers to the number of needles in a bundle sharing one attachment to the twig; not applicable
        to scalelike or awl-shaped leaves
Leaf cross-section: fl = flat, ob = oblong, rd = round, sq = square

ANALYSIS

The most efficient way to approach this problem is to photocopy the table and cut the copy into strips with one species (one row) on each slip. Then choose a characteristic that you think is the most important and arrange the strips in groups according to that characteristic. Then rearrange the strips within each group according to another characteristic. Eventually, you will perceive certain patterns. The final scheme you devise depends on which characteristic you choose as being more important than others. With the limited information given in the table, it might be very difficult to duplicate the commonly accepted version below. The importance of this exercise lies in working with a variety of observations to make a rational classification, not in "learning" the current version of conifer classification. The following table is not the *right* classification or a fact itself; it is better than yours only in that it is based on more facts.

Family	Species	Common Name	Number
Cupressaceae	*Calocedrus decurrens*	Incense cedar	3
	Cupressus arizonica	Arizona cypress	21
	C. macrocarpa	Monterey cypress	17
	Juniperus communis	Common juniper	24
	J. deppeana	Alligator juniper	4
	J. osteosperma	Utah juniper	12
	J. scopulorum	Rocky Mountain juniper	15
	Thuja plicata	Giant cedar	2
Taxodiaceae	*Sequoia sempervirens*	Coastal redwood	16
	Sequoiadendron gigantium	Giant sequoia	7
Pinaceae	*Abies concolor*	White fir	19
	A. lasiocarpa	Subalpine fir	20
	A. magnifica	Red fir	5
	Picea engelmannii	Engelmann spruce	18
	P. pungens	Blue spruce	8
	Pinus contorta	Lodgepole pine	11
	P. edulis	Two-needle pinyon	6
	P. flexilis	Limber pine	26
	P. lambertiana	Sugar pine	9
	P. longaeva	Bristlecone pine	1
	P. monophylla	One-needle pinyon	23
	P. ponderosa	Ponderosa pine	22
	P. sabiniana	Digger pine	10
	P. torreyana	Torrey pine	14
	Pseudotsuga menziesii	Douglas fir	13
	Tsuga heterophylla	Western hemlock	25

Answers

Answers to Interactive Exercises

SYSTEMATICS (20-I)
1. Systematics assesses the patterns of organismal diversity.
2. Microevolution deals primarily with the process of evolution (the origin and success of given genotypes and their associated phenotypes); macroevolution deals with large-scale patterns of diversity through time.
3. Observations of living organisms and the available fossil record.
4. Classification; 5. taxonomy; 6. phylogenetic reconstruction; 7. systematics.

TAXONOMY (20-II)
1. G; 2. I; 3. J; 4. H; 5. B; 6. A; 7. E; 8. C; 9. D; 10. F.

PHYLOGENY (20-III)
1. a. Reconstructs the tree of life by looking in a subjective way at both the similarities and the differences between organisms. Fossils were important in order to identify ancestors of living forms from the fossil record. Two researchers with the same data could arrive at different conclusions. b. Organisms are grouped according to their similarities. The problem is that similarities are not always due to a shared evolutionary history; cases of morphological convergence could deceive a researcher. c. Organisms are grouped according to similarities that are derived from a common ancestor. 2. Monophyletic lineages are lineages that share a common evolutionary heritage; 3. A phylogeny is a pattern of evolutionary relationships; 4. Phylogenetic reconstruction depends

on the identification of homologous structures; 5. The cladogram is a branching diagram (a type of family tree) that represents the patterns of relationships of organisms, based on their derived traits; 6. A derived trait is any evolutionary change that deviates from an outgroup characteristic; 7. Cladograms portray relative relationships among organisms. Distribution patterns of many homologous structures are examined. Taxa closer together on a cladogram share a more recent common ancestor than those that are farther apart. Cladograms do not convey direct information about ancestors and descendants; 8. a. The numbered traits could be discrete morphological, physiological, and behavioral traits. b. trait 8; c. trait 4; d. trait 5; e. Traits that are closer together share a more recent ancestor than those that are farther apart. f. An outgroup is the cladogram taxon exhibiting the fewest derived characteristics. g. taxon A; h. A monophyletic taxon is one in which all members have the same derived traits and share a single node on the cladogram; i. The monophyletic taxa and their derived traits are BCDEFG (derived trait 2), CDEFG (derived trait 4), CD (derived trait 5), EFG (derived trait 8), EF (derived traits 9 and 10).

CLASSIFICATION (20-IV)

1. Researchers arrive at generalizations after comparing similar attributes among a variety of organisms. Inferences about one organism can be based on observations or experiments with another organism that is closely related and is apt to have similar responses. For example, testing of antibiotics for humans can be done on close relatives of humans that will respond similarly as seen through observations or testing situations. 2. kingdom, phylum (or division), class, order, family, genus, species. 3. kingdom: Plantae; division: Anthophyta; class: Dicotyledonae; order: Asterales; family: Asteraceae; genus: *Archibaccharis*; species: *lineariloba*. 4. B; 5. E; 6. A; 7. D; 8. C.

Answers to Self-Quiz

1. phylogeny; 2. lineage; 3. gradualism; 4. punctuation; 5. genus; 6. species; 7. inclusive; 8. kingdoms; 9. c; 10. a; 11. b; 12. d; 13. e; 14. b; 15. e; 16. c; 17. d; 18. d.

21

HUMAN EVOLUTION: A CASE STUDY

THE MAMMALIAN HERITAGE
THE PRIMATES
      Primate Classification
      Trends in Primate Evolution
         Locomotion
         Modification of Hands
         Enhanced Daytime Vision
         Changes in Dentition
         Brain Expansion and Elaboration
         Behavioral Evolution

PRIMATE ORIGINS
THE HOMINIDS
      Australopiths
      Stone Tools and Early *Homo*
      Homo erectus
      Homo sapiens
      Doing Science: Mitochondria and Human Evolution

Interactive Exercises

THE MAMMALIAN HERITAGE / THE PRIMATES / PRIMATE ORIGINS (21-I, pp. 331–336)

Fill-in-the-Blanks

(1)_____ are warm-blooded animals with hair that began their evolution more than 200 million years ago. Humans are (2)_____ ; members of this order rely less on their sense of (3)_____ and more on daytime vision. Their (4)_____ became larger and more complex; this trend was accompanied by refined technologies and the development of (5)_____: the collection of behavior patterns of a social group, passed from generation to generation by learning and by symbolic behavior (language). Modification in the (6)_____ led to increased dexterity and manipulative skills. Humans demonstrate remarkable (7)_____: they remain flexible and adapt to a wide range of challenges in unpredictable, complex environments. Changes in primate (8)_____ indicate that there was a shift from eating insects to fruit and leaves and on to a mixed diet. Primates evolved from ancestral mammals more than (9)_____ million years ago, during the Paleocene. The first primates resembled small (10)_____ or tree shrews; they foraged at night for (11)_____ , seeds, buds, and eggs on the forest floor, and they could clumsily climb trees searching for safety and sleep.

    During the (12)_____ era, birds, mammals, and flowering plants rose to dominate Earth's assemblage of organisms. Humans, apes, monkeys, and prosimians are all (13)_____ ; they have excellent (14)_____ perception as a result of their forward-directed eyes, and their fingers and toes are adapted for (15)_____ instead of running. About (16)_____ million years ago, the continents began to assume their current positions, and climates became cooler and (17)_____. Under these conditions, forests began to give way to (18)_____ ; perhaps speciation was favored as subpopulations of apes became reproductively isolated within the shrinking stands of trees. The "tree apes," or (19)_____ , originated during this time; eventually, they ranged throughout Africa, Europe, and (20)_____. Between (21)_____ million and (22)_____ million years ago, some of their descendants may have given rise to the ancestors of modern gorillas, chimpanzees, and humans.

Matching

Fill in the blanks of the evolutionary diagram below with the appropriate letter from the choice list.

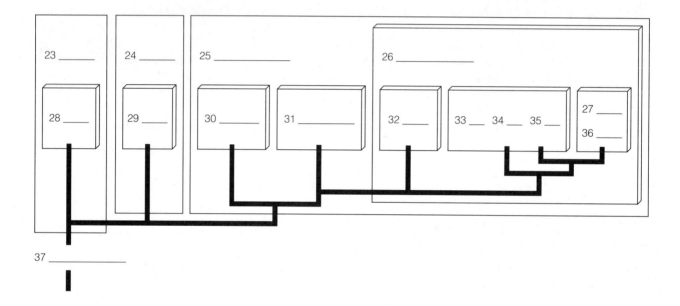

A. Anthropoids
B. Chimpanzee
C. Gibbon, siamang
D. Gorilla
E. Hominids

F. Hominoids
G. Humans and their most recent ancestors
H. Lemurs, lorises

I. New World monkeys (spider monkeys, etc.)
J. Old World monkeys (baboons, etc.)
K. Orangutan

L. Prosimians
M. Tarsiers
N. Tarsioids
O. Rodentlike primate of the Paleocene

THE HOMINIDS (21-II, pp. 337–342)

Fill-in-the-Blanks

The hominid group (which includes all species on the genetic path leading to humans since the time when that path diverged from the path leading to the (1)_____) emerged between (2)_____ million and (3)_____ million years ago, during the late Miocene. (4)_____-million-year-old fossils of humanlike forms have been discovered in Africa, and they all were bipedal and omnivorous and had an expanded brain. Lucy was one of the earliest (5)_____ , a collection of forms that combined ape and human features; they were fully two-legged, or (6)_____ , with essentially human bodies and ape-shaped heads. The oldest fossils of the genus *Homo*, makers of stone tools, date from approximately (7)_____ million years ago. Between 1.5 million and 300,000 years ago, during the Pleistocene periods of glaciation, there were also intermittent periods of warming; during these interglacial times, a larger-brained human species, (8)_____ _____ , migrated out of Africa and into China, Southeast Asia, and Europe. The (9)_____ were a distinct hominid population that appeared about 100,000 years ago in Europe and Asia; their cranial capacity was indistinguishable from our own, and they had a complex culture. By 30,000 years ago, there was only one remaining hominid species: (10)_____ _____ _____. In 1988, geneticists reported that all living humans appeared to be descended from a female who lived in Africa about (11)_____ years ago: the Mitochondrial Eve. To the geneticists who were studying human populations, the (12)_____ mutations in (13)_____ _____ seemed like an excellent choice for constructing a molecular clock. A molecular clock makes sense only when we assume that these mutations accumulate at a (14)_____ rate over evolutionary time. A slow or inconsistent rate would place the time when modern *Homo sapiens* diverged from *Homo erectus* stock at about (15)_____ years ago; a faster, more constant rate would place the ancestor common to both species at about 150,000 years ago.

Matching

Suppose you are a student who wants to be chosen to accompany a paleontologist who has spent forty years teaching and roaming the world in search of human ancestors. Above the desk in her office, she keeps reconstructions of the seven skulls shown below, and you have heard that she chooses the graduate students who accompany her on her summer safaris on the basis of their ability on a short matching quiz. The quiz is presented below. *Without consulting your text (or any other)* while you are taking the quiz, match up all seven skulls with *all* applicable letters. Sixteen correctly placed letters win you a place on the expedition. Fourteen correct answers put you on a waiting list. Thirteen or fewer and she suggests that you may wish to investigate dinosaur fossils instead of human ancestry. Which will it be for you?

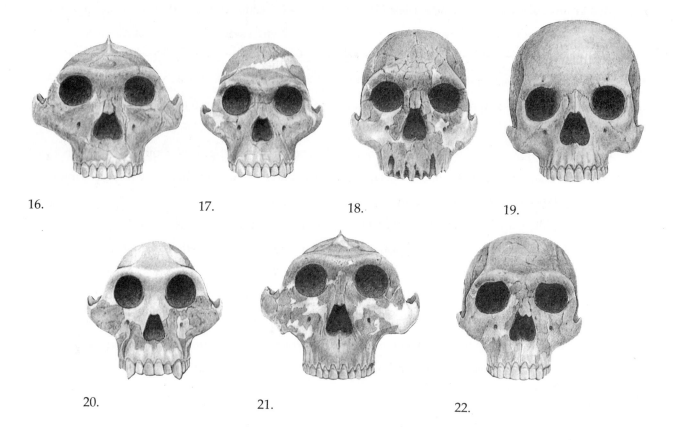

16.   17.   18.   19.

20.   21.   22.

16. _____
17. _____
18. _____
19. _____
20. _____
21. _____
22. _____

A. gracile forms of australopiths
B. robust forms of australopiths
C. fashioned stone tools and used them first
D. has a chin
E. first controlled use of fire
F. lived 4 million years ago
G. from a lineage that lived from approximately 3 million years ago until about 1.25 million years ago
H. lived 2 million years ago
I. lived 1.5 million years ago until at least 100,000 years ago
J. The most recent hominid to appear

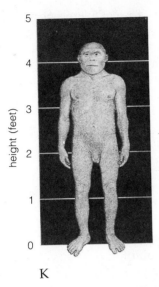

K

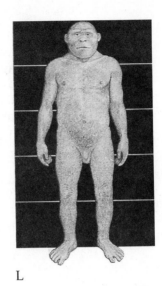

L

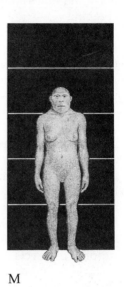

M

height (feet)

Chapter Terms

The following page-referenced terms are important; they were mostly in boldface type in the chapter. Refer to the instructions given in Chapter 1, p. 4.

dentition (332)
prosimians (332)
anthropoids (332)
hominoids (333)
hominids (333)

primate (334)
bipedalism (334)
prehensile (335)
opposable (335)
culture (335)

Aegyptopithecus (336)
"dryopiths" (336)
plasticity (337)
australopiths (338)
early *Homo* (339)

Homo erectus (340)
Homo sapiens (341)
Neandertal (341)
Mitochondrial DNA (341)

Self-Quiz

Multiple-Choice

Select the best answer.

___ 1. Which of the following is *not* considered to have been a key character in early primate evolution?
   a. eyes adapted for discerning color and shape in a three-dimensional field
   b. body and limbs adapted for tree climbing
   c. bipedalism and increased cranial capacity
   d. eyes adapted for discerning movement in a three-dimensional field

___ 2. Primitive primates generally live _____.
   a. in tropical and subtropical forest canopies
   b. in temperate savanna and grassland habitats
   c. near rivers, lakes, and streams in the East African Rift Valley
   d. in caves where there are abundant supplies of insects

___ 3. All the ancestral placental mammals apparently arose from ancestral forms of a group _____.
   a. that includes omnivorous shrews and moles
   b. that includes dogs, cats, and seals
   c. that includes mice and beavers
   d. that includes the koala ("teddy bear") and flying phalanger (resembles the flying squirrel)

___ 4. The hominid evolutionary line stems from a divergence (fork in a phylogenetic tree) from the ape line that apparently occurred _____.
   a. somewhere between 8 million and 5 million years ago
   b. about 3 million years ago
   c. during the Pliocene epoch
   d. less than 2 million years ago

___ 5. _____ was an Oligocene anthropoid that probably predated the divergence leading to Old World monkeys and the apes, with dentition more like that of dryopiths and less like that of the Paleocene primates with rodentlike teeth.
   a. *Aegyptopithecus*
   b. *Australopithecus*
   c. *Homo erectus*
   d. *Plesiadapis*

___ 6. Johanson's Lucy was a(n) _____.
   a. dryopith
   b. australopith
   c. Miocene
   d. prosimian

___ 7. A hominid of Europe and Asia that became extinct nearly 30,000 years ago was _____.
   a. a dryopith
   b. *Australopithecus*
   c. *Homo erectus*
   d. Neandertal

Matching

Choose the one most appropriate answer for each.

8. ___ anthropoids

9. ___ australopiths

10. ___ Cenozoic

11. ___ dentition

12. ___ dryopiths

13. ___ hominids

14. ___ hominoids

15. ___ Miocene

16. ___ Neandertal

17. ___ plasticity

18. ___ Pliocene

19. ___ primates

20. ___ prosimians

A. A group that includes apes and humans

B. A population of *Homo sapiens* that lived from at least 100,000 to as recently as 30,000 years ago; tool users and artisans

C. The ability to adapt to a wide range of demands

D. Organisms in a suborder that includes New World and Old World monkeys, apes, and humans

E. An era that began 65 to 63 million years ago; characterized by the evolution of birds, mammals, and flowering plants

F. The type, number, and size of teeth

G. A group that includes humans and their most recent ancestors

H. An epoch of the Cenozoic era lasting from 25 million to 17 million years ago; characterized by the appearance of primitive apes, whales, and grazing animals of the grasslands

I. Organisms in a suborder that includes tree shrews, lemurs, and others

J. An epoch of the Cenozoic era lasting from 7 million to 2 million years ago; characterized by the appearance of distinctly modern plants and animals

K. A group that includes prosimians, tarsioids, and anthropoids.

L. Bipedal organisms living from about 4 million to 1 million years ago, with essentially human bodies and ape-shaped heads; brains no larger than those of chimpanzees

M. Transitional apelike forms that could climb about in trees and walk on the ground

Chapter Objectives/Review Questions

This section lists general and detailed chapter objectives that can be used as review questions. You can make maximum use of these items by writing answers on a separate sheet of paper. Fill in answers where blanks are provided. To check for accuracy, compare your answers with information given in the chapter or glossary.

Page   *Objectives/Questions*

(332–334)   1. Describe the general physical features and behavioral patterns attributed to early primates.

(336)   2. Trace primate evolutionary development through the Cenozoic era. Describe how Earth's climates were changing as primates changed and adapted. Be specific about times of major divergence.

(334–341)   3. State which anatomical features underwent the greatest changes along the evolutionary line from early anthropoids to humans.

(333)   4. Where do prosimian survivors dwell today on Earth?

(334–335)   5. The two *key* characters of primate evolution are _____ and _____.

(332)   6. Beginning with the primates most closely related to humans, list the main groups of primates in order by decreasing closeness of relationship to humans.

(340–341)   7. Explain how you think *Homo sapiens sapiens* arose. Make sure your theory incorporates existing paleontological (fossil), biochemical, and morphological data.

Integrating and Applying Key Concepts

Suppose someone told you that some time between 12 million and 6 million years ago dryopiths were forced by larger predatory members of the cat family to flee the forests and take up residence in estuarine, riverine, and sea coastal habitats where they could take refuge in the nearby water to evade the tigers. Those that, through mutations, became naked, developed an upright stance, developed subcutaneous fat deposits as insulation, and developed a bridged nose that had advantages in watery habitats (features that other dryopiths that remained inland never developed) survived and expanded their populations. As time went on, predation by the big cats and competition with other animals for available food caused most of the terrestrial dryopiths to become extinct, but the water-habitat varieties survived as scattered remnant populations, adapting to easily available shellfish and fish, wild rice and oats, and various tubers, nuts, and fruits. It was in these aquatic habitats that the first food-getting tools (baskets, nets, and pebble tools) were developed, as well as the first words that signified different kinds of food. How does such a story fit with current speculations about and evidence of human origins? How could such a story be shown to be true or false?

Critical Thinking Exercises

1. The brain of the members of the human lineage enlarged dramatically as the genus *Homo* evolved. Which of the following is the strongest evolutionary explanation for this phenomenon?

   a. The lineage developed large brains in order to produce humans.
   b. The prehuman ancestors developed large brains because they needed them.
   c. The early humans developed large brains in order to use their new bipedal locomotion, stereoscopic vision, and opposable thumb.
   d. Early humans developed large brains because they had begun to develop culture.
   e. Because early humans practiced cooperative hunting, individuals with larger brains were more successful and left more offspring.

ANALYSIS

   a. Evolution has no predetermined outcome or "goal." It proceeds according to the dictates of selective pressures applied by the environment and to the historical accidents of mutation, genetic drift, and extinction.
   b. Evolution does not proceed according to the "needs" of the population. Selection is a pragmatic force that favors whatever accidental developments reproduce more rapidly.
   c. Again, this statement implies a predetermined direction or purpose for evolution. Evolution has none.
   d. This is a sort of evolutionary explanation. It identifies a force that favors mutations of a certain type. However, it is a very general and vague statement, and it fails to link large brains directly with reproductive success.
   e. This is a clear, specific evolutionary hypothesis. Cooperative hunting would be more successful for individuals who could communicate and do abstract planning of future movements instead of simply reacting as individuals to the immediate situation. Assuming that these abilities are improved by increasing brain size, they would constitute a strong selective force for larger brains.

2. According to the fossil evidence, the human brain reached its present average size about 200,000 years ago and has not increased since. Which of the following is the strongest evolutionary explanation?

   a. Evolutionary development of the human brain was complete at that time.
   b. Human skulls could not enlarge any further to accommodate larger brains.
   c. Individuals with larger brains would have heads too large to allow them to be born.
   d. When humans quit cooperative hunting and took up agriculture, there was no further selection pressure for large brains.
   e. The increasing size of human breeding population groups reduced the relative reproductive success of larger-brained individuals.

ANALYSIS

   a. Evolution is never "complete," because it has no goal or end.
   b. This would stop evolutionary enlargement of brains, but there is no evidence that this statement is true. There is very large variability in the size of the human skull, including developmental abnormalities that cause extreme skull enlargement and cultural practices that grossly distort the skull's growth pattern. The skull bones are apparently sufficiently plastic to accommodate larger brains.

c. This statement assumes that the pelvis cannot evolve to larger size and that the developmental program cannot be delayed even further so that more of the brain's growth takes place after birth.

d. Agriculture requires even greater cooperation and conceptualization than hunting. This selection pressure would still be present.

e. In a small group, the most successful male individual leaves a large percentage of the offspring, sometimes all of them in a polygynous social structure. As the size of the breeding group grows, however, he is unable to control as large a segment of the group, and his reproductive advantage, expressed as the percentage of the group's offspring that are his, diminishes. Assuming that larger brains were the source of the reproductive dominance, increasing populations would then relieve the selective pressure for larger and larger brains.

3. Examine the data in Figure 21.5. Is the trend toward longer periods of infant dependency a real change in the developmental program, or is it simply a result of the trend toward longer life spans?

ANALYSIS

If an animal spends twenty-one days developing before birth and its develomental program is simply slowed so that its life span doubles, it will still spend the same percentage of its life in gestation. The gestation time will double to forty-two days. Calculate the three periods of gestation, infancy, and subadult as a percentage of life span in all these animals. Use the reproductive years rather than total life span for humans. The data become approximately as follows:

Species	Percent of Lifespan Devoted to:		
	Gestation	Infancy	Subadult
Lemur	1.8	5.3	10.5
Macaque	1.8	6.0	12.5
Gibbon	1.9	10.0	16.7
Chimpanzee	1.6	11.8	17.5
Human	1.5	14.0	22.0

Without data on variation among individuals, we cannot assess the probability that the differences among species are significant. However, the data do suggest that the proportion of the life span devoted to prenatal development has not changed. The longer gestation period in "higher" primates is a result of their longer overall life spans. On the other hand, postnatal development apparently has been extended. Another way to say this is that sexual maturity is completed proportionately later in the life span of some existing primates than in that of others. This reflects a real evolutionary change in the developmental program.

Answers

Answers to Interactive Exercises

THE MAMMALIAN HERITAGE/THE PRIMATES/PRIMATE ORIGINS (21-I)

1. Mammals; 2. primates; 3. smell; 4. brains; 5. culture; 6. hands; 7. plasticity; 8. dentition; 9. 60; 10. rodents; 11. insects; 12. Cenozoic; 13. primates; 14. depth; 15. grasping; 16. 23 to 20; 17. drier; 18. grasslands; 19. dryopiths; 20. Eurasia; 21. 10; 22. 6; 23. L; 24. N; 25. A; 26. F; 27. E; 28. H; 29. M; 30. I; 31. J; 32. C; 33. K; 34. D; 35. B; 36. G; 37. O.

THE HOMINIDS (21-II)

1. apes; 2. 8; 3. 5; 4. Four; 5. australopiths; 6. bipedal; 7. 2.5; 8. *Homo erectus*; 9. Neandertals; 10. *Homo sapiens sapiens*; 11. 200,000; 12. neutral; 13. mitochondrial DNA; 14. constant; 15. 500,000; 16. B, G, H; 17. A, G, K; 18. C, H; 19. D, J; 20. A, F, M; 21. B, G, (H), L; 22. E, I.

Answers to Self-Quiz

1. c; 2. a; 3. a; 4. a; 5. a; 6. b; 7. d; 8. D; 9. L; 10. E; 11. F; 12. M; 13. G; 14. A; 15. H; 16. B; 17. C; 18. J; 19. K; 20. I.

22

VIRUSES, BACTERIA, AND PROTISTANS

Interactive Exercises

VIRUSES (22-I, pp. 347–351)

1. a. State the principal characteristics of viruses._____

_____

  b. Describe the structure of viruses._____

_____

  c. Distinguish between the ways viruses replicate themselves._____

_____

_____

_____

_____

2. a. List five specific viruses that cause human illness._____

_____

   b. Describe how each virus in (a) does its dirty work._____

_____

_____

_____

_____

_____

_____

_____

_____

_____

Fill-in-the-Blanks

A(n) (3)_____ is a noncellular, nonliving infectious agent, each of which consists of a central

(4)_____ _____ core surrounded by a protective (5)_____ _____. (6)_____

contain the blueprints for making more of themselves but cannot carry on metabolic activities. Chickenpox

and shingles are two infections caused by DNA viruses from the (7)_____ category. Naked strands or

circles of RNA that lack a protein coat are called (8)_____. (9)_____ are RNA viruses that infect

animal cells, cause AIDS, and follow (10)_____ pathways of replication. (11)_____ are the usual

units of measurement with which to measure viruses, while microbiologists measure bacteria and

protistans in terms of (12)_____. A bacterium 86 micrometers in length is (13)_____ nanometers

long. During a period of (14)_____ , viral genes remain inactive inside the host cell and any of its

descendants. Pathogenic protein particles are called (15)_____.

Identifying

Identify the virus that causes the following illnesses by writing the name of the virus in the blank preceding
the disease.

_____ 16. Common colds

_____ 17. AIDS, leukemia

_____ 18. Fever blisters, chickenpox

Matching

Match each item below with the correct lettered description.

19. ___ antibiotic

20. ___ antiviral drug

21. ___ capsid

22. ___ endemic

23. ___ host

24. ___ pathogen

25. ___ sporadic

26. ___ vector

27. ___ viroid

28. ___ virulence

29. ___ epidemic

A. An agent that transports to other creatures or temporarily houses pathogens coming from an infected organism
B. The part of a virus outside the core
C. Chemical substance that interferes with gene expression or other normal functions of bacteria
D. Severity, degree of malignancy, and rapidity of disease development
E. Organism that lets a pathogen invade and multiply inside its own cells and tissues
F. Acyclovir and AZT, for example
G. Disease abruptly spreads through large portions of a population
H. Disease that breaks out irregularly, affects few people
I. Disease that occurs continuously, but is localized to a relatively small portion of the population
J. Strands or circles of RNA with no protein coat
K. Any disease-causing organism or agent

MONERANS—THE BACTERIA (22-II, pp. 353–360)

Fill-in-the-Blanks

Photosynthetic autotrophs produce their own organic compounds from simple (1)_____ _____ , using (2)_____ as an energy source. Endospore formation by bacteria can kill humans if they enter the food supply and are not killed by high temperature and high (3)_____ . Anaerobic bacteria can live in canned food, reproduce, and produce deadly toxins. Two examples of pathogenic (disease-causing) bacteria that form endospores harmful to humans are (4)_____ _____ and (5)_____ _____ . (6)_____ autotrophs extract energy from inorganic molecules. Eubacterial cell walls are composed of (7)_____ , a substance that never occurs in eukaryotes. Heterocysts are cells in *Anabaena* that carry out (8)_____ _____ . When environmental conditions become adverse, many bacteria form (9)_____ , which resist moisture loss, irradiation, disinfectants, and even acids. Most bacteria reproduce by (10)_____ _____ . Bacteria differ from all other kinds of organisms in being (11)_____ . Spherical bacteria are (12)_____ , rod-shaped bacteria are (13)_____ , and helical bacteria are (14)_____ . (15) Gram-_____ bacteria retain the purple stain when washed with alcohol.

(16)_____ is a colonial organism that straddles the fence between protistans and (17)_____ . (18)_____ is a tiny blastulalike marine animal that may resemble the simple multicelled animals that made their entrance during the Proterozoic. In a (19)_____ , each cell benefits from a loose association with other cells, but each acts independently. In a (20)_____ organism, labor is divided up among the cells, and there is interdependence to the extent that one type of cell cannot exist without the other.

Matching

Match each of the items below with a lowercase letter designating its principal bacterial group and an uppercase letter denoting its best descriptor from the right-hand column.

a. Archaebacteria
b. Chemosynthetic eubacteria
c. Heterotrophic eubacteria
d. Photosynthetic eubacteria
e. None of the above

21. ___ , ___ *Anabaena, Nostoc*

22. ___ , ___ *Bacillus, Clostridium*

23. ___ , ___ *Desulfovibrio*

24. ___ , ___ *Escherichia coli*

25. ___ , ___ *Lactobacillus*

26. ___ , ___ *Methanobacterium*

27. ___ , ___ *Nitrobacter, Nitrosomonas*

28. ___ , ___ *Paramecium*

29. ___ , ___ *Rhizobium, Agrobacterium*

30. ___ , ___ *Rhodospirillum*

31. ___ , ___ *Salmonella, Shigella*

32. ___ , ___ *Spirochaeta, Treponema*

33. ___ , ___ *Staphylococcus, Streptococcus*

34. ___ , ___ *Streptomyces*

35. ___ , ___ *Thermoplasma, Sulfolobus*

A. Lives in anaerobic sediments of lakes and in animal gut; chemosynthetic; used in sewage treatment facilities

B. Purple; generally in anaerobic sediments of lakes or ponds; do not produce oxygen

C. Endospore-forming rods and cocci that live in the soil and in the animal gut; some major pathogens

D. Gram-positive cocci that live in the soil and in the skin and mucous membranes of animals; some major pathogens

E. Gram-positive nonsporulating rods that ferment plant and animal material; some are important in dairy industry; others contaminate milk, cheese

F. In acidic soil, hot springs, hydrothermal vents on seafloor; may use sulfur as a source of electrons for ATP formation

G. A protistan shaped like the sole of a human foot; not a bacterium

H. Gram-negative aerobic rods and cocci that live in soil or aquatic habitats or are parasites of animals and/or plants; some fix nitrogen

I. Nitrifying bacteria that live in the soil, fresh water, and marine habitats; play a major chemosynthetic role in the nitrogen cycle

J. Gram-negative anaerobic rod that inhabits the human colon where it produces vitamin K

K. Major gram-negative pathogens of the human gut that cause specific types of food poisoning

L. Mostly in lakes and ponds; cyanobacteria; produce O_2 from water as an electron donor

M. Sulfur-, sulfate-reducing bacteria that live in the anaerobic muds of bogs and marshes

N. Major producer of antibiotics; an actinomycete that lives in soil and some aquatic habitats

O. Helically coiled, motile parasites of animals; some are major pathogens

THE RISE OF EUKARYOTIC CELLS (22-III, p. 360)

Fill-in-the-Blanks

Comparisons of nucleotide sequences of the (1)_____ and RNA from different species suggest that (2)_____ , eubacteria, and the forerunners of eukaryotes diverged from a common ancestor long before (3)_____ evidence of eukaryotes began accumulating; recent estimates place this divergence as occurring between (4)_____ billion and 2.5 billion years ago. Protistans, plants, fungi, and animals are all (5)_____ ; they are like each other in having a nucleus and (6)_____-_____ organelles. Bacteria have no membrane-bound (7)_____ or nucleus. Lynn Margulis and others believe that the eukaryotic cell arose by the merger of two (8)_____ cells and the maintenance of their (9)_____ arrangement. Although the basic genetic code is identical for all living species (which suggests that the genetic code was in effect before the first cells on Earth diverged into separate evolutionary lines), the (10)_____ of several species follow a slightly altered code. These organelles resemble bacteria in size and structure; each has its own (11)_____ , which is replicated independently of the nuclear DNA. (12)_____ are like photosynthetic bacteria in their metabolism and overall DNA sequences, and they (13)_____ independently of the cell they inhabit.

True-False

____ 14. At present, there is no direct evidence of how either the single-celled eukaryotes or the multicelled eukaryotes originated.

____ 15. Among prokaryotes, purple photosynthetic bacteria resemble chloroplasts strongly, and prochlorobacteria are the most likely candidates to be the ancestors of mitochondria.

Labeling

16. _____   19. _____   22. _____

17. _____   20. _____   23. _____

18. _____   21. _____   24. _____

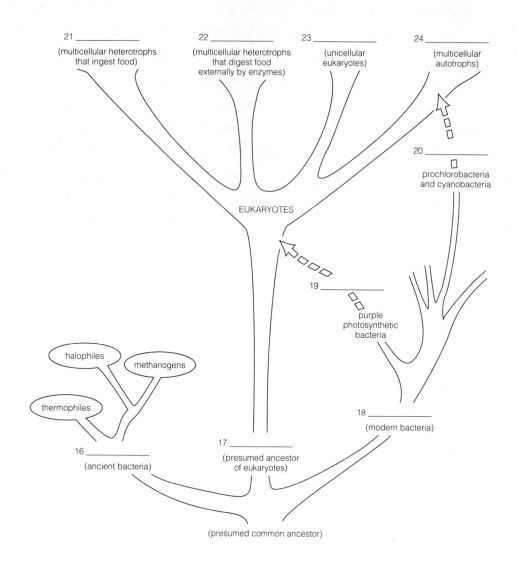

21 _____
(multicellular heterotrophs
that ingest food)

22 _____
(multicellular heterotrophs
that digest food
externally by enzymes)

23 _____
(unicellular
eukaryotes)

24 _____
(multicellular
autotrophs)

20 _____
prochlorobacteria
and cyanobacteria

EUKARYOTES

19 _____
purple
photosynthetic
bacteria

halophiles

methanogens

thermophiles

18 _____
(modern bacteria)

16 _____
(ancient bacteria)

17 _____
(presumed ancestor
of eukaryotes)

(presumed common ancestor)

PROTISTANS (22-IV, pp. 360–370)

Fill-in-the-Blanks

The cells of some (1)_____ _____ differentiate and form (2)_____ _____ , stalked
structures bearing spores at their tips; in this manner, they resemble (3)_____. Some slime-mold
spores resemble the spores of many (4)_____. Slime molds also spend part of their life creeping about
like (5)_____ and engulfing food. Euglenoids reproduce by (6)_____ _____ , a reproductive

mode that is common among all flagellated protistans; the cell grows in circumference while all (7)_____ are being duplicated and then divides along its long axis. Some strains of *Euglena* can be converted from photosynthetic, chloroplast-containing forms to strains that are (8)_____.

(9)_____ undergo explosive population growth and color the seas red or brown, causing a red tide that may kill hundreds or thousands of fish and, occasionally, people. (10)_____ include 450 species of "yellow-green algae," about 500 species of "golden algae," and more than 5,000 species of golden-brown (11)_____. Photosynthetic chrysophytes contain xanthophylls and (12)_____ _____; those pigments mask the green color of chlorophyll in golden algae and diatoms. Diatom cells have external thin, overlapping shells of (13)_____ that fit together like a pill box. 270,000 metric tons of (14)_____ _____ are extracted annually from a quarry near Lompoc, California, and are used to make abrasives, (15)_____ materials, and insulating materials. Dinoflagellates are photosynthetic members of marine (16)_____ and freshwater ecosystems; some forms are also heterotrophic.

Amoebas move by sending out (17)_____, which surround food and engulf it. (18)_____ secrete a hard exterior covering of calcareous material that is peppered with tiny holes through which sticky, food-trapping pseudopods extend. (19)_____ is a famous (20)_____ that causes malaria. When a particular mosquito draws blood from an infected individual, (21)_____ of the parasite fuse to form zygotes, which eventually develop within the mosquito. *Paramecium* is a ciliate that lives in (22)_____ environments and depends on (23)_____ _____ for eliminating the excess water constantly flowing into the cell. *Paramecium* has a (24)_____, a cavity that opens to the external watery world. Once inside the cavity, food particles become enclosed in (25)_____-_____ _____, where digestion takes place.

Matching

Put as many letters in each blank as are applicable.

26. _____ *Entamoeba histolytica*	A. Ciliophora	
27. _____ foraminiferans	B. Mastigophora	
	C. Rhizopoda	
28. _____ *Ptychodiscus brevis*	D. Sporozoa	
29. _____ *Paramecium*	E. Amoeboid protozoans	
30. _____ *Plasmodium*	F. Flagellated protozoans	
	G. Colonial photosynthetic flagellate	
31. _____ *Trichomonas vaginalis*	H. African sleeping sickness	
32. _____ *Trypanosoma brucei*	I. Malaria	
	J. Traveler's diarrhea	
33. _____ *Volvox*	K. Red tide	
	L. Primary component of many ocean sediments	

34.

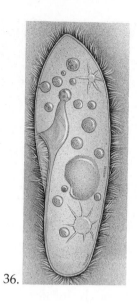

36.

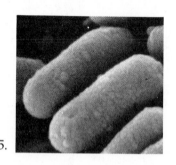

35.

37.

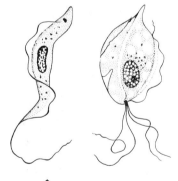

38.

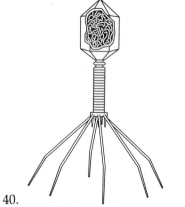

39.

40.

41.

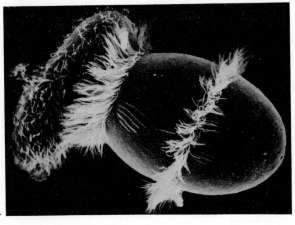
42.

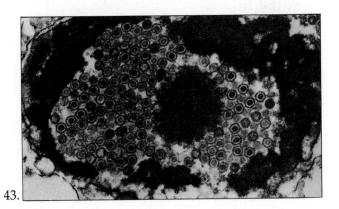
43.

Matching

Match the pictures on page 238 and above with the names below.

34. ___

35. ___

36. ___

37. ___

38. ___

39. ___

40. ___

41. ___

42. ___

43. ___

A. Bacillus
B. Bacteriophage
C. Cyanobacterium
D. Diatom
E. *Didinium,* a ciliate
F. *Euglena*
G. Flagellated protozoans
H. Foraminiferans
I. Herpesvirus
J. *Paramecium*

Chapter Terms

The following page-referenced terms are important; they were mostly in boldface in the chapter. Refer to the instructions given in Chapter 1, p. 4.

virus (347)
capsid (348)
bacteriophage (348)
lytic pathway (349)
temperate pathways (349)
latency (349)
eubacteria (353)
archaebacteria (353)
Gram stain (354)
binary fission (355)
plasmid (355)

methanogens (357)
extreme halophiles (357)
extreme thermophiles (357)
photosynthetic
    eubacteria (358)
heterocysts (358)
chemosynthetic
    eubacteria (358)
heterotrophic
    eubacteria (358)
endospore (359)

fruiting bodies (360)
eukaryotes (360)
protistans (360)
symbiosis (361)
cellular slime molds (362)
plasmodial slime molds (362)
euglenoids (364)
longitudinal fission (364)
chrysophytes (364)
dinoflagellates (365)
plankton (365)

red tides (365)
protozoans (366)
flagellated protozoans (366)
cysts (367)
amoeboid protozoans (367)
amoebas (367)
foraminiferans (367)
radiolarians (367)
heliozoans (367)
ciliated protozoans (367)
sporozoans (368)

Self-Quiz

Multiple-Choice

Select the best answer.

___ 1. Which of the following diseases is *not* caused by a virus?
   a. smallpox
   b. polio
   c. influenza
   d. syphilis

___ 2. Bacteriophages are _____.
   a. viruses that parasitize bacteria
   b. bacteria that parasitize viruses
   c. bacteria that phagocytize viruses
   d. composed of a protein core surrounded by a nucleic acid coat

___ 3. Which of the following specialized structures is not correctly paired with a function?
   a. gullet—ingestion
   b. cilia—food gathering
   c. contractile vacuole—digestion
   d. anal pore—waste elimination

___ 4. _____ form a group of related organisms that suggests how lineages of single-celled organisms might have progressed through a colonial stage to multicellularity.
   a. Ciliates such as *Paramecium, Didinium,* and *Vorticella*
   b. Volvocales such as *Chlamydomonas* and *Volvox*
   c. Golden algae and diatoms
   d. Sporozoans such as *Plasmodium, Neisseria,* and the spirochetes

___ 5. Population "blooms" of _____ cause "red tides" and extensive fish kills.
   a. *Euglena*
   b. specific dinoflagellates
   c. diatoms
   d. *Plasmodium*

___ 6. Exposure to free oxygen is lethal for all _____.
   a. obligate anaerobes
   b. bacterial heterotrophs
   c. chemosynthetic autotrophs
   d. facultative anaerobes

___ 7. For an organism to be considered truly multicellular, _____.
   a. its cells must be heterotrophic
   b. there must be division of labor and cellular specialization
   c. the organisms cannot be parasitic
   d. the organisms must at least be motile

___ 8. _____ all transform energy into usable forms and have complete genetic systems that maintain and reproduce themselves; hence, they are unquestionably alive.
   a. Bacteria, blue-green algae, and prions
   b. Bacteria and viroids
   c. Bacteria and rickettsias
   d. Bacteria and viruses

___ 9. When nutrients are scarce, many bacteria _____.
   a. engage in conjugation
   b. switch to photosynthesis
   c. form endospores
   d. become pathogenic

___ 10. Which of the following play an important role in the cycling of nitrogen-containing substances?
   a. cyanobacteria
   b. prions
   c. viruses
   d. photosynthetic flagellates

Matching

Match all applicable letters with the appropriate terms. A letter may be used more than once, and a blank may contain more than one letter.

11. _____ *Amoeba proteus*

12. _____ *Anabaena, Nostoc*

13. _____ *Clostridium botulinum*

14. _____ diatoms

15. _____ *Dictyostelium*

16. _____ *Escherichia coli*

17. _____ foraminifera

18. _____ *Ptychodiscus brevis* (red tide)

19. _____ *Herpesvirus*

20. _____ HIV

21. _____ *Lactobacillus*

22. _____ *Paramecium*

23. _____ *Plasmodium*

24. _____ *Staphylococcus*

25. _____ *Volvox*

A. Bacteria
B. Protista
C. Virus
D. Slime mold
E. Cyanobacteria
F. Photosynthetic flagellates
G. Dinoflagellates
H. Gram-positive eubacteria
I. Obtain food by using pseudopodia
J. Causes malaria
K. A sporozoan
L. A ciliate
M. Cause cold sores and a type of venereal disease
N. Live in "glass" houses
O. Live in hardened shells that have thousands of tiny holes, through which pseudopia protrude
P. Associated with AIDS, ARC

Chapter Objectives/Review Questions

This section lists general and detailed chapter objectives that can be used as review questions. You can make maximum use of these items by writing answers on a separate sheet of paper. Fill in answers where blanks are provided. To check for accuracy, compare your answers with information given in the chapter or glossary.

Page	Objectives/Questions
(348; 353–355)	1. Describe the principal body forms of monerans (inside and outside).
(348–349; 357–359)	2. Explain how, with no nucleus or membrane-bound organelles, monerans reproduce themselves and obtain energy to carry on metabolism.
(358)	3. Distinguish chemosynthesis from photosynthesis.
(357–358)	4. State the ways in which archaebacteria differ from eubacteria.
(362)	5. Trace a sequence of events that may have transformed heterotrophic prokaryotes into photosynthetic eukaryotes.
(369)	6. Outline a possible route from unicellularity to a multicellular state with division of labor that could have been traveled by protistan ancestors.
(369)	7. Distinguish between colonial organisms and truly multicellular organisms.
(361)	8. Explain how heterotrophic protistans could have acquired the capacity for photosynthesis, and state the evidence to support your explanation.
(365)	9. Explain what causes red tides.
(364)	10. How do golden algae resemble diatoms?
(366)	11. Two flagellated protozoans that cause human misery are _____ and _____.
(367)	12. State the principal characteristics of the amoebas, radiolarians, and foraminiferans. Indicate how they generally move from one place to another and how they obtain food.
(367–368)	13. List the features common to most ciliates.
(368)	14. Characterize the sporozoan group, identify the group's most prominent representative, and describe the life cycle of that organism.

Integrating and Applying Key Concepts

The textbook (Figure 48.14) identifies natural gas as a nonrenewable fuel resource, yet there is a group of archaebacteria that produce methane, the burning of which can serve as a fuel for heating and/or cooking. Recall or imagine how these bacteria could be incorporated into a system that could serve human societies by generating methane in a cycle that is renewable. Why did your text categorize natural gas as a nonrenewable resource? Is methane a constituent of natural gas? Why or why not?

Critical Thinking Exercises

1. The text notes how some bacteria "sense and move toward higher concentrations of nutrients." To test this hypothesis, an apparatus is built consisting of two chambers separated by a porous membrane. Bacteria are placed in the medium on one side; the same medium containing a test nutrient is placed in the opposite chamber. The control is identical except that it contains no test nutrient. Test nutrient can diffuse through the membrane, but bacteria that touch the membrane stick to it. After a period of incubation, the membranes are removed, and the number of bacteria attached to each is determined. If the membrane with the test nutrient has more bacteria than the control membrane, the conclusion is that the test nutrient causes the bacteria to move toward areas where it is more concentrated. Which of the following is the best criticism of this experiment?

   a. The membrane itself might have attracted the bacteria.
   b. Gravity might have caused the movement.
   c. The nutrient might have stimulated random, rather than directed, movement of the bacteria.
   d. The test solution might have contained impurities.
   e. The bacteria might grow faster where the nutrient is more concentrated.

   ANALYSIS

   a. This objection is met by the control. The membrane would have the same attraction in the control as in the experimental chamber. The protocol calls for measurement of the *difference* between the two membranes.
   b. Once again, gravity would have an equal effect on both chambers and would not produce a difference between them. Also, this objection would be met by placing the chambers side by side rather than one on top of the other.
   c. This alternative hypothesis would predict the same results. Even random movement, if it was stimulated, would cause more bacteria to accidentally bump into the membrane and stick.
   d. This is a worrisome possibility whenever reagents are used in experiments (or kitchens, for that matter). However, the presence of an impurity would not affect the conclusion that the test solution attracted the bacteria. It would complicate the determination of what kind of substance was attractive.
   e. This is also a valid interpretation of the results. If they did grow (and divide) faster near the membrane, where the nutrient was at higher concentration, more of them would be expected to stick to the membrane. However, it is reasonable to expect an investigator to stop the incubation before bacterial growth is significant or to include a growth inhibitor in the medium.

2. When bacteria are grown in culture, the number of cells begins at some level and steadily increases. Viruses, on the other hand, are inoculated in a culture at some level, then disappear for a short time, reappear in the culture, and then increase steadily in number. Why was this growth pattern so puzzling to early virologists? With current knowledge, how do you explain it?

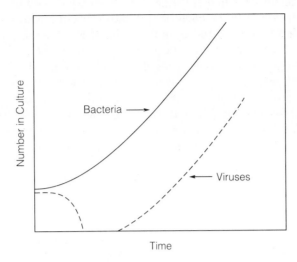

ANALYSIS

Early virologists viewed viruses as living things. As such, the virsues would have to reproduce themselves from existing individuals. What, then, could account for their reappearance in a culture that contained none of them?

We no longer view viruses as living things, and we know that production of viruses is fundamentally different from reproduction of cells. The initial inoculum of viruses first attaches to the outside of the host cells. Then the viruses inject their nucleic-acid component into the host cells and fall apart. At this point, no viruses are present in the culture. Later, new viruses, produced by the host cells under the direction of the injected viral nucleic acid, are continually released and extend the infection to more cells.

3. Some bacteriophages have alternative life cycles, lytic or lysogenic, as described in Figure 22.3. Which of the following hypotheses is most likely to be the mechanism that determines which pathway will follow infection of a given cell?

   a. Some bacteria are resistant to the lytic enzymes.
   b. Some phages inject only the genes for the lytic cycle; others inject only those for the lysogenic cycle.
   c. The whole viral genome is injected, and molecules present in some bacterial cells induce the genes for the lysogenic cycle.
   d. The whole viral genome is injected. The lysogenic genes produce inducers for the lytic enzymes, but bacterial enzymes destroy the inducers, and the lytic enzymes are not produced.
   e. The whole viral genome is injected. Bacterial proteins act as inducers of the genes for the lytic enzymes.

ANALYSIS

   a. This is possible, just as some bacteria are resistant to some antibiotics. But how is it reversible? The latent lytic genes can be activated in a subsequent generation of bacteria. If the latency were due to bacterial molecules being inappropriate substrates for the phage enzymes, reversing the situation would take precisely the right mutation in the bacterial DNA.
   b. In a sense, this begs the question. You would still not know what mechanism determines the choice. Also, this mechanism would not be reversible. The lytic genes could not be activated in a lysogenic cell, because they would not be present.

c. This mechanism would make all the phage infections lytic. Inducing the lysogenic pathway would not repress the lytic pathway. As long as the lytic enzymes are made, the cell will lyse.

d. Suppose that among the genes for the lysogenic enzymes is a gene for a protein that induces the genes for the lytic enzymes. As long as that inducer was present, the cycle would be lytic. But if the bacterial cell contained protein-digesting enzymes—not an unusual idea—the inducer would be digested and a lysogenic cycle would result. This makes the choice between lytic and lysogenic depend on conditions within the bacterial cell (whether the bacterial proteases are induced) and provides a mechanism to reverse the lysogenic choice. Recent studies indicate that this *is* the mechanism.

e. This would account for the choice. Bacterial cells with the inducers would have lytic infections; those without would have lysogenic infections. These bacterial proteins could be repressed and induced in turn according to conditions. This hypothesis might be especially attractive if you thought, with some biologists, that viruses are small portions of nucleic acid that were originally part of a cell's genome, somehow escaped, and became infectious toward the original cell. Such genes would likely be controlled by repressors and inducers in the original cells. However, you would also expect selection to favor mutations that inactivated the inducers and bacterial resistance to rapidly appear.

Answers

Answers to Interactive Exercises

VIRUSES (22-I)

1. a. Nonliving, infectious agents, smaller than the smallest cells; require living cells to act as hosts for their replication; not acted upon by antibiotics. b. The core can be DNA or RNA; the capsid can be protein and/or lipid. c. Bacteriophage viruses may use the lytic pathway, in which the virus quickly subdues the host cells and replicates itself and descendants are released as the cell undergoes lysis; or they may use a temperate pathway, in which viral genes remain inactive inside the host cell during a period of latency, which may be a long time, before activation and lysis.

2. There can be multiple answers (see Table 22.1 in text). a. Possible answers include Herpes simplex (a Herpesvirus), Varicella-zoster (a Herpesvirus), Rhinovirus (a Picornavirus), Poliovirus (an enterovirus of the Picorna group), and HTLV III (a Retrovirus). b. Herpes simplex: DNA virus. Initial infection is a lytic cycle that causes Herpes (sores) on mucous membranes on mouth or genitals. Recurrent infections are temperate. Most cells are in nerves and skin. No immunity. No cure. Varicella-zoster: DNA virus. Initial infection is a lytic cycle that causes sores on skin. Generally, immunity is conferred by one infection, but in some people subsequent infections follow temperate cycles and cause "shingles." Rhinovirus: RNA virus. Causes the *common cold*. Host cells are generally mucus-producing cells of respiratory tract. Poliovirus: RNA virus causes *polio*. Host cells are in motor nerves that lead to the diaphragm and other important muscles. Destruction of these nerve cells causes paralysis that may be temporary or permanent. Recurrences can occur. Immunize your children! HTLV III (or HIV): RNA virus. Host cells are specific white blood cells. Temperate cycle has a latency period that may last longer than a year before host tests positive for HIV. As white blood cells are destroyed, the host's immune system is progressively destroyed (*AIDS*). No cure exists.

3. virus; 4. nucleic acid; 5. protein coat (viral capsid); 6. Viruses; 7. Herpesvirus (or Varicella); 8. viroids; 9. Retroviruses (MTLV III or HIV); 10. temperate; 11. Nanometers; 12. micrometers; 13. 86,000; 14. latency; 15. prions; 16. Rhinoviruses; 17. Retroviruses; 18. Herpesviruses; 19. C; 20. F; 21. B; 22. I; 23. E; 24. K; 25. H; 26. A; 27. J; 28. D; 29. G.

MONERANS—THE BACTERIA (22-II)

1. inorganic compounds; 2. sunlight; 3. pressure; 4. *Clostridium botulinum (Clostridum tetani);* 5. *Clostridium tetani (Clostridium botulinum);* 6. Chemosynthetic; 7. peptidoglycan; 8. nitrogen fixation; 9. endospores; 10. binary fission; 11. prokaryotic; 12. cocci; 13. bacilli; 14. spirilla; 15. positive; 16. *Volvox;* 17. plants; 18. *Trichoplax;* 19. colony; 20. multicellular; 21. d, L; 22. c, C; 23. c, M; 24. c, J; 25. c, E; 26. a, A; 27. b, I; 28. e, G; 29. c, H; 30. d, B; 31. c, K; 32. c, O; 33. c, D; 34. c, N; 35. a, F.

THE RISE OF EUKARYOTIC CELLS (22-III)

1. DNA; 2. archaebacteria; 3. fossil; 4. 3.8; 5. eukaryotes; 6. membrane-bound; 7. organelles; 8. prokaryotic; 9. symbiotic; 10. mitochondria; 11. DNA; 12. Chloroplasts; 13. replicate; 14. T; 15. F; 16. archaebacteria; 17. urkaryotes; 18. eubacteria; 19. mitochondria; 20. chloroplasts; 21. animals; 22. fungi; 23. protistans; 24. plants.

PROTISTANS (22-IV)

1. slime molds; 2. fruiting bodies; 3. myxobacteria (monerans); 4. fungi; 5. animals; 6. longitudinal fission; 7. organelles; 8. heterotrophic; 9. Dinoflagellates; 10. Chrysophytes; 11. diatoms; 12. beta carotene; 13. silica; 14. diatomaceous earth; 15. filtering; 16. plankton; 17. pseudopodia; 18. Foraminiferans;

19. *Plasmodium*; 20. sporozoan (parasite); 21. gametes;
22. freshwater; 23. contractile vacuoles; 24. gullet;
25. enzyme-filled vesicles; 26. C, E, J; 27. C, E, L;
28. B, F, K; 29. A; 30. D, I; 31. B, F; 32. B, F, H; 33. G;
34. D; 35. A; 36. J; 37. C; 38. F; 39. G; 40. B; 41. H; 42. E;
43. I.

Answers to Self-Quiz

1. d; 2. a; 3. c; 4. b; 5. b; 6. a; 7. b; 8. c; 9. c; 10. a; 11. B, I;
12. A, E; 13. A, H; 14. B, N; 15. B, D; 16. A; 17. B, I, O;
18. B, F, G; 19. C, M; 20. C, P; 21. A, H; 22. B, L; 23. B, J,
K; 24. A, H; 25. B, F.

Answer to Integrating and Applying Key Concepts

The text identifies natural gas as a fossil fuel along with coal and petroleum because most supplies of it are produced as a result of drilling and mining rather than through methane generation by bacteria.

$2 CH_4$ (for cooking and heating) $+ 4 O_2$

$\longrightarrow 4H_2O + 2 CO_2 + d$

Plants then photosynthesize CO_2 to produce carbon-based organismal flesh. Methanogens generate methane (CH_4) from dead organisms or dung.

23

FUNGI

Interactive Exercises

GENERAL CHARACTERISTICS OF FUNGI (23-I, pp. 373–374)

Fill-in-the-Blanks

Many fungi have cells merged lengthwise, forming tubes that have thin transparent walls reinforced with

(1)_____ ; (2)_____ formation is also an important part of their life cycle. Some fungal species,

such as late blight, are (3) [choose one] ( ) parasitic, ( ) saprophytic. The vegetative body of most true

fungi is a (4)_____ , which is a mesh of branched, tubular filaments called (5)_____. Fungi have

been on Earth for at least (6)_____ million years. Through all that time, their metabolic activities

enable them to act as (7)_____ in ecosystems. Fungi secrete (8)_____ into their surroundings,

where large organic molecules are broken down into smaller components that the fungal cells then absorb.

While most fungi are (9)_____ (obtaining their nutrients from nonliving organic matter), some are

(10)_____ and get their nutrients directly from their living host's tissues. A common form of asexual

reproduction is the growth of a new fungal body from a(n) (11)_____. Sexual reproduction usually

involves the cytoplasmic fusion of two (12)_____ , followed immediately by nuclear fusion to form

the (13)_____. Many fungal species extend this event by having a (14)_____ stage in which cells

of the fungal body contain two distinct (15)_____ nuclei, and the stage may persist for years before

the two nuclei in each cell fuse and form a zygote.

Labeling

Identify each life cycle stage below by writing its name in the appropriate blank.

16. _____

17. _____

18. _____ _____

19. _____

20. _____

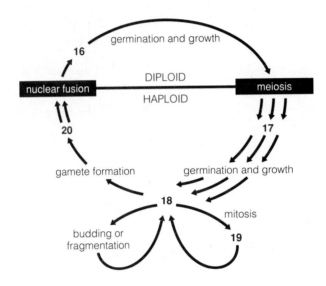

Matching

Match each item listed below with the correct lettered description.

21. ___ dikaryotic stage
22. ___ fungal spore
23. ___ hypha
24. ___ mycelium
25. ___ parasite
26. ___ saprobe

A. A mass of filaments that compose a fungal body
B. Part of the life cycle that is a walled resistant cell or multicelled structure
C. An organism that obtains its nutrients from dead organic material
D. A single fungal filament
E. Part of the fungal life cycle in which each cell has two haploid nuclei
F. An organism that obtains its nutrients from the living tissues of another creature

MAJOR GROUPS OF FUNGI (23-II, pp. 374–381)

Fill-in-the-Blanks

(1)_____ and (2)_____ _____ are the only fungi that produce motile spores; this is a primitive trait that may resemble ancestral fungi that lived several hundred million years ago in watery habitats. Water molds are only distantly related to other fungi and are seen to have evolved from (3)_____ algae. A reproductive structure that produces gametes is known as a (4)_____ ; structures that produce spores are called (5)_____. *Pilobolus* and *Rhizopus stolonifer* are examples of (6)_____-_____ fungi. Sac fungi bear spores in saclike structures called (7)_____ in a complex spore-forming body called a(n) (8)_____. Morels, truffles, and yeasts are examples of (9)_____ fungi. Rusts, smuts, puffballs, and shelf fungi are examples of (10)_____ fungi. The groups of fungi are mostly named according to their spore-producing structures; fungi of the Zygomycetes group produce (11)_____. (12)_____ attach a mycelium to a substrate. (13)_____ are imperfect fungi that parasitize humans. (14)_____ is a blue mold that flavors Roquefort cheese. A commercially important (15)_____ , *Saccharomyces cerevisiae*, produces the CO_2 that makes bread rise.

16. Complete the following table, which classifies fungi.

Group	Common Name	Typical Habitats
a. Chytridiomycetes		Aquatic; some parasitic
b. Oomycetes		
c. Zygomycetes		Soil, decaying plant parts; most saprobic, a few parasitic
d. Ascomycetes		Soil, decaying plant parts
e. Basidiomycetes		
f. Fungi Imperfecti		Diverse (for example, soil, human body)

Labeling

Identify the life-cycle stages illustrated below by writing the name in the corresponding blank.

17. _____   20. _____

18. _____   21. _____

19. _____   22. _____

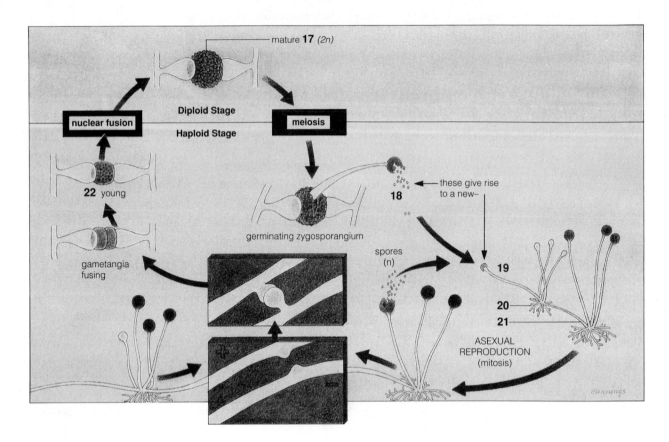

Matching

Match the following pathogens with the correct fungus group. A letter may be used more than once.

23. ___ *Rhizopus*, black bread mold

24. ___ *Phytophthora infestans*, late blight of potato

25. ___ *Amanita muscaria*, the fly agaric

26. ___ *Puccinia graminis*, wheat rust

27. ___ *Verticillium*, wilt on grapevines

28. ___ *Cryphonectria parasitica*, Chestnut blight

29. ___ *Ophiostoma ulmi*, Dutch elm disease

A. Club fungi
B. Imperfect fungi
C. Sac fungi
D. Water molds
E. Zygosporangium-forming fungi

Labeling

Complete the life cycle by labeling the numbered parts on the appropriate blanks.

30. _____

31. _____

32. _____

33. _____

34. _____

club-shaped cell (2n)
that will bear spores

30 _____
stage

fertilization

meiosis

34 _____
stage

31 _____
stage

33 _____
(stalk and cap)

32 _____ (n)

germination

cytoplasmic
fusion of two
compatible hyphae

mycelium

35. Complete the following table, which lists some fungal pathogens.

Group Name	Description
a. _____	Late blight of potato, tomato, downy mildew of grapes
Zygospore-forming fungi	*Rhizopus*, causes food spoilage
b. _____	Dutch elm disease, Chestnut blight, ergot of rye, apple scab
c. _____	Smut of corn, black stem wheat rust
d. _____	Ringworms, athlete's foot, "yeast" infection of mucous membranes

LICHENS / MYCORRHIZAE (23-III, pp. 381–382)

Fill-in-the-Blanks

(1)_____ help many complex land plants absorb certain vital nutrients from the soil. (2)_____ are composite organisms that comprise an alga and fungus living interdependently. The fungus involved in the relationship is a sac fungus or a (3)_____ _____ , and the algae are usually (4)_____ or green algae. Lichens can live on bare rocks, which they help break apart and begin the formation of (5)_____. In the absence of mycorrhizae, many plants cannot readily absorb mineral ions, particularly (6)_____. Mycorrhizae are highly susceptible to damage from (7)_____ _____.

Chapter Terms

The following page-referenced terms are important; they were mostly in boldface in the chapter. Refer to the instructions given in Chapter 1, p. 4.

Fungi (373)
saprobes (373)
parasites (373)
extracellular digestion
    and absorption (373)
mycelium, -lia (374)
hypha, -phae (374)
fungal spore (374)

dikaryotic stage (374)
chytrids
    (Chytridiomycetes) (375)
water molds
    (Oomycetes) (375)
rhizoids (375)
sporangium, -gia (375)
gametangium, -gia (375)

Rhizopus stolonifer (376)
zygosporangium-forming fungi
    (Zygomycetes) (377)
sac fungi (Ascomycetes) (378)
ascus, asci (378)
asocarps (378)
club fungi
    (Basidiomycetes) (378)

basidium, basidia (380)
basidiocarp (380)
imperfect fungi (380)
Penicillium (381)
conidium, -dia (381)
symbiotic relationship (381)
lichens (381)
mycorrhiza, -zae (382)

Self-Quiz

___ 1. Most true fungi send out cellular filaments called _____.
   a. mycelia
   b. hyphae
   c. mycorrhizae
   d. asci

___ 2. Fungi _____.
   a. are producers
   b. are generally saprobic
   c. usually have life cycles in which the diploid phase dominates
   d. include *Fucus* and liverworts

For questions 3–11, choose from the following.

   a. club fungi
   b. imperfect fungi
   c. sac fungi
   d. water molds
   e. zygosporangium-forming fungi

___ 3. The group that includes downy mildew of grapes and late blight of potato is _____.

___ 4. The group that includes *Rhizopus stolonifer*, the notorious black bread mold, is _____.

___ 5. The group that includes *Saprolegnia*, a cottony growth often parasitic on goldfish or tropical fish, is _____.

___ 6. The group that includes delectable morels and truffles but also includes bakers' and brewers' yeasts is _____.

___ 7. The group that includes shelf fungi, which decompose dead and dying trees, and mycorrhizal symbionts that help trees extract mineral ions from the soil, is _____.

___ 8. The group that includes the commercial mushroom *Agaricus brunnescens*, as well as the death cap mushroom *Amanita phalloides*, is _____.

___ 9. The group that is distantly related to other fungi and is thought to be evolutionarily linked to the red algae is _____.

___ 10. The group that includes the parasite of rye, which produces substances that can cause hysteria, hallucinations, convulsions, vomiting, diarrhea, dehydration, and gangrene in arms and legs, as well as death, is _____.

___ 11. The group that includes *Penicillium*, which has a variety of species that produce penicillin and substances that flavor Camembert and Roquefort cheeses, is _____.

Identifying-Match

In the blanks beneath each illustration below (12–21), *identify* the organism, either by common name or by scientific name. Then *match* the numbers pointing to structures (22–32) with the lettered choices on the next page.

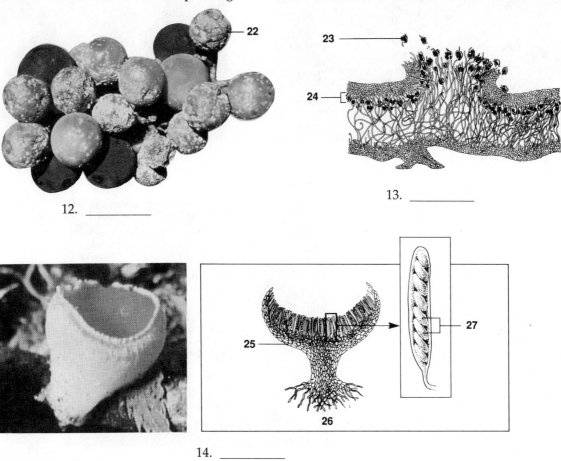

12. _____

13. _____

14. _____

15. _____

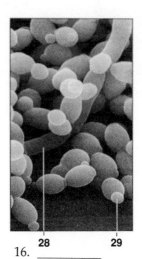

16. _____

17. _____

18. _____

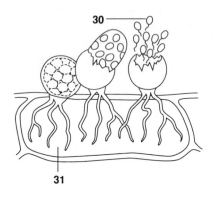

19. _____

20. _____

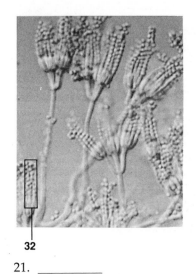

21. _____

22. _____

23. _____

24. _____

25. _____

26. _____

27. _____

28. _____

29. _____

30. _____

31. _____

32. _____

A. Algal layer
B. Asexual reproductive body
C. Ascocarp structure
D. Budding cell
E. Conidia
F. Fungal spore
G. Downy mildew mycelium
H. Hypha
I. Plant cell
J. Spores inside ascus
K. Spore-bearing hyphae

Chapter Objectives/Review Questions

This section lists general and detailed chapter objectives that can be used as review questions. You can make maximum use of these items by writing answers on a separate sheet of paper. Fill in answers where blanks are provided. To check for accuracy, compare your answers with information given in the chapter or glossary.

Page   *Objectives/Questions*

(372)   1. The resurgence of life each spring at Dragon Run could not happen without the _____ activities of fungi that make nutrients available that can also be absorbed by plants.
(373)   2. Are fungi autotrophs, heterotrophs, or producers?
(373–375)   3. Describe the general structure of fungi and its relation to their method of obtaining nutrients.
(372,   4. Describe the type of environment in which fungi are most likely to be found.
374–378)
(373)   5. Distinguish between *parasitic* and *saprobic* fungi. Mention one way in which parasitic fungi harm humans and one way in which saprobic fungi benefit humans.
(374–375,   6. List the ways that fungi can reproduce.
380)
(375–376)   7. List the five principal groups of fungi, and give one example of an organism in each group.
(376)   8. Give two examples of parasitic fungi that have played havoc with the production of crop plants.
(380)   9. Explain why the imperfect fungi are viewed as "imperfect."
(381)   10. Give two examples of fungi that participate in symbiotic relationships. Describe the separate contributions to the relationship that are made by the fungus and by the other participant.
(381–383) 11. State the fundamental contribution of fungi to ecosystems. Which fungi are the first organisms to act as pioneer "settlers" on bare rock?

Integrating and Applying Key Concepts

Suppose humans acquired a few well-placed fungal genes that caused them to reproduce in the manner of a "typical" fungus (see Figure 23.4). Try to imagine the behavioral changes that humans would likely undergo. Would their food supplies necessarily be different? Table manners? Stages of their life cycle? Courtship patterns? Habitat? Would the natural limits to population increase be the same? Would their body structure change? Would there necessarily have to be separate sexes? Compose a descriptive science-fiction tale about two mutants who find each other and set up "housekeeping" together.

Critical Thinking Exercises

1. Suppose you are studying an oomycete you isolated from potato tubers. You have developed a liquid medium in which it can grow, using potato starch as the source of carbon and energy. You decide to test the text's assertion that fungi secrete enzymes that digest food in the environment, then absorb the small product molecules into their cells. Which of the following would be the best test of that hypothesis?

    a. Measure the rate of starch disappearance from the medium as the fungus grows.
    b. Allow the fungus to grow for a while, then remove all the cells from the medium and measure the rate at which starch continues to disappear from the medium.
    c. Omit the starch from the medium and determine whether the fungus can utilize glucose as the source of carbon and energy.
    d. Use radioactive carbon to trace carbon atoms from starch to carbon dioxide.
    e. Omit the starch from the medium and determine whether the fungus can utilize cellulose as the source of carbon and energy.

ANALYSIS

   a. This will give you quantitative information about the metabolism of the fungus, but it will not tell you whether it digests the starch inside or outside of the cells.
   b. If the cells secrete digestive enzymes into the medium, the enzyme molecules will remain in the medium after the cells are removed, and starch will continue to be removed. If the hypothesis is not valid and digestion is intracellular, no more starch will disappear after the cells are removed.
   c. Glucose is the monomer formed when starch is digested. If the fungus does digest starch in the medium, it must be able to absorb glucose. However, it might also have this ability and also move the starch into the cell before digesting it.
   d. By this method, you could determine whether the fungal cells metabolize their carbohydrate by the same pathways as other cells, but it would not allow you to make any conclusions about where the metabolism was accomplished. You would predict the same results whether or not the hypothesis was valid.
   e. If the cells have the appropriate enzymes, they will be able to metabolize cellulose. This will be true whether the enzymes function inside the cells or are secreted into the medium.

2. The generalized fungal life cycle (Figure 23.4) shows production of spores by meiosis. At the next stage in the life cycle, spores are again produced, this time by mitosis. Which of the following would be the most significant advantage of producing spores two different ways in succession like this?

   a. Producing spores by mitosis uses less metabolic effort than producing spores by meiosis.
   b. Producing spores by mitosis allows the storage of larger chemical resources in the spores and thereby gives the spores a greater probability of survival.
   c. Producing spores by mitosis yields more spores per zygote.
   d. The more different stages there are in the life cycle, the more opportunities the organism has to survive temporary adverse conditions.
   e. The spores produced by mitosis disperse farther and more rapidly.

ANALYSIS

   a. The chromosomes are the essential part of a spore. Both types of spore are haploid and contain the same complement of chromosomes, produced at the same metabolic cost. The other components are also similar in both types of spore. In any case, the fungus uses both mechanisms of spore formation, so it expends whatever metabolic effort is required.
   b. Mitosis and meiosis are processes of distribution of cell components to daughter cells. How much the daughter cells get is determined by events before cell division, not by the type of cell division. Even the nuclear contents of the two types of spore are the same; both are haploid.
   c. When the zygote divides by meiosis, a maximum of four spores is produced. If a spore divides by mitosis, many cells are the result, and each can then yield two spores by mitosis. These can be dispersed over a wider area, with a greater probability of at least some of them surviving, than can the four spores produced by meiosis.
   d. This statement assumes that each of the different life-cycle stages can survive different types of adverse conditions and that some individuals at a certain place will always be in each life-cycle stage. Given these assumptions, any change in the environment would be more likely to encounter some individuals capable of surviving. But spores, however they are formed, are better adapted for surviving hostile environments than any of the other life-cycle stages. The advantageous mechanism is one that maximizes spore production.
   e. Both types of spore are single, nonmotile cells that are dispersed by the same environmental factors.

3. You are interested in the growth rate of a zygomycete. You sprinkle some spores on the surface of a solid medium in a culture dish. When the round colonies are established and big enough to see, you choose some and measure their areas at regular time intervals. Your average data are given in the table below.

Time (hrs)	Area (relative units)
2	2.2
3	2.2
4	2.3
5	3.5
6	4.2
8	4.3
9	6.1

As often happens, due to circumstances beyond your control, several points are missing that are important to you. What values would you estimate for average colony area at zero hours, 7 hours, and 10 hours?

ANALYSIS

The first step is to plot the data on a graph or put the numbers into a computer and have the computer plot them. Next draw the line that best fits the data. This is a matter of judgment. Sometimes it is best to draw line segments between each pair of consecutive points. In this case, there is a hypothesis about the growth pattern—namely, that it is exponential—and you should draw the smooth exponential curve that comes closest to the points. If you are using a computer, that is what it will do; use the equation you select and calculate the line of that form that best fits the data. This line is your best interpretation of the value of area at times you didn't measure. In this case, a few points deviate from the curve, but the overall fit is good.

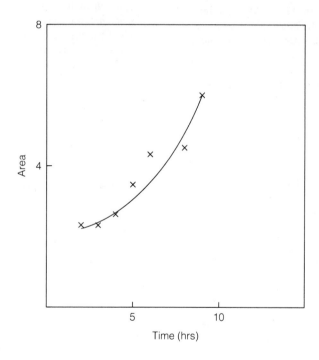

To estimate the area at 7 hours, then, simply find 7 hours on the horizontal axis, move upward until you reach the curve, and find the area value for that point, about 4.1 in this case. If you simply take a value for the area halfway between that for 6 hours and that for 8 hours (the average of these two values), you are ignoring all the other data. In effect, you are using a graph that connects adjacent data points with straight-line segments and discarding the hypothesis.

To estimate the areas at zero and 10 hours, you have to extrapolate or extend the curve beyond the range of the data. This requires that you assume that the same hypothesis applies at these other times. That is, you have to assume that the same factors determine growth rate at these times. This is dangerous, because the assumption is not always valid. Sometimes there is a lag between setting up a culture and the onset of steady growth, in which case your estimate for the starting time would be too low. In this study, you could argue that the colonies had already begun to grow steadily before you defined the beginning time and started measuring. Extrapolating to later times is also dangerous. Often, increasing size of a population causes increases in factors that reduce growth rate. For example, toxic products of metabolism accumulate faster and faster as the population grows. If something like this is happening to your colonies and you simply extend the smooth curve based on your data, your estimate for 10 hours will be too high.

Answers

Answers to Interactive Exercises

GENERAL CHARACTERISTICS OF FUNGI (23-I)
1. chitin; 2. spore; 3. parasitic; 4. mycelium; 5. hyphae; 6. 570; 7. decomposers; 8. enzymes; 9. saprobes; 10. parasites; 11. spore; 12. gametes; 13. zygote; 14. dikaryotic; 15. haploid; 16. zygote; 17. spores; 18. absorptive body; 19. spores; 20. gametes; 21. E; 22. B; 23. D; 24. A; 25. F; 26. C.

MAJOR GROUPS OF FUNGI (23-II)
1. Chytrids; 2. water molds (Oomycetes); 3. red; 4. gametangium; 5. sporangia; 6. zygospore-forming; 7. asci; 8. ascocarp; 9. sac; 10. club; 11. zygosporangia; 12. Rhizoids; 13. Ringworms; 14. *Penicillium*; 15. yeast; 16. a. Chytrids; b. water molds; Aquatic; some parasitic; c. zygosporangium-forming fungi; d. Sac fungi; e. Club fungi; soil, decaying plant parts; f. Imperfect fungi; 17. zygospore; 18. spores; 19. mycelium; 20. rhizoids;

21. stolon; 22. zygosporangium; 23. E; 24. D; 25. A; 26. A; 27. B; 28. C; 29. C; 30. diploid; 31. haploid; 32. spores; 33. basidiocarp; 34. dikaryotic; 35. a. water molds; b. sac fungi; c. club fungi; d. imperfect fungi.

LICHENS/MYCORRHIZAE (23-III)
1. Mycorrhizae; 2. Lichens; 3. club fungus; 4. cyanobacteria; 5. soil; 6. phosphorus; 7. acid rain.

Answers to Self-Quiz

1. b; 2. b; 3. d; 4. e; 5. d; 6. c; 7. a; 8. a; 9. d; 10. c; 11. b.; 12. grapes; 13. lichen; 14. *Sarcoscypha*, cup fungus; 15. *Cladonia*, reindeer moss; 16. *Candida albicans*, "yeast"; 17. shelf fungus; 18. *Pilobolus*; 19. chytrid; 20. *Amanita muscaria*, fly agaric mushroom; 21. *Penicillium*; 22. G; 23. B; 24. A; 25. K; 26. C; 27. J; 28. H; 29. D; 30. F; 31. I; 32. E.

24

PLANTS

Interactive Exercises

GENERAL CHARACTERISTICS OF PLANTS (24-I, pp. 385–386)

1. What characteristics distinguish the vascular plants?_____

_____

_____

2. What major groups of plants are the most familiar vascular plants?_____

_____

3. What characteristic distinguishes the nonvascular plants? Cite examples of nonvascular plants._____

_____

EVOLUTIONARY TRENDS AMONG PLANTS (24-II, pp. 386–388)

1. As plants evolved, several key evolutionary events occurred that solved the problems of living in new land environments. Complete the following table to summarize these events.

Evolutionary Event	Definition(s)	Survival Problem Solved
a. Xylem and phloem	_____	Vascular tissue forms important parts of root and shoot systems; allowed transition to land life

Evolutionary Event	Definition(s)	Survival Problem Solved
b. Strong, thick-walled support cells		Allowed plants to grow taller on land; allowed leaf development and display; allowed higher spore-launching platforms
c.	Waxy coating on stems and leaves	Allowed water conservation for plants living on land
d. Stomata	Many tiny passageways on leaves and young stems whose openings and closings can be controlled	
e.	Gametes that do not depend on water for movement	Allowed plants to live and reproduce in higher, drier parts of the world
f. Sporophyte dominance	Development of a large, diploid plant body, the sporophyte	
g.	Development of plants that produce two types of spores	A starting point for the evolution of pollen grains and seeds
h. Pollen grains	Mature spores, inside which male gametophytes develop	
i. Seeds		A package for plant survival on dry land

Matching

Match each of the following sketches with the appropriate description.

2. ___   A. Typical life cycle for some algae
3. ___   B. Typical life cycle for bryophytes
             C. Typical life cycle for vascular plants
4. ___

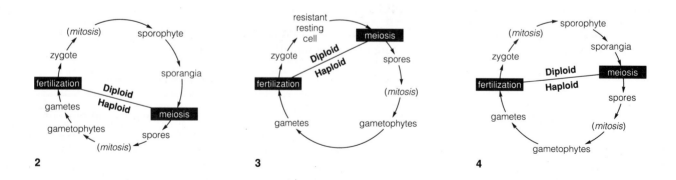

2   3   4

True-False

If false, explain why.

___ 5. The haploid sporophyte phase is dominant in bryophyte life cycles.

___ 6. In some algal life cycles, the only diploid phase is a resting cell, the zygote.

___ 7. In vascular plant life cycles, meiosis in sporangia on the sporophyte phase results in haploid gametes.

___ 8. All the life cycles pictured above show a haploid spore developing by mitosis into a haploid gametophyte.

___ 9. In the vascular plant life cycle, two haploid gametes fuse to form the haploid gametophyte phase.

THE "ALGAE" (24-III, pp. 388–392)

Fill-in-the-Blanks

The term (1) "_____" no longer has formal classification significance, because organisms once lumped under that term are now assigned to different kingdoms. Red algae secrete (2)_____ as part of their cell walls. Most red algae live in (3)_____ habitats. Some red algae have stonelike cell (4)_____ , participate in coral reef building, and are major producers. In the red algae life cycle, the (5)_____ develops spore-producing structures in which (6)_____ occurs. Haploid spores formed by (7)_____ give rise to gametophytes. Gametophytes then produce nonmotile (8)_____ and specialized egglike structures by mitosis. The (9)_____ algae contain xanthophylls, chlorophylls, and other pigments; they live offshore or in intertidal zones and have many representatives with large sporophytes known as kelps; some species produce (10)_____ , a valuable thickening agent. The brown algae life cycle includes a multicelled (11)_____ stage as well as a multicelled (12)_____ stage. Green algae are thought to be ancestral to more complex plants, because they have the same types and proportions of (13)_____ pigments, have (14)_____ in their cell walls, and store their carbohydrates as (15)_____ . The life cycles of green algae show diverse patterns of (16)_____ and (17)_____ reproduction.

Label-Match

Identify each indicated part of the illustration below by entering its name in the appropriate numbered blank. Choose from the following terms: cytoplasmic fusion, asexual reproduction, resting spore, fertilization, zygote, meiosis and germination, spore mitosis, gametes meet. Complete the exercise by matching from the list below, entering the correct letter in the parentheses following each label.

18. _____ ( )

19. _____ _____ ( )

20. _____ and _____ ( )

21. _____ _____ ( )

22. _____ _____ ( )

23. _____ _____ ( )

24. _____ _____ ( )

25. _____ ( )

A. Fusion of two haploid cells
B. A device to survive unfavorable environmental conditions
C. No genetic variation
D. Fusion of two haploid nuclei
E. Haploid spores dividing to form smaller haploid gametes
F. Formed after fertilization; the only diploid structure in this life cycle
G. Two haploid cells coming together
H. Reduction of the chromosome number

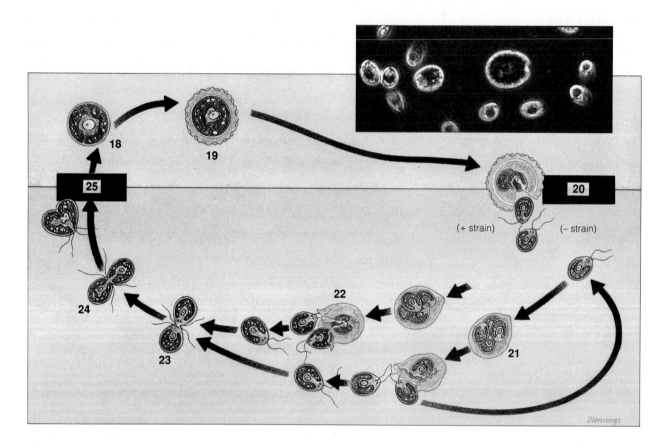

BRYOPHYTES (24-IV, p. 392)

1. List three bryophyte features that certainly were adaptive when ancient plants made the transition to life on land.

   a. _____

   b. _____

   _____

   c. _____

Fill-in-the-Blanks

Bryophytes include the mosses, (2)_____ , and (3)_____. All bryophytes are (4) [choose one] ( ) small, ( ) large plants. Bryophytes have leaflike, stemlike, and rootlike parts, but they lack the conducting tissues, (5)_____ , and (6)_____. Most species have (7)_____ , elongated cells or filaments that attach (8)_____ to the soil as well as absorb water and minerals. The most common bryophytes are (9)_____. (10)_____ give rise to threadlike (11)_____ that grow into the familiar moss plants. A (12)_____ is a saclike, jacketed structure borne at the shoot tips in which eggs and sperms develop. Sperm reach the eggs by (13)_____ through a film of water on plant parts. Following fertilization, zygotes develop into mature (14)_____ , each consisting of a stalk and a (15)_____. The (16)_____ is a saclike, jacketed structure in which spores develop. Sporophytes initially depend on the parent (17)_____ for food but continue to depend on them for nutrients and water. The bryophytes are distinct from all other land plants in that they have independent (18)_____ and dependent (19)_____.

Labeling

Identify each indicated part of the accompanying illustration.

20. _____   24. _____   27. _____

21. _____   25. _____   28. _____

22. _____   26. _____   29. _____

23. _____

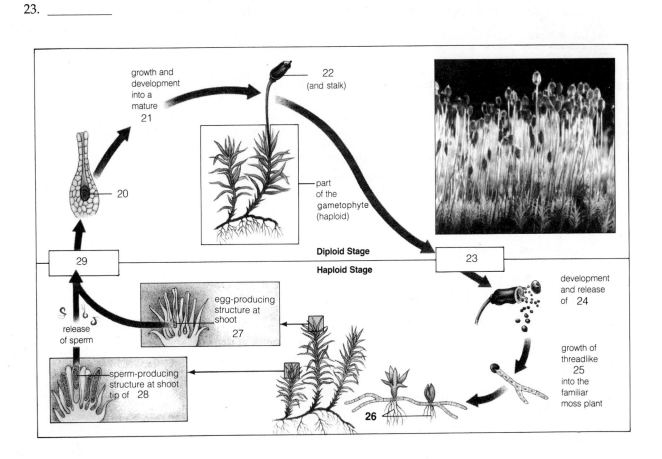

SEEDLESS VASCULAR PLANTS (24-V, pp. 393–397)

1. Based on evidence in the fossil record, how long ago were the seedless vascular plants established?

   _____

   _____

2. Some lineages of seedless vascular plants have survived to the present; list these groups._____

   _____

3. How do the existing representatives of the seedless vascular plant groups differ from the bryophytes?

   a. _____

   b. _____

   c. _____

   d. _____

4. Although the sporophytes of the seedless vascular plants are adapted for life on land, they are confined largely to wet, humid regions. Give reasons for this confinement._____

   _____

   _____

Identifying

Match each of the following characteristics with its appropriate seedless vascular plant group by writing "P" if it applies to psilophytes, "L" for lycophytes, "H" for horsetails, and "F" for ferns.

___ 5. scalelike leaves arranged in whorls about an aboveground, hollow, photosynthetic stem

___ 6. *Lycopodium* and *Selaginella*

___ 7. rust-colored sori

___ 8. outer cells of the stem photosynthetic

___ 9. *Psilotum*

___ 10. frond

___ 11. aboveground branches with scalelike projections

___ 12. grow in moist soil along streams and in disturbed habitats

___ 13. often have cone-shaped leaf clusters that bear spore sacs above these leaves

___ 14. small, green, heart-shaped gametophyte

___ 15. sporophytes lacking roots and leaves

___ 16. some tropical species the size of trees

___ 17. microscopic sporangia that look somewhat like baby rattles

___ 18. sporophytes with true roots, stems, and small leaves with a single strand of vascular tissue

___ 19. stems used by pioneers to scrub cooking pots

___ 20. leaf blades featherlike and finely divided into segments

___ 21. sporangium that snaps open and causes spores to catapult through the air

___ 22. members of one genus homosporous, those of other genera heterosporous

___ 23. ancient "fungus-roots" that performed root functions

___ 24. a single genus, *Equisetum*

Labeling

Identify each indicated part of the illustration below.

25. _____

26. _____

27. _____

28. _____

29. _____

30. _____

31. _____

32. _____

33. _____

34. _____

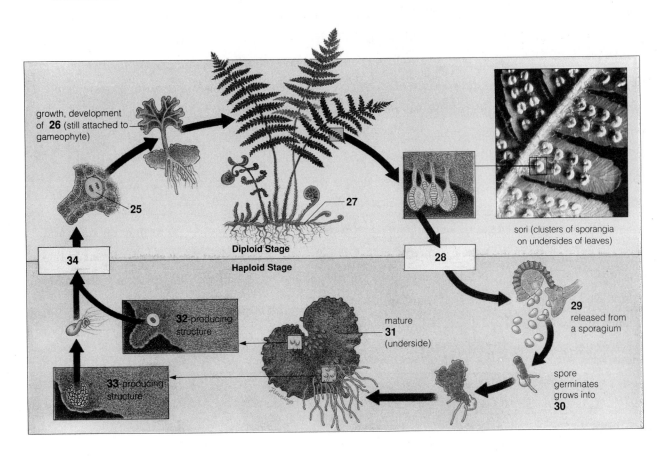

growth, development of **26** (still attached to gameophyte)

25

Diploid Stage

27

sori (clusters of sporangia on undersides of leaves)

34

Haploid Stage

28

29 released from a sporagium

32-producing structure

mature **31** (underside)

spore germinates grows into **30**

33-producing structure

EXISTING SEED PLANTS (24-VI, pp. 397–402)

Fill-in-the-Blanks

The most successful (1)_____ plants are those that produce seeds. Each seed starts out as an (2)_____. Each contains a female (3)_____, with an egg cell surrounded by tissues and a jacket of protective cell layers. A fertilized egg develops into an embryo (4)_____, and the outer ovule tissue develops into a seed coat. Ancestors of seed plants first appeared during (5)_____ times. Naked seed plants are known as (6)_____ because there are no protective tissues around the ovules. Seed plants that have tissues that surround and protect ovules are known as (7)_____. In all species of gymnosperms, the small female (8)_____ are not free-living but are enclosed by (9)_____ tissues. Small male gametophytes develop in (10)_____ _____. (11)_____ are gymnosperms that flourished with the dinosaurs, resemble small palm trees, and have large strobili that bear either pollen or ovules. The (12)_____ is a gymnosperm that diversified during the Mesozoic, became nearly extinct, and survives today only as a single species. One group of gymnosperms, the (13)_____, is composed of three genera. (14)_____ includes both trees and leathery leafed vines in moist, tropical regions. (15)_____ grows in desert regions of the world. (16)_____ is a bizarre-looking genus that lives in the hot deserts of south and west Africa. Pines, spruces, firs, hemlocks, junipers, cypresses, and redwoods are woody trees and shrubs with needlelike or scalelike leaves belonging to the gymnosperm group known as the (17)_____. (18)_____ are clusters of fertile leaves that bear sporangia. A pine tree produces two kinds of (19)_____ in two kinds of (20)_____. The scales of male cones bear sporangia in which spore mother cells undergo meiosis and give rise to haploid (21)_____, which develop into pollen grains, each containing a male (22)_____. The scales of female cones bear (23)_____, inside of which a mother cell undergoes meiosis. Only one of the resulting haploid spores, the (24)_____, survives and develops into the female (25)_____. Each spring, air currents lift millions of (26)_____ _____ from their cones; some will land on female cones. (27)_____ has occurred when pollen grains land on female reproductive parts. (28)_____ form in the tube, and (29)_____ form in the female gametophyte. (30)_____ occurs when the pollen tube reaches the egg. A pine (31)_____ includes the embryo, female gametophyte, and surrounding coats. The embryo is protected from drying by the seed coats, and the (32)_____ _____ tissue serves as food reserve for the embryo.

Labeling

Identify each indicated part of the illustration below.

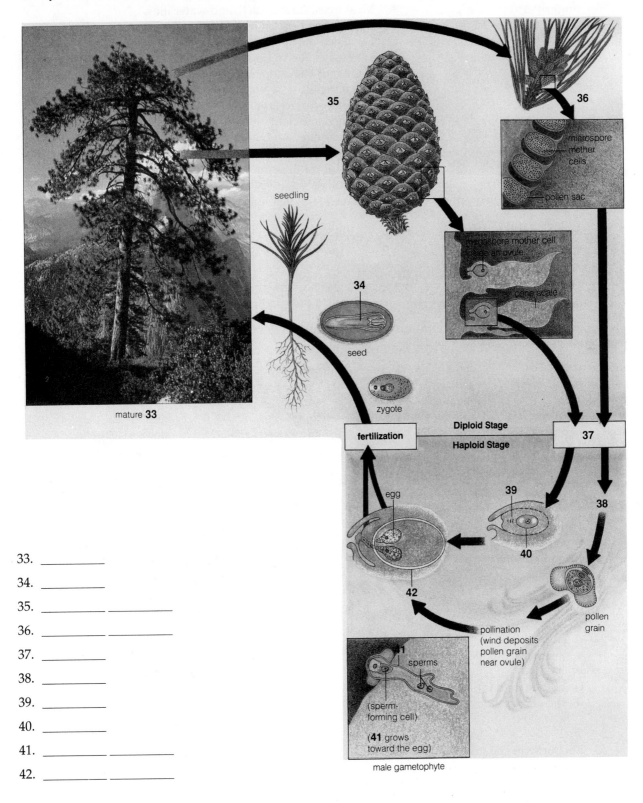

33. _____
34. _____
35. _____ _____
36. _____ _____
37. _____
38. _____
39. _____
40. _____
41. _____ _____
42. _____ _____

Matching

Choose all the appropriate answers for each.

43. ___ angiosperms

44. ___ pollen grains

45. ___ dicot plant examples

46. ___ major monocot crop plants

47. ___ endosperm

48. ___ seeds

49. ___ monocot plant examples

50. ___ flower

51. ___ land-dwelling sporophytes

52. ___ diploid sporophyte

53. ___ angiosperm gametophyte

A. Most trees, shrubs, cacti, and water lilies
B. Unique angiosperm reproductive structure
C. Packaged in fruits
D. Dominates the angiosperm life cycle
E. Have root and shoot systems to take up and conserve water and minerals
F. Wheat, corn, rice, rye, sugar cane, and barley
G. Retained and nourished by the sporophyte
H. The most successful and diverse division of plants
I. Released gymnosperms and angiosperms from dependence on water for fertilization
J. Grasses, palms, lilies, and orchids
K. Unique tissue that nourishes angiosperm embryos

Chapter Terms

The following page-referenced terms are important; they were in boldface type in the chapter. Refer to the instructions given in Chapter 1, p. 4.

vascular plants (386)
red, brown, and green
  algae (386)
nonvascular plants (386)
haploid (*n*) phase (387)
diploid (*2n*) phase (387)

sporophyte (387)
gametophyte (387)
sporangia (388)
gametangium (392)
ovule (397)

seed (397)
gymnosperm (397)
angiosperm (397)
cones (398)
microspores (398)

megaspore (398)
pollination (399)
monocots (402)
dicots (402)
flowers (402)

Self-Quiz

___ 1. Plants possessing xylem and phloem are called _____ plants.
   a. gametophyte
   b. nonvascular
   c. vascular
   d. seedless

___ 2. A gametophyte is _____.
   a. a gamete-producing plant
   b. haploid
   c. both (a) and (b)
   d. the plant produced by the fusion of gametes

___ 3. Red, brown, and green "algae" are found in the kingdom _____.
   a. Plantae
   b. Monera
   c. Protista
   d. all of the above

___ 4. Red algae _____.
   a. are primarily marine organisms
   b. are thought to have developed from green algae
   c. contain xanthophyll as their main accessory pigments
   d. all of the above

___ 5. Stemlike structure, leaflike blades, and gas-filled floats are found in the species of _____.
   a. red algae
   b. brown algae
   c. bryophytes
   d. green algae

___ 6. Because of pigmentation, cellulose walls, and starch storage similarities, the _____ algae are thought to be ancestral to more complex plants.
   a. red
   b. brown
   c. blue-green
   d. green

___ 7. Bryophytes _____.
   a. have vascular systems that enable them to live on land
   b. include lycopods, horsetails, and ferns
   c. include mosses, liverworts, and hornworts
   d. have true roots but not stems

___ 8. In horsetails, lycopods, and ferns, _____.
   a. spores give rise to gametophytes
   b. the main plant body is a gametophyte
   c. the sporophyte bears sperm- and egg-producing structures
   d. all of the above

___ 9. _____ are seed plants.
   a. Cycads and ginkgos
   b. Conifers
   c. Angiosperms
   d. All of the above

___ 10. An ovule of a gymnosperm develops into the _____ after fertilization.
   a. flower
   b. fruit
   c. seed
   d. sporangium

___ 11. In complex land plants, the diploid stage is resistant to adverse environmental conditions such as dwindling water supplies and cold weather. The diploid stage progresses through this sequence: _____.
   a. gametophyte → male and female gametes
   b. spores → sporophyte
   c. zygote → sporophyte
   d. zygote → gametophyte

___ 12. The rapid expansion of angiosperms late in the Mesozoic era appears to be related to their coevolution with _____.
   a. dinosaurs
   b. gymnosperms
   c. insects
   d. mammals

___ 13. Monocots and dicots are groups of _____.
   a. gymnosperms
   b. club mosses
   c. angiosperms
   d. horsetails

Chapter Objectives/Review Questions

This section lists general and detailed chapter objectives that can be used as review questions. You can make maximum use of these items by writing answers on a separate sheet of paper. Fill in answers where blanks are provided. To check for accuracy, compare your answers with information given in the chapter or glossary.

Page   *Objectives/Questions*

(386)   1. Most plant species are _____ plants with well-developed root and shoot systems.
(386)   2. The most familiar vascular plants are _____ (conifers) and _____ (flowering plants).
(386)   3. _____ plants such as bryophytes have either simple internal transport systems or none at all.
(387)   4. Be able to state the differences in sporophyte/gametophyte emphasis when comparing algae, bryophyte, and vascular plant life cycles.
(386–387) 5. Describe the evolution of root and shoot systems.
(386)   6. One vascular tissue, _____, distributes water and dissolved mineral ions through the plant body; another kind, _____, distributes the products of photosynthesis.
(386)   7. State the advantages of strong, thick-walled cells for plants living on land.
(387)   8. Land plants have a waxy _____ covering their young stems and leaves.

Page	Objectives/Questions
(387)	9. The life cycle of complex land plants is dominated by a large, diploid body, the _____.
(387)	10. A plant spore is a reproductive cell that can develop into a _____ , which produces gametes.
(387)	11. Distinguish homospory from heterospory; explain the evolutionary significance of heterospory.
(388–391)	12. State the outstanding characteristics of organisms of the red, brown, and green algae divisions.
(390–391)	13. Be able to trace the life cycles of *Chlamydomonas, Spirogyra,* and *Ulva.*
(392)	14. What are the three adaptive bryophyte features that allowed ancient plants to make the transition to land?
(392)	15. Know the moss life cycle.
(393)	16. List four respects in which bryophytes differ from the seedless vascular plants.
(393–395)	17. Discuss the general characteristics of psilophytes, lycophytes, horsetails, and ferns.
(395)	18. Know the fern life cycle.
(397)	19. An _____ is a structure containing a female gametophyte, with egg cell, surrounded by tissues and a protective cellular jacket.
(397)	20. _____ are the plants with naked seeds; _____ have protected ovules.
(397–399)	21. Be able to generally describe cycads, ginkgos, gnetophytes, and conifers.
(399)	22. Know the pine life cycle.
(398)	23. Scales of male pine cones bear sporangia in which spore mother cells undergo meiosis and give rise to haploid _____.
(398)	24. Inside pine ovules, a mother cell undergoes meiosis; one spore, the _____ , survives and develops into a female _____.
(399)	25. The arrival of a pollen grain on female reproductive parts is called _____.
(399)	26. A pine _____ includes the embryo, female gametophyte, and outer coats.
(400)	27. The _____ are the most successful plant division.
(402)	28. The two classes of angiosperms are the _____ and the _____.
(402)	29. The evolution of pollen grains freed gymnosperms and angiosperms from dependence on free water for _____.

Integrating and Applying Key Concepts

Explain why totally submerged aquatic plants that live in deep water never developed heterosporous life cycles.

Critical Thinking Exercises

1. The text argues that the appearance of strong-walled cells in plants conferred an advantage because it allowed plants to become taller and to compete better for light. A similar argument has been applied to a case involving two species of duckweed, a very small green plant that floats in or on water, forming a thin layer of plants. Both species could be grown successfully in aquaria, but when both species were placed in a single tank, *Lemna polyrhiza* inevitably died out, and *L. gibba* survived. The reason is thought to be that *L. gibba* has tiny air sacs that cause it to float higher in the water and absorb all the light before it penetrates to the competing species. Which of the following observations would add the most strength to this interpretation?

   a. *L. polyrhiza* does not survive in water removed from a tank where *L. gibba* is growing.
   b. Both species survive in water to which nitrogen and phosphorus fertilizer has been added.
   c. *L. polyrhiza* does survive in the same tank with *L. gibba* if *L. polyrhiza* is introduced first and allowed to grow for a while before the other species is added.
   d. *L. gibba* dies out and *L. polyrhiza* survives in a tank illuminated from the bottom only.
   e. Both species die when the tank is illuminated with green light only.

ANALYSIS

To answer this question, it is necessary to think of alternative hypotheses. How else could one plant species inhibit the growth of another? What other resources might two plant species compete for?

a. This observation would be consistent with the hypothesis that one species secretes into the water a substance that is toxic to the other species. If the basis for differential survival is competition for light, then either species should survive alone in a medium conditioned by growth of the other species as long as illumination is satisfactory.

b. Plants might compete for nitrogen and phosphorus nutrients, and one species might take them up faster than the other species. If this is the reason why one species dies out, survival of that species is predicted if the nutrients are supplied in high enough amounts that the plants no longer compete. This observation would support an alternative hypothesis and indicate that competition for light does not account for the differential survival.

c. Sometimes organisms compete for space, and one survives better because it grows faster. If the slower-growing species is started first and allowed to become established, it can survive in the same area as the other species. Once again, this observation is predicted by an alternative hypothesis and not by the one being evaluated.

d. The hypothesis of competition for light assumes that light comes from above. If the hypothesis is valid and the direction of light is reversed, the survival pattern should also be reversed. The plants that float lower would receive the light first and would shade out the plants above them.

e. This would be predicted no matter what the basis for differential survival was. All green plants depend on light, and since they all use chlorophyll, which does not absorb green light, none of them grow very well without red and/or blue light.

2. The text states that ginkgos are valued for horticultural purposes because they are resistant to insects, disease, and air pollutants. If they are so hardy, why did they become extinct in the wild? Which of the following hypotheses offers the best explanation?

a. The global climate changed so that ginkgos could not survive.
b. Human activity eliminated all the suitable ginkgo habitat.
c. More advanced plants evolved that were better competitors for ginkgo habitat.
d. Ginkgos were unable to disperse to new areas of suitable habitat as their original areas became unsuitable.
e. A fungus killed all the wild ginkgos and then became extinct along with them.

ANALYSIS

a. This clearly did not happen, because the climate is still suitable for ginkgos—they grow quite well when they are planted all around the world.

b. Ginkgos are able to grow well in a variety of habitats around the world. It is unlikely that human activity could have destroyed such a wide variety of habitats. This explanation could have been valid if ginkgos required very specific conditions, found in strictly limited areas.

c. Ginkgos are relatively primitive plants. As evolution produced gymnosperms and angiosperms that reproduce rapidly and with high success, ginkgos might well have been at a competitive disadvantage. Suitable habitat is obviously available worldwide, but ginkgos are able to occupy it only when the competition is managed by humans.

d. This would explain their extinction in nature and their survival under cultivation. However, to evaluate this hypothesis, we need much more information from the fossil record about the previous distribution of ginkgos and about their modes of dispersal. If they were once widespread or if their seeds disperse readily, this hypothesis would be less likely.

e. This scenario is possible. A highly specific, lethal pathogen species evolves, infects its host population and drives it to extinction, and then disappears itself for lack of a host organism. In the case of the ginkgo, however, the host is still extant. Why did the hypothetical fungus not infect the trees in cultivation as well as in the wild?

3. The text says that "heterospory was a starting point for the evolution of pollen grains and seeds." No plants have been discovered that do not fit this hypothesis. Which of the following hypothetical plant discoveries would not be accounted for by the text's evolutionary sequence?

   a. Heterosporous plants with large, free-living gametophytes
   b. Heterosporous plants with only one type of gamete
   c. Homosporous plants with cryptic (small and inconspicuous) gametophytes
   d. Homosporous plants with seeds
   e. Homosporous plants with two different types of gamete

ANALYSIS

   a. The seed plants have very small, inconspicuous gametophytes, and the trend in their evolution has been to steadily reduce the gametophyte to a largely internal structure within the sporophyte. However, this does not preclude a stage in which heterospory had already appeared before the reduction of the gametophyte began.
   b. The hypothesis is that heterospory appeared before the gametophyte differentiated enough to produce pollen and seeds. This allows heterospory to appear before the gametophyte and gametes have differentiated at all.
   c. The hypothesis does not require any specific timing of the reduction of the gametophyte, only that heterospory preceded the appearance of pollen and seeds. The gametophyte could have been reduced before heterospory appeared. In this case, the sporophyte would have to develop from the unprotected, unsupported spore.
   d. According to the hypothesis, heterospory preceded seeds. This predicts no homosporous seed-bearers.
   e. According to the hypothesis, full differentiation of the gametophytes and the spores that produce them had to occur before the zygote could be packaged into a seed, but not before gametes could differentiate into male and female types.

Answers

Answers to Interactive Exercises

GENERAL CHARACTERISTICS OF PLANTS (24-I)
1. Vascular plants have a well-developed root system that absorbs water and nutrients from the soil; they have a well-developed shoot system of stems and leaves.
2. Angiosperms and gymnosperms.
3. Very simple internal transport systems or none at all; bryophytes are examples of nonvascular plants.

EVOLUTIONARY TRENDS AMONG PLANTS (24-II)
1. a. Vascular tissues; xylem distributes water and dissolved mineral ions through the plant body; phloem distributes the photosynthetic products; b. Part of structural support tissues; c. Cuticle; d. A way for land plants to absorb CO_2, release O_2, and control evaporative water loss; e. Nonmotile gametes; f. Have extensive shoot and root systems to obtain nutrients and water, even on dry land; developed the capacity to hold and nourish spores and gametophytes; g. Heterospory; h. Allows movement of male gametophytes to female gametophytes on dry land; sperm production; i. Contains an embryo sporophyte plant and stored food and is covered by protective tissues; 2. C; 3. A; 4. B; 5. F; 6. T; 7. F; 8. T; 9. F.

THE "ALGAE" (24-III)
1. algae; 2. agar; 3. marine; 4. walls; 5. sporophyte; 6. meiosis; 7. meiosis; 8. sperm; 9. brown; 10. algin; 11. haploid; 12. diploid; 13. photosynthetic; 14. cellulose; 15. starch; 16. sexual (asexual); 17. asexual (sexual); 18. zygote (F); 19. resting spore (B); 20. meiosis and germination (H); 21. asexual reproduction (C); 22. spore mitosis (E); 23. gametes meet (G); 24. cytoplasmic fusion (A); 25. fertilization (D).

BRYOPHYTES (24-IV)
1. a. Aboveground parts have a water-conserving cuticle; b. A protective cellular jacket surrounds sperm-producing and egg-producing parts of the plant and keeps them from drying out; c. The sporophyte begins its early development as an embryo inside female gametophyte tissues; 2. liverworts (hornworts); 3. hornworts (liverworts); 4. small; 5. xylem (phloem); 6. phloem (xylem); 7. rhizoids; 8. gametophytes; 9. mosses; 10. Spores; 11. gametophytes; 12. gametangium; 13. swimming; 14. sporophytes; 15. sporangium; 16. sporangium; 17. gametophytes; 18. gametophytes; 19. sporophytes; 20. zygote; 21. sporophyte; 22. sporangium; 23. meiosis; 24. spores; 25. gametophyte; 26. rhizoids; 27. gametophyte; 28. gametophyte; 29. fertilization.

SEEDLESS VASCULAR PLANTS (24-V)

1. By the late Silurian, some 420 million years ago; they flourished for about 60 million years before most kinds became extinct; 2. Whisk ferns, lycophytes, horsetails, and ferns; 3. a. Their sporophytes develop independently of the gametophytes; b. Their sporophytes have well-developed vascular tissues; c. The sporophyte is the larger, longer-lived part of the life cycle; gametophytes are very small and lack chlorophyll; d. No seeds are produced by the members of these four plant divisions; 4. Gametophytes lack vascular tissues for water transport and male gametes must have water to reach the eggs; 5. H; 6. L; 7. F; 8. P; 9. P; 10. F; 11. P; 12. H; 13. L; 14. F; 15. P; 16. F; 17. F; 18. L; 19. H; 20. F; 21. F; 22. L; 23. P; 24. H; 25. zygote; 26. sporophyte; 27. rhizome; 28. meiosis; 29. spores; 30. gametophyte; 31. gametophyte; 32. egg; 33. sperm; 34. fertilization.

EXISTING SEED PLANTS (24-VI)

1. vascular; 2. ovule; 3. gametophyte; 4. sporophyte; 5. Devonian; 6. gymnosperms; 7. angiosperms; 8. gametophytes; 9. sporophyte; 10. pollen grains; 11. Cycads; 12. ginkgo; 13. gnetophytes; 14. *Gnetum*; 15. *Ephedra*; 16. *Welwitschia*; 17. conifers; 18. Cones; 19. spores; 20. cones; 21. microspores; 22. gametophyte; 23. ovules; 24. megaspore; 25. gametophyte; 26. pollen grains; 27. Pollination; 28. Sperm; 29. eggs; 30. Fertilization; 31. seed; 32. female gametophyte; 33. sporophyte; 34. embryo; 35. female cone; 36. male cones; 37. meiosis; 38. microspores; 39. ovule; 40. megaspore; 41. pollen tube; 42. female gametophyte; 43. H (A, J); 44. I; 45. A; 46. F; 47. K; 48. C; 49. J (F); 50. B; 51. E; 52. D; 53. G.

Self-Quiz

1. c; 2. c; 3. a; 4. a; 5. b; 6. d; 7. c; 8. a; 9. d; 10. c; 11. c; 12. c; 13. c.

25

ANIMALS: THE INVERTEBRATES

Interactive Exercises

OVERVIEW OF THE ANIMAL KINGDOM / SPONGES / CNIDARIANS / COMB JELLIES (25-I, pp. 406–413)

1. Complete the table below by filling in the appropriate phylum or representative group name.

Phylum	Some Representatives	Number of Known Species
Placazoa	*Trichoplax adhaerens*	1
a. _____	Sponges	8,000
b. _____	i. _____ , jellyfishes, corals, sea anemones	11,000
c. _____	Turbellarians, flukes, tapeworms	15,000

Nemertea	Species shaped like ribbons, rubber bands, shoelaces	800
d. _____	Pinworms, hookworms	20,000
Rotifera	Species with crown of cilia	1,800
e. _____	j. _____ , slugs, clams, squids, octopuses	110,000
f. _____	k. _____ , leeches, polychaetes	15,000
g. _____	Crabs, lobsters, spiders, insects	1,000,000+
h. _____	Sea stars, sea urchins, sea cucumbers	6,000
Chordata	Invertebrate chordates: Tunicates, lancelets	2,100
	l. _____	
	Fishes	21,000
	Amphibians	3,900
	Reptiles	7,000
	Birds	8,600
	Mammals	4,500

Fill-in-the-Blanks

Ninety-seven percent of all animals on Earth are (2)_____ . Multicellular animals are called

(3)_____ . (4)_____ form the most primitive major group of multicellular animals. They are

nourished by microscopic organisms extracted from the water that flows in through pores in the body wall

by sticky (5)_____ _____ . (6)_____ between sponge cells and integration of activities are

poorly developed. Sponges lack (7)_____ cells, muscles, and a gut. Cnidarians are (8)_____

symmetrical and have stinging cells called (9)_____ , which aid in defense and food capture. Most

cnidarian life cycles have a (10)_____ larval stage. Of the two body types, the (11)_____ is the

sexual stage, in which simple sex organs produce eggs or sperms. (12)_____ _____ are biradial

predatory animals that appear to be made of jelly.

Labeling

Identify each indicated part of the illustration on page 277.

13. _____ _____   15. _____ _____

14. _____ _____   16. _____

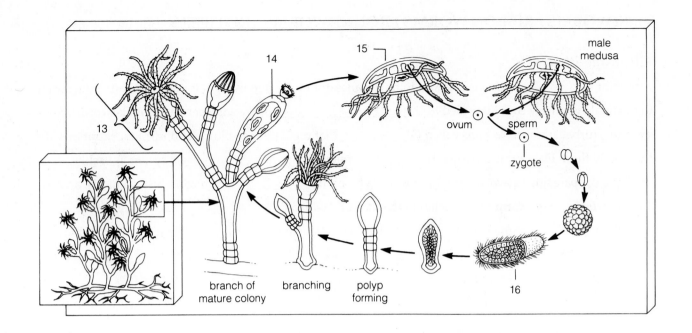

Identify the groups in the family tree shown below by writing the group names in the appropriate blank.

17. _____

18. _____

19. _____

20. _____

21. _____

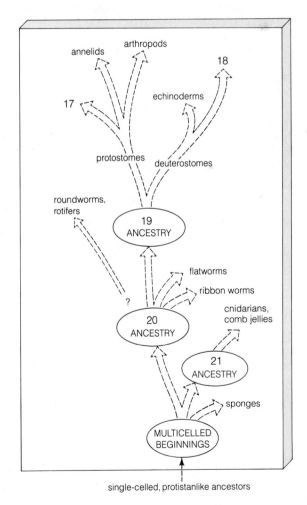

FLATWORMS / RIBBON WORMS / ROUNDWORMS / ROTIFERS (25-II, pp. 414–419)

Fill-in-the-Blanks

A mutant (1)_____ larva is believed to be the ancestor of the free-living flatworms. A shift from radial to bilateral symmetry could have led to (2)_____ _____ of the sort seen in many flatworms; for example, turbellarians have units called (3)_____ that regulate the volume and salt concentrations of their body fluid. Flatworms have no (4)_____ systems, and their (5)_____ system is saclike. Examples of parasitic flatworms are *Schistosoma*, a blood (6)_____ that causes schistosomiasis, and *Taenia saginata*, a tapeworm that attaches to the intestine with a (7)_____ and releases eggs that develop in proglottid segments. (8)_____ have complete digestive tracts, no circular muscles, and only a few longitudinal muscles. Between the gut and body wall of a nematode is a (9)_____ , which contains (10)_____ organs and serves as both a circulatory system and a hydrostatic (11)_____. (12)_____ are tiny abundant aquatic pseudocoelomates that have ciliary crowns attracting them to their food, single-celled organisms.

Labeling

Identify the parts of the animals shown dissected in the accompanying drawings.

13. _____ _____
14. _____
15. _____
16. _____ _____
17. _____
18. _____

Answer exercises 19–22 for the drawing of the accompanying dissected animal.

19. What is the common name (or genus) of the animal dissected? _____

20. Is the animal parasitic? _____

21. Is the animal hermaphroditic? _____

22. Does the animal have a coelom? _____

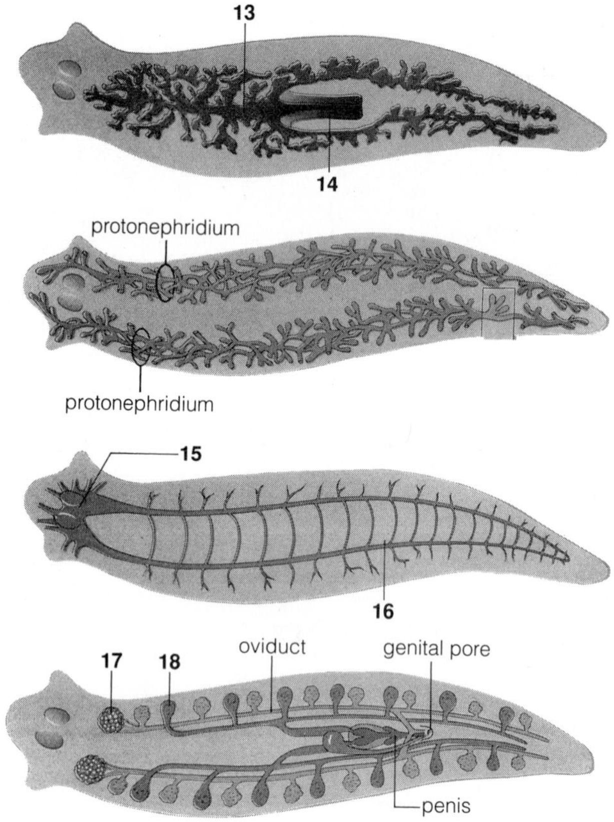

Identify the parts of the animal shown dissected in the drawing below.

23. _____

24. _____

25. _____ _____ _____

26. _____ _____

27. _____ _____ _____

Answer exercises 28–30 for the drawing of the dissected animal above.

28. What is the common name of the animal dissected? _____

29. Is the animal hermaphroditic? _____

30. Does the animal have any kind of coelom? _____

TWO MAIN EVOLUTIONARY ROADS / MOLLUSKS (25-III, pp. 420–425)

Fill-in-the-Blanks

(1)_____ include echinoderms and vertebrates; in this group, the first opening to the gut becomes the

(2)_____ , and the second one to appear becomes the (3)_____. The situation is reversed in the

(4)_____ , which includes annelids (such as (5)_____), arthropods (such as (6)_____ and

crabs), and (7)_____ (such as abalones, limpets, squids, and chambered nautiluses). The most highly

evolved invertebrates are generally considered to be the (8)_____ , which include squids and

octopuses; in terms of sheer size and complexity, the (9)_____ of these animals approach those of

mammals. Like vertebrates, these animals have acute (10)_____ and refined (11)_____ control,

which is well integrated with the activities of the nervous system. In less highly evolved mollusks, a

structure known as the (12)_____ secretes one or more pieces of calcareous armor that protect these

soft-bodied animals from predation. In the cephalopods, the mantle has become a conical cloak that

surrounds the internal organs and the much-reduced shell; seawater moves in and out of the

(13)_____ _____ in a jet-propulsive manner.

Matching

Identify the animals pictured below and on the next page by matching each with the appropriate description.

14. _____ Animal A   I.   Bivalve
15. _____ Animal B   II.   Cephalopod
                                     III.   Gastropod
16. _____ Animal C

Labeling

Identify each numbered part in the drawings below and on the next page by writing its name in the appropriate blank.

17. _____   23. _____   29. _____
18. _____   24. _____   30. _____
19. _____   25. _____   31. _____ _____
20. _____   26. _____   32. _____
21. _____   27. _____   33. _____
22. _____   28. _____   34. _____

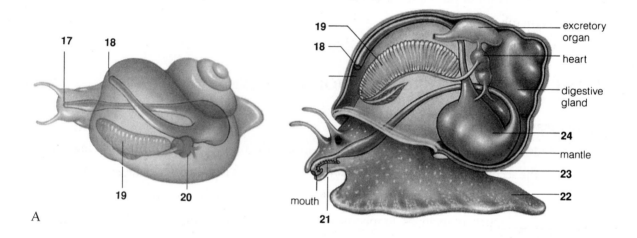

A

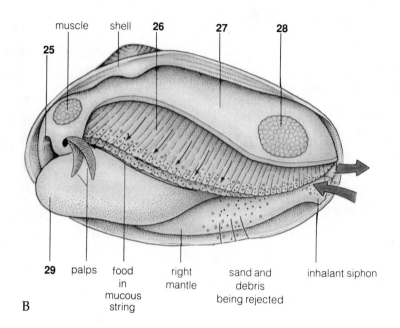

muscle   shell   **26**   **27**   **28**

25

29   palps   food in mucous string   right mantle   sand and debris being rejected   inhalant siphon

B

ANTERIOR

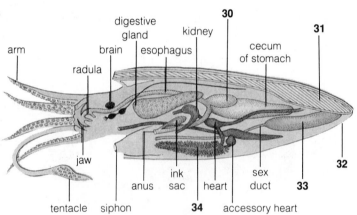

digestive gland   **30**

brain   kidney   **31**

arm   esophagus   cecum of stomach

radula

jaw   **32**

anus   ink sac   heart   sex duct   **33**

tentacle   siphon   **34**   accessory heart

C   POSTERIOR

ANNELIDS (25-IV, pp. 425–427)

Fill-in-the-Blanks

(1)_____ include truly segmented worms such as earthworms, (2)_____ , and leeches; they differ from flatworms in having (first) a complete digestive system with a mouth and (3)_____ and (second) a (4)_____ , a fluid-filled space between the gut and body wall. In a circulatory system, (5)_____ provides a means for transporting materials between internal and external environments. In some annelids, (6)_____ , a protein component of blood, dramatically increases the blood's oxygen-carrying capacity. (7)_____ , which is a repeating series of body parts, is well developed in annelids. In each segment, there are swollen regions of the (8)_____ _____ that control local activity and a pair of (9)_____ that act as kidneys, as well as bristles embedded in the body wall. Many polychaetes (marine worms) have fleshy, paddle-shaped lobes called (10)_____ , which project from the body wall.

Labeling

Identify each indicated part of the illustration below.

11. _____   14. _____   16. _____

12. _____   15. _____ _____   17. _____

13. _____ _____

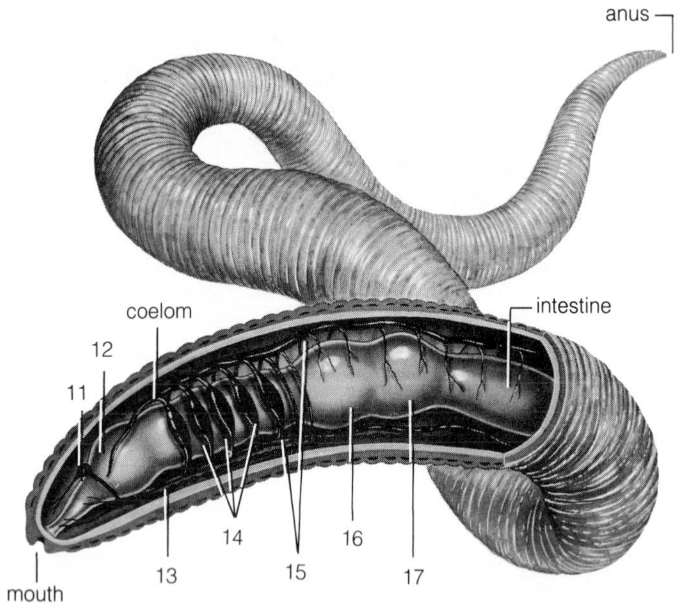

ARTHROPODS (25-V, pp. 428–435)

Fill-in-the-Blanks

Arthropods developed a thickened (1)_____ and a hardened (2)_____. In freshwater and

saltwater environments, arthropods known as (3)_____ came to be well represented. The first land

arthropods were the (4)_____. Their descendants—(5)_____, scorpions, ticks, and mites—are

still living. (6)_____ and millipedes arose later. (7)_____ are thought to have evolved from

centipedelike ancestors. The larger aquatic arthropods extract oxygen from water with (8)_____,

whereas most insects utilize (9)_____ systems, which provide the basis for the highest (10)_____

rates known. Arthropods are thought to have evolved from (11)_____.

Labeling

Identify each numbered part of the animal pictured below.

12. _____

13. _____

14. _____

15. _____

16. _____

17. _____

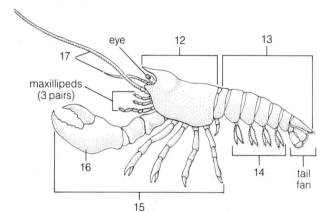

Answer exercises 18–22 for the animal pictured above.

18. Name the animal pictured above. _____

19. Name the subgroup of arthropods to which this animal belongs. _____

20. Does this animal have a coelom? _____

21. Is this animal a protostome or deuterostome? _____

22. Is this animal segmented? _____

Identify each indicated body part in the illustration below.

23. _____   25. _____

24. _____ _____   26. _____

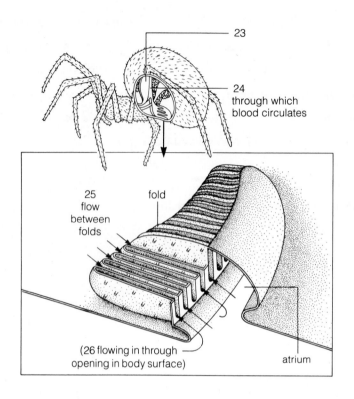

Answer exercises 27–29 for the animal pictured above.

27. Name the subgroup of arthropods to which the animal above belongs. _____

28. Does this animal have one or more gills? _____

29. Name the structure shown enlarged in the lower part of the picture above. _____ _____

ECHINODERMS (25-VI, pp. 436–438)

Fill-in-the-Blanks

Only two prominent groups of (1)_____ have survived to the present—the (2)_____ and the chordates. In echinoderms, (3)_____ symmetry has been overlaid on an earlier bilateral heritage; most echinoderms still go through a free-swimming (4)_____ symmetrical larval stage. Echinoderm locomotion is based on constant circulation of seawater through a (5)_____-_____ system of canals and (6)_____ _____. (7)_____ _____ have a rounded, globose body with bristling spines. (8)_____ have many long feather-duster arms; both mouth and anus lie within the circlet of arms.

Labeling

Identify each indicated part of the illustration below.

9. _____ _____
10. _____ _____
11. _____
12. _____

13. _____
14. _____ _____
15. _____
16. _____ _____

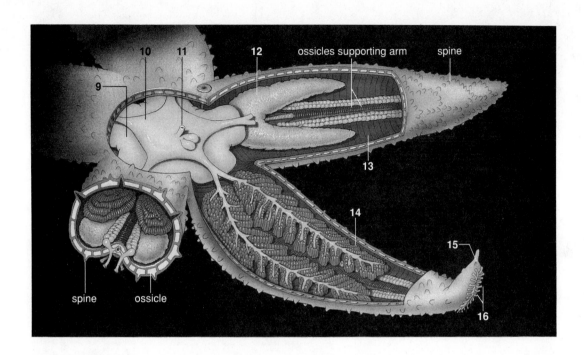

Answer exercises 17–19 for the animal pictured above.

17. Name the animal shown above. _____

18. Does it have a true coelom? _____

19. Is it a protostome or a deuterostome? _____

Identifying

20. *Name the system* shown below.

_____

21. Identify each creature below by its common name.

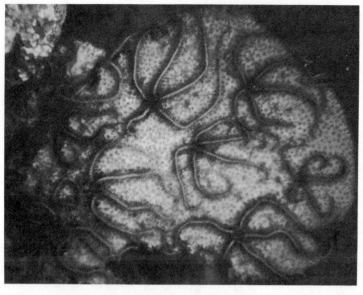

a. _____

b. _____

c. _____

d. _____

Chapter Terms

The following page-referenced terms are important; they were mostly in boldface type in the chapter. Refer to the instructions given in Chapter 1, p. 4.

vertebrates (405)
invertebrates (405)
radial symmetry (406)
body symmetry (406)
bilateral symmetry (406)
cephalization (406)
gut (406)
coelom (406)
body cavity (406)
segmentation, segmented (407)
sponges (408)
larva, -ae (409)
cnidarians (410)
medusa, -ae (410)
polyp (410)
epithelium, -ia (410)

nerve cells (410)
sensory cells (410)
colonial organisms (413)
gonads (413)
comb jellies (413)
flatworms (414)
pharynx (414)
turbellarians (414)
hermaphrodites (414)
flukes (trematodes) (414)
parasite (414)
tapeworms (cestodes) (415)
ribbon worms (418)
proboscis (418)
roundworms (418)
cuticle (419)
rotifers (419)

plankton (419)
protostomes (420)
deuterostomes (420)
radial cleavage (420)
spiral cleavage (420)
mollusks (420)
gills (421)
chitons (421)
gastropods (422)
sea slugs (nudibranchs) (422)
bivalves (423)
cephalopods (424)
annelids (425–427)
earthworms (425)
leeches (425)
polychaetes (426)
seta, -tae (426)

parapodium, -ia (426)
cuticle (426)
nerve cord (426)
ganglion, -lia (426)
nephridium, -dia (427)
arthropods (428)
exoskeleton (428)
molting (428)
specialization (428)
metamorphosis (429)
chelicerates (429–431)
crustaceans (431)
millipedes (432)
centipedes (432)
insects (432)
echinoderms (436)

Self-Quiz

Multiple-Choice

Select the best answer.

___ 1. Which of the following is *not* true of sponges? They have no _____.
   a. distinct cell types
   b. nerve cells
   c. muscles
   d. gut

___ 2. Which of the following is *not* a protostome?
   a. earthworm
   b. crayfish or lobster
   c. sea star
   d. squid

___ 3. Bilateral symmetry is characteristic of _____.
   a. cnidarians
   b. sponges
   c. jellyfish
   d. flatworms

___ 4. Flukes and tapeworms are parasitic _____.
   a. leeches
   b. flatworms
   c. nematodes
   d. annelids

___ 5. Insects include _____.
   a. spiders, mites, and ticks
   b. centipedes and millipedes
   c. termites, aphids, and beetles
   d. all of the above

___ 6. Creeping behavior and a mouth located toward the "head" end of the body may have led, in some evolutionary lines, to _____.
   a. development of a circulatory system with blood
   b. sexual reproduction
   c. feeding on nutrients suspended in the water (filter feeding)
   d. concentration of sense organs in the head region

___ 7. Which of the following is associated with the shift from radial to bilateral body form?
   a. a circulatory system
   b. a one-way gut
   c. paired organs
   d. the development of a water-vascular system

___ 8. The _____ body plan is characterized by simple gas-exchange mechanisms, two-way traffic through a relatively unspecialized gut, and a thin body with all cells fairly close to the gut.
   a. annelid
   b. nematode
   c. echinoderm
   d. flatworm

___ 9. The _____ have a tough cuticle, longitudinal muscles, and a complete digestive system, and they can live under anaerobic conditions.
   a. nematodes
   b. cnidarians
   c. flatworms
   d. echinoderms

___ 10. ___ insulates various internal organs from the stresses of body-wall movement and bathes them in a liquid through which nutrients and waste products can diffuse.
   a. A coelom
   b. Mesoderm
   c. A mantle
   d. A water-vascular system

___ 11. The annelid _____ may resemble the ancestral structure from which the vertebrate kidney evolved.
   a. trachea
   b. nephridium
   c. mantle
   d. parapodia

Matching

Match each phylum below with the corresponding characteristics (a–k) and representatives (A–P). A phylum may match with more than one letter from the group of representatives.

___ , ___ 12. Annelida

___ , ___ 13. Arthropoda

___ , ___ 14. Chordata

___ , ___ 15. Cnidaria

___ , ___ 16. Ctenophora

___ , ___ 17. Echinodermata

___ , ___ 18. Mollusca

___ , ___ 19. Nematoda

___ , ___ 20. Platyhelminthes

___ , ___ 21. Porifera

___ , ___ 22. Rotifera

a. radial (biradial) symmetry + no stinging cells + comb plates
b. choanocytes (= collar cells) + spicules
c. jointed legs + an exoskeleton
d. gill slits in pharynx + dorsal, tubular nerve cord + notochord
e. pseudocoelomate + wheel organ + soft body
f. soft body + mantle; may or may not have radula or shell
g. bilateral symmetry + blind-sac gut
h. radial symmetry + blind-sac gut; stinging cells
i. body compartmentalized into repetitive segments; coelom containing nephridia (= primitive kidneys)
j. tube feet + calcium carbonate structures in skin
k. complete gut + bilateral symmetry + cuticle; includes many parasitic species, some of which are harmful to humans

A. Dinosaurs
B. Corals, sea anemones, and *Hydra*
C. Salamanders and toads
D. Whales and opossums
E. Tapeworms and *Planaria*
F. Insects
G. Jellyfish and the Portuguese man-of-war
H. Sand dollars and starfishes
I. Earthworms and leeches
J. Lobsters, shrimp, and crayfish
K. Organisms with spicules and choanocytes
L. Scorpions and millipedes
M. Octopuses and oysters
N. Flukes
O. Hookworm, trichina worm
P. Comb jellies, sea gooseberries

Chapter Objectives/Review Questions

This section lists general and detailed chapter objectives that can be used as review questions. You can make maximum use of these items by writing answers on a separate sheet of paper. Fill in answers where blanks are provided. To check for accuracy, compare your answers with information given in the chapter or glossary.

Page   *Objectives/Questions*

(406–407) 1. Distinguish radial symmetry from bilateral symmetry, and acoelomate from pseudocoelomate.

(406–407) 2. Define coelom, and list two benefits that the development of a coelom brings to an animal.

(408)   3. Be able to reproduce from memory a phylogenetic tree that expresses the relationships between the major groups of animals.

(408)   4. List two characteristics that distinguish sponges from other animal groups.

(410)   5. Describe the two cnidarian body types.

(410)   6. Explain how radial symmetry might be more advantageous to floating or sedentary animals than bilateral symmetry.

(413)   7. State what nematocysts are used for and explain how they operate.

(410–413) 8. Tell how cnidarians obtain and digest food and tell what they do with food they cannot digest.

(414)   9. List the three main types of flatworms.

(418–419) 10. Describe the body plan of roundworms, comparing its various systems with those of the flatworm body plan.

(420)   11. Define protostome and deuterostome and give examples of each group.

(420, 425) 12. Define mantle and tell what role it plays in the molluscan body.

(425)   13. Explain why you think cephalopods came to have such well-developed sensory and motor systems and are able to learn.

(425)   14. Name the three groups of annelids and give a specific example from each group.

(426)   15. Define segmentation and explain how it is related to the development of muscular and nervous systems.

(428)   16. Explain how the development of a thickened cuticle and a hardened exoskeleton affected the ways that arthropods lived.

(428–429) 17. State which ancestors are thought to have given rise to the arthropods and tell whether you think any other major groups of animals descended from the arthropods.

(428)   18. List six different groups of arthropods.

(436–437) 19. Describe how locomotion occurs in echinoderms.

Integrating and Applying Key Concepts

Scan Table 25.2 to verify that most highly evolved animals have a complete gut, a closed blood-vascular system, both central and peripheral nervous systems, and are dioecious. Why do you suppose having two sexes in separate individuals is considered to be more highly evolved than the monoecious condition utilized by earthworms? Wouldn't it be more efficient if *all* individuals in a population could produce both kinds of gametes? Cross-fertilization would then result in both individuals being able to produce offspring.

Critical Thinking Exercises

1. The text mentions the commonly accepted idea that bilateral and radial animals diverged from a planula-like common ancestor. If this is the case, why do only Cnidaria, and no bilateral phyla, produce nematocysts?

    a. Nematocysts would not be advantageous to bilateral animals.
    b. The genetic changes necessary to form nematocysts could not occur in the bilateral line.
    c. Nematocysts appeared in the Cnidaria line after divergence from the bilateral line.

d. Nematocysts were lost early from the bilateral line.

e. The common ancestor produced nematocysts.

ANALYSIS

a. It seems reasonable that nematocysts would be equally advantageous to any soft-bodied, aquatic animal. In fact, some animals have developed the ability to conserve the nematocysts from the cnidarians they eat and install the nematocysts in the integument of the predators. Clearly, this is advantageous to these bilateral animals.

b. If both lines have a common ancestor, they have a common genetic background, and a change that occurred in one line could occur in the other.

c. If this happened, then the bilateral line would have to develop nematocysts independently; the bilateral animals could not inherit the organs from the common ancestor or from the parallel line of radial animals. While many examples of such convergent evolution exist, evolution is based on accidental genetic changes and does not produce all possible developments.

d. This is a possible explanation. Many examples exist of structures that were lost from an evolutionary lineage. However, nematocysts would seem to be highly advantageous, and explaining their loss might be difficult. Furthermore, the hypothesis that nematocysts appeared only once, after the two lines diverged, is simpler.

e. If the common ancestor already had nematocysts, then the bilateral line must have lost them after divergence from the radial line. This hypothesis makes it more, not less, difficult to explain why **no bilateral phyla produce nematocysts.**

2. The text mentions an estimate that the progeny of a single female fly over six generations would number more than 5 trillion. The text points out that this calculation requires the assumption that all the progeny survive and reproduce. Which of the following additional assumptions is also required for this kind of calculation?

a. All flies produce the same number of progeny.

b. All progeny are female.

c. All flies have the same life span.

d. None of the progeny have died by the end of the time period.

e. None of the progeny mate more than once.

ANALYSIS

a. This assumption is usually made, at least with respect to *average* number of progeny per fly. The calculation has to say that one fly has so many offspring and each of those has so many and each of those has so many, and so forth, and add up all the offspring at the end. You could use a different number for each generation or even for each individual, but then you would be making many, more complicated assumptions.

b. This assumption is not necessary, but if the progeny include males and females, it is necessary to assume that none mated with each other. If they did, then the total number of progeny calculated would have to be smaller.

c. This assumption is usually made, at least with respect to reproductive life span, when the average number of progeny per fly is calculated. You multiply the average number of offspring per mating by the average number of matings per fly and assume they are both the same for every fly. Sometimes this kind of calculation results in an estimate of the total mass of offspring, or the total area covered by offspring, or the total food consumed by offspring. In these cases, the assumption that all flies are the same average weight or size or eat the same average amount is also necessary.

d. It is not necessary to assume this, as long as any dead flies all lived long enough to meet the average production rate.

e. As long as each fly produces the average number of offspring, the number of matings required is irrelevant.

3. The pattern of water circulation through a sponge is inward through the pores and outward through the osculum. Why doesn't it go the other way?

   a. The pores have a greater total cross-sectional area than the osculum, so more water can move through them.

   b. Pumping of water by flagella is mechanically more efficient in small tubes than in large tubes.

   c. Small tubes have a greater total surface area than large tubes.

   d. The flow pattern is a secondary result of the evolutionary history of sponges.

   e. Flow inward through pores brings the water into contact with the collar cells before any other cell type, facilitating removal of food particles.

ANALYSIS

   a. The pores do have greater total cross-sectional area than the osculum, but the same amount of water leaves through the osculum as enters through the pores. If it did not, the sponge would continually change volume.

   b. Pumping by flagella does move water more efficiently in a small tube, like the pores, than in a large one, like the osculum. But the flagella could be in the pores and pump in either direction.

   c. Small tubes do have a greater surface area, which facilitates diffusion of oxygen from the water into the cells, but it would do so with water flow in either direction.

   d. This may not be the actual explanation, but it is the only one of the choices that *could* be valid. The ancestor of sponges was likely a solid organism like a sponge larva. It would have had to consume whatever food particles fell on its surface. Its feeding efficiency could have been increased by forming surface pits and then lining the pits with flagellated cells to circulate water into them. The deeper the pits became, the more efficient feeding would become, until the pits opened through into a single tube leading back out. This hypothetical scenario would produce sponges with flagella that beat a current of water inward through the pores, but the direction would no longer have adaptive significance. Not *every* single structural feature produced in evolution increases fitness or is maximally adapted.

   e. As shown in Figure 25.7, the incoming water actually comes in contact with other cell types before it arrives at the collar cells.

Answers

Answers to Interactive Exercises

OVERVIEW OF THE ANIMAL KINGDOM/ SPONGES/CNIDARIANS/COMB JELLIES (25-I)
1. a. Porifera; b. Cnidaria; c. Platyhelminthes; d. Nematoda; e. Mollusca; f. Annelida; g. Arthropoda; h. Echinodermata; i. *Hydra* (*Obelia*, Portuguese man-of-war); j. Snails (nudibranchs, oysters); k. earthworms (oligochaetes); l. Vertebrates; 2. invertebrates; 3. metazoans; 4. Sponges; 5. collar cells; 6. Communication; 7. nerve; 8. radially; 9. nematocysts; 10. planula; 11. medusa; 12. Comb jellies; 13. feeding polyp; 14. reproductive polyp; 15. female medusa; 16. planula; 17. mollusks; 18. chordates; 19. coelomate; 20. bilateral; 21. radial.

FLATWORMS/RIBBON WORMS/ ROUNDWORMS/ROTIFERS (25-II)
1. planuloid; 2. paired organs; 3. protonephridia; 4. respiratory (circulatory); 5. digestive; 6. fluke; 7. scolex; 8. Nematodes (Roundworms); 9. pseudocoel; 10. reproductive; 11. skeleton; 12. Rotifers; 13. branching gut; 14. pharynx; 15. brain; 16. nerve cord; 17. ovary;

18. testis; 19. *Dugesia* (planarian); 20. no; 21. yes; 22. no; 23. pharynx; 24. gut; 25. male reproductive organ; 26. false coelom; 27. sperm storage vesicle; 28. roundworm; 29. no; 30. yes (a "false" coelom).

TWO MAIN EVOLUTIONARY ROADS/MOLLUSKS (25-III)
1. Deuterostomes; 2. anus; 3. mouth; 4. protostomes; 5. earthworms; 6. insects (spiders); 7. mollusks; 8. cephalopods; 9. brains (eyes); 10. vision; 11. motor (movement; muscular); 12. mantle; 13. mantle cavity; 14. III; 15. I; 16. II; 17. mouth; 18. anus; 19. gill; 20. heart; 21. radula; 22. foot; 23. shell; 24. stomach; 25. mouth; 26. gill; 27. mantle; 28. muscle; 29. foot; 30. stomach; 31. internal shell; 32. mantle; 33. gonad; 34. gill.

ANNELIDS (25-IV)
1. Annelids; 2. polychaetes (marine worms); 3. anus; 4. coelom; 5. blood; 6. hemoglobin; 7. Segmentation; 8. nerve cord; 9. nephridia; 10. parapodia; 11. brain; 12. pharynx; 13. nerve cord; 14. hearts; 15. blood vessel; 16. crop; 17. gizzard.

ARTHROPODS (25-V)

1. cuticle; 2. exoskeleton; 3. crustaceans; 4. arachnids;
5. spiders; 6. Centipedes; 7. Insects; 8. gills; 9. tracheal;
10. metabolic; 11. annelids; 12. cephalothorax
(carapace); 13. abdomen; 14. swimmerets; 15. legs;
16. cheliped; 17. antennae; 18. lobster; 19. crustaceans;
20. yes; 21. protostome; 22. yes; 23. heart; 24. body
cavity; 25. "blood" (body fluids); 26. air; 27. chelicerates;
28. no; 29. book lung.

ECHINODERMS (25-VI)

1. deuterostomes; 2. echinoderms; 3. radial;
4. bilaterally; 5. water-vascular; 6. tube feet; 7. Sea
urchins; 8. Crinoids; 9. lower stomach; 10. upper
stomach; 11. anus; 12. gonad; 13. coelom; 14. digestive
gland; 15. eyespot; 16. tube feet; 17. sea star, starfish;
18. yes; 19. deuterostome; 20. water-vascular system;
21. a. brittle stars; b. sea urchin; c. sea cucumber;
d. feather star (crinoid).

Answers to Self-Quiz

1. a; 2. c; 3. d; 4. b; 5. c; 6. d; 7. c; 8. d; 9. a; 10. a; 11. b;
12. i, I; 13. c, F, J, L; 14. d, A, C, D; 15. h, B, G; 16. a, P;
17. j, H; 18. f, M; 19. k, O; 20. g, E, N; 21. b, K; 22. e.

26

EVOLUTION OF VERTEBRATES

Interactive Exercises

ON THE ROAD TO VERTEBRATES / FISHES (26-I, pp. 443–453)

Fill-in-the-Blanks

Four major features distinguish chordates from all other animals: a hollow dorsal (1)_____ _____ , a (2)_____ with slits in its wall, a (3)_____ , and a tail that extends past the anus at least during part of its life. In some chordates, the (4)_____ chordates, the notochord is *not* divided into a skeletal column of separate, hard segments; in others, the (5)_____ , it is. Invertebrate chordates living today are represented by tunicates and (6)_____ , which obtain their food by (7)_____ - _____ ; they draw in plankton-laden water through the mouth and pass it over sheets of mucus, which trap the particulate food before the water exits through the (8)_____ _____ in the pharynx. (9)_____ are among the most primitive of all living chordates; when they are tiny, they look and swim like (10)_____ . A rod of stiffened tissue, the (11)_____ , runs the length of the larval body; it was the forerunner of the chordate's (12)_____. The ancestors of the chordates existed before the (13)_____ era began (prior to 570 million years ago).

Even though the adult forms of acorn worms, echinoderms, and chordates look very different, their embryonic development has much in common: Their eggs cleave radially, the mouth forms from a second opening into the embryo, and their coeloms are formed from outpouchings of the gut. This pattern of development is said to be (14)_____. Chordates are hypothesized to have developed from a mutated deuterostome (15)_____ of a sessile, filter-feeding adult. A (16)_____ is an immature, motile form of an organism, but if a (17)_____ occurred that caused (18)_____ _____ to become functional in the larval body, then the motility of a larva that could reproduce could have been more advantageous in finding food and avoiding (19)_____ than the sessile nature of the "adult," and in time the old adult stages in the species would be dispensed with. Is there any evidence to suggest that such an event could have happened? Among some species of (20)_____—even among some amphibians—are larvae with functional sex organs, and these larvae can reproduce generation after generation!

The ancestors of the vertebrate line may have been mutated forms of their closest relatives, the (21)_____, in which the notochord became segmented and the segments became hardened (22)_____. The vertebral column was the foundation for fast-moving (23)_____, some of which were ancestral to all other vertebrates. The evolution of (24)_____ intensified the competition for prey and the competition to avoid being preyed upon; animals in which mutations expanded the nerve cord into a (25)_____ that enabled the animal to compete effectively survived more frequently than their duller-witted fellows and passed along their genes into the next generations. Fins became (26)_____; the paired fins in some fishes became (27)_____ and equipped with skeletal supports; these forms set the stage for the development of legs, arms and wings in later groups.

Labeling

Name the structures numbered on page 295.

28. _____

29. _____ _____ _____

30. _____

31. _____ _____ _____

32. _____

33. _____

34. _____

35. _____ _____

36. _____ _____

37. _____ _____

38. _____

39. _____ _____ _____

40. _____ _____ _____

41. _____

42. _____

43. _____ _____ _____

Exercises 44–47 refer to the illustrations on page 295.

44. Name the creature in illustration A._____

45. Name the creature in illustration B._____

46. Name the creature in illustration C._____

47. Name the creature in illustration D._____

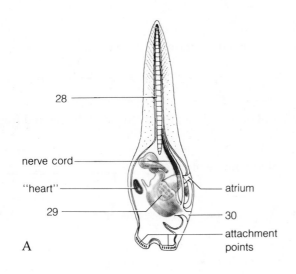

28

nerve cord

"heart"

29

atrium

30

attachment points

A

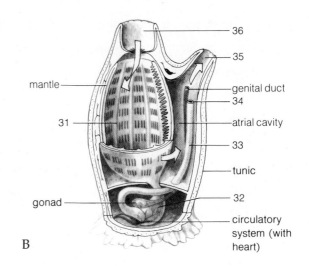

36

35

mantle

genital duct

34

31

atrial cavity

33

tunic

gonad

32

circulatory system (with heart)

B

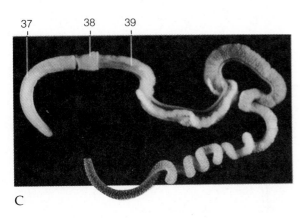

37   38   39

C

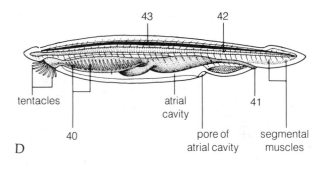

43   42

tentacles

atrial cavity

pore of atrial cavity

segmental muscles

40   41

D

48. Name the creature whose head is illustrated at the right._____

49. What structure occupies most of the head space of the creature at the right?_____

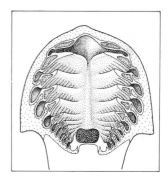

Exercises 50–60 refer to the evolutionary diagram illustrated below.

50. What single feature do the lampreys, hagfishes, and extinct ostracoderms have in common that is different from the placoderms?_____

51. How did ostracoderms feed?_____

52. A mutation in ostracoderm stock led to the development of what in all organisms that descended from ①?_____

53. Mutation at ② led to the development of an endoskeleton made of what?_____

54. Mutations at ③ led to an endoskeleton of what?_____

55. Mutations at ④ led to which spectacularly diverse fishes that have delicate fins originating from the dermis?_____

56. Mutations at ⑤ led to which fishes whose fins incorporate fleshy extensions from the body?_____
_____

57. Which branch, ④ or ⑤, gave rise to the amphibians?_____

58. Which branch gave rise to the modern bony fishes?_____

59. In which period was this (#58) thought to have occurred?_____

60. Approximately how many million years ago did the fork in the evolutionary path that led to the amphibians occur?_____

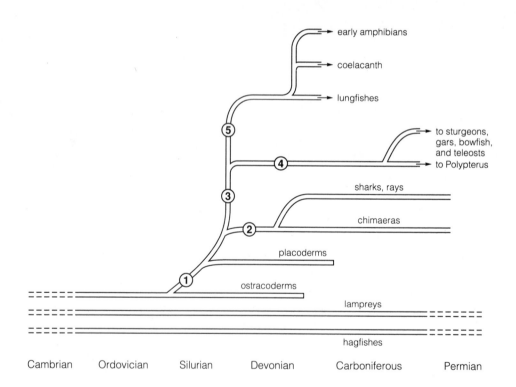

AMHIBIANS (26-II, pp. 453–455)

Fill-in-the-Blanks

Natural selection acting on lobe-finned fishes during the Devonian period favored the evolution of ever

more efficient (1)_____ used in gas exchange and stronger (2)_____ used in locomotion.

Without the buoyancy of water, an animal traveling over land must support its own weight against

the pull of gravity. The (3)_____ of early amphibians underwent dramatic modifications that

involved evaluating incoming signals related to vision, hearing, and (4)_____. Bones that

support (5)_____ in fishes became smaller and translated weak airborne sound waves into fluid

vibrations in the inner ear. Although fish have (6)_____-chambered hearts, amphibians have

(7)_____-chambered hearts. Early in amphibian evolution, mutations may have created the third

chamber, which added a second (8)_____ in addition to the already existing atrium and ventricle.

The Carboniferous period brought humid, forested swamps with an abundance of aquatic invertebrates

and (9)_____—ideal prey for amphibians.

There are three groups of existing amphibians: (10)_____ , frogs and toads, and caecilians.

Amphibians require free-standing (11)_____ or at least a moist habitat to (12)_____. Amphibians

that spend most of the time in water use (13)_____ , lungs, and skin for gas exchange; those that spend

most of their time on land use lungs, skin, and the (14)_____ _____ _____ _____.

Amphibian skin generally lacks protective (15) _____ but contains many glands, some of which

produce (16)_____. Salamanders are notable in that the adults of some species retain many

(17)_____ features, and some groups have larvae that are fully able to (18)_____.

In the late Carboniferous, (19)_____ began a major adaptive radiation into the lush habitats on

land, and only amphibians that mutated and developed certain (20)_____ features were able to follow

them and exploit an abundant food supply. Several features helped: modification of (21)_____ bones

favored swiftness, modification of teeth and jaws enabled them to feed efficiently on a variety of prey

items, and the development of a (22)_____ _____ egg protected the embryo inside from drying

out, even in dry habitats.

REPTILES (26-III, pp. 456–460)

Fill-in-the-Blanks

Today's reptiles include (1)_____ , crocodilians, snakes, and (2)_____. All rely on (3)_____ fertilization, and most lay leathery eggs, but some (4)_____ _____ and (5)_____ give birth to fully formed young. Although crocodilians have a lizardlike body, their circulatory system more closely resembles that of (6)_____. Although amphibians originated during Devonian times, ancestral "stem" reptiles appeared during the (7)_____ period. Reptilian groups living today that have existed on Earth longest are the (8)_____ ; their ancestral path diverged from that of the "stem" reptiles during the (9)_____ period.

Crocodilian ancestors appeared in the early (10)_____ period, about 220 million years ago; they have a (11)_____-chambered heart, unlike other reptiles, which have an incompletely formed septum dividing the two ventricles. Snake and lizard stocks diverged from the tuatara line during the late (12)_____ period, about 140 million years ago. (13)_____ are more closely related to extinct dinosaurs and crocodiles than to any other existing vertebrates; they, too, have a (14)_____-chambered heart. Mammals have descended from therapsids, which in turn are descended from the (15)_____ group of reptiles, which diverged earlier from the stem reptile group during the (16)_____ period, approximately 320 million years ago.

17. Consult the evolutionary diagram below and imagine what sort of mutation(s) occurred at the numbered places.

a. What may have occurred at ①? _____

b. What may have occurred at ②? _____

c. What may have occurred at ③? _____

d. What may have occurred at ④? _____

e. What may have occurred at ⑤? _____

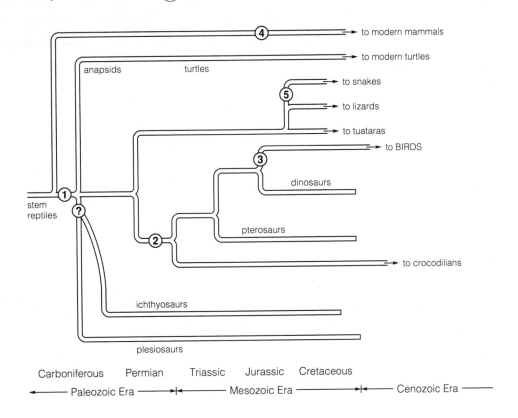

Labeling

Label the structures pictured at the right.

18. _____

19. _____

20. _____ _____

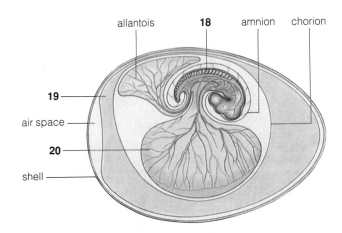

BIRDS (26-IV, pp. 460–462)

Fill-in-the-Blanks

Birds descended from (1)_____ that ran around on two legs some 160 million years ago. All birds have (2)_____ that insulate and help get the bird aloft. Generally, birds have a greatly enlarged (3)_____ to which flight muscles are attached. Bird bones contain (4)_____ _____ , and air flows *through* sacs, not into and out of them, for gas exchange in the lungs. Birds also lay (5)_____ _____ eggs, have complex courtship behaviors, and generally nurture their offspring. Birds are able to regulate their body (6)_____ , which is generally higher than that of mammals. Existing birds do not have socketed (7)_____ in their beaks.

Labeling

Name the structures pictured at the right.

8. _____

9. _____

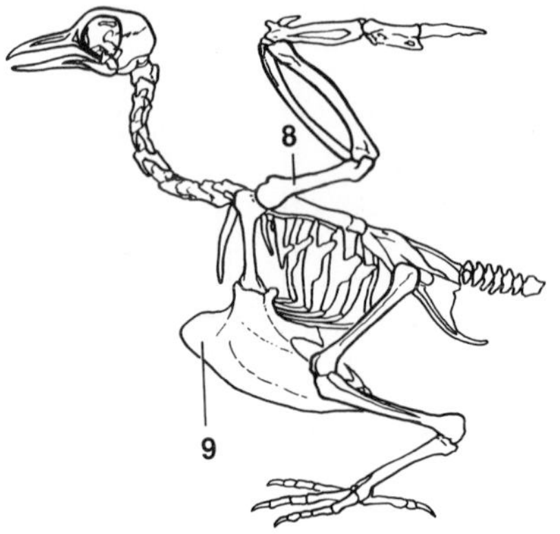

MAMMALS / SUMMARY (26-V, pp. 463–466)

Fill-in-the-Blanks

There are three groups of existing mammals: those that lay eggs (examples are the (1)_____ and the spiny anteater), those that are (2)_____ (examples are the opossum and the kangaroo), and those that are (3)_____ mammals (there are more than 4,500 species of these). Mammals regulate their body temperature, have a (4)_____-chambered heart, and show a high degree of parental nurture. Most mammals have (5)_____ as a means of insulation, and mammalian mothers generally suckle their young with milk.

Chapter Terms

The following page-referenced terms are important; they were mostly in boldface type in the chapter. Refer to the instructions given in Chapter 1, p. 4.

chordates, Chordata (444)
vertebrates (444)
invertebrate chordates (444)
notochord (444)
nerve cord (444)
pharynx (444)
gill slits (444)
urochordates (445)
larva (445)
cephalochordates (446)
endoskeleton (447)

fishes (448)
ostracoderms (448)
agnathans (448)
placoderms (448)
swim bladder (449)
fins (449)
scales (449)
jawless fishes (450)
lamprey (450)
cartilaginous fishes,
    (Chondrichthyes) (450)

bony fishes (450)
amphibian (453)
salamanders (454)
frogs (454)
toads (454)
caecilians (455)
reptiles (456)
shelled amniote egg (456)
turtles (458)
lizards (458)
snakes (458)

tuataras (460)
crocodilians (460)
birds (460)
feathers (461)
bird wing (462)
mammals (463)
synapsid reptiles (463)
egg-laying mammals (463)
pouched mammals (464)
marsupials (464)
placental mammals (464)

Self-Quiz

Multiple-Choice

Select the best answer.

___ 1. Filter-feeding chordates rely on _____ , which have cilia that create water currents and mucous sheets that capture nutrients suspended in the water.
   a. notochords
   b. differentially permeable membranes
   c. filiform tongues
   d. gill slits

___ 2. In true fishes, the gills serve primarily _____ function.
   a. a gas-exchange
   b. a feeding
   c. a water-elimination
   d. both a feeding and a gas-exchange

___ 3. The heart in amphibians _____ .
   a. pumps blood more rapidly than the heart of fish
   b. is efficient enough for amphibians but would not be efficient for birds and mammals
   c. has three chambers (ventricle and two atria)
   d. all of the above

___ 4. The feeding behavior of true fishes selected for highly developed _____ .
   a. parapodia
   b. notochords
   c. sense organs
   d. gill slits

Identifying

Provide the common name and the major chordate group to which each creature pictured below and on pages 302–304 belongs.

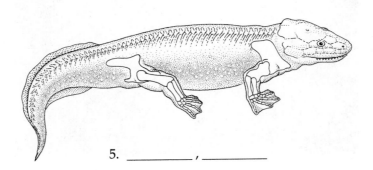

5. _____ , _____

6. _____ , _____

7. _____ , _____

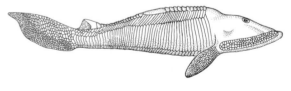

8. _____ , _____

9. _____ , _____

10. _____ , _____

11. _____ , _____

12. _____ , _____

13. _____ , _____

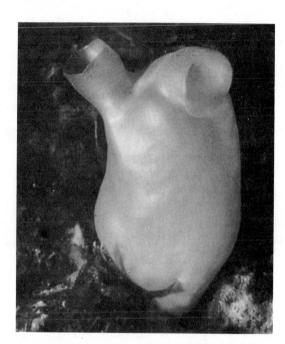

14. _____ , _____

15. _____ , _____

Matching

Match the following groups and classes with the corresponding characteristics (a–i) and representatives (A–I).

16. _____ Agnatha

17. _____ Amphibia

18. _____ Aves

19. _____ Cephalochordata

20. _____ Chondrichthyes

21. _____ Mammalia

22. _____ Osteichthyes

23. _____ Reptilia

24. _____ Urochordata

a.   hair + vertebrae
b.   feathers + hollow bones
c.   jawless + cartilaginous skeleton (in existing species)
d.   two pairs of limbs (usually) + glandular skin + "jelly"-covered eggs
e.   amniote eggs + scaly skin + bony skeleton
f.   invertebrate + sessile adult
g.   jaws + cartilaginous skeleton + vertebrae
h.   invertebrates; notochord stretches from head to tail
i.   bony skeleton + skin covered with scales and mucus

A.   lancelet
B.   loons, penguins, and eagles
C.   tunicates, sea squirts
D.   sharks and manta rays
E.   lampreys and hagfishes (and ostracoderms)
F.   true eels and sea horses
G.   lizards and turtles
H.   caecilians and salamanders
I.   platypuses and opossums

Chapter Objectives/Review Questions

This section lists general and detailed chapter objectives that can be used as review questions. You can make maximum use of these items by writing answers on a separate sheet of paper. Fill in answers where blanks are provided. To check for accuracy, compare your answers with information given in the chapter or glossary.

Page	Objectives/Questions
(443–444)	1. List three characteristics found only in chordates.
(445–447)	2. Describe the adaptations that sustain the sessile or sedentary life-style seen in primitive chordates such as tunicates and lancelets.
(447)	3. State what sort of changes occurred in the primitive chordate body plan that could have promoted the emergence of vertebrates.
(448–453)	4. Describe the differences between primitive and advanced fishes in terms of skeleton, jaws, special senses, and brain.
(452–454)	5. Describe the changes that enabled aquatic fishes to give rise to land dwellers.
(455, 460, 462)	6. State what kind of heart each of the four groups of four-limbed vertebrates has and list the principal skin structures that each produces.
(462–464)	7. Discuss the effects that increased parental nurture of offspring in birds and mammals has had on courtship behavior and reproductive physiology.

Integrating and Applying Key Concepts

Birds and mammals both have four-chambered hearts, high metabolic rates, and regulate their body temperatures efficiently. Both groups evolved from reptiles, so one would think that those same traits would have developed in ancestral reptiles. Data suggest that most reptiles have a heart intermediate between three and four chambers, lower metabolic rates, and body temperatures that are not well regulated and tend to rise and fall in accord with the environmental temperature. If the three traits mentioned in the first sentence had developed in reptilian groups, how might their lives have been different?

Critical Thinking Exercises

1. The text mentions that the sex of a turtle is determined by the temperature at which the embryo develops. Which of the following molecular mechanisms would best account for this observation?

   a. The DNA that codes for one pathway of sexual differentiation breaks down at higher temperatures.
   b. Higher temperatures inhibit editing of RNA transcripts.
   c. RNA polymerase is more active at higher temperatures.
   d. The repressor of genes for one pathway of sexual differentiation is inactivated at higher temperatures.
   e. The inducer of genes for one pathway of sexual differentiation is inactivated at higher temperatures.

ANALYSIS

   a. This mechanism could account for the pattern of sexual differentiation; however, because all bonds are the same between adjacent nucleotides, it is very difficult to propose a mechanism to make one nucleotide sequence more heat-labile than another.
   b. Again, specificity is a problem. Inhibition of editing would affect all genes equally.
   c. The explanation has to produce a differential effect on two sets of genes. Any effect on the rate of transcription mediated through RNA polymerase would affect all genes equally.
   d. Repressors are specific; each one affects only certain genes. Furthermore, repressors are proteins and as such are temperature-sensitive. Some repressors could easily be inhibited more than others by elevated temperatures and result in expression of one set of genes while another set remained inactivated by a different repressor.
   e. Inducers also are specific and could be differentially activated by the chemical and physical environment. However, inducers are often small molecules that are not affected by temperature changes within the range in which the embryos could survive.

2. Recent evidence has been interpreted to mean that at least some dinosaurs were homeothermic. If this interpretation is valid, which of the following hypotheses does it support most strongly?

   a. The ancestor of the dinosaurs was also homeothermic.
   b. Birds belong within the reptile class.
   c. Dinosaurs gave rise to mammals.
   d. Therapsids gave rise to mammals.
   e. Dinosaurs were driven to extinction by the drop in temperature caused by ash clouds from a giant meteor that struck earth.

ANALYSIS

   a. Organisms produce offspring like themselves, so the characteristics of ancestors can, to some extent, be inferred from those of descendants. But evolution also produces populations with new characteristics, so the inference is very limited.
   b. It is clear that birds evolved from a dinosaurian ancestor. The question is whether they have diverged enough to be classified as a new class or are still a subdivision of the reptiles. The more important characteristics shared by two groups, the closer their taxonomic affinity. The existence of homeothermic dinosaurs, clearly belonging among the reptiles, would reduce the number of important differences between birds and reptiles and make it more likely that they belong to the same class.
   c. This hypothesis might seem to be supported, because descendants should share characteristics with their ancestors. However, many characteristics have appeared independently more than once in evolutionary lineages. Mammals and dinosaurs, even homeothermic ones, are still less alike than mammals and therapsids.
   d. This hypothesis is not supported by the possibility of homeothermic dinosaurs. If anything, as discussed in (c), it is weakly contradicted.
   e. This hypothesis is also slightly weakened if dinosaurs were homeothermic, because warm-blooded organisms would be expected to survive better through a period of cold. The meteor hypothesis, however, does not depend solely on cold as its effector.

3. Suppose you found a fossil mid-Silurian placoderm with bony fins like a coelacanth. How would you have to modify the lineages shown in Figure 26.6?

ANALYSIS

This finding would force a revision of the divergence of fish lineages occurring at the end of the Silurian. This fossil would show that lobe-fin bone patterns were already present in ancestral placoderm populations. It would indicate that the lobe-fin lineage diverged from the other lineages *before*, not after, the sharks diverged from the teleosts. This interpretation would also mean that the development of the jaw had to occur independently in both lineages.

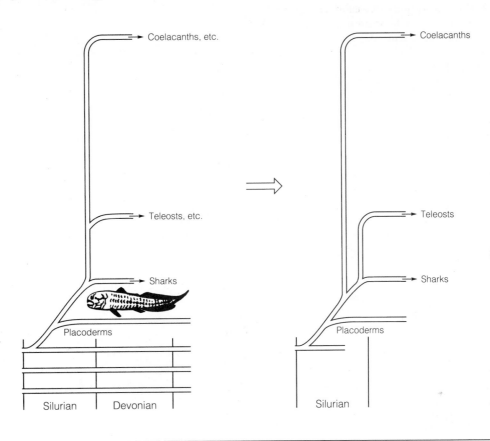

Answers

Answers to Interactive Exercises

ON THE ROAD TO VERTEBRATES/FISHES (26-I)

1. nerve cord; 2. pharynx; 3. notochord; 4. invertebrate; 5. vertebrates; 6. lancelets; 7. filter-feeding; 8. gill slits; 9. Tunicates (sea squirts); 10. tadpoles; 11. notochord; 12. spine (backbone); 13. Paleozoic; 14. deuterostome; 15. larva; 16. larva; 17. mutation; 18. sex organs; 19. predators; 20. tunicates; 21. cephalochordates (lancelets); 22. vertebrae; 23. predators; 24. jaws; 25. brain; 26. paired; 27. fleshy; 28. notochord; 29. pharynx with perforations; 30. mouth; 31. pharynx with perforations; 32. stomach; 33. intestine; 34. anus; 35. excurrent siphon; 36. incurrent siphon; 37. conelike proboscis; 38. collar; 39. gill pore region; 40. pharyngeal gill slits; 41. anus; 42. notochord; 43. dorsal, tubular nerve cord; 44. tunicate larva; 45. tunicate adult (sea squirt); 46. acorn worm; 47. lancelet; 48. ostracoderm; 49. food-straining pharynx; 50. jawless; 51. filter-feeders; 52. jaws; 53. cartilage; 54. bone; 55. ray-finned fishes; 56. lobe-finned fishes; 57. branch 5; 58. branch 4; 59. late Silurian; 60. 350 million years ago.

AMPHIBIANS (26-II)

1. lungs; 2. fins; 3. brain; 4. balance; 5. jaws; 6. two; 7. three; 8. atrium; 9. insects; 10. salamanders; 11. water; 12. reproduce; 13. gills; 14. lining of the pharynx; 15. scales; 16. toxins; 17. larval; 18. breed; 19. insects; 20. reptilian; 21. limb; 22. shelled amniote.

REPTILES (26-III)

1. turtles (lizards); 2. lizards (turtles); 3. internal;
4. garter snakes; 5. lizards; 6. birds; 7. Carboniferous,
(Mississippian); 8. turtles; 9. Carboniferous; 10. Triassic;
11. four; 12. Jurassic; 13. Birds; 14. four; 15. synapsid;
16. Carboniferous; 17. a. dry, scaly skin; b. four-
chambered heart; c. feathers; d. hair; e. loss of limbs;
18. embryo; 19. albumin; 20. yolk sac.

BIRDS (26-IV)

1. reptiles; 2. feathers; 3. sternum (breastbone); 4. air
cavities; 5. shelled amniote; 6. temperature; 7. teeth;
8. humerus; 9. breastbone (sternum).

MAMMALS/SUMMARY (26-V)

1. platypus; 2. pouched (marsupials); 3. placental;
4. four; 5. hair.

Answers to Self-Quiz

1. d; 2. a; 3. d; 4. c; 5. early amphibian, Amphibia;
6. Arctic fox, Mammalia; 7. soldier fish, Osteichthyes;
8. Ostracoderm, Agnatha; 9. owl, Aves; 10. sea turtle,
Reptilia; 11. shark, Chondrichthyes; 12. coelacanth,
Osteichthyes (lobe-finned fish); 13. reef ray,
Chondrichthyes; 14. tunicate, Urochordata; 15. lancelet,
Cephalochordata. 16. c, E; 17. d, H; 18. b, B; 19. h, A;
20. g, D; 21. a, I; 22. i, F; 23. e, G; 24. f, C.

27

PLANT TISSUES

Interactive Exercises

THE PLANT BODY: AN OVERVIEW (27-I, pp. 469–474)

1. Briefly cite the distinguishing features of gymnosperms and angiosperms. _____

_____

_____

_____

2. Xylem and phloem are vascular tissues composed of fibers, parenchyma cells, and some specialized cells. Summarize information about the specialized cells by completing the table below.

Cell Type	In Vascular Tissue?	Description	Function
a. Tracheids		Dead at maturity; all or part of their walls lignified; long cells with tapered, overlapping ends	
b. _____	Xylem		Conduct water and dissolved nutrients from the soil to photosynthetic areas
_____		_____	

Cell Type	In Vascular Tissue?	Description	Function
c. Sieve tube members		Alive at maturity; cluster of pores in the walls allows connection of cytoplasmic contents of adjacent cells	
d. _____	Phloem		Help sieve tube members load sugars produced by leaves and unload them in other plant regions

Label-Match

Identify each part of the illustration below. Choose from the following: dermal tissues, root system, ground tissues, shoot system, and vascular tissues. Complete the exercise by matching and entering the letter of the proper description in the parentheses following each label.

3. _____ _____ ( )

4. _____ _____ ( )

5. _____ _____ ( )

6. _____ _____ ( )

7. _____ _____ ( )

A. Typically consists of stems, leaves, and reproductive structures

B. Usually grows below ground, absorbs soil water and minerals, stores food, anchors the plant, and sometimes lends support

C. Plant tissues other than epidermal and vascular tissues

D. Protective covering for the plant body

E. The conduction tissues of the plant, xylem and phloem

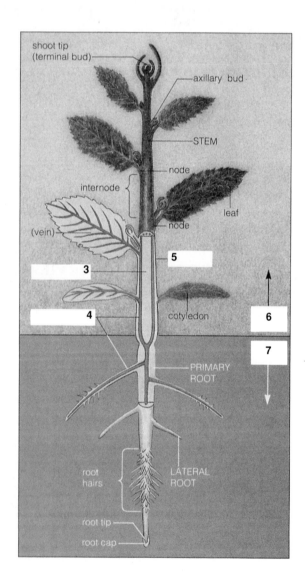

Identify each part of the illustration below. Choose from the following: vascular cambium, apical meristem, ground meristem, vascular bundle, protoderm, cork cambium, procambium. Complete the exercise by matching from the list below and entering the correct letter in the parentheses following each label.

8. _____ _____ ( )

9. _____ ( )

10. _____ _____ ( )

11. _____ ( )

12. _____ _____ ( )

13. _____ _____ ( )

14. _____ _____ ( )

A. Originates secondary growth; increases diameter of older roots and stems
B. Cell lineage of apical meristem that gives rise to epidermis
C. A cord of primary xylem and phloem threading lengthwise through ground tissue
D. Cell lineage of apical meristem giving rise to primary xylem and phloem
E. Dome-shaped cell masses at root and shoot tips
F. Cell lineage of apical meristem giving rise to ground tissues
G. Lateral meristem forming periderm in stems and roots

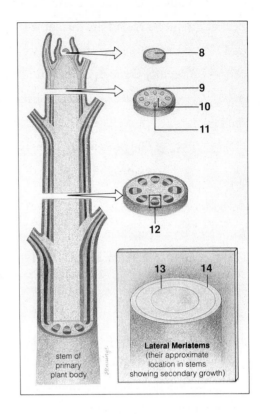

stem of primary plant body

Lateral Meristems
(their approximate location in stems showing secondary growth)

Fill-in-the-Blanks

All new plant cells develop a (15)_____ cell wall, composed of bundled cellulose strands. Many types of plant cells later deposit more cellulose and other materials inside the first wall, forming a (16)_____ wall. (17)_____ , another type of wall material, is a polysaccharide found abundantly in the middle (18)_____ layer that cements adjacent primary cell walls.

Flowering plants have well-developed (19)_____ systems, which include stems and leaves, and (20)_____ systems, which usually grow underground. (21)_____ is the most abundant cell type in ground tissue. (22)_____ and (23)_____ are ground tissues that provide mechanical support for plant parts. Vascular tissues consist of (24)_____ , which conducts water and ions from the roots to the photosynthetic areas, and (25)_____ , which conducts the products of photosynthesis away to storage areas; both help support the plant. The main types of cells in xylem are (26)_____ and (27)_____ members. Both types are dead at maturity with all or parts of their walls (28)_____ .

(29)_____ are long cells with tapered, overlapping ends. Vessel members are shorter cells joined end to end to form a (30)_____ , a tube through which water flows freely. The specialized cells of phloem that conduct sugars and solutes rapidly through the plant are called (31)_____ _____ members. (32)_____ cells in the phloem assist sieve tube members in moving sugars from photosynthetic regions to other plant parts. Dermal tissues include the outer cuticle-covered (33)_____ cells of the primary plant body and the thicker (34) _____ that forms as roots and stems grow in diameter and become woody. Root and shoot tips have dome-shaped (35)_____ meristems where new cells form rapidly through mitotic divisions. (36)_____ growth originates at root and shoot tips. (37)_____ growth occurs primarily in (38)_____ plants and originates at areas of dividing cells, the (39)_____ meristems. The two classes of flowering plants are the (40)_____ and the (41)_____. The presence of one (42)_____ on the flowering-plant embryo identifies (43) _____ plants; the presence of two (44)_____ on the flowering-plant embryo identifies (45)_____ plants.

Labeling

Parenchyma, collenchyma, and sclerenchyma are the types of ground tissue. For each of the following statements, enter "P" if it applies to parenchyma, "C" if it applies to collenchyma, and "S" if it applies to sclerenchyma.

46. ___ Cells alive at maturity; primary walls often thickened at the corners

47. ___ Sclereids and fibers

48. ___ Helps strengthen the plant body, found just beneath the dermal tissue of stem and leaf stalks

49. ___ Gives plant parts mechanical support and protection

50. ___ The most abundant ground tissue

51. ___ Generally has thin primary walls

52. ___ Secondary walls impregnated with lignin

53. ___ Different types active in photosynthesis, storage, secretion, and other tasks

54. ___ Cells still alive at maturity and retain the capacity for division

55. ___ Ample air spaces between cells

Place one of these three labels under each of the following illustrations: collenchyma, parenchyma, sclerenchyma.

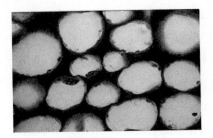

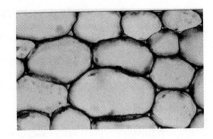

56. _____

57. _____

58. _____

For each of the following items, place a "D" in the blank if it is a dicot characteristic and an "M" in the blank if it is a monocot characteristic.

59. ___ One cotyledon

60. ___ Two cotyledons

61. ___ Floral parts usually occur in threes or multiples of three

62. ___ Floral parts usually occur in fours or fives or multiples thereof

63. ___ Leaf veins usually parallel

64. ___ Leaf veins usually netlike

65. ___ Pollen grains basically have one pore or furrow

66. ___ Pollen grains basically have three pores or furrows

67. ___ Bundles of vascular tissue distributed throughout ground tissue of the stem

68. ___ Bundles of vascular tissue positioned in a ring in the stem

SHOOT SYSTEM (27-II, pp. 474–478)

Fill-in-the-Blanks

A (1)_____ bundle is a strandlike arrangement of primary xylem and phloem. Cross sections of (2)_____ stems generally show vascular bundles scattered throughout the ground tissue. The stems of most (3)_____ and conifers have vascular bundles arranged in a ring that divides the ground tissue into an outer cortex and an inner (4)_____. For most vascular plants, (5)_____ are the major photosynthetic sites. A (6)_____ is the point on a stem where one or more leaves attach. The stem region between two successive nodes is an (7)_____. The bud at the stem tip is the (8)_____ bud; buds at the sides of the stem are called (9)_____ buds. Buds can develop into (10)_____ or (11)_____ or both. The blade of a (12)_____ leaf has its blade divided into smaller leaflets. Plants that drop their leaves each fall are (13)_____ species. Vascular bundles that form a network through leaves are known as (14)_____. (15)_____ are tiny openings usually located in lower leaf surfaces through which water vapor and oxygen move out of leaves and carbon dioxide enters them.

Label-Match

Identify each indicated part of the accompanying illustration. Complete the exercise by matching and entering the letter of the proper function description in the parentheses following each label.

16. _____ _____ ( )
17. _____ _____ ( )
18. _____ _____ ( )
19. _____ _____ ( )
20. _____ ( )

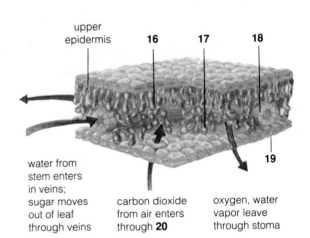

upper epidermis   16   17   18

water from stem enters in veins; sugar moves out of leaf through veins

carbon dioxide from air enters through **20**

oxygen, water vapor leave through stoma

19

A. Lowermost cuticle-covered cell layer
B. Loosely packed photosynthetic parenchyma cells just above the lower epidermal layer
C. Allows movement of oxygen and water vapor out of leaves and allows carbon dioxide to enter
D. Photosynthetic parenchyma cells just beneath the upper epidermis
E. Move water and solutes to photosynthetic cells and carry products away from them

ROOT SYSTEM (27-III, pp. 479–481)

Label-Match

Identify each indicated part of the accompanying illustration. Complete the exercise by matching the letter of the proper description in the parentheses following each label. Some choices are used more than once.

1. _____ _____ ( )
2. _____ ( )
3. _____ ( )
4. _____ ( )
5. _____ ( )
6. _____ _____ ( )
7. _____ _____ ( )
8. _____ _____ _____ ( )
9. _____ _____ ( )
10. _____ ( )
11. _____ ( )
12. _____ _____ ( )

A. Dome-shaped cell mass produced by the apical meristem
B. Part of the vascular column; gives rise to lateral roots
C. Part of the vascular column; transports photosynthetic products
D. Ground tissue region
E. The absorptive interface with the root's environment
F. Part of the vascular column; transports water and minerals
G. The region of dividing cells
H. Innermost part of the root cortex; helps control water and mineral movement into the vascular column
I. Greatly increases the surface available for taking up water and solutes

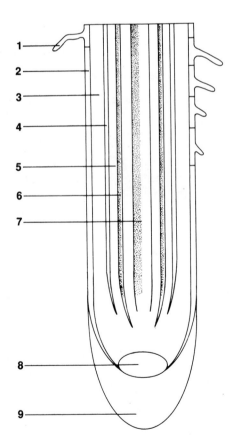

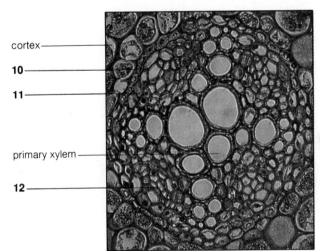

Vascular Cylinder

Fill-in-the-Blanks

Cytoplasm of all adjacent living plant cells is interconnected at cell junctions, the (13)_____. Carrots and dandelions are examples of plants whose primary root and its lateral branchings represent a (14)_____ system. In monocots such as grasses, the primary root is short-lived; in its place, numerous (15)_____ roots arise from the stem of the young plant. Such roots and their branches are somewhat alike in length and diameter and form a (16)_____ root system.

WOODY PLANTS (27-IV, pp. 481–483)

Fill-in-the-Blanks

(1)_____ plants show little or no secondary growth during their life cycle. (2)_____ plants show secondary growth during two or more growing seasons. Plants that complete their life cycle in two growing seasons are (3)_____ plants. (4)_____ are plants in which vegetative growth and seed formation continue year after year. The (5)_____ cambium produces secondary xylem and phloem. Xylem forms on the (6) [choose one] ( ) outer, ( ) inner face of the vascular cambium, and phloem forms on the (7) [choose one] ( ) outer, ( ) inner face. (8)_____ cambium produces the (9)_____ , a corky replacement for epidermis. (10) "_____" refers to all living and nonliving tissues between the vascular cambium and the stem or root surface. A tree can be killed by stripping off a band of phloem completely around the tree, an activity called (11)_____. In regions with prolonged dry spells or cool winters, the vascular cambium of a woody plant's stems and roots becomes (12)_____ during parts of the year. Xylem cells produced earliest in the growing season have thin walls and large diameters; they represent (13)_____ wood. Later in the season, xylem cell diameters become smaller, and walls thicken; these cells represent the (14)_____ wood. Trees in many wet tropical regions have poorly defined xylem growth layers due to a (15)_____ growing season.

Label-Match

Identify each indicated part of the accompanying illustration. Complete the exercise by matching and entering the letter of the proper description in the parentheses following each label.

16. _____ _____ ( )

17. _____ ( )

18. _____ ( )

19. _____ ( )

20. _____ ( )

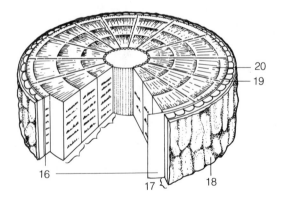

A. Corky replacement for epidermis
B. Vascular tissue that conducts water and dissolved minerals absorbed from soil; gives mechanical support to the plant
C. Has meristematic cells that give rise to secondary xylem and phloem tissues
D. The vascular tissue that transports sugars and other solutes through the plant body
E. All living and nonliving tissues between the vascular cambium and the stem or root surface

Chapter Terms

The following page-referenced terms are important; they were in boldface type in the chapter. Refer to the instructions given in Chapter 1, p. 4.

angiosperms (469)	xylem (472)	dicots (474)	fibrous root system (479)
gymnosperms (469)	phloem (472)	vascular bundle (474)	root hairs (480)
shoot system (470)	epidermis (473)	cortex (474)	vascular cylinder (480)
root system (470)	cuticle (473)	pith (474)	lateral roots (481)
parenchyma (470)	periderm (473)	leaves (476)	vascular cambium (482)
collenchyma (470)	apical meristem (473)	bud (476)	cork cambium (482)
sclerenchyma (470)	lateral meristems (473)	veins (478)	early wood (483)
lignification (471)	monocots (474)	taproot system (479)	late wood (483)

Self-Quiz

___ 1. _____ develops into the plant's surface layers.
   a. Ground tissue
   b. Dermal tissue
   c. Vascular tissue
   d. Pericycle

___ 2. Which of the following is *not* considered a ground cell type?
   a. epidermis
   b. parenchyma
   c. collenchyma
   d. sclerenchyma

___ 3. The _____ produces secondary xylem growth.
   a. apical meristem
   b. vascular cambium
   c. cork cambium
   d. endodermis

___ 4. The _____ is a leaflike structure that is part of the embryo; monocot embryos have one, dicot embryos have two.
   a. shoot tip
   b. root tip
   c. cotyledon
   d. apical meristem

___ 5. Leaves are differentiated and buds develop at specific points along the stem called _____.
   a. nodes
   b. internodes
   c. vascular bundles
   d. cotyledons

___ 6. Which of the following structures is *not* considered to be meristematic?
   a. vascular cambium
   b. lateral meristem
   c. cork cambium
   d. endodermis

___ 7. Which of the following statements about monocots is *false*?
   a. They are usually herbaceous.
   b. They develop one cotyledon in their seeds.
   c. Their vascular bundles are scattered throughout the ground tissue of their stems.
   d. They have a single central vascular cylinder in their stems.

___ 8. Of the following tissues, the ground meristem would *not* give rise to _____.
   a. epidermis
   b. leaf mesophyll
   c. cortex
   d. pith

___ 9. Vascular bundles called _____ form a network through a leaf blade.
   a. xylem
   b. phloem
   c. veins
   d. stomata

___ 10. A primary root and its lateral
branchings represent a _____
system.
a. lateral root
b. adventitious root
c. taproot
d. branch root

___ 11. Plants whose vegetative growth and seed
formation continue year after year are
_____ plants.

a. annual
b. perennial
c. biennial
d. herbaceous

___ 12. The _____ layer of a root divides to
produce lateral roots.
a. endodermis
b. pericycle
c. xylem
d. cortex

Chapter Objectives/Review Questions

This section lists general and detailed chapter objectives that can be used as review questions. You can make maximum use of these items by writing answers on a separate sheet of paper. Fill in answers where blanks are provided. To check for accuracy, compare your answers with information given in the chapter or glossary.

Page   *Objectives/Questions*

(469)   1. _____ are flowering plants, such as oak and apple trees; conifers and other "naked seed" plants are _____.
(470)   2. Define shoot system and root system.
(470)   3. Name and define the three tissue systems that extend throughout the plant body.
(470–471) 4. _____ with thin primary walls forms the most abundant ground tissue; _____ is a ground tissue that strengthens the plant body; sclereids and fibers are _____ cells with lignified secondary walls that give mechanical support and protection to mature plant parts.
(472)   5. Distinguish xylem from phloem in structure and function.
(473)   6. Distinguish cuticle, epidermis, and periderm from one another in structure, plant location, and function.
(473)   7. _____ growth is initiated at root and shoot tips in dome-shaped cell masses called _____ _____.
(473)   8. List three cell lineages that descend from apical meristems and the mature plant tissues they give rise to.
(473)   9. Secondary growth originates at self-perpetuating tissue masses called _____ _____.
(474)   10. Flowering plants with one embryonic cotyledon are called _____ ; those with two embryonic cotyledons are known as _____.
(474)   11. List the principal characteristics that distinguish monocots from dicots.
(474)   12. Describe a vascular bundle.
(474)   13. Most dicot and conifer stems have a ring of vascular bundles that divides ground tissue into two zones, an outer _____ and an inner _____.
(476)   14. _____ are the main sites of photosynthesis.
(476)   15. Distinguish nodes from internodes, tell where buds are located, and indicate how cross sections of monocot stems differ from those of dicot stems.
(476)   16. List the types of buds found on stems; give their general functions.
(476–478) 17. Briefly describe how leaves develop and how they are arranged on stems.
(476–478) 18. Describe how leaves might vary from one species to another; define deciduous and evergreen.
(478)   19. Vascular bundles called _____ form a lacy network through the leaf; give their function.
(478)   20. Generally describe the internal structure of a leaf; give the function of stomata.
(480)   21. A root tip has a dome-shaped mass, called the root _____.
(480)   22. Root _____ are epidermal cell extensions that increase absorptive surface area.
(480)   23. Define the vascular cylinder.
(481)   24. _____ are cell junctions connecting the cytoplasm of adjacent cortex cells.
(481)   25. Explain the location and function of the endodermis and pericycle.

Page	Objectives/Questions
(479)	26. Distinguish between the taproot system and the fibrous root system in terms of structure and developmental origins; define adventitious root.
(481)	27. Distinguish between herbaceous plants and woody plants; give examples of each.
(481)	28. Define and distinguish among annuals, biennials, and perennials.
(482–483)	29. Describe the activity of the vascular cambium and the cork cambium.
(481)	30. _____ is a process of stripping off a band of phloem all the way around the tree.
(483)	31. Describe the cells of early wood and late wood and tell the cause of their differences.

Integrating and Applying Key Concepts

Try to imagine the specific behavioral restrictions that might be imposed if the human body resembled the plant body in having (1) open growth with apical meristematic regions, (2) stomata in the epidermis, (3) cells with chloroplasts, (4) excess carbohydrates stored primarily as starch rather than as fat, and (5) dependence on the soil as a source of water and inorganic compounds.

Critical Thinking Exercises

1. The general functional plan of a potato plant involves photosynthesis in the leaves, transport of the carbohydrate downward in the phloem, and storage in the tuber. Which of the following observations would provide the greatest support for this model?

   a. Starch is present in both the leaves and the tubers during daylight.
   b. Tubers contain a greater concentration of starch than leaves do.
   c. When plants are exposed to radioactive carbon dioxide, radioactive starch can later be extracted from the tuber.
   d. When plants are prepared with radioactive starch only in the tubers, radioactive starch can later be found in the leaves.
   e. When plants are exposed to radioactive carbon dioxide, radioactive carbon can be detected in the stem.

ANALYSIS

   a. The mere presence of a substance in parts of an organism does not indicate where it is synthesized or how it is moved. Because we can assume that starch is synthesized only in leaves, its presence in tubers is most easily explained by movement of carbohydrate from the leaves, but it is not conclusive and does not indicate anything about the pathway.
   b. It is expected that a substance will be at highest concentration where it is synthesized and will move passively away. However, if there is a mechanism to actively carry the substance to another location, the substance could readily reach a higher concentration at the storage location. This observation does not support or contradict the model.
   c. This is a clear demonstration of movement of material from one location to another. The key is the conversion of the label from carbon dioxide to starch. Fixation of the inorganic carbon into organic molecules must occur in the leaves. The molecules then must be transported to the tuber. Notice that the exact chemical form in which they are transported is not indicated; it could be as starch, as glucose monomers, or as something that could be converted to starch. Furthermore, only a sample of phloem fluid containing radioactive carbon would implicate phloem as the route of movement.
   d. This observation would contradict the hypothesis. It would indicate movement of organic compounds upward from tuber to leaf. This could happen via the xylem as glucose metabolites were redistributed. It could also happen outside the plant as the carbon dioxide released by respiration in the tuber diffused upward in the air, reentered the leaf, and was once again incorporated into carbohydrate by photosynthesis.
   e. If we assume that the only entry of radioactive carbon dioxide is through the stomata into the leaf, this observation would indicate internal circulation of material within the plant. It would not demonstrate the involvement of photosynthesis unless it could be shown not to happen in the dark. It would not

even show whether the molecules move in one of the fluid compartments or move through the air spaces among the cells.

2. You drive a nail into a tree trunk. Twenty years later the nail is still at the same height above the ground, although the tree has grown many meters taller. Which of the following is the best explanation?

    a. Tree growth is mainly secondary growth of vascular cambium.
    b. The functional vascular tissue in a tree is a thin layer near the surface.
    c. Tree elongation is by growth of the apical meristems.
    d. Tree elongation is by expansion of the junction zone between the root and the shoot.
    e. Vascular bundles elongate as the shoot elongates.

ANALYSIS

    a. Secondary growth is important in trees, but it only enlarges the diameter of the shoot system parts. If this were the only form of growth, it would explain why the nail didn't rise but not why the tree became taller.
    b. This is true, but it does not explain the growth pattern. Puncturing this layer with a nail would not significantly reduce the flow of water and solutes up and down the trunk of a tree and thus would not affect growth.
    c. Apical meristems add cells and length to the tips of the shoot (and the root) system. Parts closer to the center of the plant remain in the same place. This explains both observations.
    d. If this happened, tree growth would be by addition of height below the nail and would move the segment containing the nail to a higher position. It would not explain the observation, and it does not occur.
    e. They must do so, but they, like the other tissue types, grow by addition of cells at their tips.

3. The text says, "with upright growth . . . flowers are favorably situated for pollination." This conclusion is based on some assumptions. What mode of growth and flower position might be optimum for pollination by each of the following?

    a. flying insects
    b. crawling insects
    c. wind
    d. splashing raindrops
    e. direct contact

ANALYSIS

Obviously, there are many different ways to answer this question. The optimum arrangement will be different depending on details relating to each individual situation. Some possibilities that would *not* be favored by upright growth with flowers high above the ground include pollination by crawling insects, especially if they are not highly agile and mobile, pollination by splashing raindrops, which might have a higher success rate if the flowers were spread in an array on the ground (so splashes would be less likely to drop below the flowers), pollination by direct contact, in which the optimum arrangement would be to place the flowers on long, flexible horizontal elements, and even pollination by flying insects, which are subject to predation from above and would be more successful if the flowers were low and hidden while the insects foraged and pollinated.

Answers

Answers to Interactive Exercises

THE PLANT BODY: AN OVERVIEW (27-I)

1. Angiosperms produce flowers; their seeds are completely enclosed in protective layers. Gymnosperms produce "naked" seeds that are borne on surfaces of reproductive structures. 2. a. Xylem; Water flows from cell to cell in pits; b. Vessel members; Dead at maturity; short cells joined end to end to form a vessel; have pits and perforation plates; c. Phloem; Main sucrose-conducting cells; d. Companion cells; Living cells found adjacent to sieve tube members; 3. ground tissues (C); 4. vascular tissues (E); 5. dermal tissues (D); 6. shoot system (A); 7. root system (B); 8. apical meristem (E); 9. protoderm (B); 10. ground meristem (F); 11. procambium (D); 12. vascular bundle (C); 13. vascular cambium (A); 14. cork cambium (G); 15. primary; 16. secondary; 17. Pectin; 18. lamella; 19. shoot; 20. root; 21. Parenchyma; 22. Collenchyma (Sclerenchyma); 23. sclerenchyma (collenchyma); 24. xylem; 25. phloem; 26. tracheids; 27. vessel; 28. lignified; 29. Tracheids; 30. vessel; 31. sieve tube; 32. Companion; 33. epidermis; 34. periderm; 35. apical; 36. Primary; 37. Secondary; 38. woody; 39. lateral; 40. monocots (dicots); 41. dicots (monocots); 42. cotyledon; 43. monocot; 44. cotyledons; 45. dicot; 46. C; 47. S; 48. C; 49. S; 50. P; 51. P; 52. S; 53. P; 54. P; 55. P; 56. collenchyma; 57. sclerenchyma; 58. parenchyma; 59. M; 60. D; 61. M; 62. D; 63. M; 64. D; 65. M; 66. D; 67. M; 68. D.

SHOOT SYSTEM (27-II)

1. vascular; 2. monocot; 3. dicots; 4. pith; 5. leaves; 6. node; 7. internode; 8. terminal; 9. lateral; 10. leaves (flowers); 11. flowers (leaves); 12. compound; 13. deciduous; 14. veins; 15. Stomata; 16. palisade mesophyll (D); 17. spongy mesophyll (B); 18. minor vein (E); 19. lower epidermis (A); 20. stomata (C).

ROOT SYSTEM (27-III)

1. root hair (I); 2. epidermis (E); 3. cortex (D); 4. endodermis (H); 5. pericycle (B); 6. primary phloem (C); 7. primary xylem (F); 8. root apical meristem (G); 9. root cap (A); 10. endodermis (H); 11. pericycle (B); 12. primary phloem (C); 13. plasmodesmata; 14. taproot; 15. adventitious; 16. fibrous.

WOODY PLANTS (27-IV)

1. Herbaceous; 2. Woody; 3. biennial; 4. Perennials; 5. vascular; 6. inner; 7. outer; 8. Cork; 9. periderm; 10. Bark; 11. girdling; 12. inactive (dormant); 13. early; 14. late; 15. continuous; 16. vascular cambium (C); 17. bark (E); 18. periderm (A); 19. phloem (D); 20. xylem (B).

Answers to Self-Quiz

1. b; 2. a; 3. b; 4. c; 5. a; 6. d; 7. d; 8. a; 9. c; 10. c; 11. b; 12. b.

28

PLANT NUTRITION
AND TRANSPORT

NUTRITIONAL REQUIREMENTS
UPTAKE OF WATER AND NUTRIENTS
      Root Nodules
      Mycorrhizae
      Root Hairs
      Controls over Nutrient Uptake
WATER TRANSPORT AND CONSERVATION
      Transpiration
      Control of Water Loss

TRANSPORT OF ORGANIC SUBSTANCES
      Storage and Transport Forms
      of Organic Compounds
      Translocation
      Pressure Flow Theory

Interactive Exercises

NUTRITIONAL REQUIREMENTS (28-I, pp. 487–488)

1. Refer to Table 28.1 in the text; list the sixteen essential elements generally required by plants._____

    _____

    _____

2. What three elements do plants use as the building blocks for carbohydrates, lipids, proteins, and

    nucleic acids?_____

3. Define and distinguish between macronutrients and micronutrients._____

    _____

    _____

4. Refer to Table 28.2 in the text; list the six mineral ions known as macronutrients._____

    _____

5. Refer to Table 28.2 in the text; list the mineral ions known as micronutrients._____

    _____

Fill-in-the-Blanks

The three essential elements that plants use as their main metabolic building blocks are oxygen, carbon,

and (6)_____. The thirteen essential elements available to plants as dissolved salts are known as

(7)_____ _____. Six dissolved salts become significantly incorporated in plant tissues and are known as (8)_____. The remainder of the dissolved salts occur in very small amounts in plant tissues and are known as (9)_____.

Matching

Refer to Table 28.2 in the text; a blank may have more than one letter, and letters may be used more than once.

10. ___ boron
11. ___ calcium
12. ___ chlorine
13. ___ copper
14. ___ iron
15. ___ magnesium
16. ___ manganese
17. ___ molybdenum
18. ___ nitrogen
19. ___ phosphorus
20. ___ potassium
21. ___ sulfur
22. ___ zinc

A. Macronutrient
B. Micronutrient
C. Component of nucleic acids, ATP, and phospholipids
D. Component of chlorophyll molecule
E. Activation of enzymes, role in maintaining water-solute balance
F. Component of enzyme used in nitrogen metabolism
G. Component of proteins, nucleic acids, coenzymes
H. Involved in chlorophyll synthesis and electron transport
I. Needed in cementing cell walls, and in forming mitotic spindles

UPTAKE OF WATER AND NUTRIENTS (28-II, pp. 488–491)

Fill-in-the-Blanks

The availability of water and dissolved mineral ions profoundly affects (1)_____ development, which in turn affects growth of the entire plant. (2)_____ _____ are specialized extensions of root epidermal cells that promote the uptake of water and mineral ions. "Fungus roots" known as (3)_____ also enhance water and mineral ion absorption. This permanent association is called (4)_____ because both species benefit. String beans, soybeans, peas, alfalfa, and clover are examples of (5)_____ plants. In many agricultural regions, soils suffer from (6)_____ scarcity. There is ample gaseous (7)_____ in the atmosphere, but it is not metabolically available to plants. Legume plants have a (8)_____ relationship with the (9)_____-_____ bacteria that live in localized root swellings called root (10)_____. These bacteria convert the nitrogen to forms they and the plants can use.

Once water and dissolved nutrients pass through the cortex cells and enter the vascular cylinder, their distribution and uptake in different tissue regions is coordinated in ways that affect (11)_____. Energy from (12)_____ drives the active transport membrane pumps which moves solutes into cells. The pumps are transport (13)_____ embedded in the plasma membrane. ATP necessary for membrane pump operation in photosynthetic cells is formed during both (14)_____ and aerobic respiration. In (15)_____ cells, ATP necessary for active transport is formed nearly entirely through aerobic respiration.

Label-Match

Identify each indicated part of the illustration below. Choose from the following: water movement, cytoplasm, vascular cylinder, endodermal cell wall, exodermis, endodermis, and Casparian strip. Complete the exercise by matching from the list below and entering the correct letter in the parentheses following each label.

16. _____ _____ ( )

17. _____ ( )

18. _____ ( )

19. _____ ( )

20. _____ _____ ( )

21. _____ _____ ( )

22. _____ _____ _____ ( )

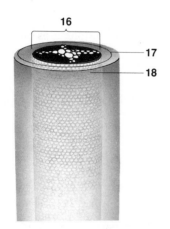

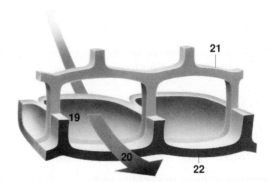

A. Cellular area through which water and dissolved nutrients must move due to Casparian strips
B. A layer of cortex cells just inside the epidermis; also equipped with Casparian strips
C. Waxy band acting as an impermeable barrier between the walls of abutting endodermal cells; forces water and dissolved nutrients through the cytoplasm of endodermal cells
D. Specific location of the waxy strips known as Casparian strips
E. Substance whose diffusion occurs through the cytoplasm of endodermal cells due to Casparian strips
F. Sheetlike layer of single cortex cells wrapped around the vascular cylinder
G. Tissues include the xylem, phloem, and pericycle

WATER TRANSPORT AND CONSERVATION (28-III, pp. 492–494)

1. Define transpiration._____

2. What plant tissues are involved in moving water to the tops of tall trees?_____

3. If the cells are not pulling water upward in the plant, what force is involved?_____

_____

Sequence

Arrange the continuum of events involved in transpiration in correct hierarchical order with the initial event first.

4. ___

5. ___

6. ___

    A. When water molecules move out of veins, replacements are drawn in from stem xylem. The pulling action puts water inside the xylem in a state of tension.

    B. Replacement water moves into the roots—and more soil water is drawn into the plant, following its osmotic gradient (from higher to lower concentration). This inward movement continues until the soil becomes so dry that an osmotic gradient no longer exists.

    C. Water evaporates from the walls of photosynthetic cells inside leaves. As water molecules escape, they are replaced by others from the cell cytoplasm. Then water from xylem in the leaf veins replaces the water being lost from the cells.

True-False

If false, explain why.

___ 7. Transpiration is the evaporation of water from leaves and other exposed plant parts.

___ 8. Due to transpiration, water in phloem is placed in a continuous state of tension.

___ 9. As water is transpired from the plant, hydrogen bonds between water molecules break.

___ 10. Water columns rise in xylem cells due, in part, to the collective strength of hydrogen bonds between water molecules.

___ 11. The continuous escape of water molecules from the plant reduces the tendency for more water molecules to be pulled up to replace them.

Fill-in-the-Blanks

More than 90 percent of the water moving into a leaf is lost through (12)_____. (13)_____ of the plant will occur when water loss by transpiration exceeds water uptake by roots. The (14)_____ is the waxy covering that reduces the rate of water loss from aboveground plant parts. Transpiration occurs mostly at (15)_____ , the small passageways across the cuticle-covered leaf and stem epidermis. The gas (16)_____ _____ also enters the leaf through these passageways. Two kidney-shaped (17)_____ cells flank each stomatal opening; when they are swollen with water, (18)_____ pressure changes their shape to move them apart. When the water content of the guard cells decreases, this pressure drops, and the stomatal opening (19)_____. The amount of (20)_____ and carbon dioxide in the guard cells determines the opening and closing of stomata. When the sun comes up, carbon dioxide is used in (21)_____. Eventually, carbon dioxide levels (22) [choose one] ( ) rise, ( ) drop in cells, including the guard cells. (23)_____ wavelengths in the sun's rays and the decrease in carbon dioxide concentration trigger active transport of (24)_____ ions into the guard cells. This is followed by inward water movement due to (25)_____. Guard cells swell and the stomata (26) [choose one] ( ) open, ( ) close. Now water vapor can (27)_____ the leaf, and carbon dioxide can (28)_____. Photosynthesis ceases when the sun goes down, but (29)_____ _____ accumulates in the cells as a by-product of aerobic respiration. Now (30)_____ ions leave the guard cells, followed by (31)_____. The guard cells then collapse and (32) [choose one] ( ) open, ( ) close. At (33)_____ , transpiration is reduced and water is conserved. As long as soil is moist, the stomata of plants growing in it can remain (34) [choose one] ( ) open, ( ) closed during daylight. Stressful conditions such as drought trigger production of a plant hormone called (35)_____ acid in roots that travels to leaves. When this hormone accumulates in leaves, it causes guard cells to give up (36)_____ ions, thus closing the stomata. CAM plants such as cacti and other succulents open stomata during the (37) [choose one] ( ) day, ( ) night, when they fix carbon dioxide by way of a special C4 metabolic pathway. The fixed carbon dioxide is used in photosynthesis the next (38) [choose one] ( ) day, ( ) night when stomata are closed.

TRANSPORT OF ORGANIC SUBSTANCES (28-IV, pp. 495–498)

1. How are carbohydrates stored in most plant cells?_____

2. In which plant organs are proteins and fats likely to be stored?_____

   _____

3. Describe the mechanism by which plants are able to transport compounds that are large and insoluble.

   _____

4. What is the main form in which sugars are transported throughout most plants?_____

Fill-in-the-Blanks

Sucrose and other organic compounds resulting from (5)_____ are used throughout the plant. Most plant cells store their carbohydrates as (6)_____. Quantities of (7)_____ become stored in some fruits; (8)_____ store proteins and fats. (9)_____ molecules are too large to cross cell membrane and too insoluble to be transported to other regions of the plant body. (10)_____ are largely insoluble in water and cannot be transported from storage sites. (11)_____ proteins do not lend themselves to transport. Specific chemical reactions such as (12)_____ convert storage forms of organic compounds to their subunits, which are transportable forms.

Labeling

Identify each part of the accompanying illustration.

13. _____ _____

14. _____ _____

15. _____ _____ _____ _____

16. _____ _____ _____

17. _____ _____

18. _____ _____ _____ _____

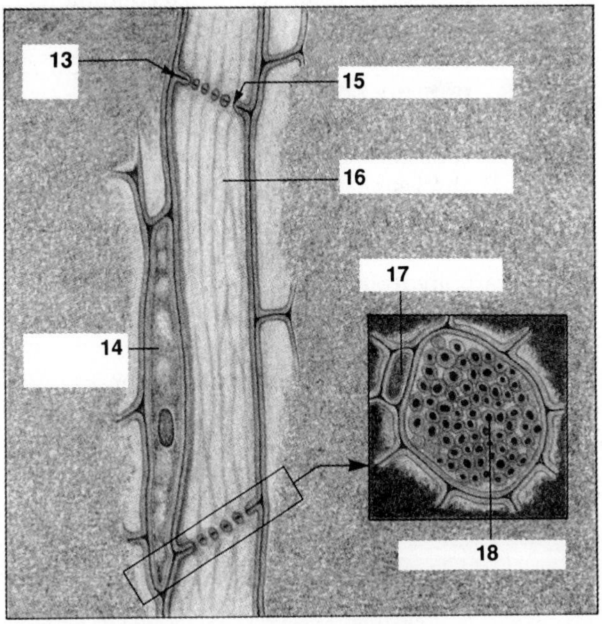

Matching

Match the appropriate letter with its numbered partner.

19. ___ translocation

20. ___ sieve tube members

21. ___ companion cells

22. ___ aphids

23. ___ source

24. ___ sink

25. ___ pressure flow theory

A. Any region where organic compounds are being loaded into the sieve tube system

B. Nonconducting cells adjacent to sieve tube members that supply energy to load sucrose at the source

C. Any region of the plant where organic compounds are being unloaded from the sieve tube system and used or stored

D. Process occurring in phloem that distributes sucrose and other organic compounds through the plant

E. States that pressure builds up at the source end of a sieve tube system and pushes solutes toward a sink, where they are removed

F. Passive conduits for translocation within vascular bundles; water and organic compounds flow rapidly through large pores on their end walls

G. Insects used to verify that in most plant species sucrose is the main carbohydrate translocated under pressure

Chapter Terms

The following page-referenced terms are important; they were in boldface type in the chapter. Refer to the instructions given in Chapter 1, p. 4.

root nodules (489)
mycorrhizae (490)
root hairs (490)
endodermis (490)
Casparian strip (490)

exodermis (490)
transpiration (492)
cohesion-tension theory
   of water transport (493)
stomata (493)

abscisic acid (494)
CAM plants (494)
translocation (495)
sieve tube members (495)

companion cells (495)
source (496)
sink (496)
pressure flow theory (496)

Self-Quiz

___ 1. The _____ theory of water transport states that hydrogen bonding allows water molecules to maintain a continuous fluid column as water is pulled from roots to leaves.
   a. pressure flow
   b. evaporation
   c. cohesion-tension
   d. abscission

___ 2. The three elements that are present in carbohydrates, lipids, proteins, *and* nucleic acids are _____.
   a. oxygen, carbon, and nitrogen
   b. oxygen, hydrogen, and nitrogen
   c. oxygen, carbon, and hydrogen
   d. carbon, nitrogen, and hydrogen

___ 3. Macronutrients are the six mineral ions that _____.
   a. play vital roles in photosynthesis and other metabolic events
   b. occur in only small traces in plant tissues
   c. become heavily incorporated in plant tissues
   d. can function only without the presence of micronutrients
   e. both a and c

___ 4. Without _____ , plants would rapidly wilt and die during hot, dry spells.
   a. a cuticle
   b. a mycorrhiza
   c. phloem
   d. cotyledons

5. Water inside all the xylem cells is being pulled upward by _____.
   a. turgor
   b. tension and negative pressure
   c. osmotic gradients
   d. pressure-flow

6. Gaseous nitrogen is converted to a plant-usable form by _____.
   a. root nodules
   b. mycorrhizae
   c. nitrogen-fixing bacteria
   d. Venus flytraps

7. Symbiotic relationships between fungi and young roots are _____.
   a. root nodules
   b. mycorrhizae
   c. nitrogen-fixing bacteria
   d. Venus flytraps

8. _____ prevent(s) water from moving past the abutting walls of the root endodermal cells.
   a. Cytoplasm
   b. Plasma membranes
   c. Casparian strips
   d. Osmosis

9. Most of the water moving into a leaf is lost through _____.
   a. osmotic gradients being established
   b. transpiration
   c. pressure-flow forces
   d. translocation

10. _____ causes transpiration.
    a. Hydrogen bonding
    b. The drying power of air
    c. Cohesion
    d. Turgor pressure

11. Stomata remain _____ during daylight, when photosynthesis occurs, but remain _____ during the night when carbon dioxide accumulates through aerobic respiration
    a. open; open
    b. closed; open
    c. closed; closed
    d. open; closed

12. By control of _____ levels inside the guard cells of stomata, the activity of stomata is controlled when leaves are losing more water than roots can absorb.
    a. oxygen
    b. potassium
    c. carbon dioxide
    d. ATP

13. Large pressure gradients arise in sieve tube systems by means of _____.
    a. vernalization
    b. abscission
    c. osmosis
    d. transpiration

14. Leaves represent _____ regions; growing leaves, stems, fruits, seeds, and roots represent _____ regions.
    a. source; source
    b. sink; source
    c. source; sink
    d. sink; sink

Chapter Objectives/Review Questions

This section lists general and detailed chapter objectives that can be used as review questions. You can make maximum use of these items by writing answers on a separate sheet of paper. Fill in answers where blanks are provided. To check for accuracy, compare your answers with information given in the chapter or glossary.

Page	Objectives/Questions
(487)	1. Plants generally require _____ (number) essential elements.
(487)	2. What three elements are the main building blocks for carbohydrates, lipids, proteins, and nucleic acids?
(487–488)	3. Distinguish between macronutrients and micronutrients in relation to their role in plant nutrition.

Integrating and Applying Key Concepts

How do you think maple syrup is made from maple trees? Which specific systems of the plant are involved, and why are maple trees tapped only at certain times of the year?

Critical Thinking Exercises

1. A plant cell is placed in a solution of NaCl that is more concentrated than the sodium and chloride ions inside the cell. After thirty minutes, the cell has not changed volume. Which of the following is the best explanation of this observation?

   a. The cell membrane is not permeable to either sodium or chloride.
   b. The cell is surrounded by a rigid wall.
   c. The cell transported sodium out of the cytoplasm into the external solution.
   d. The cell died after only five minutes of exposure to the salt solution.
   e. The cell interior contains a great variety of impermeable solutes.

ANALYSIS

   a. If the cell membrane was permeable to sodium or chloride, these ions would enter the cell by diffusion, and water would tend to accompany them by osmosis. However, if the membrane was not permeable to the ions, the net movement of water would be outward because the ions in the solution would lower the water potential. This would produce visible shrinkage of the cell.
   b. A rigid wall would prevent visible swelling under conditions that favored movement of water into the cell. But these conditions favor net movement of water out of the cell, and the cell would be expected to shrink within the wall. This should be visible.

c. If the cell transported sodium outward, the water potential would become greater inside and less outside, and the tendency to shrink would be increased. The cell might equalize its internal water potential to the outside potential by transporting sodium *inward*.

d. Death of the cell should not affect the structure of the membrane within the time span of this experiment, and, if it did, the change should result in visible damage to the cell.

e. A tendency for water to leave a cell results from the greater water potential inside the cell relative to outside. In this example, the outside water potential is kept low by the dissolved sodium and chloride ions. If the cell contained an equivalent concentration of other impermeable molecules, such as proteins, the cytoplasmic water potential would be equal to the external water potential, net water movement across the cell membrane would be zero, and the cell would not change volume.

2. The Casparian strip is clearly visible in the microscope and forms a continuous band among the cells of the endodermis. Which of the following observations provides the strongest support for the hypothesis that the Casparian strip is impermeable to water and solutes?

   a. Fluid in the xylem vessels flows upward with water and ions.
   b. The composition of the intercellular fluid in the cortex is different from the composition of the water in the vascular column.
   c. Radioactive water supplied to the outside of the roots can later be detected in the intercellular fluid of the cortex, the cytoplasm of the endodermis cells, and the xylem fluid.
   d. Photosynthesis takes place in the leaves, not in the roots.
   e. Glucose, containing radioactive carbon provided to the plant as carbon dioxide gas around the leaves, can be found in the cortex cells of the roots.

ANALYSIS

   a. This observation shows what happens to water and ions *after* they pass the endodermis. It neither supports nor refutes a hypothesis about the route of substances moving between the root cortex and the xylem.
   b. Fluid moves freely among the cells of the cortex. If the Casparian strip was permeable, water and ions would diffuse through it, and the solutions on both sides would have the same composition. The observation that the solutions are different means that a permeability barrier exists between them. In the root, that barrier has two components—the membranes of the endodermis cells and the Casparian strip among the cells.
   c. This observation would directly support the idea that water is absorbed from the roots into the vascular column and distributed to cells. It does not indicate whether the water moved through the endodermal cells, among them and then into them from the inside of the vascular column, or both.
   d. This observation would lead us to expect products of photosynthesis to move downward in the vascular column to the roots. It would not lead to any conclusion as to whether the molecules moved through or among the endodermal cells.
   e. This observation would be direct evidence of distribution of photosynthetic products downward within the plant, but it, too, would not support any conclusion regarding the pathway of movement.

3. Suppose you are studying the effects of a nutrient substance on plant growth. You want to know the optimum concentration of the substance. You grow a series of plants on liquid medium to which you add various concentrations of the nutrient. You measure the weight of the plants after a set time of growth. The data are shown in the table below.

Concentration of Nutrient (mM)	Weight (gm)
50	6.8
75	9.3
100	12.1
125	15.8
150	15.9

Plot the data on a graph and make a conclusion. On the basis of these observations, you decide to repeat the experiment using smaller increments of concentration as well as a slightly wider range of concentrations. This time the data are as follows.

Concentration (mM)	Weight (gm)
0	2.0
10	2.0
20	1.9
30	2.2
40	5.0
50	6.8
60	7.9
70	8.9
80	10.0
90	11.0
100	12.1
110	13.3
120	14.8
130	16.7
140	19.6
150	15.9
160	5.4
170	1.8
180	2.0

How do the new data affect your interpretations?

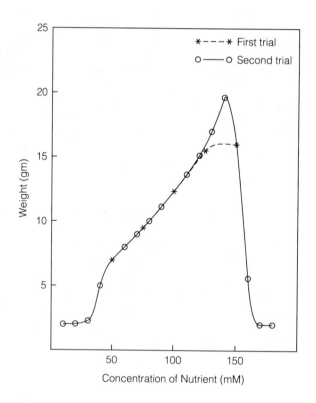

ANALYSIS

The graph shows the two data sets and curves that fit them. The first trial looks like a typical saturation curve. The nutrient stimulates growth at all concentrations, reaching maximum effectiveness at about 125 mM. Supplying more nutrient has no additional effect on growth. The data do not extend below 50 mM, but the curve might reasonably be expected to extrapolate smoothly to zero concentration at a plant weight of about 2 gm. This would be the weight of a seed that did not grow.

The additional data of the second trial require changes of interpretation at both ends of the concentration range. At the low end, it is clear that a threshold concentration of the nutrient around 35 or 40 mM is required before any growth can commence. Growth stimulation increases with concentration to a maximum around 140 mM or perhaps a little higher. Then the substance inhibits growth very effectively, preventing growth totally at concentrations above 160 mM. This kind of biphasic effect is not at all uncommon in biological systems. A substance may activate an enzyme or stimulate some other process at low concentrations. The same substance often can also inhibit a different enzyme or process, but only at a higher concentration. Thus, at low concentrations the substance is beneficial, perhaps even essential, but at high concentrations it is toxic. Effects like this can easily be missed by studies with too few data points.

Answers

Answers to Interactive Exercises

NUTRITIONAL REQUIREMENTS (28-I)

1. carbon, oxygen, hydrogen, nitrogen, potassium, calcium, magnesium, phosphorus, sulfur, chlorine, iron, boron, manganese, zinc, copper, and molybdenum; 2. oxygen, carbon, and hydrogen; 3. Macronutrients are those six mineral ions that become incorporated in plant tissues in significant amounts; micronutrients are those seven mineral ions that occur in plant tissues in only small traces; 4. nitrogen, potassium, calcium, magnesium, phosphorus, and sulfur; 5. chlorine, iron, boron, manganese, zinc, copper, and molybdenum; 6. hydrogen; 7. mineral ions; 8. macronutrients; 9. micronutrients; 10. B; 11. A, I; 12. B; 13. B; 14. B, H; 15. A, D; 16. B; 17. B, F; 18. A, G; 19. A, C; 20. A, E; 21. A; 22. B.

UPTAKE OF WATER AND NUTRIENTS (28-II)

1. root; 2. Root hairs; 3. mycorrhizae; 4. mutualism; 5. legume; 6. nitrogen; 7. nitrogen; 8. mutualistic; 9. nitrogen-fixing; 10. nodules; 11. growth; 12. ATP; 13. proteins; 14. photosynthesis; 15. nonphotosynthetic; 16. vascular cylinder (G); 17. exodermis (B); 18. endodermis (F); 19. cytoplasm (A); 20. water movement (E); 21. Casparian strip (C); 22. endodermal cell wall (D).

WATER TRANSPORT AND CONSERVATION (28-III)

1. Water evaporation from stems, leaves, and other plant parts; 2. The conducting cells of xylem; 3. Water is pulled up by the drying power of air, which creates continuous negative pressures that extend downward from the leaves to the roots. 4. C; 5. A; 6. B; 7. T; 8. F; 9. T; 10. T; 11. F; 12. transpiration; 13. Dehydration; 14. cuticle; 15. stomata; 16. carbon dioxide; 17. guard; 18. turgor; 19. closes; 20. water; 21. photosynthesis; 22. drop; 23. Blue; 24. potassium; 25. osmosis; 26. open; 27. leave; 28. enter; 29. carbon dioxide; 30. potassium; 31. water; 32. close; 33. night; 34. open; 35. abscisic; 36. potassium; 37. night; 38. day.

TRANSPORT OF ORGANIC SUBSTANCES (28-IV)

1. Carbohydrates are stored as starch. 2. Fats can be stored in fruits; proteins and fats can be stored in seeds. 3. Hydrolysis is used to convert storage forms of organic compounds to transportable forms (smaller subunits). 4. sucrose; 5. photosynthesis; 6. starch; 7. fats; 8. seeds; 9. Starch; 10. Fats; 11. Storage; 12. hydrolysis; 13. sieve plate; 14. companion cell; 15. pore of sieve plate; 16. sieve tube member; 17. companion cell; 18. pore of sieve plate; 19. D; 20. F; 21. B; 22. G; 23. A; 24. C; 25. E.

Answers to Self-Quiz

1. c; 2. c; 3. c; 4. a; 5. b; 6. c; 7. b; 8. c; 9. b; 10. b; 11. d; 12. b; 13. c; 14. c.

29

PLANT REPRODUCTION

REPRODUCTIVE MODES
GAMETE FORMATION IN FLOWERS
    Floral Structure
    Microspores to Pollen Grains
    Megaspores to Eggs
POLLINATION AND FERTILIZATION
    Pollination
    Commentary: Coevolution of Flowering
    Plants and Their Pollinators

SEED AND FRUIT FORMATION
    Seed Formation
    Fertilization and Endosperm Formation
    Fruit Formation and Seed Dispersal
    Doing Science: Why So Many Flowers?
ASEXUAL REPRODUCTION
OF FLOWERING PLANTS

Interactive Exercises

REPRODUCTIVE MODES (29-I, pp. 501–502)

1. Distinguish between sexual and asexual reproduction in terms of a comparison of the genetics of
   parents and offspring._____

   _____

   _____

Label-Match

Study the generalized life cycle for flowering plants as shown in Figure 29.2 in the text. Identify each indicated part of the accompanying illustration. Complete the exercise by matching and entering the letter of the proper description in the parentheses following each label.

2. _____ ( )

3. _____ ( )

4. _____ _____ ( )

5. _____ _____ ( )

6. _____ ( )

A. An event that produces a young sporophyte
B. A reproductive shoot produced by the sporophyte
C. The "plant"; a vegetative body that develops from a zygote
D. Produce eggs
E. Produce sperm

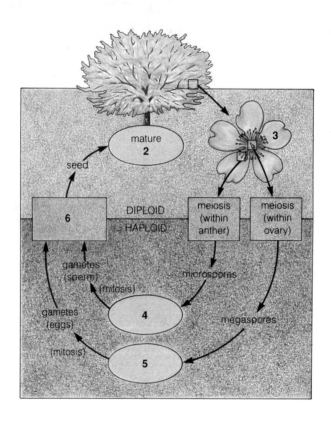

Fill-in-the-Blanks

(7)_____ reproduction requires the formation of gametes, followed by fertilization. (8)_____ reproduction occurs by mitosis and produces individuals that are genetically (9)_____. Familiar plants such as radishes, cactus plants, and elm trees are examples of (10)_____ ; they produce reproductive shoots called (11)_____. Some cells within flowers undergo (12)_____ to produce (13)_____. Male gametophytes produce (14)_____ ; female gametophytes produce

(15)_____. Female gametophytes are usually tiny multicelled bodies embedded within

(16)_____ tissues. Immature male gametophytes are released from flowers as small (17)_____

_____. (18)_____ may also reproduce asexually by several means.

GAMETE FORMATION IN FLOWERS (29-II, pp. 502–505)

Labeling

Identify each indicated part of the accompanying illustration.

1. _____
2. _____
3. _____
4. _____
5. _____
6. _____
7. _____

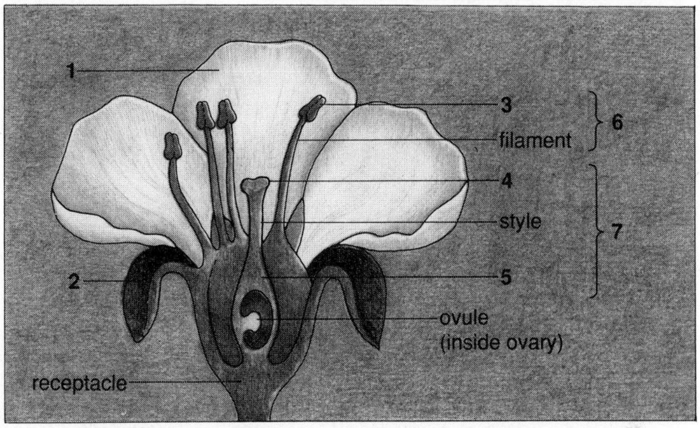

For each listed structure or event, enter "O" in the blank if it is associated with the ovary and "A" if it is associated with the anther.

8. ___ Embryo sac

9. ___ Pollen grain

10. ___ Endosperm mother cell

11. ___ Pollen tube

12. ___ Integuments

13. ___ Microspores

14. ___ Sperm nuclei

15. ___ Nucellus

16. ___ Pollen sac

17. ___ Megaspores

18. ___ Endosperm

19. ___ Megaspore mother cell

20. ___ Embryo sac

Matching

Match each letter with its appropriate numbered partner.

21. ___ sepals

22. ___ petals

23. ___ stamens

24. ___ pollen sacs

25. ___ carpels

26. ___ ovaries

27. ___ perfect flowers

28. ___ imperfect flowers

A. Have both male and female parts
B. Four chambers within an anther where pollen grains develop
C. Collectively, the flower's "corolla"
D. Have male or female parts, but not both
E. Female reproductive parts
F. Chambers in the carpels where egg formation, fertilization, and seed development occur
G. Outermost whorl of floral organs; the "calyx"
H. Male reproductive parts; a slender stalk capped by an anther

Label-Match

Identify each indicated part of the accompanying illustration. Choose from the following: microspores, male gametophyte, pollen sac, tube cell, microspore mother cell, and sperm-producing cell. Complete the exercise by matching and entering the letter of the proper description in the parentheses following each label.

29. _____ _____ ( )

30. _____ _____ _____ ( )

31. _____ ( )

32. _____ _____ ( )

33. _____ - _____ _____ ( )

34. _____ _____ ( )

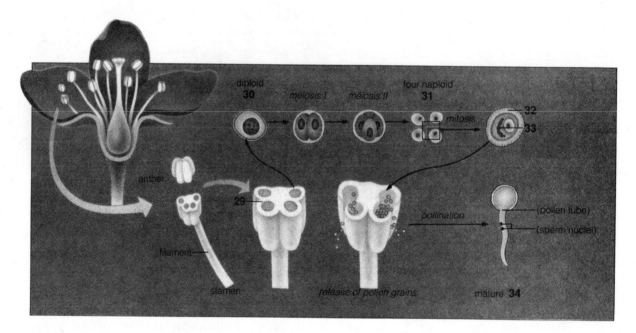

A. Will develop into a pollen tube
B. Haploid cells resulting from meiosis
C. Pollen tube with included haploid nuclei
D. Produced by mitotic divisions while an anther develops in a flower bud
E. Produces two sperm nuclei
F. Chamber in which pollen develops

Identify each indicated part of the accompanying illustration. Choose from the following: megaspore mother cell, micropyle, ovule, nucellus, and integuments. Complete the exercise by matching and entering the letter of the proper description in the parentheses following each label.

35. _____ ( )

36. _____ ( )

37. _____ _____

    _____ ( )

38. _____ ( )

39. _____ ( )

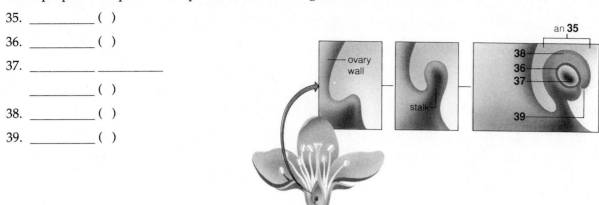

A. A diploid cell undergoing meiosis to form megaspores
B. A tiny gap in the nucellus for pollen tube penetration
C. Structure within the ovary that becomes a seed
D. Inner ovular tissue inside integuments
E. Protective tissue layers outside the nucellus

Fill-in-the-Blanks

Each diploid mother cell inside a pollen sac undergoes (40)_____ and produces four haploid (41)_____. Each of these cells undergoes mitotic cell division to produce two-celled haploid bodies, the (42)_____ _____. After one of these haploid bodies lands on a stigma, one cell will give rise to (43)_____ nuclei, and the other will develop into a (44)_____ _____. The latter structure will grow through carpel tissues and transport sperm nuclei to the (45)_____. (46)_____ begin their development as dome-shaped cell masses that develop on the inner ovary wall. Ovules may become (47)_____. As a young ovule develops, some cells form a (48)_____ that attaches it to the ovary wall. The remainder develop into an inner tissue, the (49)_____ , and one or two protective layers called (50)_____ form around it. The (51)_____ is a tiny gap in the nucellus that does not become covered with integuments. This tiny gap is a place where the pollen (52)_____ may penetrate the ovule. Within the inner cell mass of the ovule, a diploid (53)_____ cell divides by meiosis to form four haploid (54)_____. Typically, three of the four disintegrate, but the remaining one undergoes three successive (55)_____ divisions without cytoplasmic division. A single cell is formed with (56)_____ nuclei. The cytoplasm divides after each nucleus migrates to a specific cell location; this forms a seven-celled (57)_____ _____. This structure is the equivalent of the female (58)_____. One cell within the embryo sac is the (59) _____ ; another cell is the (60)_____ mother cell, which will help form (61)_____ , a nutritive tissue around the embryo.

POLLINATION AND FERTILIZATION (29-III, pp. 505–508)

1. Distinguish between pollination and fertilization._____

   _____

   _____

Fill-in-the-Blanks

Pollen grains are transferred from (2)_____ to (3)_____ ; this is called (4)_____. Flowering plants and their (5)_____ have coevolved. After a pollen grain is deposited on a stigma, the (6)_____ _____ begins to grow toward the ovules. (7)_____ sperm cells form inside the pollen tube. The sperm cells are released after the pollen tube penetrates the (8)_____ _____ inside the ovule. (9)_____ fertilization occurs in flowering plants. One sperm nucleus fuses with the egg nucleus; this forms a diploid (10)_____. The other sperm nucleus fuses with the two nuclei of the endosperm mother cell, forming a triploid (11)_____ nucleus that produces the (12)_____ tissues. These tissues will nourish the young (13)_____ until its leaves form. Endosperm forms only in (14)_____ plants.

Label-Match

Identify each indicated part of the illustration below. Complete the exercise by matching and entering the letter of the proper description in the parentheses following each label.

15. _____ ( )

16. _____ ( )

17. _____ ( )

18. _____ ( )

19. _____ _____ ( )

20. _____ ( )

21. _____ _____ _____ ( )

22. _____ _____ _____ ( )

23. _____ ( )

24. _____ _____ ( )

25. _____ _____ ( )

26. _____ ( )

27. _____ ( )

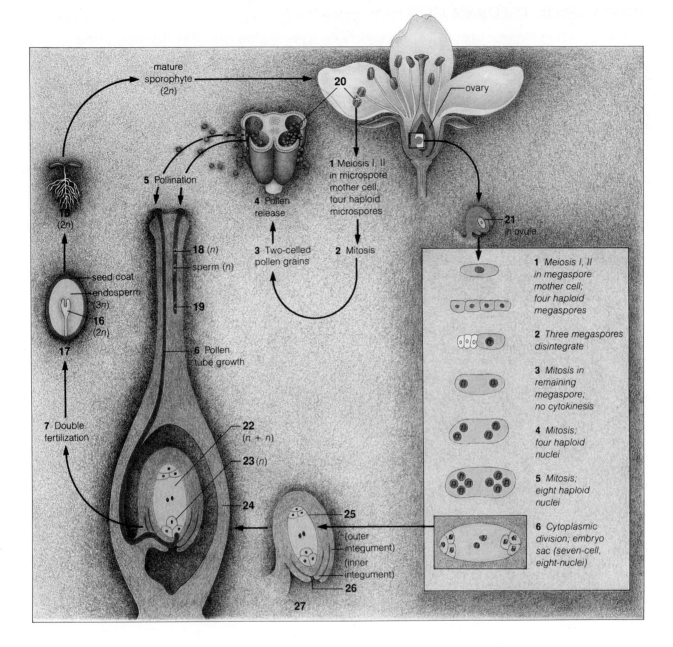

A. Portion of an immature fruit
B. Female gametophyte
C. Will fuse with one sperm nucleus to form a triploid cell
D. Site of meiosis to form microspores
E. Developing sporophyte
F. Male gamete
G. New sporophyte in a seed
H. Ovule pore that allows entry of pollen tube
I. Haploid nucleus in the pollen tube that does not act as a sperm
J. Undergoes meiosis to form four megaspores
K. Immature seed
L. Female gamete
M. A mature ovule

SEED AND FRUIT FORMATION (29-IV, pp. 509–513)

1. Complete the following table, which summarizes the types of fruits formed by flowering plants.

Examples	Fruit Type	Characteristics
a. Pear, apple		A single ovary surrounded by receptacle tissue
b.	Simple	Fruit wall dry; split at maturity
c. Blackberry, raspberry	Aggregate	
d.	Accessory	Swollen, fleshy receptacle with dry fruits on its surface
e. Sunflower, wheat, rice, maple	Simple	
f. Pineapple, fig, mulberry		Matured ovaries, grown together in a mass; may include accessory structures
g. Grape, banana, lemon, cherry		Fruit wall fleshy, sometimes with leathery skin

Labeling

Identify each indicated part of the accompanying illustration.

2. _____
3. _____
4. _____
5. _____
6. _____
7. _____

8. _____ _____
9. _____
10. _____
11. _____ _____
12. _____
13. _____

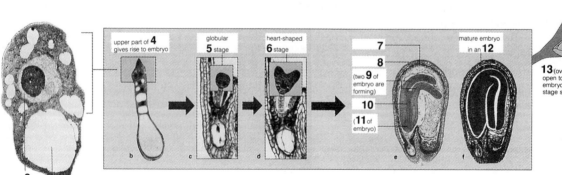

Sequence

Arrange the following developmental events in correct chronological sequence from first to last.

14. ___
15. ___
16. ___
17. ___
18. ___
19. ___

A. Embryo continues to develop, endosperm expands, integuments of ovule harden and thicken.
B. A nutrient-transferring suspensor forms.
C. The mature ovule is a seed; integuments become the seed coat.
D. The zygote forms with the nucleus and most of the organelles residing in its top half.
E. Cotyledons develop as a part of the embryo.
F. The upper part of the zygote gives rise to a multicelled embryo.

Fill-in-the-Blanks

A fully matured ovule is a (20)_____ ; the (21)_____ surrounding the seed(s) develops into most or all of a fruit. The (22)_____ of the ovule harden and thicken to become the seed coats.

(23)_____ are embryo structures that function to provide the embryo with nutrition. (24)_____ and nuts are dry fruits; apples and tomatoes are (25)_____ fruits. Pineapples are formed from (26)_____ flowers whose ovaries remain together. Fruits function in (27)_____ protection and (28)_____ in specific environments. Many (29)_____ have hooks, spines, hairs, and sticky surfaces that allow animals to move them to new locations. Fleshy fruits such as blueberries and cherries have their seeds dispersed as they are eaten and expelled by (30)_____. Digestive enzymes in the animal guts remove enough of the hard seed coats to increase the chance of successful (31)_____ after expulsion from the body.

ASEXUAL REPRODUCTION OF FLOWERING PLANTS (29-V, pp. 513–514)

Matching

Match the following asexual reproductive modes of flowering plants.

1. ___ asexual reproduction involving modified stems
2. ___ parthenogenesis
3. ___ vegetative propagation
4. ___ tissue culture propagation

A. New plants develop from tissues or organs that drop or separate from parent plants.
B. New plants arise from cells in parent plant that were not irreversibly differentiated (a laboratory technique).
C. A mechanism that involves asexual reproduction utilizing runners, rhizomes, corms, tubers, and bulbs.
D. The embryo develops without nuclear or cellular fusion.

Chapter Terms

The following page-referenced terms are important; they were in boldface type in the chapter. Refer to the instructions given in Chapter 1, p. 4.

sporophyte (502)	carpels (503)	megaspores (504)	coevolution (506)
flowers (502)	ovary (503)	embryo sac (504)	cotyledons (509)
gametophytes (502)	pollen grain (503)	endosperm (504)	double fertilization (509)
stamens (503)	microspores (504)	pollination (505)	fruit (510)
pollen sacs (503)			

Self-Quiz

___ 1. A stamen is _____.
   a. composed of a stigma
   b. the mature male gametophyte
   c. the site where microspores are produced
   d. part of the vegetative phase of an angiosperm

___ 2. A gametophyte is _____.
   a. a gamete-producing plant
   b. haploid
   c. both a and b
   d. the plant produced by the fusion of gametes

___ 3. The phase in the life cycle of plants that gives rise to spores is known as the _____.
   a. gametophyte
   b. embryo
   c. sporophyte
   d. seed

___ 4. Fertilization and seed maturation occurs in _____.
   a. the style of the carpel
   b. the ovary of the carpel
   c. pollen sacs of anthers
   d. microspores

___ 5. A characteristic of a seed is that it _____.
   a. contains an embryo sporophyte
   b. represents an arrested growth stage
   c. is covered by hardened and thickened integuments
   d. all of the above

___ 6. An immature fruit is a(n) _____ and an immature seed is a(n) _____.
   a. ovary; megaspore
   b. ovary; ovule
   c. megaspore; ovule
   d. ovule; ovary

___ 7. The joint evolution of flowers and their pollinators is known as _____.
   a. adaptation
   b. coevolution
   c. joint evolution
   d. covert evolution

___ 8. In flowering plants, one sperm nucleus fuses with that of an egg, and a zygote forms that develops into an embryo. Another sperm fuses with _____.
   a. a primary endosperm cell to produce three cells, each with one nucleus
   b. a primary endosperm cell to produce one cell with one triploid nucleus
   c. both nuclei of the endosperm mother cell, forming a primary endosperm cell with a single triploid nucleus
   d. one of the smaller megaspores to produce what will eventually become the seed coat

___ 9. Simple, aggregate, multiple, and accessory refer to types of _____.
   a. carpels
   b. seeds
   c. fruits
   d. ovaries

___ 10. "When a leaf falls or is torn away from a jade plant, a new plant can develop from the leaf, from meristematic tissue." This statement refers to _____.
   a. parthenogenesis
   b. runners
   c. tissue culture propagation
   d. vegetative propagation

Chapter Objectives/Review Questions

This section lists general and detailed chapter objectives that can be used as review questions. You can make maximum use of these items by writing answers on a separate sheet of paper. Fill in answers where blanks are provided. To check for accuracy, compare your answers with information given in the chapter or glossary.

Page   *Objectives/Questions*

(501)   1. _____ reproduction requires formation of gametes, followed by fertilization.
(501)   2. The production of individuals in clones describes _____ reproduction.
(502)   3. Give the relationship of a flower to a sporophyte.
(502)   4. Distinguish between sporophytes and gametophytes.
(502)   5. _____ have several means of reproducing themselves asexually.
(502–503)   6. Be able to identify the various parts of a typical flower and give their functions.
(503)   7. The male floral reproductive parts are called _____.
(503)   8. _____ are the female reproductive parts of a flower.
(503)   9. Give the distinctions between a flower that is perfect and one that is imperfect.
(503–504)   10. Relate the sequence of events that give rise to microspores and megaspores.
(504)   11. Describe the development of the male gametophyte.
(504)   12. Describe the development of the embryo sac.
(504)   13. What represents the male gametophyte and female gametophyte in flowering plants?
(504)   14. The gamete made in the embryo sac is the _____.
(504)   15. In the ovule, the four haploid cells that form after meiosis are called _____.
(505)   16. _____ is the transfer of pollen grains to a stigma.
(509)   17. Describe the double fertilization that occurs uniquely in the flowering-plant life cycle.
(509)   18. How is endosperm formed? What is the function of endosperm?
(509)   19. _____ , or "seed leaves," develop as part of the embryo.
(509)   20. An ovule is an immature _____.
(510)   21. An ovary containing ovule(s) develops into a _____.
(510)   22. From the plant's perspective, what is the function of fruits?
(510)   23. Review the general types of fruits produced by flowering plants (Table 29.1).
(513)   24. Review Table 29.2, which lists major types of asexual reproductive modes among flowering plants.
(513–514)   25. Distinguish between parthenogenesis, vegetative propagation, and tissue culture propagation.

Integrating and Applying Key Concepts

Unlike the growth pattern of animals, the vegetative body of vascular plants continues growing throughout its life by means of meristematic activity. Explain why most representatives of the plant kingdom probably would not have survived and grown as well as they do if they had followed the growth pattern of animals.

Critical Thinking Exercises

1. It has long been thought that the growth of the pollen tube into the style is stimulated and directed by a chemical substance released from the ovule. This substance is thought to produce a chemotropic response in the growing pollen tube. Experiments with extracts of ovules indicated that the chemotropic substance was small, stable to heating, and readily soluble in water. These observations indicated that the substance is an inorganic ion. In one type of experiment on chemotropic responses of pollen tubes, pollen grains are placed on the surface of an agar gel. Calcium ions are placed in a shallow well nearby in the gel. After several hours, the pollen tubes are examined. The data are given in the table below.

Calcium in Well (micrograms)	Pollen Tubes Growing Toward Well
0.0	21
0.8	46
1.6	70
2.4	90
4.0	110
6.0	198
8.0	226
9.6	260
20.0	269

The authors concluded from these observations that calcium does exert a chemotropic effect on pollen tubes. Which of the following additional observations would be most important in evaluating the conclusion?

a. The concentration of calcium in the agar at the pollen tubes
b. The length of the pollen tubes
c. The results achieved with a longer incubation time
d. The number of pollen tubes growing away from the wells
e. The incubation temperature

ANALYSIS

a. In order to conclude that calcium mediates the response, it is important to show that the response is concentration-dependent. However, it is reasonable to assume that calcium diffuses through the agar gel and that the concentration at some distance from the well is dependent on the concentration in the well. This additional observation would be a reasonable check on the validity of the assumption, but it is not necessary in order to interpret the data.

b. If calcium stimulates growth of pollen tubes, they should be longer given higher concentrations of calcium. This observation would be useful to confirm the interpretation, but the data on numbers alone can evaluate the hypothesis.

c. As long as the response is clear-cut at the time used, longer times might only confound the results. The test of chemotropism depends on applying the calcium from one side only and showing that the pollen tubes grow nonrandomly to that side. After a longer incubation, the calcium would diffuse beyond the pollen grains and destroy the gradient.

d. The first well, containing no calcium, shows that calcium is not absolutely essential for pollen-tube growth in one direction. The conclusion now depends on showing that *more* tubes grow toward calcium than away from it. If equal numbers of tubes grow in both directions, the conclusions would simply be that calcium stimulates growth, not that it is chemotropic.

e. Knowing the effects of temperature on the response and the rate of growth might be valuable. Assuming there is a temperature effect, you would need to know the temperature in order to repeat the experiment, but the evaluation of the hypothesis on the basis of the present data does not require knowledge of the temperature.

2. Which of the following independent observations would provide the greatest support for the hypothesis that calcium functions as a chemotropic substance for the pollen-tube growth in the living flower?

a. In the experiment in exercise 1, pollen tubes that start growing away from the calcium well sooner or later turn back and grow toward the calcium.
b. Extracts prepared from ovule tissue have chemotropic effects.
c. Ovules contain calcium.
d. Nonchemotropic parts of the plant do not contain calcium.
e. Ovule extracts are not chemotropic when the agar contains optimum concentrations of calcium.

ANALYSIS

a. This observation would strengthen the conclusion that calcium has a chemotropic effect on pollen, but it would not link the ion to the growth in the plant. Many biological processes can be stimulated or inhibited by substances that are not part of the normal control mechanisms in intact organisms.

b. This is the observation that started the whole study. The question is *what substance* in the extract mediates the effect.

c. This is a necessary link in the chain of evidence. Unless calcium is found in the ovule, it cannot be the chemotropic substance *in vivo*. However, this is not conclusive evidence. Other substances in the cells might mediate the response.

d. This observation would strengthen the conclusion, once it is known that calcium *is* present in the ovule. Even better evidence would be to observe that the chemotropic effectiveness of extracts of a variety of plant parts varies as their calcium content varies.

e. This would tell you that a stimulatory background of calcium could overcome any chemotropic effect of the extract, but it would not indicate what component of the extract mediated its chemotropism. Just as the sound of a jet engine can drown out any music, so the effect of calcium can stimulate pollen-tube growth to the maximum rate, and no other addition can add measurably to the effect.

3. In a more detailed study, the investigators mixed uniform concentrations of calcium into the agar medium. The pollen grains were planted on the surface, and the pattern of pollen-tube growth was observed several hours later. The data follow.

Calcium Concentration (M)	Total Number of Tubes Formed	Average Length of Tubes (mm)
0	2	0.3
10^{-6}	0	—
10^{-5}	8	0.3
10^{-4}	69	0.4
0.001	270	0.7
0.005	500	0.8
0.01	700	1.0
0.02	320	1.0
0.03	283	0.6
0.04	235	0.6
0.05	0	—
0.10	0	—

Given these observations, which of the following hypotheses best explains the mechanism of the chemotropic effect of calcium?

a. Calcium is essential for pollen-tube growth.
b. Calcium stimulates elongation of pollen tubes.
c. Calcium stimulates growth of pollen tubes, causing them to bend toward the side where the calcium concentration is higher.
d. Calcium inhibits growth of pollen tubes, causing them to bend toward the side where the calcium concentration is higher.
e. Calcium attracts the pollen tubes.

ANALYSIS

a. Apparently, calcium is essential, because no significant number of tubes grow until calcium concentration reaches a minimum level. This alone, however, would not make the tubes grow toward a source of calcium. It would simply be a necessary condition for the chemotropic substance to work.

b. The data support this conclusion, and this could be a chemotropic mechanism. If calcium causes more rapid elongation of tubes, any tubes that happen to start growing toward the source of calcium would grow faster. As they grew closer to the calcium source, the concentration would become higher, and they would grow still faster. On the other hand, any tubes that happened to start out growing away from the source would encounter lower and lower concentrations of calcium and would grow slower and slower, unless they happened to turn up the calcium gradient.

c. Calcium stimulates growth in a concentration-dependent fashion. If a pollen tube grows perpendicular to a calcium gradient, the tube will experience greater calcium on one side than on the other and, assuming the gradient is steep enough, will grow faster on one side than on the other. This will cause the tube to bend. However, it will bend *away from* the side of greater growth.

d. The data show that calcium at high enough concentrations inhibits tube growth. That could cause a tube that was growing across the calcium gradient to turn toward the calcium source. However, the tube would then encounter increasing concentrations of calcium and would stop growing.

e. None of the studies presented in these three questions support this hypothesis. In all cases, the calcium either did or could be assumed to contact the pollen tube directly and influence its growth. Without such direct contact, it is very difficult to propose a mechanism by which a pollen tube could detect the presence of calcium at a distance and respond by turning toward the calcium. It is necessary to propose some direct effect of the calcium on the molecular mechanisms of the pollen tube.

Answers

Answers to Interactive Exercises

REPRODUCTIVE MODES (29-I)
1. Sexual reproduction involves two sets of genetic instructions from gametes of two parents; genetics of parents and offspring vary. Asexual reproduction produces offspring genetically identical to the single parent. 2. sporophyte (C); 3. flower (B); 4. male gametophyte (E); 5. female gametophyte (D); 6. fertilization (A); 7. Sexual; 8. Asexual; 9. identical; 10. sporophytes; 11. flowers; 12. meiosis; 13. meiospores; 14. sperm; 15. eggs; 16. sporophyte; 17. pollen grains; 18. Sporophytes.

GAMETE FORMATION IN FLOWERS (29-II)
1. petal; 2. sepal; 3. anther; 4. stigma; 5. ovary; 6. stamen; 7. carpel; 8. O; 9. A; 10. O; 11. A; 12. O; 13. A; 14. A; 15. O; 16. A; 17. O; 18. O; 19. O; 20. O; 21. G; 22. C; 23. H; 24. B; 25. E; 26. F; 27. A; 28. D; 29. pollen sac (F); 30. microspore mother cell (D); 31. microspores (B); 32. tube cell (A); 33. sperm-producing cell (E); 34. male gametophyte (C); 35. ovule (C); 36. nucellus (D); 37. megaspore mother cell (A); 38. integuments (E); 39. micropyle (B); 40. meiosis; 41. microspores; 42. pollen grains; 43. sperm; 44. pollen tube; 45. egg; 46. Ovules; 47. seeds; 48. stalk; 49. nucellus; 50. integuments; 51. micropyle; 52. tube; 53. mother; 54. megaspores; 55. mitotic; 56. eight; 57. embryo sac; 58. gametophyte; 59. egg; 60. endosperm; 61. endosperm.

POLLINATION AND FERTILIZATION (29-III)
1. Pollination refers to the transfer of pollen grains to a stigma; wind, insects, birds, or other agents make the transfer. Fertilization generally means the fusion of a

sperm nucleus and an egg nucleus. 2. anthers; 3. stigmas; 4. pollination; 5. pollinators; 6. pollen tube; 7. Two; 8. embryo sac; 9. Double; 10. zygote; 11. endosperm; 12. endosperm; 13. sporophyte; 14. flowering; 15. seedling (sporophyte) (E); 16. embryo (G); 17. seed (M); 18. sperm (F); 19. tube nucleus (I); 20. anther (including pollen sacs) (D); 21. megaspore mother cell (J); 22. endosperm mother cell (C); 23. egg (L); 24. ovary wall (A); 25. embryo sac (B); 26. micropyle (H); 27. ovule (K).

SEED AND FRUIT FORMATION (29-IV)
1. a. Accessory; b. Pea, mustard; c. Cluster of matured ovaries, all attached to receptacle; d. Strawberry; e. Fruit wall dry; intact at maturity; f. Multiple; g. Simple; 2. nucleus; 3. vacuole; 4. zygote; 5. embryo; 6. embryo; 7. endosperm; 8. seed coat; 9. cotyledons; 10. embryo; 11. root tip; 12. ovule; 13. fruit; 14. D; 15. B; 16. F; 17. E; 18. A; 19. C; 20. seed; 21. ovary; 22. integuments; 23. Cotyledons; 24. Grains; 25. fleshy; 26. multiple; 27. seed; 28. dispersal; 29. fruits; 30. animals; 31. germination.

ASEXUAL REPRODUCTION OF FLOWERING PLANTS (29-V)
1. C; 2. D; 3. A; 4. B.

Answers to Self-Quiz

1. c; 2. c; 3. c; 4. b; 5. d; 6. b; 7. b; 8. c; 9. c; 10. d.

30

PLANT GROWTH AND DEVELOPMENT

SEED GERMINATION
PATTERNS OF EARLY GROWTH
PLANT HORMONES
    Auxins
    Gibberellins
    Cytokinins
    Abscisic Acid
    Ethylene
    Other Plant Hormones
PLANT RESPONSES TO THE ENVIRONMENT
    The Many and Puzzling Tropisms
        Phototropism
        Gravitropism
        Thigmotropism

Response to Mechanical Stress
Biological Clocks and Their Effects
    Circadian Rhythms
    Photoperiodism
The Flowering Process
    Photoperiodic Responses
    Vernalization
Senescence
Dormancy
Commentary: From Embryo to Mature Oak

Interactive Exercises

SEED GERMINATION (30-I, pp. 517–518)

1. What events occur at germination?_____

    _____

2. Name the factors that influence germination._____

    _____

3. Why does germination often coincide with the return of spring rains?_____

    _____

4. Explain the process of imbibition._____

    _____

Fill-in-the-Blanks

Before or after (5)_____ of seeds, embryonic growth idles. The embryo absorbs water, resumes

growth, and breaks through the seed coat at (6)_____. In a process called (7)_____ , water

molecules move into the seed, being especially attracted to (8)_____ groups of the stored proteins.

As more water moves inside, the seed (9)_____ , and the coat ruptures. Embryonic metabolism increases when oxygen is available and (10)_____ respiration occurs. Cells divide and continuously elongate to produce the (11)_____. Inside the germinating seed, (12)_____ cells of the embryo are the first to grow and give rise to the (13)_____ root of the new plant. When this first root emerges through the seed coat, (14)_____ is completed.

PATTERNS OF EARLY GROWTH (30-II, pp. 518–519)

Labeling

Label the development patterns illustrated below as dicot or monocot.

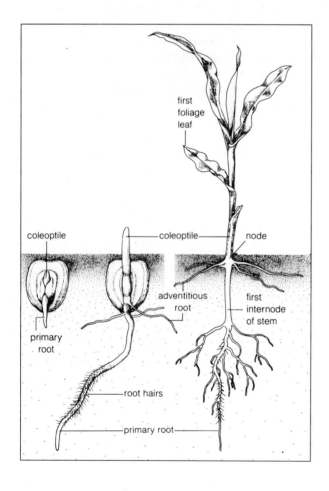

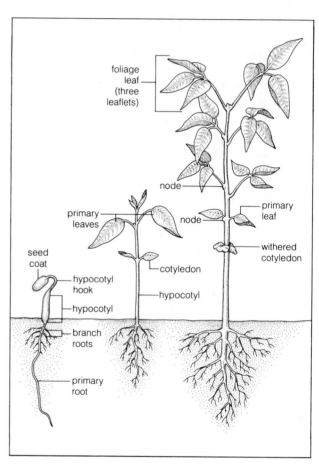

1. _____

2. _____

True-False

If false, explain why.

___ 3. The cell divisions in a growing plant are equal.

___ 4. Differences such as unequal distribution of cytoplasm affect how cells interact with their neighbors during growth.

___ 5. Activation of genes governing synthesis of growth-stimulating hormones in some cells and not in others is called selective gene expression.

___ 6. Plant growth and development are under environmental controls.

___ 7. Selective gene expression in plant embryo cells influences the direction of cell divisions and how they will differentiate.

___ 8. In a growing plant, all the dividing cells undergo enlargement and retain the capacity to divide.

___ 9. Cell enlargement in a growing plant is driven by water uptake.

___ 10. As plant cells grow, their walls become thinner and discourage plant growth.

PLANT HORMONES (30-III, pp. 520–522)

1. What is the role of hormones in the life of complex plants?_____

2. Define hormone and target cell._____

_____

_____

3. List the five types of plant hormones known to be produced by most flowering plants._____

_____

4. Which hormone has been named but is as yet not identified?_____

Identifying

Name the hormone that is primarily responsible for the effects described after each blank.

5. _____ Promotes stem elongation (especially in dwarf plants); might help break dormancy of seeds and buds

6. _____ Arbitrary designation for as-yet-unidentified hormone (or hormones) thought to cause flowering

7. _____ Promote cell elongation in coleoptiles and stems; thought to be involved in phototropism and gravitropism

8. _____ Promotes stomatal closure; might trigger bud and seed dormancy

9. _____ Promotes fruit ripening; promotes abscission of leaves, flowers, and fruits

10. _____ Promotes cell division; promotes leaf expansion and retards leaf aging

Matching

Match the appropriate letter with its numbered partner.

11. ___ 2,4-D
12. ___ hormone
13. ___ apical dominance
14. ___ herbicide
15. ___ abscission
16. ___ IAA
17. ___ gibberellins
18. ___ coleoptile
19. ___ 2,4,5-T
20. ___ target cell

A. Agent Orange; used to defoliate during the Vietnam conflict
B. Have little effect on stem elongation of pines, firs, and other conifers
C. Hollow, protective cylinder protecting new leaves of monocot embryos
D. Dropping of flowers, fruits, and leaves
E. A cell with receptors for a given signaling molecule
F. Hormonal effect that inhibits lateral bud growth, which promotes stem elongation
G. Synthetic auxin used as an herbicide
H. A signaling molecule released from one cell that changes the activity of target cells
I. The most important naturally occurring auxin
J. Any compound used to selectively kill plants

PLANT RESPONSES TO THE ENVIRONMENT (30-IV, pp. 522–527)

1. Complete the following table with information about plant growth responses called tropisms.

Tropism	Definition	Example
a. Phototropism		Many plant leaves will turn until their flat upper surface faces light.
b.	A plant growth response to the earth's gravitational force	
c.		While growing, a plant stem curls around a fencepost.

Matching

Choose the one most appropriate answer for each.

2. ___ gravitropism
3. ___ phototropism
4. ___ thigmotropism

A. Negative response in shoot; positive response in root due to growth inhibitor produced by the root cap
B. A response to physical contact evident in climbing vines; may involve auxin and ethylene
C. Known to be controlled by auxin and light

Labeling

Identify each indicated part of the accompanying illustration.

5. _____

6. _____

7. _____

8. _____

9. _____

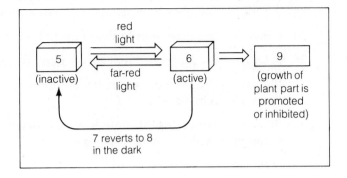

Fill-in-the-Blanks

Strong winds, rainstorms, grazing animals, and even farm machinery can (10)_____ a plant so much that its overall (11)_____ will be inhibited. Like other organisms, plants have internal time-measuring mechanisms called biological (12)_____ that function in daily and seasonal adjustments. (13)_____ rhythms refer to plant activities that occur regularly in cycles of about twenty-four hours. For example, many plants hold their (14)_____ in horizontal positions during the day but fold them in sleep movements at night. (15)_____ is defined as any biological response to a change in the relative length of daylight and darkness in a twenty-four hour cycle. A blue-green pigment molecule, (16)_____ , serves as a switching mechanism in the photoperiodism response. Phytochrome is converted to an active form (Pfr) at sunrise, when (17)_____ wavelengths dominate the sky. Phytochrome reverts to an inactive form (Pr) at sunset, at night, or in the shade, where (18)_____ wavelengths predominate. Pfr may control which types of (19)_____ are being produced in particular cells for different growth responses. (20)_____ influences seed germination, stem elongation and branching, leaf expansion, and the formation of flowers, fruit, and seeds. Photoperiodism is especially apparent in the (21)_____ process, which is often keyed to daylength changes throughout the year. "Long-day plants" flower in spring when daylength becomes (22) [choose one] ( ) shorter, ( ) longer than some critical value. "Short-day plants" flower in late summer or early autumn when daylength becomes (23) [choose one] ( ) shorter, ( ) longer than some critical value. "Day-neutral plants" flower whenever they become (24)_____ enough to do so without regard to daylength. Low-temperature stimulation of plants to induce flowering is called (25)_____. Dropping of plant parts such as leaves, flowers, or fruits is called (26)_____. (27)_____ is the total of the processes that lead to the death of plant parts or the entire plant. When a plant ceases growth under suitable growing conditions, it is said to enter (28)_____. (29)_____ days and (30)_____ nights are strong cues for dormancy. Depending on the species, two compounds that may play a role in breaking dormancy are (31)_____ and (32)_____ acid.

If false, explain why.

___ 33. Frits Went named auxin and demonstrated that it moves from the coleoptile tip into cells less exposed to light; this stimulates these cells to shorten.

___ 34. Growing plant stems curve toward the light source.

___ 35. A horizontally positioned root curves downward if its root cap is surgically removed.

___ 36. Plants of the same species growing outdoors are shorter and have somewhat thicker stems than those grown indoors.

___ 37. Botanists suspect that Pfr controls which enzymes are produced in particular cells; this in turn determines which growth responses occur.

___ 38. Spinach, a short-day plant, will not flower and produce seeds unless it is exposed to fourteen hours of light each day for two weeks.

___ 39. Cocklebur, a short-day plant, requires a single night longer than 8-1/2 hours to trigger flowering. A minute or two of artificial light during that dark period will block flowering.

Chapter Terms

The following page-referenced terms are important; they were in boldface type in the chapter. Refer to the instructions given in Chapter 1, p. 4.

germination (517)	abscisic acid (520)	phototropism (522)	photoperiodism (524)
hormone (520)	ethylene (521)	gravitropism (522)	vernalization (526)
auxins (520)	abscission (521)	thigmotropism (523)	senescence (527)
gibberellins (520)	apical dominance (522)	biological clocks (524)	dormancy (527)
cytokinins (520)			

Self-Quiz

___ 1. Promoting fruit ripening and abscission of leaves, flowers, and fruits is a function ascribed to _____.
   a. gibberellins
   b. ethylene
   c. abscisic acid
   d. auxins

___ 2. Auxins _____.
   a. cause flowering
   b. promote stomatal closure
   c. promote cell division
   d. promote cell elongation in coleoptiles and stems

___ 3. _____ is demonstrated by a germinating seed whose first root always curves down while the stem always curves up.
   a. Phototropism
   b. Photoperiodism
   c. Gravitropism
   d. Thigmotropism

___ 4. Light of _____ wavelengths is the main stimulus for phototropism.
   a. blue
   b. yellow
   c. red
   d. green

___ 5. Plants whose leaves are open during the day but fold at night are exhibiting _____.
   a. a growth movement
   b. a circadian rhythm
   c. a biological clock
   d. both b and c

___ 6. 2,4-D, a potent dicot weed killer, is a synthetic _____.
   a. auxin
   b. gibberellin
   c. cytokinin
   d. phytochrome

7. All the processes that lead to the death of a plant or any of its organs are called

_____.

a. dormancy
b. vernalization
c. abscission
d. senescence

___ 8. Phytochrome is converted to an active form, _____, at sunrise and reverts to an inactive form, _____, at sunset, at night, or in the shade.

a. Pr; Pfr
b. Pfr; Pfr
c. Pr; Pr
d. Pfr; Pr

___ 9. Which of the following is not promoted by the active form of phytochrome?

a. seed germination
b. flowering
c. leaf expansion
d. none of the above

___ 10. When a perennial or biennial plant stops growing under conditions suitable for growth, it has entered a state of

_____.

a. senescence
b. vernalization
c. dormancy
d. abscission

Chapter Objectives/Review Questions

This section lists general and detailed chapter objectives that can be used as review questions. You can make maximum use of these items by writing answers on a separate sheet of paper. Fill in answers where blanks are provided. To check for accuracy, compare your answers with information given in the chapter or glossary.

Page   *Objectives/Questions*

(517)   1. The plant embryo absorbs water, resumes growth, and breaks the seed coat at _____.
(518)   2. Define imbibition and its role in germination.
(518)   3. What portion of the embryo emerges from the seed first?
(518)   4. Plant growth and development are under _____ controls.
(518–519) 5. What determines the directions in which plant embryo cells will divide and how they will differentiate?
(520)   6. Describe the general role of plant hormones and target cells.
(520)   7. Be able to list the five types of hormones found in most flowering plants.
(520)   8. _____ promote cell elongation in coleoptiles and stems and are involved in phototropism and gravitropism.
(520)   9. _____ promote stem elongation, might help break dormancy of seeds and buds, and stimulate starch breakdown.
(520)   10. _____ promote cell division, promote leaf expansion, and retard leaf aging.
(520–521) 11. _____ promotes stomatal closure and bud and seed dormancy.
(521)   12. _____ promotes fruit ripening and abscission of leaves, flowers, and fruits.
(522)   13. _____ is an arbitrary designation for an as-yet-unidentified hormone (or hormones) thought to cause flowering.
(520)   14. Define herbicide and give examples.
(522)   15. _____ dominance is the hormonal effect that inhibits lateral bud growth while plant resources are diverted to stem elongation.
(522)   16. Through interactions with the _____, a flowering plant adjusts its growth patterns.
(522–523) 17. Describe and give examples of phototropism, gravitropism, and thigmotropism.
(523)   18. List possible stresses for plants and describe some responses.
(524)   19. Plants have internal time-measuring mechanisms called biological _____.
(524)   20. What are circadian rhythms? Give an example.
(524)   21. _____ is a biological response to a change in the relative length of daylight and darkness in a twenty-four-hour cycle.

Integrating and Applying Key Concepts

An oak tree has grown up in the middle of a forest. A lumber company has just cut down all the surrounding trees except for a narrow strip of woods that includes the oak. How is the oak likely to respond as it adjusts to its changed environment? To what new stresses will it be exposed? Which hormones will most probably be involved in the adjustment?

Critical Thinking Exercises

1. In a study of flowering in a short-day plant, a light-tight barrier was used to expose one half of the plant to constant darkness and the other half to short days. At the beginning of the study, the plant had no flowers; at the end, both halves had flowers. When both halves of another specimen were exposed to constant darkness, no flowers were formed. Which of the following is the best conclusion made from these observations?

   a. A flower-initiating hormone is produced by the leaves when they are exposed to short days.
   b. On short days, the plant can produce a flower-initiating hormone that can move to other parts of the plant.
   c. When exposed to long days, the plant produces a flower-inhibiting hormone.
   d. When exposed to long days, the plant produces a flower-initiating hormone.
   e. Constant darkness stimulates the plant to produce a flower-inhibiting hormone that can be distributed to other parts of the plant.

ANALYSIS

   a. Because the plant flowered when one half was exposed to short days, the production of a flower-initiating hormone might be indicated. However, there is no reason to identify the leaves as the site of hormone production. Furthermore, this choice does not account for flowering in the unlighted parts of the plant.
   b. This answer is similar to (a) but is significantly better. This choice does not go beyond the data to postulate a specific site of synthesis of the hormone. Also, this choice accounts for flowering in unlighted parts of the plant by postulating a hormone that moves throughout the plant.
   c. It is known that the plant does not flower when exposed to long days, but there are no data leading to the conclusion that an inhibitory hormone is involved. Moreover, this conclusion does not explain flowering under any circumstances.
   d. This choice contradicts the observation that flowers are not formed on a long-day light schedule.
   e. This mechanism could account for the lack of flowering under constant darkness. However, it does not account for the formation of flowers under any circumstances, especially in the part of the plant exposed to darkness in the split-light experiment. Such a hormone would be expected to be present, and some mechanism to overcome its inhibitory effect must be assumed.

2. In one of the first studies of phototropism in coleoptiles, Charles Darwin made the hypothesis that the tip of the coleoptile somehow senses the direction of the light source. He covered the tips of growing coleoptiles with black caps, and the plants did not bend toward the light source. The uncovered controls did. He concluded that his hypothesis was supported. Which of the following was he most likely assuming?

    a. The caps were allowing some light to reach the tips.
    b. The weight of a cap on the tip did not prevent bending.
    c. Coleoptiles bend toward a light source when their tips are not covered.
    d. The tip of a coleoptile can sense the direction of the light source.
    e. Light causes uneven distribution of a cell growth hormone within the coleoptiles and causes bending of the shoot.

ANALYSIS

    a. He could not have assumed this. His experiment depends on the assumption that the black caps do prevent light from reaching the coleoptiles. Only if no light passes the caps can he make different predictions: bending if the hypothesis is not valid, no bending if it is valid.
    b. If the weight of the caps prevents the tips from bending, then the same outcome, no bending, is predicted whether the hypothesis is valid or not.
    c. This observation was made previously. The hypothesis is an attempt to explain this observation.
    d. This is the hypothesis itself. It is being tested, not accepted without evidence.
    e. This is a more elaborate hypothesis that suggests a mechanism for the hypothesis being tested in the experiment. No assumptions about possible mechanisms are needed (or reasonable) until the hypothesis is tested and found acceptable.

3. Darwin later worried that he had not used adequate controls in his experiment on coleoptile phototropism. Which of the following would be the best control?

    a. Grow coleoptiles in complete darkness.
    b. Cut off the tips of the coleoptiles.
    c. Place a transparent cap on the tips.
    d. Illuminate the plants from all directions.
    e. Let substances from coleoptile tips diffuse into an agar block, and then apply the block to the side of an intact coleoptile.

ANALYSIS

    a. The experiment tests a hypothesis about the mechanism of phototropism. A plant is illuminated from one direction and manipulated to determine whether the response is abolished. The control has to be a plant in which the stimulus, directional illumination, is applied and the plant responds normally. Plants treated according to this choice would not have the stimulus.
    b. This is another experiment, not a control. It is a different manipulation that would predict no bending if the hypothesis is valid, bending if it is not.
    c. The experiment depends on the assumption that the caps only block the light from the tips. Some other property of the caps, such as their weight, could cause the failure of the coleoptiles to bend. This operation imposes all the conditions except blockage of light on the control shoots.
    d. This choice is subject to the same objection as choice (a). It eliminates the essential test stimulus.
    e. Like choice (b), this is another experiment, but this one tests a more elaborate hypothesis about the mechanism by which the tips detect and respond to light. There is no reason to do this experiment until an adequate control establishes the validity of the first hypothesis.

Answers

Answers to Interactive Exercises

SEED GERMINATION (30-I)

1. The embryo absorbs water, resumes growth, and breaks through the seed coat. 2. the amount of moisture and oxygen, the temperature, the number of daylight hours, and other environmental aspects. 3. Water availability is seasonal. 4. Water molecules move into the seed and are especially attracted to hydrophilic groups of the stored proteins. 5. dispersal; 6. germination; 7. imbibition; 8. hydrophilic; 9. swells; 10. aerobic; 11. seedling; 12. root; 13. primary; 14. germination.

PATTERNS OF EARLY GROWTH (30-II)

1. monocot; 2. dicot; 3. F; 4. T; 5. T; 6. F; 7. T; 8. F; 9. T; 10. F.

PLANT HORMONES (30-III)

1. Hormones govern the growth and development of complex plants. 2. A hormone is a signaling molecule released from one cell that changes the activity of target cells. A target cell is a cell that has receptors for a given signaling molecule, either within the cell or at the surface of its plasma membrane. 3. auxins, gibberellins, cytokinins, abscisic acid, and ethylene. 4. florigen; 5. gibberellins; 6. florigen; 7. auxins; 8. abscisic acid (ABA); 9. ethylene; 10. cytokinins; 11. G; 12. H; 13. F; 14. J; 15. D; 16. I; 17. B; 18. C; 19. A; 20. E.

PLANT RESPONSES TO THE ENVIRONMENT (30-IV)

1. a. The adjustment of a plant's direction and rate of growth in response to light; b. Gravitropism; Stems curve upward when a potted seedling is turned on its side in a dark room; c. Thigmotropism; Unequal plant growth resulting from physical contact with solid objects in their surroundings; 2. A; 3. C; 4. B; 5. Pr; 6. Pfr; 7. Pfr; 8. Pr; 9. response; 10. stress; 11. growth; 12. clocks; 13. Circadian; 14. leaves; 15. Photoperiodism; 16. phytochrome; 17. red; 18. far-red; 19. enzymes; 20. Pfr; 21. flowering; 22. longer; 23. shorter; 24. mature; 25. vernalization; 26. abscission; 27. Senescence; 28. dormancy; 29. Short; 30. long; 31. gibberellins; 32. abscisic; 33. F; 34. T; 35. F; 36. T; 37. T; 38. F; 39. T.

Answers to Self-Quiz

1. b; 2. d; 3. c; 4. a; 5. d; 6. a; 7. d; 8. d; 9. d; 10. c.

31

TISSUES, ORGAN SYSTEMS, AND HOMEOSTASIS

ANIMAL STRUCTURE AND
FUNCTION: AN OVERVIEW
ANIMAL TISSUES
    Tissue Formation
    Epithelial Tissue
        General Features
        Glandular Epithelium
        Cell-to-Cell Contacts in Epithelium
    Connective Tissue
        Connective Tissue Proper
        Specialized Connective Tissue

Muscle Tissue
Nervous Tissue
MAJOR ORGAN SYSTEMS
HOMEOSTASIS AND SYSTEMS CONTROL
    The Internal Environment
    Mechanisms of Homeostasis

Interactive Exercises

ANIMAL STRUCTURE AND FUNCTION: AN OVERVIEW (31-I, pp. 533–534)

True-False

If false, explain why.

___ 1. Physiology is the study of the way body parts are arranged in an organism.

___ 2. Groups of like cells that work together to perform a task are known as an organ.

___ 3. Most animals are constructed of only four types of tissue; epithelial, connective, nervous, and muscle tissues.

Fill-in-the-Blanks

Groups of *like* cells that work together to perform a task are known as a(n) (4)_____. Groups of

different types of tissues that interact to carry out a task are known as a(n) (5)_____. Each cell engages

in basic (6)_____ activities that assure its own survival. The combined contributions of cells, tissues,

organs, and organ systems help maintain a stable (7)_____ _____ that is required for individual

cell survival.

Matching

Choose the most appropriate answer to match with each of the following terms.

8. ___ circulatory system

9. ___ digestive system

10. ___ endocrine system

11. ___ immune system

12. ___ integumentary system

13. ___ muscular system

14. ___ nervous system

15. ___ reproductive system

16. ___ respiratory system

17. ___ skeletal system

18. ___ urinary system

A. Picks up nutrients absorbed from gut and transports them to cells throughout body

B. Helps cells use nutrients by supplying them with oxygen and relieving them of CO_2 wastes

C. Helps maintain the volume and composition of body fluids that bathe the body's cells

D. Provides basic framework for the animal and supports other organs of the body

E. Uses chemical messengers to control and guide body functions

F. Protects the body from viruses, bacteria, and other foreign agents

G. Produces younger, temporarily smaller versions of the animal

H. Breaks down larger food molecules into smaller nutrient molecules that can be absorbed by body fluids and transported to body cells

I. Consists of contractile parts that move the body through the environment and propel substances about in the animal

J. Serves as an electrochemical communications system in the animal's body

K. In the meerkat, served as a heat catcher in the morning and protective insulation at night

ANIMAL TISSUES (31-II, pp. 534–539)

True-False

If false, explain why.

___ 1. Mammalian skin contains squamous epithelium and other tissues.

___ 2. The more tight junctions there are in a tissue, the more permeable the tissue will be.

___ 3. Endocrine glands secrete their products through ducts that empty onto an epithelial surface.

___ 4. Endocrine-cell products include digestive enzymes, saliva, and mucus.

___ 5. Muscle bundles are identical to skeletal muscle cells.

___ 6. Both skeletal and cardiac muscle tissues are striated.

___ 7. Cardiac muscle cells are fused, end-to-end, at regions called muscle bundles.

___ 8. Smooth muscle tissue is involuntary and not striated.

___ 9. Smooth muscle tissue is located in the walls of the intestine.

___ 10. Neurons conduct messages to other neurons or to muscles or glands.

Fill-in-the-Blanks

In (11)_____ _____ (immature reproductive cells that later develop into (12)_____), a special form of cell division known as (13)_____ occurs. In all other body tissues that consist of (14)_____ cells, the usual form of cell division, (15)_____ , occurs. The life of almost any animal begins with two gametes merging to form a fertilized egg, which undergoes reorganization and then divides by mitosis to form first undifferentiated cells, then three types of (16)_____ (groups of similar cells that perform similar activities) in the early embryo. Eventually, the outer layer of skin and the tissues of the nervous system are formed from the relatively unspecialized embryonic tissue known as (17)_____ located on the embryo's surface. The inner lining of the gut and the major organs formed from the embryonic gut develop from the internal embryonic tissue known as (18)_____ . Most of the internal skeleton, muscle, the circulatory, reproductive, and urinary systems, and the connective tissue layers of the gut and body covering are formed from (19)_____ , the embryonic tissue composed of cells that can move about like amoebae.

While a specific kind of tissue (for example, simple squamous epithelium) is composed of cells that look very similar and do similar jobs, a specific (20)_____ (for example, a kidney) is composed of different tissues that cooperate to do a specific job (in this case, to remove waste products from blood, produce urine, and maintain the composition of body fluids). Kidneys, a urinary bladder, and various tubes are grouped together into a(n) (21)_____ _____ , which adds the functions of urine storage and elimination to the jobs that the kidneys do. A multicellular (22)_____ is most often composed of organ systems that cooperate to keep activities running smoothly in a coordinated fashion; this maintenance of stable operating conditions in the internal environment is known as (23)_____ .

All the diverse body parts found in different animals can be assembled from a few (24)_____ types through variations in the way they are combined and arranged. (25)_____ cells constitute the physical structure of the animal body; they become differentiated into the components of four main types of tissue—(26)_____ , (27)_____ , (28)_____ , and (29)_____ .

Cells of the connective tissues are scattered throughout an extensive extracellular (30)_____ _____ . (31)_____ connective tissue contains a weblike scattering of strong, flexible protein fibers ((32)_____) and a few highly elastic protein fibers ((33)_____) and serves as a packing material that holds in place blood vessels, nerves, and internal organs. (34)_____ and (35)_____ are examples of dense, regular connective tissue that help connect elements of the skeletal and muscular systems. (36)_____ and (37)_____ are examples of supportive connective tissue.

Label-Match

Identify each of the illustrations below by labeling it with one of the following: connective, epithelial, muscle, nervous, germ cells and/or gametes. Complete the exercise by matching and entering all appropriate letters from each group below in the parentheses following each label.

38. _____ ( )
39. _____ ( )
40. _____ ( )
41. _____ ( )
42. _____ ( )
43. _____ ( )
44. _____ ( )
45. _____ ( )
46. _____ ( )
47. _____ ( )
48. _____ ( )
49. _____ ( )
50. _____ ( )

38.

39.

40.

41.

42.

43.

44. Haversian canal

cells in spaces

45.

46.

lymphocyte

platelets

neutrophils

erythrocytes

47.

48.

49.

50.

A. Adipose
B. Bone
C. Cardiac
D. Dense, regular
E. Loose
F. Simple columnar
G. Simple cuboidal
H. Simple squamous
I. Smooth
J. Skeletal

1. Absorption
2. Maintain diploid number of chromosomes in sexually reproducing populations
3. Communication by means of electrochemical signals
4. Energy reserve
5. Contraction for voluntary movements
6. Diffusion
7. Padding
8. Contract to propel substances along internal passageways; not striated
9. Attaches muscle to bone and bone to bone
10. In vertebrates, provides the strongest internal framework of the organism
11. Elasticity
12. Secretion
13. Pumps circulatory fluid; striated
14. Insulation
15. Transport of nutrients and waste products to and from body cells

MAJOR ORGAN SYSTEMS (31-III, pp. 540–541)

Matching

Match the most appropriate function with each system shown on the next page.

1. ___ Male: production and transfer of sperm to the female. Female: production of eggs; provision of a protected nutritive environment for developing embryo and fetus. Both systems have hormonal influences on other organ systems.

2. ___ Ingestion of food, water; preparation of food molecules for absorption; elimination of food residues from the body.

3. ___ Movement of internal body parts; movement of whole body; maintenance of posture; heat production.

4. ___ Detection of external and internal stimuli; control and coordination of responses to stimuli; integration of activities of all organ systems.

5. ___ Protection from injury and dehydration; body temperature control; excretion of some wastes; reception of external stimuli; defense against microbes.

6. ___ Provisioning of cells with oxygen; removal of carbon dioxide wastes produced by cells; pH regulation.

7. ___ Support, protection of body parts; sites for muscle attachment, blood cell production, and calcium and phosphate storage.

8. ___ Hormonal control of body functioning; works with nervous system in integrative tasks.

9. ___ Maintenance of the volume and composition of extracellular fluid.

10. ___ Rapid internal transport of many materials to and from cells; helps stabilize internal temperature and pH.

11. ___ Return of some extracellular fluid to blood; roles in immunity (defense against specific invaders of the body).

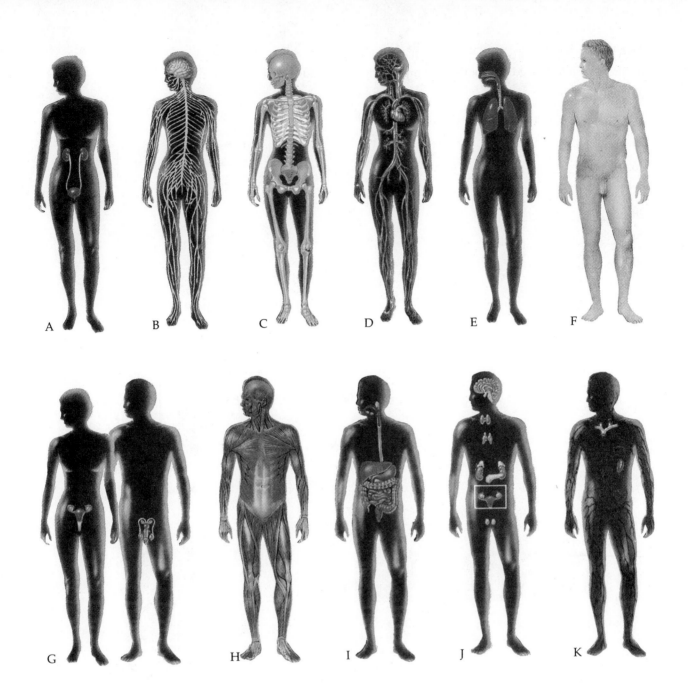

Fill-in-the-Blanks

There are four major body cavities in humans. Lungs are located in the (12)_____ cavity, the brain is in the (13)_____ cavity, and, in the female, ovaries and the urinary bladder are in the (14)_____ cavity. The (15)_____ plane divides the body into right and left halves. The (16)_____ plane divides the body into (17)_____ (front) and posterior (back) parts. The (18)_____ plane divides it into dorsal (upper) and (19)_____ (lower) parts. The urinary system of an animal is responsible for the disposal of (20)_____ wastes; fecal material is not considered in the category. The endocrine system is generally responsible for internal (21)_____ control; together with the (22)_____ system, it integrates physiological processes.

Labeling

Identify each indicated part of the accompanying illustration.

23. _____ _____

24. _____

25. _____

26. _____

27. _____

28. _____ _____

29. _____

30. _____

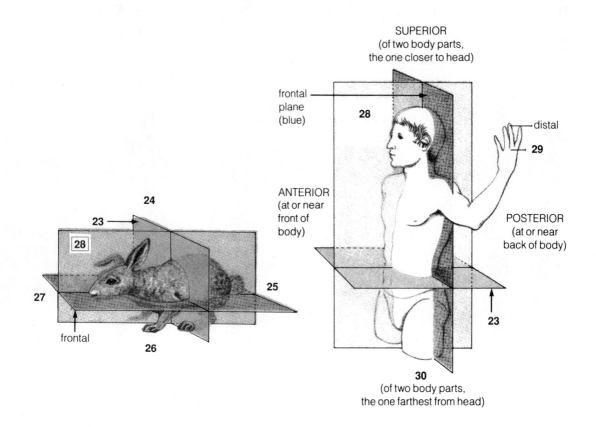

HOMEOSTASIS AND SYSTEMS CONTROL (31-IV, pp. 542–543)

True-False

If false, explain why.

___ 1. The process of childbirth is an example of a negative feedback mechanism.

___ 2. The human body's effectors are, for the most part, muscles and glands.

___ 3. An integrator is constructed in such a way that it detects specific energy changes in the environment and relays messages about them to a receptor.

___ 4. Interstitial fluid is a synonym for plasma.

Fill-in-the-Blanks

In a (5)_____ feedback mechanism, a chain of events is set in motion that intensifies the original condition; sexual arousal and childbirth are two examples. Generally, physiological controls work by means of (6)_____ feedback, in which an activity changes some condition in the internal environment, and the change causes the condition to be reversed; the maintenance of body temperature close to a "set point" is an example. Your brain is a(n) (7)_____ , a control point where different bits of information are pulled together in the selection of a response. Muscles and (8)_____ are examples of effectors. Internal temperature control of the human body is achieved by (9)_____ in the skin and elsewhere sensing a temperature change at the body's surface and relaying the neural information to an integrator. In this example, the (10)_____ in the brain compares neural input against a set point. This part of the brain then sends output signals to (11)_____. Examples of these are the (12)_____ gland, the secretions of which enhance or suppress rates of metabolic activity, the (13)_____ glands, which excrete water from the skin, (14)_____ muscles in blood vessels that regulate the distribution of blood flow, and (15)_____ muscles that may be involved in shivering and other changes in behavior.

Chapter Terms

The following page-referenced terms are the important terms in the chapter. Refer to the instructions given in Chapter 1, p. 4.

tissue (533)
organs (534)
organ systems (534)
epithelium, -lia (534)
exocrine glands (535)
endocrine glands (535)
connective tissue (536)
loose connective
   tissue (537)

dense, irregular connective
   tissue (537)
dense, regular connective
   tissue (537)
cartilage (538)
bone (538)
adipose tissue (538)
blood (538)
muscle tissue (538)

nervous tissue (539)
integumentary system (540)
muscular system (540)
skeletal system (540)
nervous system (540)
endocrine system (540)
circulatory system (540)
lymphatic system (541)
respiratory system (541)

digestive system (541)
urinary system (541)
reproductive system (541)
extracellular fluid (542)
homeostasis (542)
negative feedback
   mechanism (542)
positive feedback
   mechanisms (542)

Self-Quiz

____ 1. Which of the following is *not* included in connective tissues?
   a. bone
   b. blood
   c. cartilage
   d. skeletal muscle

____ 2. A surrounding substance within which something originates, develops, or is contained is known as a _____.
   a. lamella
   b. ground substance
   c. plasma
   d. lymph

____ 3. Blood is considered to be a(n) _____ tissue.
   a. epithelial
   b. muscular
   c. connective
   d. none of these

____ 4. Which group is arranged correctly from smallest structure to largest?
   a. muscle cells, muscle bundle, muscle
   b. muscle cells, muscle, muscle bundle
   c. muscle bundle, muscle cells, muscle
   d. none of the above

_____ 5. Muscle that is not striped and is involuntary is _____.
   a. cardiac
   b. skeletal
   c. striated
   d. smooth

_____ 6. Chemical and structural bridges link groups or layers of like cells, uniting them in structure and function as a cohesive _____.
   a. organ
   b. organ system
   c. tissue
   d. cuticle

_____ 7. A fish embryo was accidentally stabbed by a graduate student in developmental biology. Later, the embryo developed into a creature that could not move and had no supportive or circulatory systems. Which embryonic tissue had suffered the damage?
   a. ectoderm
   b. endoderm
   c. mesoderm
   d. protoderm

_____ 8. A tissue whose cells are striated and fused at the ends by intercalated disks so that the cells are not autonomous is called _____ tissue.
   a. smooth muscle
   b. dense fibrous connective
   c. supportive connective
   d. cardiac muscle

_____ 9. The secretion of tears, milk, sweat, and oil are functions of _____ tissues.
   a. epithelial
   b. loose connective
   c. lymphoid
   d. nervous

Chapter Objectives/Review Questions

This section lists general and specific chapter objectives that can be used as review questions. You can make maximum use of these items by writing answers on a separate sheet of paper. Fill in answers where blanks are provided. To check for accuracy, compare your answers with information given in the chapter or glossary.

Page   *Objectives/Questions*

(532–533) 1. Explain how the meerkat maintains a rather constant internal environment in spite of changing external conditions.

(533–534) 2. Cells are the basic units of life; in a multicellular animal, like cells are grouped into a _____.

(534–539) 3. Know the characteristics of the various types of tissues. Know the types of cells that compose each tissue type and be able to cite some examples of organs that contain significant amounts of each tissue type.

(534–536) 4. _____ tissues cover the body surface of all animals and line internal organs from gut cavities to vertebrate lungs; this tissue always has one _____ surface; the opposite surface adheres to a _____ _____.

(536)   5. Explain the nature of three different cell-to-cell junctions and state the types of tissues in which these junctions occur.

(534–536) 6. List the functions carried out by epithelial tissue and state the general location of each type.

(535)   7. Explain the meaning of the term *gland*, cite three examples of glands, and state the extracellular products secreted by each.

(536–538) 8. Describe the basic features of connective tissue, and explain how they enable connective tissue to carry out its various tasks.

(536)   9. Connective tissue cells and fibers are surrounded by a _____ _____.

(538)   10. List three functions of blood.

(538)   11. Distinguish among skeletal, cardiac, and smooth muscle tissues in terms of location, structure, and function.

(538)   12. Muscle tissues contain specialized cells that can _____.

(539)   13. Neurons are organized as lines of _____.

Interpreting and Applying Key Concepts

Explain why, of all places in the body, marrow is located on the interior of long bones. Explain why your bones are remodeled after you reach maturity. Why does your body not keep the same mature skeleton throughout life?

Critical Thinking Exercises

1. Suppose you are studying microscope slides of tissue from a previously undescribed animal. You find a small patch of cells that look empty, like the adipose cells in Figure 31.5. Which of the following interpretations would be satisfactory?

   a. The slide contains a patch of adipose tissue.
   b. The cells contained some material that dissolved in the solutions used to prepare the tissue for microscopy.
   c. The cells contained some material that fell out during handling of the sections before they were mounted on the slide.
   d. The cells contain some material that does not react with the stain used on these slides.
   e. The cells really are hollow in the animal and help control the density of its body, like the gas bladder of a fish.

   ANALYSIS

   a. This interpretation is certainly possible and might be a reasonable hypothesis to guide further investigation. But it is by no means the only possibility. The appearance of slides is sometimes an inaccurate reflection of the actual structure of the tissue, and various tissues can have the same appearance. Cross sections of some nerve tissues look like these adipose cells, too.
   b. This is the reason the adipose cells look hollow and empty. The lipid is dissolved by the solvents used to process the tissue. Even water-soluble materials can be removed from cells this way, however, because aqueous solutions are also used in the preparation of tissues for microscopy.
   c. Material can easily fall out of the very thin sections used for microscope slides, leaving cells or other structures looking hollow and empty. Usually, however, this happens only to some of the cells in a section, and some still show their contents. In text Figure 31.5, all the cells are empty.
   d. This is always a possibility. Stains are chemicals, and staining is a chemical reaction. Not all stains react with all cell components. It is possible to cut sections from frozen pieces of tissue and prepare them for microscopy without using lipid solvents, but the adipose cells still look hollow and empty unless stains that can react with lipid are used.
   e. This would be an unexpected, but not impossible, cell structure, and it would explain the hollow appearance of the cells. Even in this decade, novel structures are still being discovered in organisms.

2. When a human voluntarily stops breathing, the pulse rate increases. In order to conclude that this is an example of a negative feedback system, which of the following assumptions must be made?

   a. The body has sensors that detect pulse rate.
   b. The body has sensors that detect carbon dioxide concentration in the blood.

c. The body has sensors that detect arterial blood pressure.
d. Increased pulse rate increases removal of carbon dioxide from the blood.
e. Increased pulse rate decreases removal of carbon dioxide from the blood.

ANALYSIS

a. Although the heart is an effector in the postulated feedback system, it is not necessary that there be sensors for its rate. A negative feedback system requires sensors for some variable and the ability to activate an effector that reverses a change in the variable. The stimulus to the effector will be removed when the original variable returns to control levels. The effector itself can function without being monitored.

b. The system is proposed to control some variable related to breathing. Blood carbon dioxide is a reasonable candidate, because normal metabolism produces carbon dioxide, and breathing removes it from the blood. Whatever variable is controlled, there must be a sensor that detects it.

c. Blood pressure is likely to increase with an increase in pulse rate, but how blood pressure would be first lowered by breath holding is not clear. There are blood pressure receptors, but they probably don't function in a feedback loop triggered by stopping breathing.

d. If an increase in blood carbon dioxide is the stimulus in the proposed feedback loop, then the response (increased pulse rate) must reduce blood carbon dioxide. One way to do this is to increase the rate of removal of carbon dioxide.

e. Even though a negative feedback loop is proposed, the mechanism does not have to be a decrease in the function of an effector. The stimulus has to be decreased, and in this case that is accomplished by increasing an effector process.

3. Suppose a person consumes the same amount of glucose every day and maintains a constant blood glucose concentration. If that person doubles the daily glucose intake, several predictions might be made about the effect on blood glucose concentration, including the following:

a. Blood glucose would constantly rise.
b. Blood glucose would rise to a new constant level about twice the former level.
c. Blood glucose would rise slightly to a new constant level.
d. Blood glucose would rise slightly, then return to the former level.
e. Blood glucose would be unchanged.

For each prediction given above, state whether it assumes that blood glucose is part of a negative feedback loop, a positive feedback loop, or is not subject to control by a feedback mechanism.

ANALYSIS

a. Although the prediction is of a constant rise, it does not assume a positive feedback mechanism. If blood glucose concentration is constant, the rate of removal of glucose is equal to the rate of consumption of glucose. If consumption increases and removal mechanisms are not adjusted, blood glucose will increase at a constant rate and will not reach a steady state. This prediction assumes there is no feedback control of blood glucose that can adjust removal mechanisms in response to changes in intake.

b. This prediction does assume that the removal mechanisms increase but does not assume a homeostatic feedback loop. Removal of glucose is accomplished by enzymes and membrane transport carriers. As the intake of glucose increases, the blood concentration increases, and the enzymes and carriers work at higher rate because their substrate concentration is increased. The system establishes a new, higher steady state without any receptors, integrators, or information transmission.

c. This prediction assumes a negative feedback control mechanism. A slight increase in blood glucose triggers an increase in the removal mechanisms that balances the increased rate of intake and maintains a constant value. The response is maintained by a constant, slightly higher value of the variable. In effect, the set point has been slightly raised.

d. This prediction also assumes a negative feedback control mechanism. A slight increase in blood glucose triggers the response, but enough extra enzymes are produced not only to balance the increased glucose intake but also to reduce the blood glucose level to the original set point. The enzyme level must then be reduced to the balance level. This is a slightly more elaborate version of (c).

e. This prediction assumes not a negative feedback system but a system that can anticipate a stimulus and prepare for it. A negative feedback system is activated only after the control variable changes, and the system then reverses the change. Here, no change at all is predicted. This can happen only if another receptor, perhaps for glucose concentration in the stomach or the mouth, activates glucose removal mechanisms before the glucose is absorbed into the blood. Unless this kind of mechanism is precisely timed, it would result in decreases in blood glucose below the set point.

Authors' note: Some of the Critical Thinking sections that follow include descriptions of real experimental procedures that manipulate organisms in various ways to test hypotheses and isolate variables. These graphic descriptions are included to inform students about part of the "real world" of science and, at the same time, lead them to question if and when ethics and morals should be applied to the treatment of organisms used in experimentation. Do any of the other organisms on Earth have the right not to be surgically altered, not to be infected with deadly viruses, not to be dosed with carcinogenic or mutagenic chemicals? Is any experimental procedure in the pursuit of knowledge that might help humans legitimate? We hope you discuss these issues.

Answers

Answers to Interactive Exercises

ANIMAL STRUCTURE AND FUNCTION: AN OVERVIEW (31-I)
1. F; 2. F; 3. T; 4. tissue; 5. organ; 6. metabolic; 7. internal environment; 8. A; 9. H; 10. E; 11. F; 12. K; 13. I; 14. J; 15. G; 16. B; 17. D; 18. C.

ANIMAL TISSUES (31-II)
1. T; 2. F; 3. F; 4. F; 5. F; 6. T; 7. F; 8. T; 9. T; 10. T; 11. germ cells; 12. gametes; 13. meiosis; 14. somatic; 15. mitosis; 16. tissues; 17. ectoderm; 18. endoderm; 19. mesoderm; 20. organ; 21. organ system (urinary system, excretory system); 22. organism; 23. homeostasis; 24. tissue; 25. Somatic; 26. epithelial; 27. connective; 28. muscle; 29. nervous; 30. ground substance; 31. Loose; 32. collagen; 33. elastin; 34. Tendons (Ligaments); 35. ligaments (tendons); 36. Bone (Cartilage); 37. cartilage (bone); 38. connective D, 9, 11; 39. epithelial G, 1, 6, 12; 40. muscle I, 8, (11); 41. muscle J, 5, (11); 42. connective E, 7, 11, (14); 43. gametes 2; 44. connective B, 10; 45. epithelial H, 1, 6, 12; 46. connective 1, 6, 15; 47. nervous 3; 48. muscle C, 13; 49. epithelial F, 1, 6, 12; 50. connective A, 4, 14.

MAJOR ORGAN SYSTEMS (31-III)
1. G; 2. I; 3. H; 4. B; 5. F; 6. E; 7. C; 8. J; 9. A; 10. D; 11. K; 12. thoracic; 13. cranial; 14. pelvic; 15. midsagittal; 16. transverse; 17. anterior; 18. frontal; 19. ventral; 20. fluid; 21. chemical/hormonal; 22. nervous; 23. transverse plane; 24. dorsal; 25. posterior; 26. ventral; 27. anterior; 28. midsagittal plane; 29. proximal; 30. inferior.

HOMEOSTASIS AND SYSTEMS CONTROL (31-IV)
1. F; 2. T; 3. F; 4. F; 5. positive; 6. negative; 7. integrator; 8. glands; 9. receptors; 10. hypothalamus; 11. effectors; 12. thyroid; 13. sweat; 14. smooth; 15. skeletal.

Answers to Self-Quiz

1. d; 2. b; 3. c; 4. a; 5. d; 6. c; 7. c; 8. d; 9. a.

32

INFORMATION FLOW
AND THE NEURON

CELLS OF THE NERVOUS SYSTEM
FUNCTIONAL ZONES OF THE NEURON
NEURAL MESSAGES
     Membrane Excitability
     Neurons "At Rest"
     Local Disturbances in Membrane Potential
A CLOSER LOOK AT ACTION POTENTIALS
     Mechanism of Excitation
     Duration of Action Potentials
     Propagation of Action Potentials
          Refractory Period
          Sheathed Axons

CHEMICAL SYNAPSES
     Effects of Transmitter Substances
     Neuromodulators
     Synaptic Integration
     Commentary: Deadly Imbalances
     at Chemical Synapses
PATHS OF INFORMATION FLOW

Interactive Exercises

CELLS OF THE NERVOUS SYSTEM / FUNCTIONAL
ZONES OF THE NEURON (32-I, pp. 547–548)

Fill-in-the-Blanks

Nerve cells that conduct messages are called (1)_____. (2)_____ cells, which support and
nurture the activities of neurons, make up about half the volume of the nervous system. (3)_____
neurons are receptors for environmental stimuli, (4)_____ connect different neurons in the central
nervous system, and (5)_____ neurons are linked with muscles or glands. All neurons have a
(6)_____ _____ that contains the nucleus and the metabolic means to carry out protein synthesis.
(7)_____ are short, slender extensions of (6), and together these two neuronal parts are the neurons'
"input zone" for receiving (8)_____. The (9)_____ is a single long cylindrical extension of the
(10)_____ _____ ; in motor neurons, the (11)_____ has finely branched (12)_____ that
terminate on muscle or gland cells and are "output zones," where messages are sent on to other cells.

Labeling

Label the parts of neurons and types of neurons in the illustration below.

13. _____

14. _____

15. _____

16. _____

17. _____

18. _____ _____

19. _____

20. _____ _____

21. _____

22. _____ _____

23. _____

24. _____

25. _____ _____

26. _____ _____

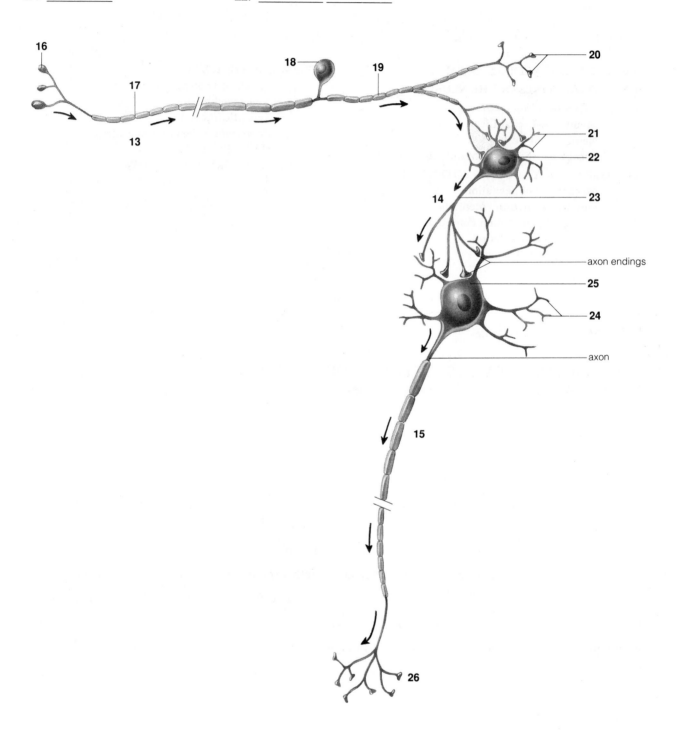

axon endings

axon

NEURAL MESSAGES (32-II, pp. 548–551)

Fill-in-the-Blanks

A neuron at rest establishes unequal electric charges across its plasma membrane, and a (1)_____ _____ is maintained. Another name for (1) is the (2)_____ _____ _____; it represents a tendency for activity to happen along the membrane. Weak disturbances of the neuronal membrane might set off only slight changes across a small patch, but strong disturbances can cause a(n) (3)_____ _____, which is an abrupt, short-lived reversal in the polarity of charge across the plasma membrane of the neuron. For a fraction of a second, the cytoplasmic side of a bit of membrane becomes positive with respect to the outside. The (4)_____ that travels along the neural membrane is nothing more than short-lived changes in the membrane potential. Any cell that can respond to stimulation by producing action potentials is said to show (5)_____ _____. When action potentials reach the end of a motor neuron, they cause (6)_____ to be released that serve as chemical signals to adjacent muscle cells. Muscles (7)_____ in response to the signals.

How is the resting membrane potential established, and what restores it between action potentials? The concentrations of (8)_____ ions (K^+), sodium ions (9)(___$^+$), and other charged substances are not the same on the inside and outside of the neuronal membrane. (10)_____ proteins that span the membrane affect the diffusion of specific types of ions across it. (11)_____ proteins that span the membrane pump sodium and potassium ions against their concentration gradients across it by using energy stored in ATP. A neuronal membrane has many more positively charged (12)_____ ions inside than out and many more positively charged (13)_____ ions outside than inside. An electrical gradient also exists across the neuronal membrane; compared with the outside, the inside of a neuron at rest has an overall (14)_____ charge. For most neurons in most animals, the difference in charge across the neuronal membrane is within the range of 60–90 (15)_____. There are about (16)_____ times more potassium ions on the cytoplasmic side as outside, and there are about (17)_____ times more sodium ions outside as inside. These ions can cross the membrane only by traveling along passages through (18)_____ proteins, which is called (19)_____ diffusion. Some channel proteins leak ions through them all the time; others have (20)_____ that open only when stimulated. Transport proteins called (21)_____-_____ _____ counter the leakage of ions across the neuronal membrane and maintain the resting membrane potential. In all neurons, stimulation at an input zone produces (22)_____ signals that do not spread very far (half a millimeter or less). (23)_____ means that signals can vary in magnitude—small or large—depending on the intensity and (24)_____ of the stimulus. When stimulation is intense or prolonged, graded signals can spread into an adjacent (25)_____ _____ of the membrane—the site where action potentials can be initiated.

Labeling

Identify the parts of the neuron illustrated below.

26. _____ _____   29. _____-_____ _____

27. _____ _____   30. _____ _____

28. _____

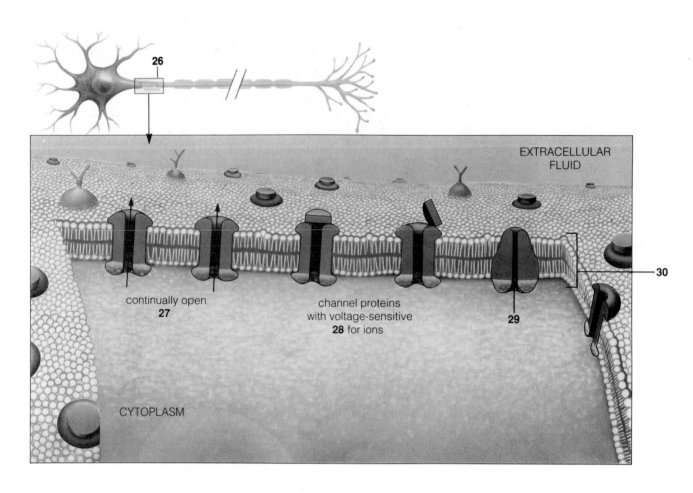

26

EXTRACELLULAR
FLUID

continually open
27

channel proteins
with voltage-sensitive
28 for ions

29

30

CYTOPLASM

A CLOSER LOOK AT ACTION POTENTIALS (32-III, pp. 552–554)

Fill-in-the-Blanks

A(n) (1)_____ _____ is an all-or-nothing, brief reversal in membrane potential; it is also

known as a(n) (2)_____ _____. Once an action potential has been achieved, it is an

(3)_____-_____-_____ event; its amplitude will not change even if the strength of the

stimulus changes. The minimum change in membrane potential needed to achieve an action potential is

the (4)_____ value. Each action potential is followed by a (5)_____ period, that is, a time of

insensitivity to stimulation. Some narrow-diameter neurons are wrapped in lipid-rich (6)_____

produced by specialized neuroglial cells called Schwann cells; each of these is separated from the next by a

(7)_____ _____ _____—a small gap where the axon is exposed to extracellular fluid. An action potential jumps from one node to the next in line and is called (8)_____ conduction; this means of conduction affords the most rapid signal (9)_____ using the least (10)_____ effort by the cell. In the largest myelinated axon, signals travel (11)_____ miles per hour.

Labeling

Identify the numbered parts of the illustrations below.

12. _____ _____

13. _____

14. _____ _____ _____

15. _____

16. _____

17. _____ _____ _____

18. _____ _____

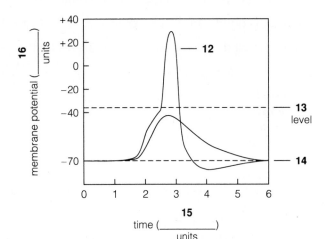

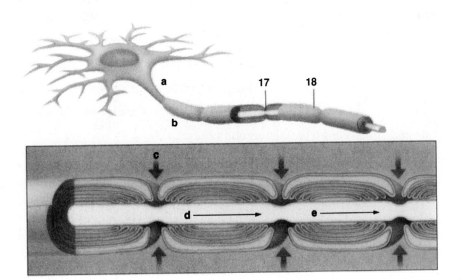

CHEMICAL SYNAPSES / PATHS OF INFORMATION FLOW (32-IV, pp. 555–560)

Fill-in-the-Blanks

The junction specialized for transmission between a neuron and another cell is called a (1)_____ _____. Usually, the signal being sent to the receiving cell is carried by chemical messengers called (2)_____ _____. (3)_____ is an example of this type of chemical messenger that diffuses across the synaptic cleft, combines with protein receptor molecules on the muscle cell membrane, and soon thereafter is rapidly broken down by enzymes. At an (4)_____ synapse, the membrane potential is driven toward the threshold value and increases the likelihood that an action potential will occur. At an (5)_____ synapse, the membrane potential is driven away from the threshold value, and the receiving neuron is less likely to achieve an action potential. A specific transmitter substance can have either excitatory or inhibitory effects depending on which type of protein channel it opens up in the (6)_____ membrane.

A (7)_____ _____ is a synapse between a motor neuron and muscle cells.

(8)_____ acts on brain cells that govern sleeping, sensory perception, temperature regulation, and emotional states. (9)_____ are neuromodulators that inhibit perceptions of pain and may have roles in memory and learning, emotional depression, and sexual behavior. (10)_____ _____ at the cellular level is the moment-by-moment tallying of all excitatory and inhibitory signals acting on a neuron. Incoming information is (11)_____ by cell bodies, and the charge differences across the membranes are either enhanced or inhibited. An (12)_____ postsynaptic potential (EPSP) brings the membrane closer to threshold and has a depolarizing effect. An inhibitory postsynaptic potential (IPSP) drives the membrane away from threshold and either has a (13)_____ effect or maintains the membrane at its resting level. In the brain and spinal cord (central nervous system), (14)_____ _____ are sets of interacting neurons confined to a single region. Cordlike communication lines called (15)_____ _____ connect neurons from one region to neurons in different regions within the brain and spinal cord.

A (16)_____ is an involuntary sequence of events elicited by a stimulus. During a (17)_____ _____ , a muscle contracts involuntarily whenever conditions cause a stretch in length; many of these help you maintain an upright posture despite small shifts in balance. Located within skeletal muscles are length-sensitive organs called (18)_____ _____ , which generate action potentials when stretched beyond a critical point; these potentials are conducted rapidly to the (19)_____ _____ , where they are communicated to motor neurons leading right back to the muscle that was stretched.

Imbalances can occur at chemical synapses; a neurotoxin produced by *Clostridium tetani* interferes with the effect of (20)_____ on motor neurons, which may cause tetanus—a prolonged, spastic paralysis that can lead to death.

Labeling

Label the parts of the nerve illustrated below.

21. _____

22. _____ _____

23. _____ _____

24. _____

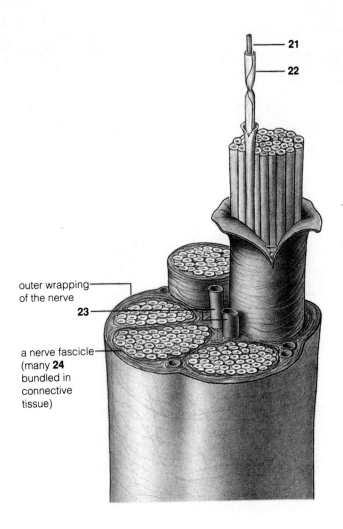

outer wrapping of the nerve

23

a nerve fascicle (many **24** bundled in connective tissue)

Matching

Match the choices below with the correct number in the diagram. Two of the numbers match with two lettered choices.

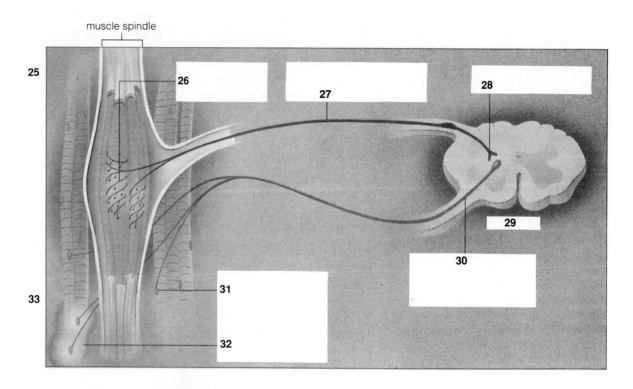

25. _____
26. _____
27. _____
28. _____
29. _____
30. _____
31. _____
32. _____
33. _____

A. Response
B. Action potentials generated in motor neuron and propagated along its axon toward muscle
C. Motor neuron synapses with muscle cells
D. Muscle cells contract
E. Local signals in receptor endings of sensory neuron
F. Muscle spindle stretches
G. Action potentials generated in all muscle cells innervated by motor neuron
H. Stimulus
I. Axon endings synapse with motor neuron
J. Spinal cord
K. Action potential propagated along sensory neuron toward spinal cord

---

Chapter Terms

The following page-referenced terms are the important terms in the chapter. Refer to the instructions given in Chapter 1, p. 4.

neuron (547)
sensory neurons (547)
interneurons (547)
motor neurons (548)
neuroglial cells (548)
dendrites (548)
axon (548)

cell body (548)
resting membrane
    potential (548)
action potential (549)
membrane excitability (549)
potassium ions (K+) (550)
sodium ions (Na+) (550)

sodium-potassium
    pumps (551)
concentration gradient (551)
electric gradient (551)
trigger zone (551)
transmitter substance (555)
chemical synapses (555)

synaptic cleft (555)
neuromuscular junctions (556)
neuromodulators (556)
synaptic integration (556)
nerves (558)
reflex arc (558)
muscle spindles (558)

Self-Quiz

___ 1. Which of the following is *not* true of an action potential?
   a. It is a short-range message that can vary in size.
   b. It is an all-or-none brief reversal in membrane potential.
   c. It doesn't decay with distance.
   d. It is self-propagating.

___ 2. The conducting zone of a neuron is the _____.
   a. axon
   b. axonal terminals
   c. cell body
   d. dendrite

___ 3. The integrative zone of a neuron is the _____.
   a. axon
   b. axonal terminals
   c. cell body
   d. dendrite

___ 4. In the nervous system, a cotransport mechanism involves _____.
   a. the inactivation of signals along the parasympathetic nerves by signals from the sympathetic division
   b. acetylcholine being reciprocally inactivated by cholinesterase in the synaptic zone
   c. the exchange of signals between the peripheral and central nervous systems
   d. two different substances being exchanged across a membrane by the same enzyme system

___ 5. An action potential is brought about by _____.
   a. a sudden membrane impermeability
   b. the movement of negatively charged proteins through the neuronal membrane
   c. the movement of lipoproteins to the outer membrane
   d. a local change in membrane permeability caused by a greater-than-threshold stimulus

___ 6. The resting membrane potential _____.
   a. exists as long as a charge difference sufficient to do work exists across a membrane
   b. occurs because there are more potassium ions outside the neuronal membrane than there are inside
   c. occurs because of the unique distribution of receptor proteins located on the dendrite exterior
   d. is brought about by a local change in membrane permeability caused by a greater-than-threshold stimulus

___ 7. The phrase "all or none" used in conjunction with discussion about an action potential means that _____.
   a. a resting membrane potential has been received by the cell
   b. an impulse does not decay or dissipate as it travels away from the stimulus point
   c. the membrane either achieves total equilibrium or remains as far from equilibrium as possible
   d. propagation along the neuron is saltatory

___ 8. Endorphins _____.
   a. are neuromodulators
   b. block perceptions of pain
   c. may play a role in causing emotional depression
   d. are involved in all of the above roles

___ 9. An action potential passes from neuron to neuron across a synaptic cleft by _____.
   a. saltatory conduction
   b. the resting membrane potential
   c. neurotransmitter substances
   d. neuromodulator substances

___ 10. _____ are responsible for integration in the nervous system.
   a. Interneurons
   b. Schwann cells
   c. Motor neurons
   d. Sensory neurons

Chapter Objectives/Review Questions

This section lists general and detailed chapter objectives that can be used as review questions. You can make maximum use of these items by writing answers on a separate sheet of paper. Fill in answers where blanks are provided. To check for accuracy, compare your answers with information given in the chapter or glossary.

Page		Objectives/Questions
(546–549)	1.	Outline some of the ways by which information flow is regulated and integrated in the human body.
(548)	2.	Draw a neuron and label it according to its three general zones, its specific structures, and the specific function(s) of each structure.
(548)	3.	Define resting membrane potential; explain what establishes it and how it is used by the cell neuron.
(550–551)	4.	Describe the distribution of the invisible array of large proteins, ions, and other molecules in a neuron, both at rest and as a neuron experiences a change in potential.
(551)	5.	Define sodium-potassium pump and state how it helps maintain the resting membrane potential.
(549)	6.	Define action potential by stating its three main characteristics.
(552–554)	7.	Explain the chemical basis of the action potential. Look at Figure 32.8 in your text and determine which part of the curve represents the following:

> a. the point at which the stimulus was applied;
> b. the events prior to achievement of the threshold value;
> c. the opening of the ion gates and the diffusing of the ions;
> d. the change from net negative charge inside the neuron to net positive charge and back again to net negative charge; and
> e. the active transport of sodium ions out of and potassium ions into the neuron.

(554)	8.	Define refractory period and state what causes it.
(554)	9.	Define Schwann cell, nodes of Ranvier, and myelin sheath and explain how each helps narrow-diameter neurons conduct nerve impulses quickly.
(551)	10.	Explain how graded signals differ from action potentials.
(560)	11.	Understand how a nerve impulse is received by a neuron, conducted along a neuron, and transmitted across a synapse to a neighboring neuron, muscle, or gland.
(556–557)	12.	Distinguish the way excitatory synapses function from the way inhibitory synapses function.
(558)	13.	Explain what a reflex is by drawing and labeling a diagram and telling how it functions.
(558–559)	14.	Explain what the stretch reflex is and tell how it helps an animal survive.

Integrating and Applying Key Concepts

What do you think might happen to human behavior if inhibitory postsynaptic potentials did not exist and if the threshold stimulus necessary to provoke an EPSP were much higher?

Critical Thinking Exercises

1. Suppose you raised the sodium concentration inside an axon to a level several times normal. Which of the following recordings would be most likely when the axon was stimulated to generate an action potential? Choice (a) represents a normal action potential, unaffected by the change.

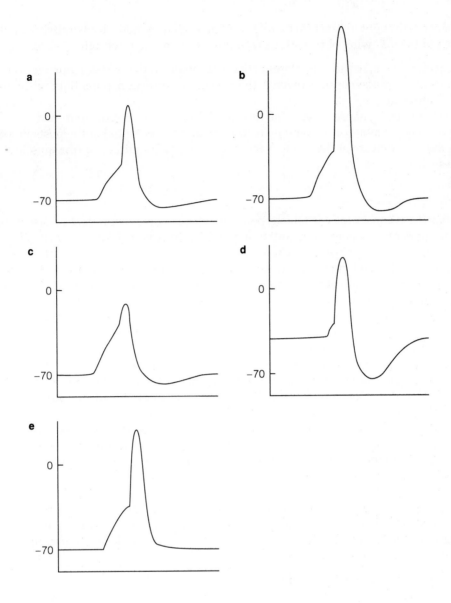

ANALYSIS

a. The action potential is mediated by movements of ions, including sodium across the membrane. Movements of substances by diffusion are determined by their concentrations. Changing the concentration of sodium would change its movements and also change the action potential.

b. The membrane potential at the peak of the action potential is determined by the ratio of sodium outside to sodium inside. Raising the inside concentration would reduce the ratio and hence result in a lower, not higher, peak of membrane potential.

c. This is the predicted result, as discussed in (b).

d. In this graph, the resting membrane potential is changed. However, the determinant of resting potential is potassium, and changing sodium would have no effect.

e. In this case, the overshoot is abolished. As is the case for the resting potential, the overshoot is mediated by potassium. The sodium channels are closed during this time, and sodium concentrations will not affect the membrane potential.

2. A certain synapse is thought to function by release of an excitatory neurotransmitter from the presynaptic neuron. Which of the following observations would *not* be consistent with this hypothesis?

   a. A drug, placed in the synapse, stimulates action potentials on the postsynaptic neuron.
   b. A drug, placed in the synapse, prevents the formation of action potentials when the presynaptic neuron is stimulated.
   c. A drug, placed in the synapse, lowers the threshold of the postsynaptic neuron.
   d. A drug, placed in the synapse, lowers the resting membrane potential of the postsynaptic neuron.
   e. Lowering the temperature of the synapse increases the time required for transmission of information across the synapse.

ANALYSIS

   a. This observation is consistent with the hypothesis. The drug may stimulate release of the neurotransmitter or be similar to the neurotransmitter and therefore activate the receptors itself.
   b. This observation is also consistent. The drug could bind to the receptors and block access by the neurotransmitter, it could prevent release of the neurotransmitter, or it could even accelerate the activity of the enzymes or carriers that remove the neurotransmitter molecules so that they never reach the postsynaptic membrane.
   c. This would be a drug that alters the structure of the postsynaptic membrane or its voltage-gated sodium channels. It would not affect the synapse and hence is consistent with any model of synaptic function.
   d. This would imply that there are inhibitory receptors on the postsynaptic membrane. This is not consistent with the hypothesis that the synapse is excitatory. Each synapse functions specifically with only one kind of neurotransmitter from the presynaptic neuron and one kind of receptor on the postsynaptic membrane.
   e. This observation is consistent with the hypothesis. In a chemical synapse, the movement of the neurotransmitters across the cleft is by simple diffusion, due to spontaneous random molecular motion. At lower temperature, movement is slower, diffusion is slower, and synaptic transmission would take longer.

3. Two neurons, A and B, were electrically stimulated, and the responses of a skeletal muscle were observed. When A alone was stimulated, the muscle contracted. When B alone was stimulated, the muscle did not contract. When A and B were stimulated simultaneously, the muscle did not contract. Which of the following is the best interpretation of these observations?

   a. Neurons A and B both innervate the muscle directly.
   b. Neuron B transmits no information in this system.
   c. Neurons A and B both innervate a third neuron that innervates the muscle.
   d. Neuron A innervates the muscle directly; neuron B innervates neuron A.
   e. Neuron A innervates the muscle directly; neuron B innervates a third neuron that innervates the muscle.

ANALYSIS

   a. Direct innervation of skeletal muscle is only excitatory. All neuromuscular junctions transmit acetylcholine and stimulate contraction of the muscle. This means that choice (a) does not account for the second and third observations. Even if you assume that the stimulation of neuron B alone was simply not strong enough to make the muscle contract, this model does not explain how simultaneous stimulation of B was able to inhibit the effects of stimulating A.
   b. Clearly, it does transmit information, as shown by the third observation, in which stimulation of B inhibits the effect of stimulating A. It apparently transmits inhibitory information.
   c. This interpretation is consistent with all the observations. The third neuron can only be excitatory to the muscle. Neuron A stimulates the third neuron and causes contraction. Neuron B inhibits the third neuron and leaves the muscle unexcited. The third neuron summates inputs from the other two as a subthreshold stimulus and again does not excite a contraction.

d. This interpretation is not consistent with the observation. It explains the first two observations but not the third. Stimulating a neuron electrically generates an action potential somewhere on the axon independently of any synaptic inputs to the integrating region on the cell body. Stimulating B, even if it exerted synaptic inhibition on A, would not prevent electrical stimulation of A.

e. This also would explain the first two observations but not the third. Stimulation of B would inhibit the third neuron and generate zero input to the muscle. Simultaneous stimulation of A would produce a net positive excitatory input to the muscle.

Answers

Answers to Interactive Exercises

CELLS OF THE NERVOUS SYSTEM/FUNCTIONAL ZONES OF THE NEURON (32-I)
1. neurons; 2. Neuroglial; 3. Sensory; 4. interneurons; 5. motor; 6. cell body; 7. Dendrites; 8. signals (stimuli); 9. axon; 10. cell body; 11. axon; 12. endings; 13. sensory; 14. inter; 15. motor; 16. receptor; 17. dendrite; 18. cell body; 19. axon; 20. axon endings; 21. dendrites; 22. cell body; 23. axon; 24. dendrites; 25. cell body; 26. axon endings.

NEURAL MESSAGES (32-II)
1. voltage differential; 2. resting membrane potential; 3. action potential (nerve impulse); 4. disturbance; 5. membrane excitability; 6. molecules (chemicals); 7. contract; 8. potassium; 9. Na; 10. Channel; 11. Transport; 12. potassium; 13. sodium; 14. negative; 15. millivolts; 16. 30; 17. 10; 18. channel; 19. facilitated; 20. gates; 21. sodium-potassium pumps; 22. localized; 23. Graded; 24. duration; 25. trigger zone; 26. axonal membrane; 27. channel proteins; 28. gates; 29. sodium-potassium pump; 30. lipid bilayer.

A CLOSER LOOK AT ACTION POTENTIALS (32-III)
1. action potential; 2. nerve impulse; 3. all-or-nothing; 4. threshold; 5. refractory; 6. myelin; 7. node of Ranvier; 8. saltatory; 9. propagation; 10. metabolic; 11. 270; 12. action potential; 13. threshold; 14. resting membrane potential; 15. milliseconds; 16. millivolts; 17. node of Ranvier; 18. Schwann cell (myelin sheath).

CHEMICAL SYNAPSES/PATHS OF INFORMATION FLOW (32-IV)
1. chemical synapse; 2. transmitter substances; 3. Acetylcholine (ACh); 4. excitatory; 5. inhibitory; 6. postsynaptic; 7. neuromuscular junction; 8. Serotonin; 9. Endorphins; 10. Synaptic integration; 11. summed; 12. excitatory; 13. hyperpolarizing; 14. local circuits; 15. nerve tracts; 16. reflex; 17. stretch reflex; 18. muscle spindles; 19. spinal cord; 20. acetylcholine; 21. axon; 22. myelin sheath; 23. blood vessels; 24. axons; 25. H, F; 26. E; 27. K; 28. I; 29. J; 30. B; 31. C; 32. G; 33. A, D.

Answers to Self-Quiz

1. a; 2. a; 3. c; 4. d; 5. d; 6. a; 7. b; 8. d; 9. c; 10. a.

33

INTEGRATION AND CONTROL: NERVOUS SYSTEMS

Interactive Exercises

INVERTEBRATE NERVOUS SYSTEMS (33-I, pp. 564–565)

Fill-in-the-Blanks

(1)_____ systems enable an animal to sense specific information about external and internal conditions, to (2)_____ or evaluate that information, and to issue commands to the muscles and glands so that a response can be made. The simplest nervous systems are the (3)_____ _____ of cnidarians, which have (4)_____ symmetry; a nerve net is arranged about a central axis like spokes of a bike wheel.

Probably the earliest response mechanism that developed in invertebrates was that of the (5)_____ , a simple, stereotyped movement made in response to a specific kind of stimulus; in the simplest of these pathways, a (6)_____ neuron directly signals a (7)_____ neuron, which acts on muscle cells that respond to the stimulus. Radially symmetrical organisms are adapted for a life that is either fixed to some surface or floating through water, but a crawling existence might get an organism more resources. In crawling animals, the leading, forward end is the part most likely to encounter (8)_____ or food; it would be more advantageous if (9)_____ _____ were located there rather than in the tail section that follows, if the animal is interested in surviving. Organisms in which sensory nerve cells became concentrated as a sort of brain in the forward, head end are said to have undergone the evolutionary process known as (10)_____ ; such animals are (11)_____ symmetrical with right and left sides. Invertebrate animals tend to have two (12)_____ _____ , which are located on the lower side and receive information from and issue commands to the right and left halves of the animal. As invertebrate body structure became more complex, the cell bodies of neurons became clustered to form integrative/evaluative centers known as (13)_____ ; the fiberlike axons of their neurons were grouped together and encased in connective tissue and became known as (14)_____. As brain regions and sense organs evolved and became more complex, they exerted more and more control over the more basic (15)_____ behaviors; brain regions were able to store and compare information about experiences and use that information to override reflex activities and initiate innovative actions. These neural pathways within the more recent evolutionary additions to brains are the bases of memory, (16)_____ , and reasoning. Octopuses, for example, can be taught symbols that guide them to a crab feast.

Labeling

Label each numbered part of the illustration below.

17. _____ _____

18. _____ _____

19. _____ _____

20. _____ _____

21. _____

22. _____ _____

23. _____

24. _____ _____

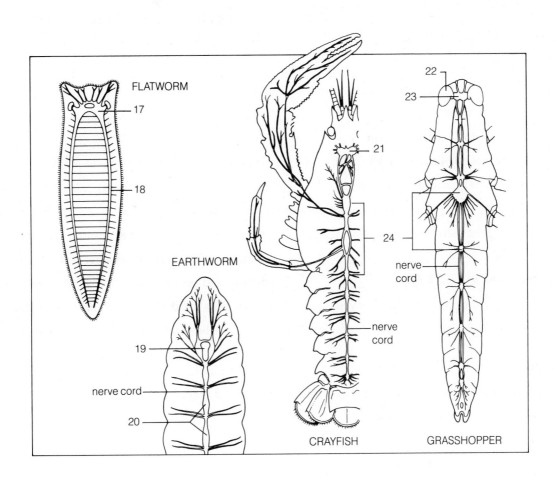

VERTEBRATE NERVOUS SYSTEMS / PERIPHERAL NERVOUS SYSTEM (33-II, pp. 566–568)

Fill-in-the-Blanks

A shift from radial to bilateral symmetry could have led, in some evolutionary lines of animals, to paired (1)_____ and (2)_____ , paired sensory structures such as eyes, and paired (3)_____ regions. The (4)_____ nervous system consists of the brain and nerve cord (or paired cords). The (5)_____ nervous system includes cell bodies of sensory neurons and all nerves (bundles of axons) that lead to and from the central nervous system.

In addition to cephalization and bilateral symmetry, the first vertebrates also had a (6)_____ _____ , a hollow, tubular structure running dorsally above the (7)_____ , and it was the forerunner of the (8)_____ _____ and brain. As time passed, the brain expanded and became divided into three specialized parts: the (9)_____ , midbrain, and hindbrain.

We call the nerve cord that develops in all vertebrate embryos the (10)_____ _____ ; it undergoes expansion and regional modification and becomes enclosed within the (11)_____ _____. Adjacent tissues in the embryo form (12)_____ that thread through all body regions and connect with the spinal cord and brain.

All motor-nerves-to-skeletal muscle pathways and all sensory pathways make up the (13)_____ nervous system. The remaining nerve tissue, which generally is not under conscious control, is collectively known as the (14)_____ nervous system; it is subdivided into two parts: (15)_____ nerves, which respond to emergency situations, and (16)_____ nerves, which oversee the restoration of normal body functioning. The (17)_____ _____ _____ consists of the brain and spinal cord. A skull encloses the brain, and the (18)_____ _____ encloses and protects the spinal cord in vertebrates. In humans, thirty-one pairs of spinal nerves connect with the spinal cord and are grouped by anatomical region; twelve pairs of (19)_____ nerves connect parts of the head and neck with brain centers.

Matching

Choose the most appropriate letter to match with each numbered blank.

20. ___ cervical
21. ___ coccygeal
22. ___ lumbar
23. ___ sacral
24. ___ thoracic

A. Chest
B. Neck
C. Pelvic
D. Tail
E. Waist

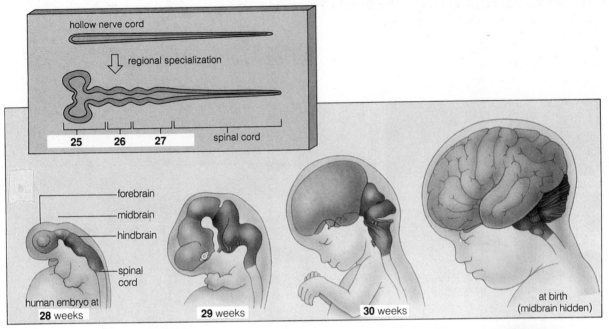

Labeling

Label each numbered part of the accompanying illustrations.

25. _____

26. _____

27. _____

28. _____

29. _____

30. _____

Identify the divisions of the nervous system in the posterior view below.

31. _____ _____

32. _____ _____

33. _____ _____

34. _____ _____

35. _____ _____

36. _____ _____

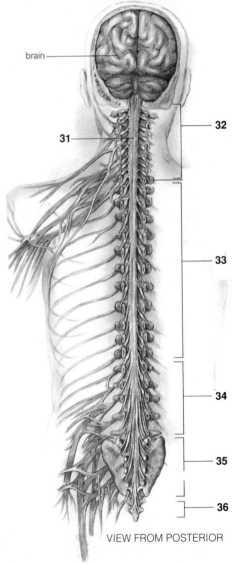

VIEW FROM POSTERIOR

Label each numbered part of the accompanying illustration.

37. _____

38. _____

39. _____

40. _____

41. _____

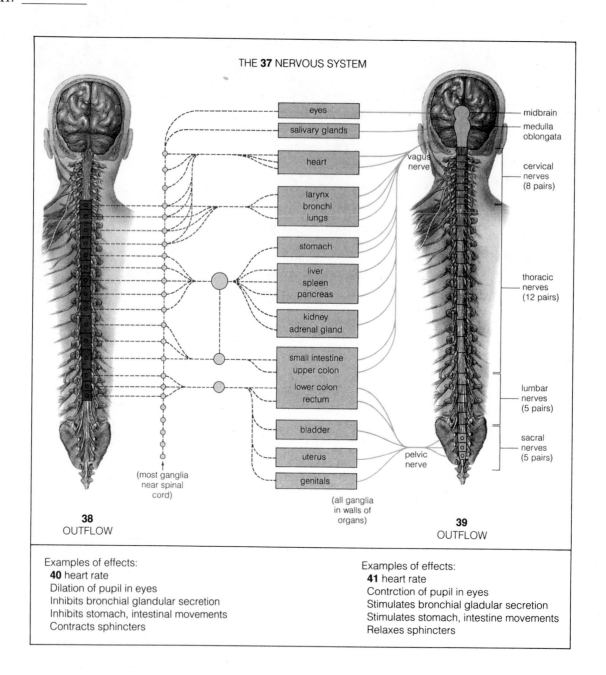

THE **37** NERVOUS SYSTEM

midbrain

medulla oblongata

vagus nerve

cervical nerves (8 pairs)

thoracic nerves (12 pairs)

lumbar nerves (5 pairs)

sacral nerves (5 pairs)

pelvic nerve

eyes

salivary glands

heart

larynx bronchi lungs

stomach

liver spleen pancreas

kidney adrenal gland

small intestine upper colon

lower colon rectum

bladder

uterus

genitals

(most ganglia near spinal cord)

(all ganglia in walls of organs)

38 OUTFLOW

39 OUTFLOW

Examples of effects:
40 heart rate
Dilation of pupil in eyes
Inhibits bronchial glandular secretion
Inhibits stomach, intestinal movements
Contracts sphincters

Examples of effects:
41 heart rate
Contrction of pupil in eyes
Stimulates bronchial gladular secretion
Stimulates stomach, intestine movements
Relaxes sphincters

CENTRAL NERVOUS SYSTEM (33-III, pp. 569–571)

Fill-in-the-Blanks

The (1)_____ _____ is a region of local integration and reflex connections with nerve pathways leading to and from the brain; its (2)_____ _____ , which contains myelinated sensory and motor axons, is the through-conducting zone. The (3)_____ _____ includes neuronal cell bodies, dendrites, nonmyelinated axon terminals, and neuroglial cells; this is the (4)_____ zone, where sensory input is linked to motor output.

The hindbrain is an extension and enlargement of the upper spinal cord; it consists of the (5)_____ _____ , which contains the control centers for the heartbeat rate, blood pressure, and breathing reflexes. The hindbrain also includes the (6)_____ , which helps coordinate motor responses associated with refined limb movements, maintenance of posture, and spatial orientation. The (7)_____ is a major routing station for nerve tracts passing between brain centers.

The (8)_____ evolved as a center that coordinates visual and auditory (hearing) input with reflex responses. The midbrain, pons, and medulla oblongata make up the brain stem; within its core, the (9)_____ _____ is a major network of interneurons that extends its entire length and governs an organism's level of alertness. The (10)_____ contains two cerebral hemispheres composed of gray matter. Underlying the cerebral hemisphere is the (11)_____ , a region that monitors internal organs and influences hunger, thirst, and (12)_____ behaviors, and the (13)_____ , where some motor pathways converge and which is the principal gateway to the cerebral hemispheres. Both the spinal cord and the brain are bathed in (14)_____ _____ . The (15)_____-_____ _____ determines which blood-borne substances are allowed to affect the neurons.

Labeling

Identify the vertebrae in the side view of the spinal cord shown below.

16. _____ vertebrae

17. _____ vertebrae

18. _____ vertebrae

19. _____ vertebrae

20. _____ vertebrae

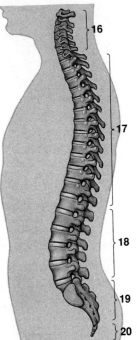

Identify the numbered parts of the illustration below.

21. _____ _____
22. _____
23. _____
24. _____ _____
25. _____ _____
26. _____
27. _____ _____
28. _____ _____
29. _____ _____
30. _____ _____

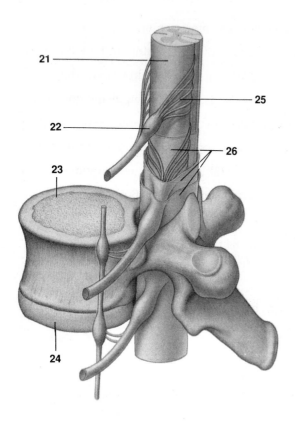

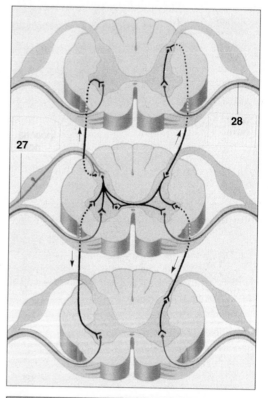

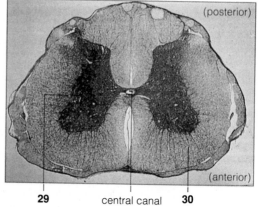

THE HUMAN BRAIN (33-IV, pp. 572–577)

Matching

Match the named part with the letter that describes its function.

1. _____ Broca's area
2. _____ cerebellum
3. _____ corpus callosum
4. _____ hypothalamus
5. _____ limbic system
6. _____ medulla oblongata
7. _____ meninges
8. _____ motor cortex
9. _____ olfactory lobes
10. _____ primary auditory cortex
11. _____ somatic visual cortex
12. _____ somatic sensory cortex
13. _____ thalamus

A. Monitors visceral activities; influences behaviors related to thirst, hunger, reproductive cycles, and temperature control
B. Coordinates muscles required for speech
C. Receives inputs from cochleas of inner ears
D. Issues commands to muscles
E. Relays and coordinates sensory signals to the cerebrum
F. Receives inputs from receptors in nasal epithelium
G. Receives and processes input from body-feeling areas
H. Broad channel of white matter that keeps the two cerebral hemispheres communicating with each other
I. Coordinates nerve signals for maintaining balance, posture, and refined limb movements
J. Receives inputs from retinas of eyeballs
K. Connects pons and spinal cord; contains reflex centers involved in respiration, stomach secretion, and cardiovascular function
L. Protective coverings of brain and spinal cord
M. Contains brain centers that coordinate activities underlying emotional expression

Match the most appropriate letter with each numbered item.

14. _____ amphetamines
15. _____ caffeine
16. _____ cocaine
17. _____ ethyl alcohol
18. _____ heroin
19. _____ lithium
20. _____ LSD
21. _____ marijuana
22. _____ nicotine
23. _____ Quaalude
24. _____ Valium

A. Antipsychotic drug
B. Depressant or hypnotic drug
C. Narcotic analgesic drug
D. Psychedelic or hallucinogenic drug
E. Stimulant

Fill-in-the-Blanks

The storage of individual bits of information somewhere in the brain is called (25)_____ ; the neural representation of such bits is known as a (26)_____ _____. Experiments suggest that at least two stages are involved in its formation. One is a (27)_____-_____ _____ period, lasting only a few minutes; information then becomes spatially and temporally organized in neural pathways. The other is a (28)_____-_____ _____ ; information then is put in a different neural representation and is permanently filed in the brain.

The principal wave pattern for someone who is relaxed, with eyes closed, is a(n) (29)_____ _____. The (30)_____-_____ _____ pattern occupies about 80 percent of the total sleeping time for adults. When individuals shift from sleep to a conscious focus on external stimuli or on their own thoughts, the pattern is called (31)_____ _____. (32)_____ _____ accompanies vivid dreaming periods. Activities in the (33)_____ _____ determine whether you are awake or asleep. (34)_____ are analgesics produced by the brain that inhibit regions concerned with our emotions and perception of (35)_____. High (36)_____ levels in the brain stem's core bring about drowsiness and sleep. Imbalances in (37)_____ _____ can produce emotional disturbances.

Labeling

Identify each numbered part of the accompanying illustration.

38. _____   42. _____ _____
39. _____ _____   43. _____
40. _____   44. _____
41. _____ _____   45. _____ _____

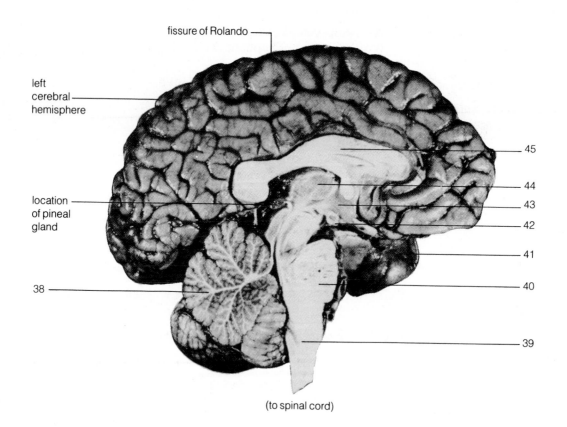

fissure of Rolando

left cerebral hemisphere

location of pineal gland

38

45
44
43
42
41
40
39

(to spinal cord)

Chapter Terms

The following page-referenced terms are the important terms in the chapter. Refer to the instructions given in Chapter 1, p. 4.

nerve net (564)
ganglion, -glia (564)
nerves (564)
nerve cords (564)
central nervous system (566)
peripheral nervous system (566)

somatic system (567)
autonomic system (567)
parasympathetic nerves (567)
sympathetic nerves (567)
spinal cord (569)
nerve tracts (569)
brain (570)

hindbrain (570)
midbrain (571)
reticular formation (571)
forebrain (571)
cerebrospinal fluid (571)
blood-brain barrier (571)
cerebrum (572)
cerebral cortex (572)

memory (574)
motor cortex (574)
stimulants (576)
depressants (576)
hypnotics (576)
analgesics (577)
psychedelics (577)
hallucinogens (577)

Self-Quiz

____ 1. All nerves that lead away from the central nervous system are _____.
a. efferent nerves
b. sensory nerves
c. afferent nerves
d. spinal nerves
e. peripheral nerves

____ 2. _____ nerves generally dominate internal events when environmental conditions permit normal body functioning.
a. Ganglia
b. Pacemaker
c. Sympathetic
d. Parasympathetic
e. All of the above

____ 3. The center of consciousness and intelligence is the _____.
a. medulla
b. thalamus
c. hypothalamus
d. cerebellum
e. cerebrum

____ 4. The _____ are the protective coverings of the brain.
a. ventricles
b. meninges
c. tectums
d. olfactory bulbs
e. pineal glands

____ 5. Broca's area is concerned with _____.
a. coordination of hands and fingers
b. speech
c. memory
d. vision
e. sense of taste and smell

____ 6. The left hemisphere of the brain is responsible for _____.
a. music
b. mathematics
c. language skills
d. abstract abilities
e. artistic ability and spatial relationships

____ 7. The part of the brain that controls the basic responses necessary to maintain life processes (breathing, heartbeat) is _____.
a. the cerebral cortex
b. the cerebellum
c. the corpus callosum
d. the medulla
e. Broca's area

____ 8. To produce a split-brain individual, an operation would need to sever the _____.
a. pons
b. fissure of Rolando
c. hypothalamus
d. reticular formation
e. corpus callosum

____ 9. The center for balance and coordination in the human brain is the _____.
   a. cerebrum
   b. pons
   c. cerebellum
   d. hypothalamus
   e. thalamus

____ 10. The sleep center of the human brain is the _____.
   a. medulla
   b. pons
   c. thalamus
   d. hypothalamus
   e. reticular formation

Chapter Objectives/Review Questions

This section lists general and detailed chapter objectives that can be used as review questions. You can make maximum use of these items by writing answers on a separate sheet of paper. Fill in answers where blanks are provided. To check for accuracy, compare your answers with information given in the chapter or glossary.

Page	Objectives/Questions
(563–566)	1. Contrast invertebrate and vertebrate nervous systems in terms of neural patterns.
(564)	2. Describe how the shift from radial to bilateral symmetry influenced the complexity of nervous systems.
(569–570)	3. Describe the basic structural and functional organization of the spinal cord. In your answer, distinguish spinal cord from vertebral column.
(566)	4. Define and contrast the central and peripheral nervous systems.
(567–568)	5. Explain how parasympathetic nerve activity balances sympathetic nerve activity. List activities of the sympathetic and parasympathetic nerves in regulating pupil diameter, rate of heartbeat, activities of the gut, and elimination of urine.
(569, 572)	6. Compare the structures of the spinal cord and brain with respect to white matter and gray matter.
(570–571)	7. List the parts of the brain found in the hindbrain, midbrain, and forebrain and tell the basic functions of each.
(570–571)	8. For each part of the brain in the list developed for objective 7, state how the behavior of a normal person would change if he or she suffered a stroke in that part of the brain.
(571–572)	9. Describe how the cerebral hemispheres are related to the other parts of the forebrain.
(573)	10. State what the results of the "split-brain" experiments suggest about the functioning of the cerebral hemispheres.
(575)	11. Explain what an electroencephalogram is and what EEGs can tell us about the levels of conscious experience. Describe three typical EEG patterns and tell which level of consciousness each characterizes.
(576)	12. Locate and identify the function of the reticular formation.
(576–577)	13. Explain the relationship between transmitter substances and analgesics.
(577)	14. List the major classes of psychoactive drugs and provide an example of each class.

Integrating and Applying Key Concepts

Suppose that anger is eventually determined to be caused by excessive amounts of specific transmitter substances in the brains of angry people. Also suppose that an inexpensive antidote to anger that neutralizes these anger-producing transmitter substances is readily available. Can violent murderers now argue that they have been wrongfully punished because they were victimized by their brain's transmitter substances and could not have acted in any other way? Suppose an antidote is prescribed to curb violent tempers in an easily angered person. Suppose also that the person forgets to take the pill and subsequently murders a family member. Can the murderer still claim to be victimized by transmitter substances?

Critical Thinking Exercises

1. In an early experiment on the localization of brain function, a rat was trained to run a maze. Symmetrical bilateral lesions were made in the cerebral cortex, and the rat was tried again in the maze. After lesioning, it ran as if it had never been in the maze before. The conclusion was that the lesioned cortex area was the site of memory. Which of the following assumptions was most likely made in reaching that conclusion?

   a. The lesion did not damage the corpus callosum.
   b. The lesion did not damage the motor cortex.
   c. The lesion did not damage the sensory cortex.
   d. Memory is a function of the lesioned part of the cortex.
   e. One part of the brain can take over function from a damaged part.

ANALYSIS

   a. Because the lesion was bilateral, it is not necessary to assume that communication between the hemispheres is intact. The same part of the cortex was destroyed on both sides of the brain.
   b. There is evidence that the motor cortex was intact—the rat ran the maze as if it had never seen it before. All that was lost was the ability to reach the goal without false turns.
   c. Many and varied sensory cues are used in running a maze. If the sensory cortex was damaged, the memory of the maze could still be intact, but it could not be matched to the sensory input, and the rat would run like a naive rat. This kind of assumption is always present when complicated responses like learning are studied by removing something from an animal. The rest of the system must be assumed to be undamaged. Furthermore, the conclusion that the lesioned part is the site of the response mechanism is not always valid. The lesion may actually be in a structure that is only secondarily necessary for the response.
   d. This is the hypothesis, not an assumption. This statement is supported by the observations, not taken as true without evidence.
   e. This statement is an alternative hypothesis about brain function. If the experiment were done to test this hypothesis, the predicted outcome would be that the lesioned rat would be able to retain its learning and run the maze just as well as before the lesion.

2. Many substances have been identified or suspected as neurotransmitters in the human brain, yet very few are used as either therapeutic or psychoactive drugs. Which of the following is the least acceptable explanation of the lack of effectiveness of exogenous neurotransmitter substances?

   a. There are enzymes in the blood that catalyze the destruction of the neurotransmitter substances.
   b. The neurotransmitter substances do not move from the blood into contact with brain neurons.
   c. Doses of neurotransmitter substances placed into the bloodstream become too dilute to be effective.
   d. One neurotransmitter spread system-wide produces antagonistic effects that cancel each other.
   e. There is a background level of activity with each kind of neurotransmitter that would mask any increment of activity caused by administering the same substance.

ANALYSIS

   a. This would make exogenous neurotransmitter ineffective, because it would be catalytically destroyed before it reached the brain. Furthermore, such enzymes would be expected, because the effect of a neurotransmitter must be limited in space to the one synapse where it is released and in time to the few milliseconds before another action potential arrives.
   b. If synapses function by release of neurotransmitter from vesicles only when an action potential arrives, the neurotransmitter must be impermeable to membranes. That means that it would not pass the blood-brain barrier unless it happened to be a substrate for a membrane carrier in the endothelial cell. If it could not leave the brain capillaries, it could have no effect on brain neurons.
   c. This could easily be true. Doses of drugs are dissolved in a large volume of blood; thus, they tend to reach low concentrations, especially if they are taken orally and thus absorbed slowly from the digestive system. In the synapse, on the other hand, they must be functioning at a fairly high concentration, because the volume is very low. Obviously, there are many serious assumptions in this

analysis about specific numbers of molecules and volumes. Different and equally reasonable assumptions could lead to an opposite evaluation.

    d. Most neurotransmitters are excitatory in some synapses and inhibitory in others, depending on the kind of receptors present in the postsynaptic neuron. This means that a single neurotransmitter applied indiscriminately around the whole nervous system would be expected to have a variety of effects, some of which might well antagonize each other. The function of the nervous system depends on the coordination of activity at some synapses and inactivity at others.

    e. While the whole nervous system is never quiet, any given synapse is active only intermittently and would be affected by application of its neurotransmitter substance from outside.

3. In a pioneering experiment on the function of the corpus callosum, an animal was trained to perform a visual discrimination task with one eye covered. The cover was then switched to the other eye, and the animal could still perform the task, this time with information coming to the opposite cerebral hemisphere. This result was taken to indicate that the corpus callosum was a pathway of communication between the hemispheres. Which of the following additional experiments would be the strongest confirmation of that conclusion?

    a. Train with one eye, section the corpus callosum, and test with the other eye.

    b. Section the corpus callosum, then train with one eye and test with the other.

    c. Train with one eye, then try to elicit the response by electrically stimulating the trained side of the brain.

    d. Train with one eye, then try to elicit the response by electrically stimulating the untrained side of the brain.

    e. Block the passage of action potentials in the corpus callosum with a drug instead of with a surgical procedure.

ANALYSIS

    a. Sectioning the corpus callosum would block the postulated pathway of transfer between the two hemispheres. It would also leave intact the other possible pathways, such as transmission downward to a lower brain region, crossover there, and transmission back up to the opposite cerebral hemisphere. In this case, if the training of one side was transferred to the other side, the corpus callosum could be eliminated as a pathway. However, because the protocol delays surgery until after training, the transfer could already have occurred, and the results would not be conclusive.

    b. This protocol would avoid the objection to (a).

    c. Electrical stimulation can provide much stronger evidence than surgical manipulation because it evokes a positive response rather than merely blocking something, which is often a secondary effect, even an artifact of surgery. But this experiment would at best only demonstrate that the trained side of the brain could be stimulated to recall its training.

    d. This protocol is not significantly better than (c). It could show that the training had been transferred to the opposite side, but it would indicate nothing at all about the pathway of the transfer.

    e. This would be conceptually the same as (a) and (b) but would avoid the procedural problems of surgery. On the other hand, the corpus callosum is a large structure, and reliably blocking all of it with a local injection presents a serious technical problem. The drug could be administered systemically and thus be sure to reach all of the corpus callosum, but then the drug would reach all the rest of the brain, too, with disruptive consequences.

Note: Please see authors' note on page 371.

Answers

Answers to Interactive Exercises

INVERTEBRATE NERVOUS SYSTEMS (33-I)
1. Nervous; 2. integrate; 3. nerve nets; 4. radial; 5. reflex; 6. sensory; 7. motor; 8. danger (trouble); 9. sensory cells (sense organs); 10. cephalization; 11. bilaterally; 12. nerve cords; 13. ganglia; 14. nerves; 15. reflex; 16. learning; 17. rudimentary "brain"; 18. nerve cord; 19. rudimentary "brain"; 20. segmental ganglia; 21. brain; 22. optic lobe; 23. brain; 24. segmental ganglia.

VERTEBRATE NERVOUS SYSTEMS/PERIPHERAL NERVOUS SYSTEM (33-II)
1. nerves (muscles); 2. muscles (nerves); 3. brain; 4. central; 5. peripheral; 6. nerve cord; 7. notochord; 8. spinal cord; 9. forebrain; 10. neural tube; 11. vertebral column; 12. nerves; 13. somatic; 14. autonomic; 15. sympathetic; 16. parasympathetic; 17. central nervous system; 18. vertebral column; 19. cranial; 20. B; 21. D; 22. E; 23. C; 24. A; 25. forebrain; 26. midbrain; 27. hindbrain; 28. 3; 29. 7; 30. 9; 31. spinal cord; 32. cervical nerves; 33. thoracic nerves; 34. lumbar nerves; 35. sacral nerves; 36. coccygeal nerves; 37. autonomic; 38. sympathetic; 39. parasympathetic; 40. increases; 41. decreases.

CENTRAL NERVOUS SYSTEM (33-III)
1. spinal cord; 2. white matter; 3. gray matter; 4. integrative; 5. medulla oblongata; 6. cerebellum; 7. pons; 8. midbrain; 9. reticular formation; 10. forebrain; 11. hypothalamus; 12. sexual; 13. thalamus; 14. cerebrospinal fluid; 15. blood-brain barrier; 16. cervical; 17. thoracic; 18. lumbar; 19. sacral; 20. coccygeal; 21. spinal cord; 22. ganglion; 23. vertebra; 24. intervertebral disk; 25. spinal nerve; 26. meninges; 27. sensory axon; 28. motor axon; 29. gray matter; 30. white matter.

THE HUMAN BRAIN (33-IV)
1. B; 2. I; 3. H; 4. A; 5. M; 6. K; 7. L; 8. D; 9. F; 10. C; 11. J; 12. G; 13. E; 14. E; 15. E; 16. E; 17. B; 18. C; 19. A; 20. D; 21. D; 22. E; 23. B; 24. B; 25. memory; 26. memory trace; 27. short-term memory; 28. long-term memory; 29. alpha rhythm; 30. slow-wave sleep; 31. EEG arousal; 32. REM sleep; 33. reticular formation (sleep centers); 34. Endorphins (Enkephalins); 35. pain; 36. serotonin; 37. neurotransmitter substances; 38. cerebellum; 39. medulla oblongata; 40. pons; 41. temporal lobe; 42. optic chiasm; 43. hypothalamus; 44. thalamus; 45. corpus callosum.

Answers to Self-Quiz

1. a; 2. d; 3. e; 4. b; 5. b; 6. c; 7. d; 8. e; 9. c; 10. e.

34

INTEGRATION AND CONTROL: ENDOCRINE SYSTEMS

Interactive Exercises

"THE ENDOCRINE SYSTEM" / HORMONES AND OTHER SIGNALING MOLECULES (34-I, pp. 581–582)

Fill-in-the-Blanks

(1)_____ _____ are substances that are secreted from nerve endings, travel only a short distance across a synaptic cleft to an adjacent cell, and then are rapidly degraded or recycled. (2)_____ _____ _____ are secreted from cells in many different tissues. They change chemical conditions in the immediate vicinity and then are degraded quickly. The scattered sources of hormones came to be known as the (3)_____ system, and it was thought that they formed a separate means of control within the body, *apart* from the nervous system. Today we know that the two systems are intimately linked. Animal (4)_____ are produced by endocrine cells and by some neurons, and transported by the bloodstream to tissues and organs some distance away, where they regulate specific cellular reactions in (5)_____ cells. Pheromones are produced by specialized (6)_____ glands, which have ducts that lead out to the body surface; pheromones may activate behavioral changes in other animals of the (7)_____ species.

Labeling

Identify each indicated part of the accompanying illustration.

8. _____

9. _____

10. _____

11. _____

12. _____

13. _____

14. _____

15. _____ _____ _____

16. _____

17. _____

18. _____

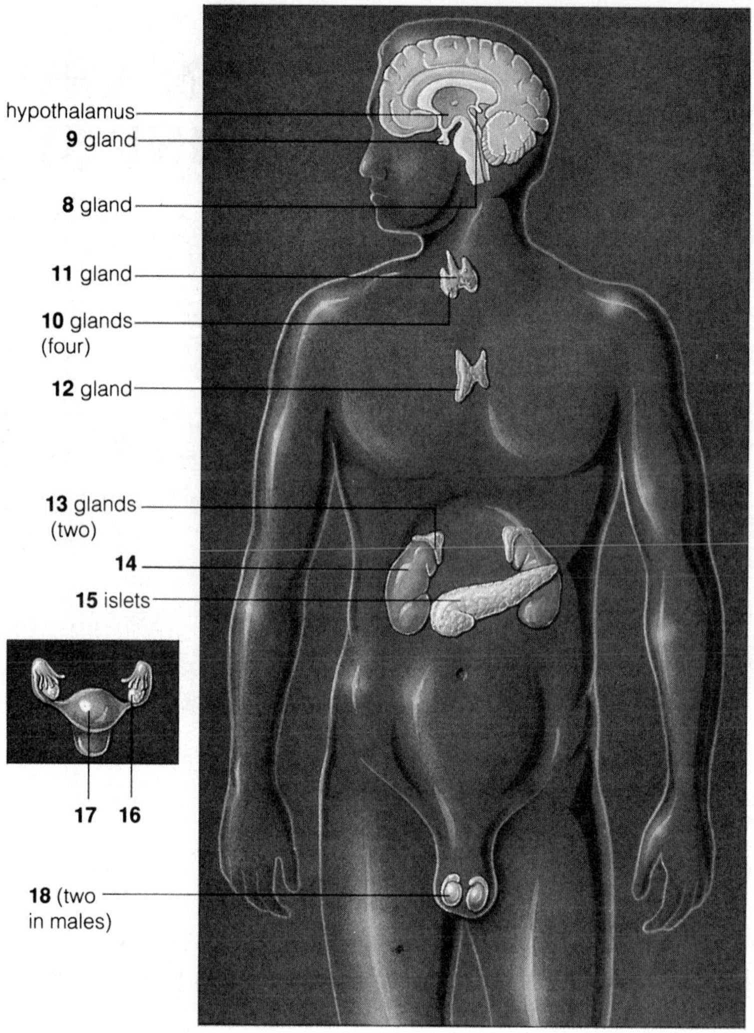

hypothalamus

9 gland

8 gland

11 gland

10 glands (four)

12 gland

13 glands (two)

14

15 islets

17   **16**

18 (two in males)

THE HYPOTHALAMUS-PITUITARY CONNECTION (34-II, pp. 583–586)

Fill-in-the-Blanks

In the (1)_____ , incoming neural messages are summed, shifts in hormonal concentrations are detected, and responses are sent out in the form of (2)_____ _____ to one of two regions of the (3)_____ . The tissues of the anterior lobe are (4)_____ in nature; the tissues of the posterior lobe are (5)_____ . Vessels that connect two distinct capillary beds are called (6)_____ vessels; such vessels link the capillary bed in the (7)_____ with appropriate capillary beds in either lobe of the pituitary. From the anterior pituitary come several hormones: (8)_____ , which stimulates the adrenal cortex; (9)_____ , which stimulates the thyroid to produce thyroxine; and (10)_____ , which stimulates milk production in mammary glands.

Labeling

Identify each indicated part of the accompanying illustrations.

11. _____   15. _____ _____

12. _____   16. _____ _____

13. _____   17. _____

14. _____ _____   18. _____

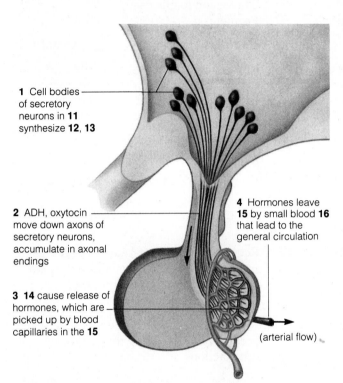

1 Cell bodies of secretory neurons in **11** synthesize **12**, **13**

2 ADH, oxytocin move down axons of secretory neurons, accumulate in axonal endings

3 14 cause release of hormones, which are picked up by blood capillaries in the **15**

4 Hormones leave **15** by small blood **16** that lead to the general circulation

(arterial flow)

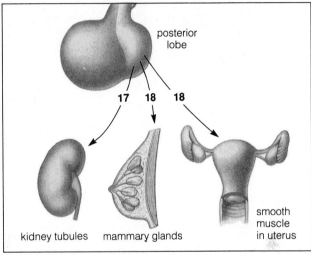

posterior lobe

17 **18** **18**

kidney tubules   mammary glands   smooth muscle in uterus

Label-Match

Label each hormone given below with an "A" if it is secreted by the anterior lobe of the pituitary, a "P" if it is released from the posterior pituitary, and an "I" if it is secreted by the intermediate lobe. Complete the exercise by entering the letter of the corresponding action in the parentheses following each label.

19. ___ ( ) ACTH

20. ___ ( ) ADH

21. ___ ( ) FSH

22. ___ ( ) GH (STH)

23. ___ ( ) LH

24. ___ ( ) MSH

25. ___ ( ) oxytocin

26. ___ ( ) PRL

27. ___ ( ) TSH

A. Acts on ovaries, testes to develop gametes
B. Acts on pigmented cells in the skin and on other surface coverings
C. Acts on mammary glands to develop milk supplies
D. Acts on ovaries, testes to release gametes and to stimulate the production of testosterone in males and formation of corpus luteum in females
E. Induces uterine contractions and milk movement into secretory ducts
F. Acts on the thyroid gland
G. Acts on the kidneys to conserve water
H. Acts on the adrenal cortex to increase adrenal steroid hormone production
I. Acts on most cells to induce protein synthesis and cell division; also plays a role in the metabolism of glucose and protein in adults

SELECTED EXAMPLES OF HORMONAL ACTION / LOCAL SIGNALING MOLECULES (34-III, pp. 586–593)

1. Complete the following table, which gives hormone sources other than the mammalian hypothalamus and pituitary.

Source	Secretions	Main Targets	Primary Actions
a. _____ _____	Glucocorticoids (including cortisol)	Most cells	Promote protein breakdown and conversion to glucose
	Mineralocorticoids (including aldosterone)	Kidney	Promotes q. _____ reabsorption; controls salt and r. _____ balance
b. _____ _____	Epinephrine (adrenalin)	Liver, muscle, adipose tissue	Raises blood level of sugar, fatty acids; increases heart rate, force of contraction
	Norepinephrine	Smooth muscle of blood vessels	Promotes constriction or dilation of blood vessel diameter
c. _____	Triiodothyronine, thyroxine	Most cells	Regulates s. _____ ; has roles in growth, development
	Calcitonin	l. _____	t. _____ calcium levels in blood
d. _____	Parathyroid hormone	m. _____ , kidney	u. _____ calcium levels in blood
e. _____	Androgens (including testosterone)	General	Required in sperm formation, development of genitals, maintenance of secondary sex traits; influences growth, development
f. _____ {	Estrogens	General	Required in egg maturation and release; prepares uterine lining for pregnancy; other actions same as above
	Progesterone	n. _____ , _____	Prepares, maintains v. _____ lining for pregnancy; stimulates breast development
Pancreatic islets	g. _____	Muscle, adipose tissue	Lowers blood sugar level
	h. _____	o. _____ Insulin-secreting cells of pancreas	Raises blood sugar level Influences carbohydrate metabolism
	i. _____		
Thymus	j. _____	p. _____	Has roles in immune responses
Pineal	k. _____	Gonads (indirectly)	Influences daily biorhythms, seasonal sexual activity

Fill-in-the-Blanks

The (2)_____ _____ produces glucocorticoids, such as (3)_____ , in response to

(4)_____ secreted by the anterior pituitary lobe. The adrenal medulla produces (5)_____ and

(6)_____ , which help regulate blood circulation and carbohydrate metabolism. When calcium levels

in the blood plasma rise, the (7)_____ secretes calcitonin, which (8) [choose one] ( ) promotes,

( ) inhibits the release of calcium from bone storage sites. Counterbalancing the effects of calcitonin, the (9)_____ _____ secrete their hormone, which (10) [choose one] ( ) promotes, ( ) inhibits the release of calcium into the bloodstream.

Three kinds of hormonal interactions are common. Becoming pregnant is an example of (11)_____ _____ , in which the lining of the uterus must first be primed by estrogens before (12)_____ can exert its influence in finishing the preparation of the lining to receive the developing blastocyst. The secretion of milk by mammals cannot occur unless two or more hormones work together in the target cell; this is an example of (13)_____ interaction. Finally, the effect of one hormone may oppose the effect of another, as in (14)_____ interactions. For example, (15)_____ promotes a decrease in the level of glucose in the blood, while (16)_____ promotes an increase.

In humans, the pineal gland takes part in (17)_____ physiology, although its precise role has not been identified. (18)_____ is a hormone that promotes the breakdown of glycogen (a storage starch) into glucose subunits; it counterbalances some aspects of (19)_____ action. Both hormones are produced by the pancreatic islets. The (20)_____ is involved in the immune response by producing a hormone that takes part in the production of specific types of infection-fighting cells.

Following a meal, (21)_____ enters the bloodstream faster than cells can use it, so blood glucose levels rise. Pancreatic (22)_____ cells are stimulated to secrete (23)_____. The hormonal targets use glucose or store it as (24)_____. Between meals, blood glucose levels drop. Pancreatic (25)_____ cells are stimulated to secrete (26)_____. The hormonal target cells convert (27)_____ back to glucose, which enters the blood. As a result, the blood glucose level is restored to its set point. (28)_____ and growth factors are examples of (29)_____ molecules, whose influence is exerted on a local level, in the immediate area where changes are occurring in a cell's surrounding chemical environment. (30)_____ promotes survival and growth of neurons in a developing embryo.

SIGNALING MECHANISMS (34-IV, pp. 593–595)

Fill-in-the-Blanks

A cell can respond to specific (1)_____ or other signaling molecules only if its cell membrane has (2)_____ for them. The receptors for (3)_____ hormones are in the (4)_____ of their target cells, while the receptors for (5)_____ hormones are on the plasma membranes of their target cells. Responses to nonsteroid hormones are often mediated by a (6)_____ _____ inside the cell. (7)_____ is one kind of second messenger that communicates information from cell surfaces to specific inner regions of the cell and amplifies the response to a (8)_____ molecule. (9)_____ hormones are synthesized from (10)_____ , are lipid-soluble, and readily cross plasma membranes; they switch certain genes on or off, and this either stimulates or inhibits (11)_____ synthesis.

Most (12)_____ hormones alter the activity of proteins already present in target cells, and these alterations in activity maintain the internal environment of the target cells or contribute to the developmental program or (13)_____ program of the cell.

Chapter Terms

The following page-referenced terms are important terms in the chapter. Refer to the instructions given in Chapter 1, p. 4.

hormones (582)
transmitter substances (582)
local signaling
    molecules (582)
pheromones (582)
neuroendocrine control
    center (583)
pituitary (583)
posterior lobe (583)
antidiuretic hormone
    (ADH) (583)
oxytocin (585)

hypothalamus (584)
releasing hormones (584)
inhibiting hormones (584)
anterior lobe (585)
intermediate lobe (585)
ACTH (585)
TSH (585)
FSH (585)
LH (585)
prolactin (585)
homeostatic feedback
    loops (588)

antagonistic
    interaction (588)
synergistic interaction (588)
permissive interaction (588)
adrenal cortex (588)
adrenal medulla (588)
thyroid gland (589)
parathyroid glands (589)
parathyroid hormone
    (PTH) (589)
gonads (590)
pancreatic islets (590)

glucagon (590)
insulin (590)
somatostatin (590)
thymus gland (592)
pineal gland (592)
prostaglandins (593)
growth factors (593)
steroid hormones (593)
nonsteroid
    hormones (593)
second messengers (595)

Self-Quiz

Multiple-Choice

___ 1. The _____ is often called the master gland, but it is controlled by the _____.
    a. pituitary, hypothalamus
    b. pancreas, hypothalamus
    c. thyroid, parathyroid glands
    d. hypothalamus, pituitary
    e. pituitary, thalamus

___ 2. Although the _____ is nervous in embryonic origin, structure, and behavior, it also secretes substances into the bloodstream.
    a. anterior pituitary
    b. adrenal cortex
    c. pancreas
    d. thyroid
    e. posterior pituitary

___ 3. If you were lost in the desert and had no fresh water to drink, the level of _____ in your blood would increase as a means to conserve water.

    a. insulin
    b. erythropoietin
    c. oxytocin
    d. antidiuretic hormone
    e. salt

For questions 4–6, choose from the following answers:
    a. estrogen
    b. progesterone
    c. FSH
    d. LH
    e. prolactin

___ 4. _____ prepares and maintains the uterine lining for pregnancy.

___ 5. Sudden high levels of _____ bring about ovulation.

___ 6. _____ is a hormone produced by the pituitary that stimulates female gametes to mature in the ovary.

For questions 7–9, choose from the following answers:

    a. adrenal medulla
    b. adrenal cortex
    c. thyroid
    d. anterior pituitary
    e. posterior pituitary

___ 7. The _____ produces steroid hormones that exert their effects on the target nucleus.

___ 8. The gland that is most closely associated with emergency situations is the _____.

___ 9. The _____ gland regulates the basic metabolic rate.

___10. If all sources of calcium were eliminated from your diet, your body would secrete more _____ in an effort to release calcium stored in your body and send it to the tissues that require it.
    a. parathyroid hormone
    b. aldosterone
    c. calcitonin
    d. mineralocorticoids
    e. none of the above

Matching

11. ___ ACTH
12. ___ ADH
13. ___ aldosterone
14. ___ calcitonin
15. ___ cortisol
16. ___ epinephrine
17. ___ estrogen
18. ___ FSH
19. ___ insulin
20. ___ melatonin
21. ___ parathyroid hormone
22. ___ STH (GH)
23. ___ testosterone
24. ___ thyroxine
25. ___ TSH

A. Produced by anterior pituitary; essential for egg maturation
B. Influences daily biorhythms, sexual activity, and sexual development
C. Essential for sperm production; secreted by gonad
D. Increases heart rate and force of contraction; the main "emergency hormone"
E. Produced by gonad; essential for egg maturation and maintenance of secondary sex characteristics in the female
F. The water conservation hormone; released from posterior pituitary
G. Lowers blood sugar by encouraging cells to take in glucose; responsible for protein synthesis and fat storage
H. Stimulates adrenal cortex to secrete hormones involved in responses to stress
I. Elevates calcium levels in blood by stimulating calcium reabsorption from bone and kidneys and calcium absorption from gut
J. Influences overall metabolic rate, growth and development, and sensitivity to temperature extremes
K. Promotes sodium reabsorption; involved in salt, water balance; produced by adrenal cortex
L. Lowers calcium levels in blood by inhibiting reabsorption from bone
M. Raises blood sugar by stimulating glucose production
N. Secreted by anterior pituitary; stimulates release of thyroid hormones
O. Secreted by anterior pituitary; enhances growth in young animals; stimulates release of somatomedins

Chapter Objectives/Review Questions

This section lists general and detailed chapter objectives that can be used as review questions. You can make maximum use of these items by writing answers on a separate sheet of paper. Fill in answers where blanks are provided. To check for accuracy, compare your answers with information given in the chapter or glossary.

Page	Objectives/Questions
(582)	1. Explain why the boundaries of the endocrine and nervous systems are not sharply defined.
(583–584)	2. Understand how the neuroendocrine center controls secretion rates of other endocrine glands and responses in nerves and muscles.

Integrating and Applying Key Concepts

Suppose you suddenly quadruple your already high daily consumption of calcium. State which body organs would be affected and tell how they would be affected. Name two hormones whose levels would most probably be affected and tell whether your body's production of them would increase or decrease. Suppose you continue this high rate of calcium consumption for ten years. Can you predict the organs that would be subject to the most stress as a result?

Critical Thinking Exercises

1. The concentration of iodine inside the cells of the thyroid gland is much higher than in any other tissue or in the extracellular fluid. Which of the following mechanisms would best explain the accumulation?

   a. Iodine is synthesized within the cells of the thyroid.
   b. Iodine is bound to TSH and carried on it into the cells.
   c. Iodine is moved into the cells by active transport.
   d. Iodine moves into the cells by diffusion, but the membrane is permeable to iodine only in the inward direction.
   e. Iodine moves into the cells by diffusion and is attached to impermeable molecules in the cells.

ANALYSIS

   a. Although compounds do tend to have higher concentrations where they are synthesized than elsewhere, iodine is an element, not a compound, and thus is not synthesized.
   b. This mechanism is akin to an active transport mechanism and would be a way to accumulate iodine if TSH were present in extracellular fluid at a higher concentration than iodine. TSH is a protein and belongs to the class of impermeable hormones.
   c. This is the obvious mechanism to maintain a substance at a higher concentration on one side of a membrane than on the other side.

d. This is not an acceptable mechanism. The membrane is essentially a lipid bilayer, the same on both sides. A substance is equally permeable or impermeable in both directions.

e. This is a common mechanism and operates in the thyroid. Thyroid hormone is a modified amino acid with iodine covalently bonded to it. The hormone is stored by being incorporated into a protein with normal peptide bonds. The stimulus to secrete causes hydrolysis (digestion) of the protein and release of the iodinated amino acid hormone.

2. Tadpole tails can be amputated and placed in a dish of pond water, where they will survive intact for many days. In contrast, if thyroid hormone is placed in the water, the tails shrink, shrivel, and lose most of their mass in a few days. Finally, actinomycin D is a compound that inhibits RNA synthesis. If actinomycin D is included with thyroid hormone in the water, amputated tails survive normally and do not shrink. Which of the following is the best conclusion about the mechanism of thyroid hormone action?

a. Thyroid hormone poisons tadpole tails.
b. Thyroid hormone stimulates translation of mRNA for digestive enzymes.
c. Thyroid hormone stimulates transcription of genes for digestive enzymes.
d. Thyroid hormone stimulates synthesis of enzymes for oxidative metabolism.
e. Thyroid hormone promotes metamorphosis of tadpoles into frogs.

ANALYSIS

a. This is one way to say that thyroid hormone causes regression of the tails *in vitro*, but it does not propose a mechanism for the action, nor does it account for the antagonism by actinomycin D, a very toxic substance in its own right.

b. If thyroid hormone did stimulate translation of digestive enzymes, it would promote regression of the cells by self-digestion. However, specific control of translation is very rare. More important, this mechanism does not explain how it could be antagonized by actinomycin D, since the mRNAs would already be present.

c. Again, this would account for the stimulation of regression by thyroid hormone. This conclusion would also account for the antagonism of the response by actinomycin D. If the response to the hormone depends on mRNA synthesis, it should be blocked by inhibitors of transcription.

d. A general stimulation of metabolism might be expected to accelerate regression of a nonfeeding piece of tissue, which might account for the observations if the increase in enzyme activity was based on stimulation of transcription. However, this is a very nonspecific mechanism that would require increases in many enzymes, and it would be a fairly slow response. This conclusion should be kept as a possible, but secondary, hypothesis.

e. This goes beyond proposing a mechanism for the action of thyroid hormone and suggests a physiological role for the hormone. It is a reasonable next hypothesis, because tail regression is a major part of metamorphosis, but it is not directly supported by this experiment.

3. In studying the role of cAMP as a second messenger, the usual experimental design is to apply the hormone to cells and then measure the increase in cAMP that results. In such experiments, a substance called theophylline or a similar molecule is almost always included in the incubation mixture. Which of the following effects would be the best reason for including theophylline?

a. Theophylline inhibits binding of cAMP to enzymes.
b. Theophylline stimulates activity of the enzymes activated by cAMP.
c. Theophylline stimulates adenylate cyclase activity.
d. Theophylline inhibits activity of enzymes that hydrolyze cAMP.
e. Theophylline blocks interaction of hormone receptor with adenylate cyclase.

ANALYSIS

a. The physiological effects of theophylline depend on its binding to enzymes. However, in this experimental design, the goal is to detect increases in cAMP concentration, not its effect. Unless the cAMP assay could not detect molecules that were bound to enzymes, there would be no value to blocking that binding.

b. Once again, the experiment is designed to detect events *before* the activation of enzymes by cAMP. A substance that potentiated the effect of cAMP would have no value in this design.

c. This would interfere with the experiment. If an increase in cAMP was detected, whether it was due to the hormone action or to the action of the theophylline would not be clear.

d. Like any biological molecule, cAMP is continually being synthesized *and* removed. The concentration at any time is the result of the relative rates of these two processes. The goal of this experiment is to detect an increase in the rate of synthesis by showing that the concentration increases. Because the extra cAMP could be removed by the hydrolytic enzymes, it is important to add something that inhibits the removal.

e. This effect would block the very process that the experiment is supposed to study.

Note: Please see authors' note on page 371.

Answers

Answers to Interactive Exercises

"THE ENDOCRINE SYSTEM"/HORMONES AND OTHER SIGNALING MOLECULES (34-I)
1. Transmitter substances; 2. Local signaling molecules; 3. endocrine; 4. hormones; 5. target; 6. exocrine; 7. same; 8. pineal; 9. pituitary; 10. parathyroid; 11. thyroid; 12. thymus; 13. adrenal; 14. kidney; 15. islets of Langerhans; 16. ovary; 17. placenta; 18. testis.

THE HYPOTHALAMUS-PITUITARY CONNECTION (34-II)
1. hypothalamus; 2. releasing hormones; 3. pituitary; 4. glandular (secretory); 5. nervous; 6. portal; 7. hypothalamus; 8. ACTH; 9. TSH; 10. prolactin (PRL); 11. hypothalamus; 12. ADH; 13. oxytocin; 14. action potentials; 15. posterior pituitary; 16. blood vessels; 17. ADH; 18. oxytocin; 19. A (H); 20. P (G); 21. A (A); 22. A (I); 23. A (D); 24. I (B); 25. P (E); 26. A (C); 27. A (F).

SELECTED EXAMPLES OF HORMONAL ACTION/LOCAL SIGNALING MOLECULES (34-III)
1. a. Adrenal cortex; b. Adrenal medulla; c. Thyroid; d. Parathyroids; e. Testis; f. Ovary; g. Insulin; h. Glucagon; i. Somatostatin; j. Thymosins; k. Melatonin; l. Bone; m. Bone; n. Uterus, ovaries, breasts; o. Liver; p. Lymphocytes; q. sodium; r. water; s. metabolism; t. Lowers; u. Elevates; v. uterine; 2. adrenal cortex; 3. cortisone (cortisol); 4. ACTH; 5. epinephrine (adrenalin); 6. norepinephrine (noradrenalin); 7. thyroid; 8. inhibits; 9. parathyroid glands; 10. promotes; 11. permissive interaction; 12. progesterone; 13. synergistic; 14. antagonistic; 15. insulin; 16. glucagon; 17. reproductive (biorhythm); 18. Glucagon; 19. insulin; 20. thymus; 21. glucose; 22. beta; 23. insulin; 24. glycogen; 25. alpha; 26. glucagon; 27. glycogen; 28. Prostaglandins; 29. signaling; 30. NGF.

SIGNALING MECHANISMS (34-IV)
1. hormones; 2. receptors; 3. steroid; 4. nucleus; 5. nonsteroid; 6. second messenger; 7. cAMP; 8. signaling; 9. Steroid; 10. cholesterol; 11. protein; 12. nonsteroid; 13. reproductive.

Answers to Self-Quiz

1. a; 2. e; 3. d; 4. b; 5. d; 6. c; 7. b; 8. a; 9. c; 10. a; 11. H; 12. F; 13. K; 14. L; 15. M; 16. D; 17. E; 18. A; 19. G; 20. B; 21. I; 22. O; 23. C; 24. J; 25. N.

35

SENSORY RECEPTION

Interactive Exercises

SENSORY SYSTEMS: AN OVERVIEW (35-I, pp. 599–603)

Fill-in-the-Blanks

Finely branched peripheral endings of sensory neurons that detect specific kinds of stimuli are

(1)_____. A (2)_____ is any form of energy change in the environment that the body actually

detects. (3)_____ detect impinging chemical energy; (4)_____ detect mechanical energy

associated with changes in pressure, position, or acceleration; (5)_____ detect the energy of visible

and ultraviolet light; (6)_____ detect radiant energy associated with temperature changes. A

(7)_____ is conscious awareness of change in internal or external conditions; this is not to be confused

with (8)_____ , which is an understanding of what sensation means. A sensory system consists of

sensory receptors for specific stimuli, (9)_____ _____ that conduct information from those

receptors to the brain, and (10)_____ _____ where information is evaluated.

Signals from receptors in the skin and joints travel to the primary (11)_____ _____ _____ , which is a strip little more than an inch wide running from the top of the (12)_____ to just above the (13)_____ on the surface of each (14)_____ _____. The largest portion of the somatic sensory cortex is A (see figure at right), which receives signals coming from the (15)_____. The second largest region is C, which receives signals coming from the (16)_____. Every type of sensation is caused by (17)_____ _____ arriving from particular nerve pathways activating specific neurons in the (18)_____. Besides sensing a stimulus, the brain also interprets variations in (19)_____ _____. Interpretation is based on the (20)_____ of action potentials propagated along single axons and the (21)_____ of axons carrying action potentials from a given tissue.

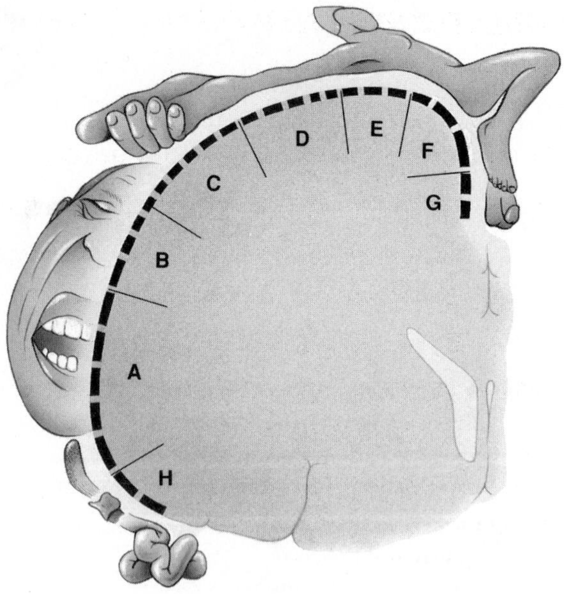

Matching

Select the best match for each item below.

22. ___ Vision is associated with _____.

23. ___ Pain is associated with _____.

24. ___ Odors are detected by _____.

25. ___ Hearing is detected by _____.

26. ___ CO_2 concentration in the blood is detected by _____.

27. ___ Environmental temperature is detected by _____.

28. ___ Internal body temperature is detected by _____.

29. ___ Touch is detected by _____.

30. ___ Rods and cones

31. ___ Hair cells in the ear's organ of Corti

32. ___ Pacinian corpuscles in the skin

33. ___ Olfactory receptors

34. ___ Any stimulus that causes tissue damage

35. ___ The movement of fluid in the inner ear is associated with _____.

A. chemoreceptors
B. mechanoreceptors
C. nociceptors
D. photoreceptors
E. thermoreceptors

SOMATIC SENSATIONS (35-II, pp. 603–604)

Fill-in-the-Blanks

The somatic sensations (awareness of (1)_____ , pressure, heat, (2)_____ , and pain) start with receptor endings that are embedded in (3)_____ and other tissues at the body's surfaces, in (4)_____ muscles, and in the walls of internal organs. All skin (5)_____ are easily deformed by pressure on the skin's surface; these make you aware of touch, vibrations, and pressure. (6)_____ nerve endings serve as "heat" receptors, and their firing of action potentials increases with increases in temperature. (7)_____ is the perception of injury to some body region; the perception begins when (8)_____ , which include free nerve endings, send signals via the thalamus to the parietal lobe of the brain, where they are interpreted. When the (9)_____ _____ pass a certain threshold, the signals generated are translated into sensations of pain. The brain sometimes gets confused and may associate perceived pain with a tissue some distance from the damaged area; this phenomenon is called (10)_____ _____. Usually the nerve pathways to both the injured and the mistaken area pass through the same segment of spinal cord. (11)_____ in skeletal muscle, joints, tendons, ligaments, and (12)_____ are responsible for awareness of the body's position in space and of limb movements.

Label-Match

Identify each indicated parts of the illustration below. Complete the exercise by entering the appropriate letter in the parentheses that follow the labels.

13. _____ _____ _____ ( )

14. _____ _____ ( )

15. _____ _____ ( )

16. _____

17. _____

18. _____ _____ ( )

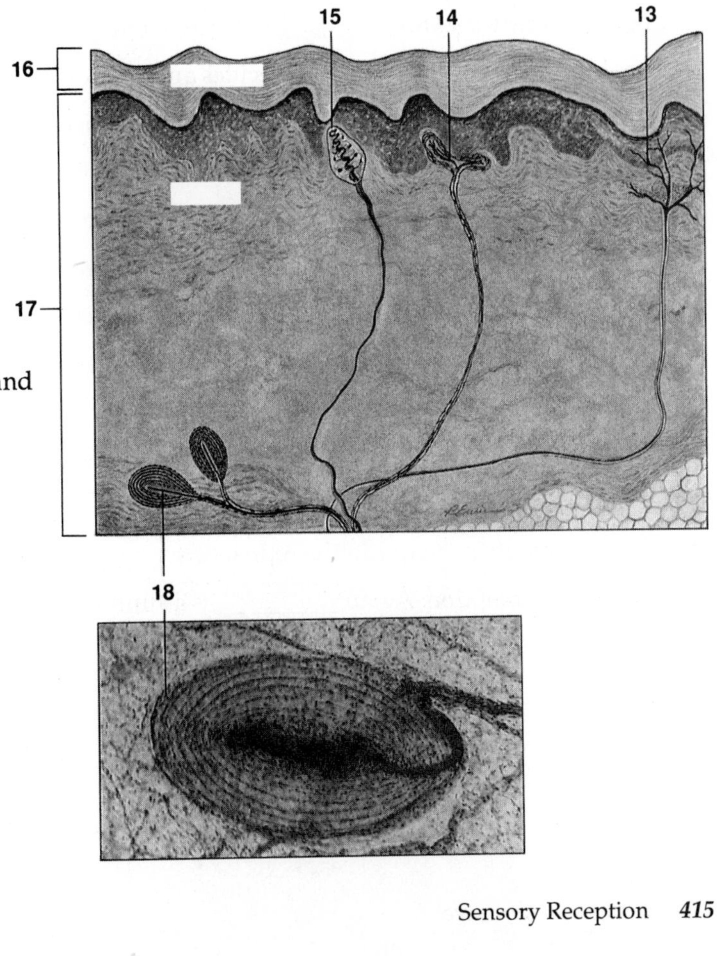

A. React continually to ongoing stimuli
B. Contribute to sensations of vibrations
C. Involved in sensing heat, light pressure, and pain
D. Stimulated at the beginning and end of sustained pressure

THE SPECIAL SENSES (35-III, pp. 604–615)

Fill-in-the-Blanks

(1)_____ receptors detect molecules that become dissolved in fluid next to some body surface. Receptors are the modified dendrites of (2)_____ neurons. In the case of taste, these receptors, when located on animal tongues, are often part of sensory organs, (3)_____ _____ , which are enclosed by circular papillae. Animals smell substances by means of (4)_____ receptors, such as the ones in your (5)_____ ; humans have about (6)_____ million of these in a nose. Sensory nerve pathways lead from the nasal cavity to the region of the brain where odors are identified and associated with their sources—the (7)_____ bulb and nerve tract. (8)_____ receptors sampling odors from food in the mouth are important for our sense of taste. When we suffer from the common cold and have a "runny" nose, odor molecules from food have difficulty reaching the olfactory receptors and contributing their information signals. Our senses of taste and (9)_____ are both dulled. The (10)_____ (perceived loudness) of sound depends on the height of the sound wave. The (11)_____ (perceived pitch) of sound depends on how fast the wave changes occur. The faster the vibrations, the (12) [choose one] ( ) higher, ( ) lower the sound. Hair cells are (13) [choose one] ( ) nociceptors, ( ) mechanoreceptors, ( ) thermoreceptors that detect vibrations. The hammer, anvil, and stirrup are located in the (14) [choose one] ( ) inner, ( ) middle ear. The (15)_____ is a coiled tube that resembles a snail shell and contains the (16)_____ _____ _____—the organ that changes vibrations into electrochemical impulses. Structures that detect rotational acceleration in humans are (17)_____ _____.

Light is a stream of (18)_____—discrete energy packets. (19)_____ is a process in which photons are absorbed by pigment molecules and photon energy is transformed into the electrochemical energy of a nerve signal. (20)_____ requires precise light focusing onto a layer of photoreceptive cells that are dense enough to sample details of the light stimulus, followed by image formation in the brain. (21)_____ are simple clusters of photosensitive cells, usually arranged in a cuplike depression in the epidermis. (22)_____ are well-developed photoreceptor organs that allow at least some degree of image formation. The (23)_____ is a transparent cover of the lens area, and the (24)_____ consists of tissue containing densely packed photoreceptors. Compound eyes contain several thousand photosensitive units known as (25)_____. In the vertebrate eye, lens adjustments assure that the (26)_____ _____ for a specific group of light rays lands on the retina. (27)_____ refers to the lens adjustments that bring about precise focusing onto the retina. (28)_____ people focus light from nearby objects posterior to the retina. (29)_____ cells are concerned with daytime vision and, usually, color perception. A (30)_____ is a funnel-shaped pit on the retina that provides the greatest visual acuity.

Labeling

Identify each indicated part of the accompanying illustrations.

31. _____ _____

32. _____

33. _____ _____

34. _____ _____

35. _____ _____

36. _____ _____

37. _____ _____

38. _____ _____

39. _____

40. _____

41. _____

42. _____ _____

43. _____

44. _____

45. _____

46. _____ _____

47. _____ _____

48. _____

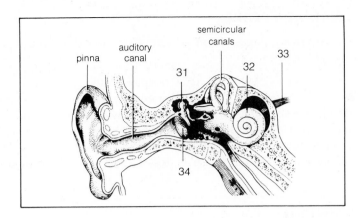

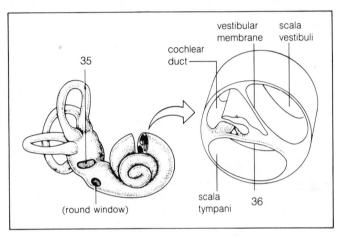

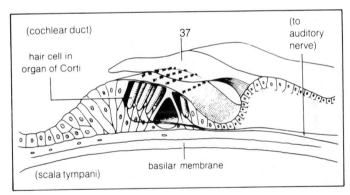

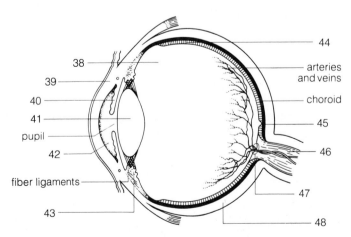

Chapter Terms

The following page-referenced terms are important terms in the chapter. Refer to the instructions given in Chapter 1, p. 4.

sensory systems (599)
sensory receptors (600)
sensation (600)
perception (600)
chemoreceptors (600)
mechanoreceptors (600)

photoreceptors (600)
thermoreceptors (600)
sensory pathways (601)
adaptation (603)
somatic sensations (603)
pain (604)

nociceptors (604)
echolocation (607)
vestibular apparatus (608)
vision (608)
eyes (608)
eyespots (608)

compound eyes (609)
mosaic theory (609)
photoreception (611)
rod cells (611)
cone cells (611)

Self-Quiz

___ 1. According to the mosaic theory,
_____.
   a. the basement membrane's pigment molecules prevent the scattering of light
   b. light falling on the inner area of an "on-center" field activates firing of the cells
   c. hair cells in the semicircular canals cooperate to detect rotational acceleration
   d. each ommatidium detects information about only one small region of the visual field; many ommatidia contribute "bits" to the total image
   e. all of the above

___ 2. The principal place in the human ear where sound waves are amplified is
_____.
   a. the pinna
   b. the ear canal
   c. the middle ear
   d. the organ of Corti
   e. none of the above

___ 3. The place where vibrations are translated into patterns of nerve impulses is
_____.
   a. the pinna
   b. the ear canal
   c. the middle ear
   d. the organ of Corti
   e. none of the above

For questions 4–8, choose from the following answers:
   a. fovea
   b. cornea
   c. iris
   d. retina
   e. sclera

___ 4. The white protective fibrous tissue of the eye is the _____.

___ 5. Rods and cones are located in the _____.

___ 6. The highest concentration of cones is in the _____.

___ 7. The adjustable ring of contractile and connective tissues that controls the amount of light entering the eye is the _____.

___ 8. The outer transparent protective covering of part of the eyeball is the _____.

___ 9. Accommodation involves the ability to _____.
   a. change the sensitivity of the rods and cones by means of transmitters
   b. change the width of the lens by relaxing or contracting certain muscles
   c. change the curvature of the cornea
   d. adapt to large changes in light intensity
   e. all of the above

___ 10. Nearsightedness is caused by _____.
   a. eye structure that focuses an image in front of the retina
   b. uneven curvature of the lens
   c. eye structure that focuses an image posterior to the retina
   d. uneven curvature of the cornea
   e. none of the above

Chapter Objectives/Review Questions

This section lists general and detailed chapter objectives that can be used as review questions. You can make maximum use of these items by writing answers on a separate sheet of paper. Fill in answers where blanks are provided. To check for accuracy, compare your answers with information given in the chapter or glossary.

Page	Objectives/Questions
(600)	1. Define and distinguish among chemoreceptors, mechanoreceptors, photoreceptors, and thermoreceptors. Name at least one example of each type that appears in an animal.
(603–604; 606–607)	2. Distinguish the types of stimuli detected by tactile and stretch receptors from those detected by hearing and equilibrium receptors.
(604–605; 603–604)	3. Explain how a taste bud works and distinguish the types of stimuli it detects from those detected by tactile or stretch receptors.
(606–607)	4. Follow a sound wave from pinna to organ of Corti; mention the name of each structure it passes and state where the sound wave is amplified and where the pattern of pressure waves is translated into electrochemical impulses.
(606)	5. State how low- and high-pitch sounds affect the organ of Corti.
(606–607)	6. State how low- and high-amplitude sounds affect the organ of Corti.
(607–608)	7. Explain how the three semicircular canals of the human ear detect changes of position and acceleration in a variety of directions.
(608)	8. Explain what a visual system is and list four of the five aspects of a visual stimulus that are detected by different components of a visual system.
(609–610; 608)	9. Contrast the structure of compound eyes with the structures of invertebrate eyespots and of the human eye.
(612–613)	10. Define nearsightedness and farsightedness and relate each to eyeball structure.
(611; 614–615)	11. Describe how the human eye perceives color and black-and-white.
(610–611; 614)	12. Define light and explain the general principles that affect how light is detected by photoreceptors and changed into electrochemical messages.

Integrating and Applying Key Concepts

How might human behavior be changed if human eyes were compound eyes composed of ommatidia and if humans perceived only vibrations—as fish do—rather than sounds?

Critical Thinking Exercises

1. A study of the muscle stretch receptor, part of the sensory organ called a muscle spindle (see text p. 604), used a muscle suspended from one end with a weight attached to the free end. Action potentials were recorded in an axon from the spindle as the weight attached to the muscle was varied. The results, shown below, were interpreted as evidence that information about amount of stretch was encoded in the frequency of the action potentials.

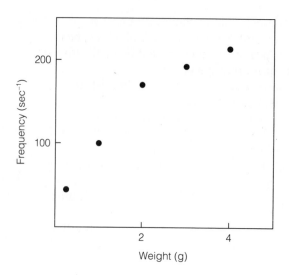

Which of the following assumptions is most important for this conclusion?

   a. Frequency of action potentials is directly proportional to amount of stretch.
   b. Amount of stretch is directly proportional to amount of weight loaded.
   c. Stretching the spindle depolarizes it.
   d. Depolarization of the spindle triggers action potentials.
   e. Action potentials in the spindle are transmitted to the muscle, which contracts in response.

ANALYSIS

   a. This is the conclusion, not an assumption. This statement is supported by the data if appropriate assumptions are made.
   b. The data are expressed as frequency as a function of weight loaded; the conclusion is about the amount of stretch. To link weight to stretch, we must make an assumption. The next experiment might check this assumption by measuring muscle length as the load is varied.
   c. This is a more detailed hypothesis about the mechanism by which the spindle responds to increasing stretch with higher frequency of action potentials. It is not necessary for this to be a valid statement in order to conclude that stretch is encoded in frequency.
   d. This choice is subject to the same analysis as choice (c). Both are aspects of the same, more elaborate explanation of the mechanism.
   e. This statement would establish a physiological context for the response, a reflex response that reverses the stretch of a muscle. However, this is not necessary in order to conclude that frequency is proportional to stretch. This response would occur only *after* the stretch information was encoded in whatever way.

Reference: Adrian, E. D., and Zotterman, Y. (1926) *J. Physiol.* 61:151–71.

2. Suppose you are studying a photoreceptor. You vary the intensity of white light shining on the receptor as you record the frequency of action potentials transmitted from the receptor. Your data are as shown on the graph.

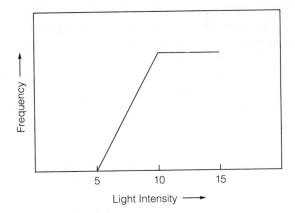

Which of the following conclusions would be valid based on these data?

a. The animal cannot detect the difference between light at intensity 3 and light at intensity 7.
b. The animal can detect the difference between light at intensity 6 and light at intensity 9.
c. The organism cannot detect the difference between light at intensity 9 and light at intensity 12.
d. The organism can detect the difference between light at intensity 11 and light at intensity 18.
e. The organism cannot detect light at intensity 4.

ANALYSIS

Information on light intensity is encoded in the frequency of action potentials. Thus, light at one intensity will be distinguishable from light at another intensity as long as the two intensities elicit different frequencies of action potentials. By this criterion, choice (b) is valid, and choices (a) and (c) are not. In all three cases, the frequency at one listed intensity is different from the frequency at the other intensity. Choice (d) requires extrapolation beyond the data. If you assume that the frequency response continues unchanged from light intensity 15 to intensity 18, then you would conclude that the animal could not differentiate these intensities. If you assume that the response will increase or decrease between 15 and 18, then you would make the opposite conclusion. Choice (e) means that the animal cannot detect the difference between intensity 4 and darkness. Since both zero intensity and intensity 4 generate no action potentials, this conclusion is valid.

3. A fine electrode was inserted into a single cell in the olfactory epithelium of an anaesthetized frog. Action potentials were recorded as vapors of different chemicals were blown over the epithelium. Recordings from two different cells with the compounds menthol and menthone are shown below.

	Cell 1	Cell 2
Unstimulated	‖‖‖‖‖‖‖‖‖‖‖‖‖‖‖‖‖‖	⎸ ⎸ ‖ ⎸⎸⎸⎸⎸⎸
Menthol	‖‖‖‖‖‖‖‖‖‖‖‖‖‖‖‖‖	⎸‖ ⎸ ⎸⎸
Menthone	‖‖‖‖‖‖‖‖‖‖‖‖‖‖‖‖‖	⎸ ⎸ ‖‖‖‖‖‖‖ ⎸⎸⎸

For each of the two cells, indicate whether it can do the following:

a. detect the presence of menthol
b. detect the presence of menthone
c. detect the difference between menthol and menthone

ANALYSIS

Cell 1 generates action potentials at the same frequency whether either substance is present or not. It can detect neither substance. The firing frequency of Cell 2 is depressed when menthol is present and increased when menthone is present. This cell can detect both substances and the difference between them. These data illustrate two forms of specificity of olfactory reception: (1) cells that are responsive to a substance versus cells that are not responsive to it and (2) cells that respond differently to two different substances. The data also show that information can be encoded in either an increase or a decrease in action potential frequency.

Reference: Gesteland, R. C. (1966) *Discovery* 27(2).

Answers

Answers to Interactive Exercises

SENSORY SYSTEMS: AN OVERVIEW (35-I)
1. receptors; 2. stimulus; 3. Chemoreceptors; 4. mechanoreceptors; 5. photoreceptors; 6. thermoreceptors; 7. sensation; 8. perception; 9. nerve pathways; 10. brain regions; 11. somatic sensory cortex; 12. head; 13. ear; 14. cerebral hemisphere; 15. mouth; 16. hand; 17. action potentials; 18. brain; 19. stimulus intensity; 20. frequency; 21. number; 22. D; 23. C; 24. A; 25. B; 26. A; 27. E; 28. E; 29. B; 30. D; 31. B; 32. B; 33. A; 34. C; 35. B.

SOMATIC SENSATIONS (35-II)
1. touch; 2. cold; 3. skin; 4. skeletal; 5. mechanoreceptors; 6. Free; 7. Pain; 8. nociceptors; 9. action potentials; 10. referred pain; 11. Mechanoreceptors; 12. skin; 13. free nerve endings (C); 14. Ruffini endings (A); 15. Meissner corpuscle (D); 16. epidermis; 17. dermis; 18. Pacinian corpuscle (B).

THE SPECIAL SENSES (35-III)
1. Taste; 2. sensory; 3. taste buds; 4. olfactory; 5. nose; 6. 10; 7. olfactory; 8. Olfactory; 9. smell; 10. amplitude; 11. frequency; 12. higher; 13. mechanoreceptors; 14. middle; 15. cochlea; 16. organ of Corti; 17. semicircular canals; 18. photons; 19. Photoreception; 20. Vision; 21. Eyespots; 22. Eyes; 23. cornea; 24. retina; 25. ommatidia; 26. focal point; 27. Accommodation; 28. Farsighted; 29. Cone; 30. fovea; 31. middle earbones (malleus, incus, stapes); 32. cochlea; 33. auditory nerve; 34. tympanic membrane/eardrum; 35. oval window; 36. basilar membrane; 37. tectorial membrane; 38. vitreous humor; 39. cornea; 40. iris; 41. lens; 42. aqueous humor; 43. suspensory ligament; 44. retina; 45. fovea; 46. optic nerve; 47. blind spot/optic disk; 48. sclera.

Answers to Self-Quiz

1. d; 2. c; 3. d; 4. e; 5. d; 6. a; 7. c; 8. b; 9. b; 10. a.

36

PROTECTION, SUPPORT, AND MOVEMENT

Interactive Exercises

INTEGUMENTARY SYSTEM (36-I, pp. 620–623)

Fill-in-the-Blanks

Human skin is an organ system that consists of two layers: the outermost (1)_____ , which contains mostly dead cells, and the (2)_____ , which contains hair follicles, nerves, tiny muscles associated with the hairs, and various types of glands. The (3)_____ layer, with its loose connective tissue and store of fat in (4)_____ tissue, lies beneath the skin.

Labeling

Label the numbered parts of the illustration below.

5. _____

6. _____ _____

   _____

7. _____ _____

8. _____ _____

9. _____ _____

10. _____ _____

11. _____ _____

12. _____

13. _____

14. _____

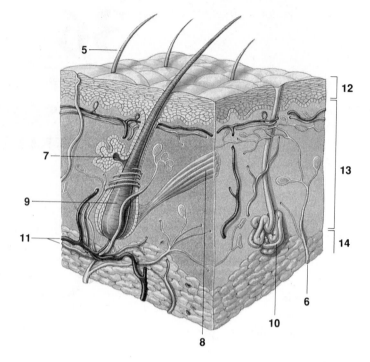

SKELETAL SYSTEMS / HUMAN SKELETAL SYSTEM (36-II, pp. 624–629)

Fill-in-the-Blanks

All motor systems are based on effector cells that are able to (1)_____ and (2)_____ , and on the presence of a medium against which the (3)_____ force can be applied. Longitudinal and radial muscle layers work as an (4)_____ muscle system, in which the action of one motor element opposes the action of another. A membrane filled with fluid resists compression and can act as a (5)_____ skeleton. Arthropods have antagonistic muscles attached to an (6) [choose one] ( ) endoskeleton, ( ) exoskeleton. (7)_____ are hard compartments that enclose and protect the brain, spinal cord, heart, lungs, and other vital organs of vertebrates. Bones support and anchor (8)_____ and soft organs, such as eyes. (9)_____ systems are found in the long bones of mammals and contain living bone cells that receive their nutrients from the blood. (10)_____ _____ is a major site of blood cell formation. Bone tissue serves as a "bank" for (11)_____ , (12)_____ , and other mineral ions; depending on metabolic needs, the body deposits ions into and withdraws ions from this "bank."

Bones develop from (13)_____ secreting material inside the shaft and on the surface of the cartilage model. Bone can also give ions back to interstitial fluid as osteoclasts dissolve out component minerals and remodel bone in response to (14)_____ signals, lack of exercise, and calcium deficiencies in the diet. Extreme decreases in bone density result in (15)_____ , particularly among older women.

The (16)_____ skeleton includes the skull, vertebral column, ribs, and breastbone; the (17)_____ skeleton includes the pectoral and pelvic girdles and the forelimbs and hindlimbs, when they exist in vertebrates.

(18)_____ joints are freely movable and are lubricated by a fluid secreted into the capsule of dense connective tissue that surrounds the bones of the joint. Bones are often tipped with (19)_____ ; as a person ages, the cartilage at (20)_____ joints may simply wear away, a condition called (21)_____. By contrast, in (22)_____ _____ , the synovial membrane becomes inflamed, cartilage degenerates, and bone becomes deposited in the joint.

Labeling

Identify each indicated part of the accompanying illustrations.

23. _____ _____
24. _____ _____ _____
25. _____ _____
26. _____ _____
27. _____ _____ _____

28. _____ _____
29. _____ _____
30. _____ _____
31. _____

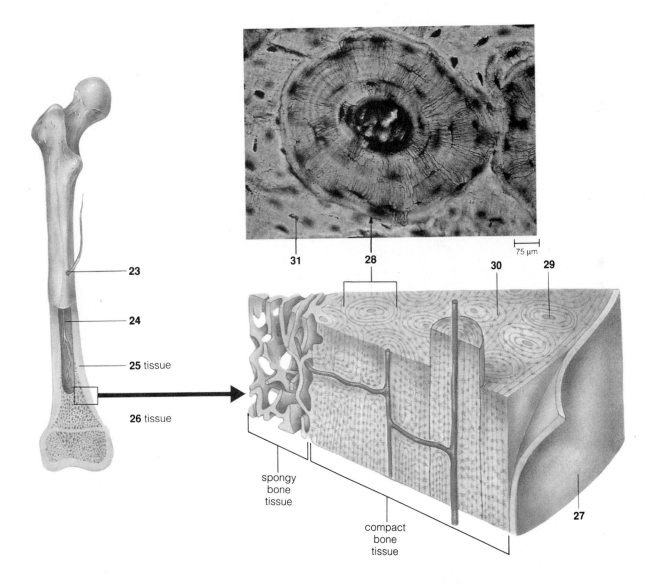

25 tissue

26 tissue

spongy bone tissue

compact bone tissue

75 μm

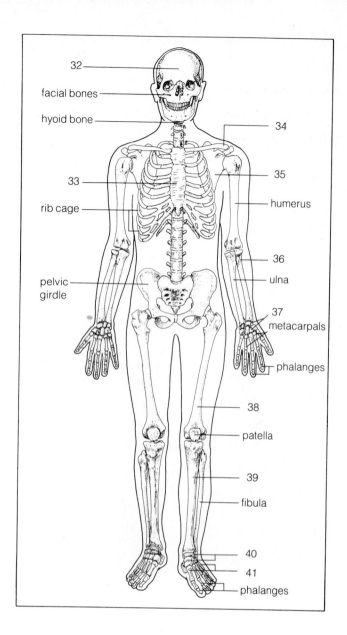

32. _____
facial bones
hyoid bone
34 _____
33 _____
35 _____
rib cage
humerus
36 _____
pelvic girdle
ulna
37 _____
metacarpals
phalanges
38 _____
patella
39 _____
fibula
40 _____
41 _____
phalanges

32. _____
33. _____
34. _____
35. _____
36. _____
37. _____   _____
38. _____
39. _____
40. _____   _____
41. _____
42. _____ vertebrae
43. _____ vertebrae
44. _____ vertebrae
45. _____ vertebrae
46. _____ vertebrae

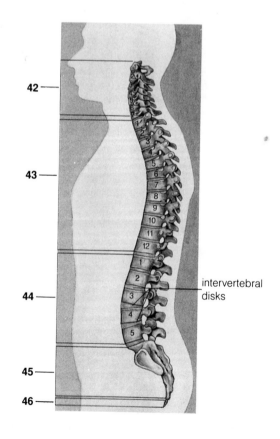

42 —
43 —
44 —
45 —
46 —

intervertebral disks

MUSCULAR SYSTEM (36-III, pp. 630–635)

Sequence

Arrange in order of decreasing size.

1. _____
2. _____
3. _____
4. _____
5. _____
6. _____

   A. Muscle fiber (muscle cell)
   B. Myosin filament
   C. Muscle bundle
   D. Muscle
   E. Myofibril
   F. Actin filament

Labeling

Identify each indicated part of the accompanying illustration.

7. _____

8. _____ _____

9. _____

10. _____

11. _____ _____

12. _____ _____

13. _____

14. _____ _____

15. _____

16. _____ _____

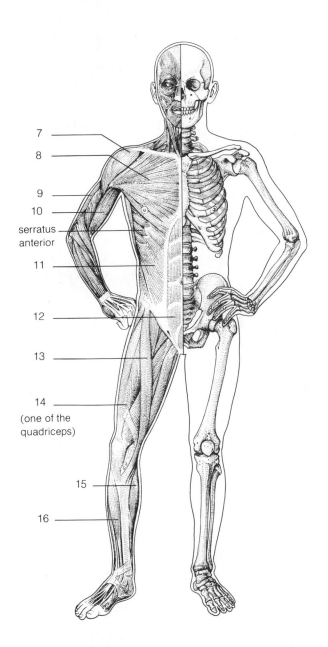

7

8

9

10

serratus anterior

11

12

13

14
(one of the quadriceps)

15

16

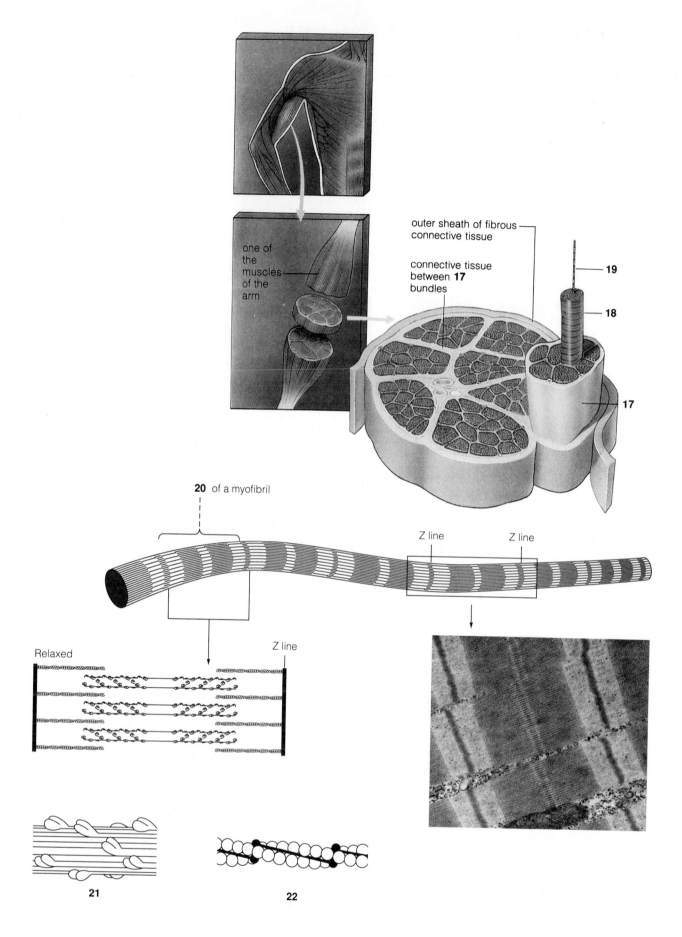

outer sheath of fibrous connective tissue

connective tissue between **17** bundles

19

18

17

one of the muscles of the arm

20 of a myofibril

Z line   Z line

Relaxed

Z line

21

22

Label the numbered parts of the accompanying illustrations on p. 428.

17. _____ bundle

18. _____ _____

19. _____

20. _____

21. _____ _____

22. _____ _____

Fill-in-the-Blanks

Each myofibril contains (23)_____ filaments, which have cross-bridges, and (24)_____ filaments, which are thin and lack cross-bridges. The repetitive fundamental unit of muscle contraction in striated muscle cells is the (25)_____. The sarcoplasmic reticulum releases (26)_____ ions as a consequence of charge reversal.

There are three types of muscle tissue: (27)_____ , which is striated, involuntary, and located in the heart; (28)_____ , which is striated, largely voluntary, and generally attached to bones or cartilage; and (29)_____ , which is involuntary, not striated, and mostly located in the wall of internal organs in vertebrates. Of these three, (30)_____ muscle interacts with the skeleton to bring positional changes of body parts and to move the animal through its environment.

Muscle cells have three properties in common: contractility, elasticity, and (31)_____. (32)_____ of an entire muscle is brought about by the combined decreases in length of the individual sarcomeres that make up the myofibrils of the muscle cells. According to the (33)_____-_____ model, (34)_____ filaments physically slide along and pull the (35)_____ filaments toward the center of a sarcomere during contraction. The energy that drives the attaching and detaching of the cross-bridges comes immediately from (36)_____. (37)_____ phosphate is then enzymatically broken apart, and its available phosphate group is then used to regenerate ATP. When supplies of available creatine run out, the muscle cells take (38)_____ from the bloodstream and from the breakdown of glycogen stored in the muscle cells and send it through lactate fermentation if the demand for muscle action is intense but brief. If the demand for muscle action is moderate, it can also be prolonged by muscle cells sending glucose through (39)_____ _____ phosphorylation, the final stage of the aerobic respiration pathway.

Muscle contracts when (40)_____ ions are released from the (41)_____ _____ where they are stored and relaxes when they are actively sent back. By controlling the (42)_____ _____ that reach the (43)_____ _____ in the first place, the nervous system controls muscle contraction by controlling calcium ion levels in muscle tissue. (44)_____ binding alters the actin filaments so that the heads of the myosin filaments bind to them.

CONTRACTION OF A SKELETAL MUSCLE / LIMB MOVEMENTS (36-IV, pp. 635–638)

Fill-in-the-Blanks

The larger the (1)_____ of a muscle, the greater its strength. A (2)_____ neuron and the muscle cells under its control are called a (3)_____ _____. A (4)_____ _____ is a response in which a muscle contracts briefly when reacting to a single, brief stimulus and then relaxes. The strength of the muscular contraction depends on how far the (5)_____ response has proceeded by the time another signal arrives. (6)_____ is the state of contraction in which a motor unit that is being stimulated repeatedly is maintained. In a (7)_____ contraction, the nervous system activates only a small number of motor units; in a stronger contraction, a larger number are activated at a high (8)_____ of stimulation. Some of the human body's more than 600 muscles work (9)_____ (in opposition) so that the action of one opposes or reverses the action of another. Some work (10)_____ (together) to enhance the action of each other.

Together, the skeleton and its attached muscles are like a system of levers in which rigid rods, (11)_____ , move about at fixed points, called (12)_____. Most attachments are close to joints, so a muscle has to shorten only a small distance to produce a large movement of some body part.

When the (13)_____ contracts, the elbow joint bends (flexes). As it relaxes and as its partner, the (14)_____ , contracts, the forelimb extends and straightens. In order to accomplish this, (15)_____ signals sent to one muscle's neurons prevent that muscle from contracting while the opposing muscle group is being stimulated (excited). This cooperative action results partly from (16)_____ _____ occurring in the spinal cord. In addition, the central nervous system uses signals from (17)_____ receptors to coordinate the contractions.

(18)_____ _____ are synthetic hormones that mimic the effects of testosterone in building greater (19)_____ mass in both men and women. It is illegal for competitive athletes to use them because of unfair advantage and because of the side effects. In men, (20)_____ , baldness, shrinking (21)_____ , and infertility are the first signs of damage. In women, (22)_____ hair becomes more noticeable, (23)_____ _____ become irregular, breasts may shrink, and the (24)_____ may become grossly enlarged. Aside from these physical side effects, some men experience uncontrollable (25)_____ , delusions, and wildly manic behavior.

Labeling

Identify the numbered parts of the accompanying illustrations.

26. _____

27. _____ _____

28. _____

29. _____ _____ _____

30. _____

31. _____ _____

32. _____ _____

33. _____ (_____)

34. _____ _____ _____

35. _____ _____

36. _____ _____

37. _____ _____

38. _____ _____

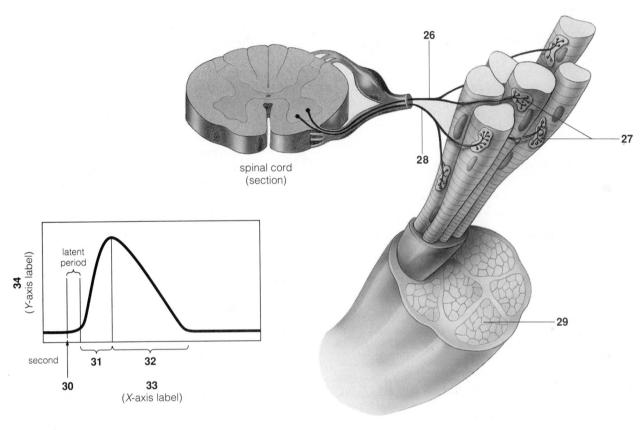

spinal cord
(section)

latent
period

34
(*Y*-axis label)

second

30   **31**   **32**

33
(*X*-axis label)

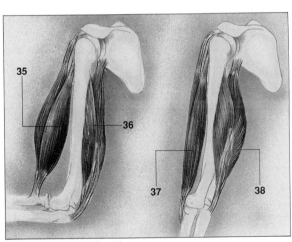

Chapter Terms

The following page-referenced terms are important; they were boldfaced in the chapter. Refer to the instructions given in Chapter 1, p. 4.

integument (620)
epidermis (621)
dermis (621)
hydrostatic skeleton (624)
exoskeleton (624)
endoskeleton (624)
bones (626)

red marrow (627)
yellow marrow (627)
axial skeleton (628)
appendicular skeleton (628)
intervertebral disks (628)
synovial joint (628)

cartilaginous joints (629)
fibrous joints (629)
myofibrils (631)
actin (632)
myosin (632)
sarcomeres (632)

sliding-filament model (632)
sarcoplasmic reticulum (634)
motor unit (635)
muscle twitch (636)
tetanus (636)
reciprocal innervation (638)

Self-Quiz

For questions 1–5, choose from the following answers:

    a. clavicle
    b. femur
    c. humerus
    d. phalanx
    e. scapula

____ 1. The bone in the upper arm is the
_____.

____ 2. Any bone in a toe or finger is called a
_____.

____ 3. The collarbone is the _____.

____ 4. Another name for the shoulder blade is the _____.

____ 5. The thigh bone is the _____.

____ 6. Reciprocal innervation of reflexes between antagonistic muscle pairs _____.
    a. is the usual basis of coordinated contractions
    b. explains the mechanism for the operation of the calcium pump
    c. refers to the adjustments in lens width that focus an image precisely on the retina
    d. causes rods to interfere with the stimulation of cones
    e. all of the above

____ 7. Muscle tone refers to _____.
    a. the length of time it takes for a muscle to contract
    b. the speed at which a motor unit recovers after contraction
    c. the fact that some motor units remain contracted even when the total muscle is considered to be relaxed
    d. the state in which a muscle is left after reciprocal innervation of antagonistic muscle pairs has occurred
    e. the state in which a muscle bundle is left after a twitch contraction

For questions 8–11, choose from the following answers:

    a. cartilaginous
    b. fibrous
    c. hinge
    d. synovial
    e. none of the above

____ 8. Vertebral disks with small amounts of movement are examples of _____ joints.

____ 9. Nonmoving joints between skull bones are examples of _____ joints.

____ 10. Freely movable bones that are separated by a fluid-filled cavity compose a _____ joint.

____ 11. Knees and elbows are _____ joints.

___ 12. Which of the following statements does *not* describe normal activity in a motor system?
   a. All motor systems require the presence of some medium or structural element against which force can be applied.
   b. In a resting muscle, energy is stored in the form of tropomyosin.
   c. In a skeletal muscle system, coordinated contraction depends on reciprocal innervation of motor neurons to antagonistic muscle pairs.
   d. Calcium ions are required for contraction in all muscle cells.
   e. In vertebrates, only skeletal muscle actually moves the body through the environment.

___ 13. Haversian systems _____.
   a. are composed of receptors, modulators, and effectors
   b. are negative feedback mechanisms
   c. are positive feedback mechanisms
   d. supply bone cells with nutrients, oxygen, and signals from elsewhere in the body

Chapter Objectives/Review Questions

This section lists general and detailed chapter objectives that can be used as review questions. You can make maximum use of these items by writing answers on a separate sheet of paper. Fill in answers where blanks are provided. To check for accuracy, compare your answers with information given in the chapter or glossary.

Page	Objectives/Questions
(621)	1. Name the four functions of human skin.
(621–623)	2. Describe the two-layered structure of human skin and identify the items located in each layer.
(624–625)	3. Compare invertebrate and vertebrate motor systems in terms of skeletal and muscular components and their interactions.
(626–627)	4. Explain the various roles of osteoblasts, osteoclasts, cartilage models, long bones, and epiphyseal plates in the development of human bones.
(630)	5. Identify human bones by name and location.
(630)	6. Refer to Figure 36.17 of your main text and indicate (a) a muscle used in sit-ups, (b) another used in dorsally flexing and inverting the foot, and (c) another used in flexing the elbow joint.
(630–632)	7. Explain in detail the structure of muscles, from the molecular level to the organ systems level. Then explain how biochemical events occur in muscle contractions and how antagonistic muscle action refines movements.
(631–632)	8. Describe the fine structure of a muscle fiber; use terms such as *myofibril, sarcomere, motor unit, actin,* and *myosin.*
(632–635)	9. List, in sequence, the biochemical and fine structural events that occur during the contraction of a skeletal muscle fiber and explain how the fiber relaxes.
(636)	10. Distinguish twitch contractions from tetanic contractions.
(638)	11. Define reciprocal innervation and explain how it helps coordinate motor elements.

Integrating and Applying Key Concepts

If humans had an exoskeleton rather than an endoskeleton, would they move differently from the way they do now? Name any advantages or disadvantages that having an exoskeleton instead of an endoskeleton would present in human locomotion.

Critical Thinking Exercises

1. Muscles and muscle cells can be prepared for experiments in a variety of ways. Which of the following preparations would contract only if calcium was present in the surrounding medium? For each preparation, how could a contraction be triggered?

   a. An intact muscle exposed in the animal
   b. A single muscle cell dissected free of all other cells
   c. A single muscle cell with all its membranes chemically stripped
   d. An artificial myofibril made of actin, myosin, and troponin
   e. An artificial myofibril made only of actin and myosin

ANALYSIS

   a. The calcium that stimulates contraction in whole muscle cells is released from the sarcoplasmic reticulum. As long as this organelle is intact, calcium is not needed in the medium. A muscle contraction can be triggered by direct electrical stimulation of the muscle or by electrical stimulation of the axons that innervate the muscle. Contraction can be triggered chemically by applying to the muscle acetylcholine or an appropriate drug that mimics acetylcholine action. Also, raising the extracellular potassium concentration depolarizes the muscle cell membrane and can induce contraction if the potential reaches threshold.
   b. The same considerations apply to contraction of a single intact cell as to a bundle of cells. In this case, however, the nerve has been removed, so the muscle fiber cannot be stimulated via electric current applied to the nerve. All the other methods would work.
   c. This kind of preparation, made by soaking muscle cells in glycerin solutions, has no sarcoplasmic reticulum, so it requires calcium in the medium in order to contract. If ATP is present in the medium, the glycerinated fiber will contract when calcium is added. Alternatively, calcium can be present from the beginning, and contraction will be triggered by addition of ATP. With no plasma membrane, this material is unaffected by electric stimulation or by neurotransmitters and synaptic drugs.
   d. Although this preparation does have the calcium-binding protein troponin, the material contracts without calcium because tropomyosin is missing. Tropomyosin is the protein that blocks the myosin-binding site on the actin filament and prevents contraction in the absence of calcium. Without tropomyosin, the sarcomere will contract when ATP is added.
   e. This preparation behaves the same as (d).

2. A muscle can be clamped in an apparatus so that its sarcomeres are all limited to a certain minimum length. When the muscle is stimulated to contract, it generates tension, which can be measured, but it becomes no shorter than the apparatus allows. The graph below shows the typical behavior of a muscle set at various lengths. Which of the following statements is the best explanation of these data?

   a. There is an optimum length for muscle function.
   b. ATP is synthesized most rapidly at a certain muscle length.
   c. More calcium is released at a certain muscle length.
   d. More myosin heads can engage the actin filaments at a certain muscle length.
   e. The amount of force the arm can exert depends on the angle at which the elbow is bent.

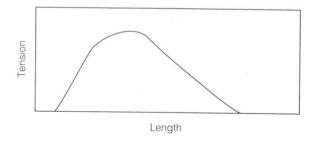

Reference: Gordon, A. M., Huxley, A. F., and Julian, F. S. (1966) *J. Phys. London* 184:170–92.

ANALYSIS

a. This is a statement of the observations, a description of the graph, not an explanation.

b. This statement assumes that the availability of ATP limits the force that can be generated by a muscle. This can be true when the contraction is prolonged, but in a single twitch the number of ATP molecules available is much greater than the number of myosin heads that are activated. Furthermore, it is very difficult to imagine a mechanism that would couple the rate of ATP synthesis to the length of the sarcomere.

c. Tension does seem to be proportional to calcium concentration, but how changing cell length could alter release of calcium from the sarcoplasmic reticulum is not at all clear. You might imagine membrane permeability to calcium increasing as the sarcoplasmic reticulum is stretched, but this does not account for the fact that calcium release is still dependent on action potentials or for the drop in tension at high sarcomere lengths.

d. The more individual forces applied to a rod, the greater the total force. Thus, the more myosin heads activated, the greater the tension generated. At very short sarcomere lengths, the ends of the thin filaments extend over the central part of the thick filament where there are no myosin heads; hence, those actin molecules do not function to generate more tension. Furthermore, the ends of the actin filaments begin to push against each other and on structures in the center of the thick filaments, reducing the total tension generated. Likewise, the ends of the thick filaments can begin to push against the ends of the sarcomere. At intermediate lengths, there is maximum overlap between the thick and thin filaments, and the greatest number of myosin heads is activated. As the length is increased, the overlap between the filaments decreases, and fewer active myosin heads generate less tension. At some point, the thick and thin filaments are pulled completely apart, and no tension can be generated.

e. This statement is a consequence of the observation, not an explanation. The angle of the elbow determines the length of the muscles that move it. If there is a length of maximum tension, there must be an angle of maximum tension.

3. One gram of contracting muscle consumes about 1 millimole (mmole) of ATP per minute. The same gram of muscle contains about 5 micromoles of ATP and 25 micromoles of creatine phosphate. There are also about 10 milligrams, or 1 percent of the tissue weight, of glycogen. Assume that the muscle uses aerobic metabolism to completely oxidize glucose. How long could this muscle contract using only its stored energy reserves?

ANALYSIS

One mole of creatine phosphate can be used in muscle to produce 1 mole of ATP. The molecular weight of glucose is 180 milligrams per millimole; thus, the glycogen, a polymer of glucose, can yield about 0.06 millimoles, or 60 micromoles, of glucose. Oxidation of one glucose molecule produces 36 ATPs, so the stored glycogen can yield 2,160 micromoles of ATP. The sum of all the ATP sources is thus 2,190 micromoles, or 2.19 millimoles, enough to sustain contraction for a little over 2 minutes. Note, however, that the supply of ATP by aerobic metabolism may be limited by the availability of oxygen. As soon as the cell begins to supplement with anaerobic glycolysis, the amount of ATP produced per glucose drops, and the time of contraction drops as well.

Note: Please see authors' note on page 371.

Answers

Answers to Interactive Exercises

INTEGUMENTARY SYSTEM (36-I)

1. epidermis; 2. dermis; 3. subcutaneous; 4. adipose;
5. hair; 6. sensory nerve ending; 7. sebaceous gland;
8. smooth muscle; 9. hair follicle; 10. sweat gland;
11. blood vessel; 12. epidermis; 13. dermis;
14. hypodermis.

SKELETAL SYSTEMS/HUMAN SKELETAL SYSTEM (36-II)

1. contract; 2. relax; 3. contractile; 4. antagonistic;
5. hydrostatic; 6. exoskeleton; 7. Bones; 8. muscles;
9. Haversian; 10. Red marrow; 11. calcium (phosphate);
12. phosphate (calcium); 13. osteoblasts; 14. hormonal;
15. osteoporosis; 16. axial; 17. appendicular;
18. Synovial; 19. cartilage; 20. synovial;
21. osteoarthritis; 22. rheumatoid arthritis; 23. nutrient
canal; 24. contains yellow marrow; 25. compact bone;
26. spongy bone; 27. connective tissue covering
(periosteum); 28. Haversian system; 29. Haversian
canal; 30. mineral deposits (calcium phosphate);
31. osteocyte (bone cell); 32. cranium; 33. sternum;
34. clavicle; 35. scapula; 36. radius; 37. carpal bones;
38. femur; 39. tibia; 40. tarsal bones; 41. metatarsals;
42. cervical; 43. thoracic; 44. lumbar; 45. sacral;
46. coccygeal.

MUSCULAR SYSTEM (36-III)

1. D; 2. C; 3. A; 4. E; 5. B; 6. F; 7. deltoid; 8. pectoralis
major; 9. triceps; 10. biceps; 11. external oblique;
12. rectus abdominis; 13. sartorius; 14. rectus femoris;
15. gastrocnemius; 16. tibalis anterior; 17. muscle;

18. muscle cell (muscle fiber); 19. myofibril;
20. sarcomere; 21. myosin filament; 22. actin filament;
23. myosin; 24. actin; 25. sarcomere; 26. calcium;
27. cardiac; 28. skeletal; 29. smooth; 30. skeletal;
31. excitability; 32. Contraction; 33. sliding-filament;
34. myosin; 35. actin; 36. ATP; 37. Creatine; 38. glucose;
39. electron transport; 40. calcium; 41. sarcoplasmic
reticulum; 42. action potentials; 43. sarcoplasmic
reticulum; 44. Calcium.

CONTRACTION OF A SKELETAL MUSCLE/ LIMB MOVEMENTS (36-IV)

1. diameter; 2. motor; 3. motor unit; 4. muscle twitch;
5. twitch; 6. Tetanus; 7. weak; 8. frequency (rate);
9. antagonistically; 10. synergistically; 11. bones;
12. joints; 13. biceps; 14. triceps; 15. inhibitory;
16. reciprocal innervation; 17. stretch; 18. Anabolic
steroids; 19. muscle; 20. acne; 21. testes; 22. facial;
23. menstrual periods; 24. clitoris; 25. aggression;
26. axon of motor neuron serving one motor unit;
27. neuromuscular junction; 28. axon of another motor
neuron serving another motor unit; 29. individual
muscle cells; 30. time that stimulus is applied;
31. contraction phase; 32. relaxation phase; 33. time
(seconds); 34. strength of contraction; 35. biceps
contracts; 36. triceps relaxes; 37. biceps relaxes;
38. triceps contracts.

Answers to Self-Quiz

1. c; 2. d; 3. a; 4. e; 5. b; 6. a; 7. c; 8. a; 9. b; 10. d; 11. c;
12. b; 13. d.

37

DIGESTION AND HUMAN NUTRITION

Interactive Exercises

TYPES OF DIGESTIVE SYSTEMS AND THEIR FUNCTIONS (37-I, pp. 641–643)

Fill-in-the-Blanks

A digestive system is some form of body cavity or tube in which food is reduced first to (1)_____ and then to small (2)_____. Digested nutrients are then (3)_____ into the internal environment. Predators and scavengers have (4)_____ feeding habits, while ruminants have (5)_____ feeding. A(n) (6)_____ digestive system has only one opening, two-way traffic, and a highly branched gut cavity that serves both digestive and (7)_____ functions. A(n) (8)_____ digestive system has a tube or cavity with regional specializations and a(n) (9)_____ at each end. (10)_____ involves the muscular movement of the gut wall, but (11)_____ is the release into the lumen of enzyme fluids and other substances required to carry out digestive functions.

HUMAN DIGESTIVE SYSTEM: AN OVERVIEW / INTO THE MOUTH, DOWN THE TUBE (37-II, pp. 643–648)

1. Complete the following table by naming the organs described.

Organ	Main Functions
a. _____	Mechanically breaks down food, mixes it with saliva
b. _____	Moisten food; start polysaccharide breakdown; buffer acidic foods in mouth
c. _____	Stores, mixes, dissolves food; kills many microoganisms; starts protein breakdown; empties in a controlled way
d. _____	Digests and absorbs most nutrients
e. _____	Produces enzymes that break down all major food molecules; produces buffers against hydrochloric acid from stomach
f. _____	Secretes bile for fat emulsification; secretes bicarbonate, which buffers hydrochloric acid from stomach
g. _____	Stores, concentrates bile from liver
h. _____	Stores, concentrates undigested matter by absorbing water and salts
i. _____	Controls elimination of undigested and unabsorbed residues

Fill-in-the-Blanks

Saliva contains an enzyme (2)_____ that hydrolyzes starch. Contractions force the larynx against a cartilaginous flap called the (3)_____ , which closes off the trachea. The (4)_____ is a muscular tube that propels food to the stomach. Any alternating progression of contracting and relaxing muscle movements along the length of a tube is known as (5)_____. The (6)_____'s most important function is to regulate the rate at which food reaches the intestine. Most digestion and absorption of nutrients occurs in the (7)_____ _____. (8)_____ is an enzyme that works in the stomach. (9)_____ is made by the liver, is stored in the gallbladder, and works in the (10)_____ _____. (11)_____ is an example of an enzyme that is made by the pancreas but works in the small intestine.

Labeling

Identify each numbered structure in the accompanying illustration.

12. _____ _____

13. _____ _____

14. _____

15. _____

16. _____

17. _____ _____

18. _____ _____

19. _____

20. _____

21. _____

22. _____

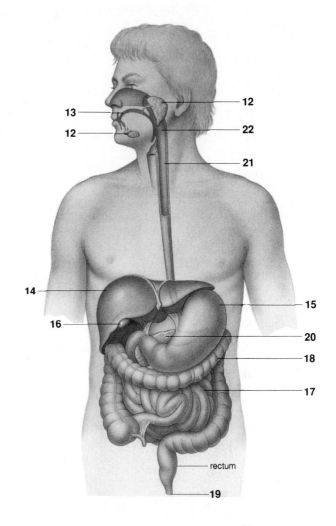

True-False

If false, explain why.

___ 23. Amylase digests starch, lipase digests lipids, and proteases break peptide bonds.

___ 24. ATP is the end product of digestion.

___ 25. The appendix has no known digestive functions.

___ 26. Water and minerals are absorbed into the bloodstream from the lumen of the large intestine.

___ 27. Gallstones are formed in the kidney when something changes the concentrations of bile salts, lecithin, and cholesterol in the bile.

HUMAN NUTRITIONAL REQUIREMENTS (37-III, pp. 648–653)

1. Complete the following table by determining how many kilocalories the people described should take in daily, given the stated exercise level, in order to *maintain* their weight. Consult pages 648–649 of the text.

	Weight	Age	Sex	Level of Physical Activity	Present Weight (lbs.)	Number of Kilocalories
a.	5'6"	25	Female	Moderately active	138	_____
b.	5'10"	18	Male	Very active	145	_____
c.	5'8"	53	Female	Not very active	143	_____

Fill-in-the-Blanks

People who are (2)_____ percent heavier than "ideal" are considered to be obese. (3)_____ _____ are the body's main sources of energy; they should make up (4)_____ to _____ percent of the human daily caloric intake. (5)_____ and cholesterol are components of animal cell membranes. Fat deposits are used primarily as (6)_____ _____ , but they also cushion many organs and provide insulation. Lipids should constitute less than (7)_____ percent of the human diet. One tablespoon a day of polyunsaturated oil supplies all (8)_____ _____ _____ that the body cannot synthesize. (9)_____ are digested to twenty common amino acids, of which eight are (10)_____ , cannot be synthesized, and must be supplied by the diet. Animal proteins such as (11)_____ and (12)_____ contain high amounts of essential amino acids (that is, they are complete). (13)_____ are organic substances needed in small amounts in order to build enzymes or help them catalyze metabolic reactions. (14)_____ are inorganic substances needed for a variety of uses.

NUTRITION AND METABOLISM (37-IV, pp. 654–655)

To answer the following questions, consult Table 37.2 and Figure 37.13.

1. What is the pool of amino acids used for in the human body?_____

   _____

2. Which breakdown products result from carbohydrate and fat digestion?_____

   _____

3. Monosaccharides, free fatty acids, and monoglycerides all have three uses; identify them._____

   _____

Fill-in-the-Blanks

Amino acid conversions in the liver form (4)_____ , which is potentially toxic to cells; the liver immediately converts this substance to (5)_____ , a much less toxic waste product that is expelled by the urinary system from the body. Beta cells of the pancreas secrete (6)_____ , and alpha cells secrete (7)_____ , a hormone that commands liver cells to convert (8)_____ (a storage starch) into glucose subunits. Under hypothalamic commands, the adrenal medulla begins secreting (9)_____ and

(10)_____ , which stop (11)_____ synthesis in the liver, stop (12)_____ uptake in muscles, and promote the shift from "burning" (13)_____ during cellular respiration to "burning" fats.

Chapter Terms

The following page-referenced terms are the important terms in the chapter. Refer to the instructions given in Chapter 1, p. 4.

digestive system (641)
incomplete digestive
   system (641)
complete digestive
   system (642)
motility (642)
secretion (642)
digestion (642)
absorption (642)

sphincter (644)
peristalsis (644)
segmentation (644)
salivary glands (645)
pharynx (645)
esophagus (645)
stomach (645)
small intestine (646)
pancreas (646)

liver (646)
bile (646)
emulsification (647)
gallbladder (647)
villus, -li (647)
microvillus, -li (647)
large intestine, colon (648)
rectum (648)
anus (648)

obesity (649)
essential amino acids (649)
NPU, net protein
   utilization (651)
essential fatty acids (652)
vitamins (652)
minerals (653)
glucagon (655)
insulin (655)

Self-Quiz

_____ 1. The process that moves nutrients into the blood or lymph is _____.
a. ingestion
b. absorption
c. assimilation
d. digestion
e. none of the above

_____ 2. The enzymatic digestion of proteins begins in the _____.
a. mouth
b. stomach
c. liver
d. pancreas
e. small intestine

_____ 3. The enzymatic digestion of starches begins in the _____.
a. mouth
b. stomach
c. liver
d. pancreas
e. small intestine

_____ 4. The greatest amount of absorption of digested nutrients occurs in the _____.
a. stomach
b. pancreas
c. liver
d. colon
e. duodenum

_____ 5. Glucose moves through the membranes of the small intestine mainly by _____.
a. peristalsis
b. osmosis
c. diffusion
d. active transport
e. bulk flow

_____ 6. Which of the following is *not* found in bile?
a. lecithin
b. salts
c. digestive enzymes
d. cholesterol
e. pigments

_____ 7. The average American consumes approximately _____ pounds of sugar per year.
a. 25
b. 50
c. 75
d. 100
e. 125

_____ 8. Of the following, _____ has (have) the highest net protein utilization.
a. milk
b. eggs
c. fish
d. meat
e. bread

___ 9. Obesity is defined as being _____ percent above the ideal weight, which has been defined by insurance companies.
  a. 5
  b. 10
  c. 15
  d. 20
  e. 25

___ 10. A deficiency of vitamin _____ causes rickets in children and osteomalacia in adults.
  a. A
  b. B
  c. C
  d. D
  e. E

___ 11. The element needed by humans for blood clotting, nerve impulse transmission, and bone and tooth formation is _____.
  a. magnesium
  b. iron
  c. calcium
  d. iodine
  e. zinc

Chapter Objectives/Review Questions

This section lists general and detailed chapter objectives that can be used as review questions. You can make maximum use of these items by writing answers on a separate sheet of paper. To check for accuracy, compare your answers with information given in the chapter or glossary.

Page   *Objectives/Questions*

(641–642) 1. Distinguish between incomplete and complete digestive systems and tell which is characterized by (a) specialized regions, (b) two-way traffic, and (c) discontinuous feeding.
(642)   2. Define and distinguish among motility, secretion, digestion, and absorption.
(643–644) 3. List all parts (in order) of the human digestive system through which food actually passes. Then list the auxiliary organs that contribute one or more substances to the digestive process.
(644–648) 4. Explain how, during digestion, food is mechanically broken down. Then explain how it is chemically broken down.
(646–648) 5. Describe how the digestion and absorption of fats differ from the digestion and absorption of carbohydrates and proteins.
(647–648) 6. List the items that leave the digestive system and enter the circulatory system during the process of absorption.
(646)   7. Tell which foods undergo digestion in each of the following parts of the human digestive system and state what the food is broken into: oral cavity, stomach, small intestine, large intestine.
(646)   8. List the enzyme(s) that act in (a) the oral cavity, (b) the stomach, and (c) the small intestine. Then tell where each enzyme was originally made.
(647–648) 9. Describe the cross-sectional structure of the small intestine and explain how its structure is related to its function.
(648)   10. State which processes occur in the colon (large intestine).
(649, 651, 11. Compare the contributions of carbohydrates, proteins, and fats to human nutrition with the
652–653)   contributions of vitamins and minerals.
(648; 650) 12. Summarize the 1979 U. S. Surgeon General's report that presented ideas for promoting health by eating properly.
(648–649) 13. Summarize the daily nutritional requirements of a 25-year-old man who works at a desk job and exercises very little. State what he needs in energy, carbohydrates, proteins, and lipids and name at least six vitamins and six minerals that he needs to include in his diet every day.
(651–653) 14. Distinguish vitamins from minerals, and state what is meant by net protein utilization.
(653)   15. Name four minerals that are important in human nutrition and state the specific role of each.
(654)   16. List four functions of the liver.
(654–656) 17. Explain how the human body manages to meet the energy and nutritional needs of the various body parts even though the person may be feasting sometimes and fasting at other times.

Integrating and Applying Key Concepts

Suppose you could not eat solid food for two weeks and you had only water to drink. List in correct sequential order the measures your body would take to try to preserve your life. Mention the command signals that are given as one after another critical point is reached, and tell which parts of the body are the first and the last to make up for the deficit.

Critical Thinking Exercises

1. A rise in blood glucose elicits a rise in blood insulin. However, the insulin response is quicker and greater when a dose of glucose is given orally than when an equivalent dose is given intravenously. Which of the following substances is most likely the cause of the augmented response to an oral dose?

   a. Gastrin
   b. GIP
   c. Glucagon
   d. Glucose
   e. Secretin

   ANALYSIS

   a. Gastrin is secreted in response to amino acids in the stomach. It would not be expected to be involved in a response to glucose.
   b. GIP is released in response to the presence of glucose in the small intestine, and it stimulates release of insulin. An oral dose of glucose would reach the small intestine before it caused an increase in blood glucose. This intestinal glucose would stimulate GIP release; GIP in turn would stimulate insulin release in addition to the later release due to the glucose that was later absorbed into the bloodstream.
   c. Glucagon release would be inhibited by glucose, and glucagon does not directly affect insulin release. Glucagon cannot account for activation of an insulin response; it antagonizes glucose.
   d. Glucose is the initial stimulus for the release of insulin in this experiment. However, the problem is to account for different effects caused by equivalent doses of glucose. This requires a third substance. In fact, the problem is even more pronounced than it may appear; an oral dose of glucose would be absorbed into the blood over a period of time, whereas an intravenous dose is delivered more quickly and hence would produce a quicker, larger rise in blood glucose concentration. This would lead to the prediction that the insulin response would be greater to an intravenous dose than to an oral dose.
   e. Secretin does stimulate insulin secretion, indicating that it could be involved in this response. However, glucose does not stimulate secretin release; HCl in the stomach does.

2. When glucose is infused into the blood, insulin secretion is stimulated. Increased insulin secretion also is the response to intravenous infusion of amino acid solutions. However, blood glucose level does not drop as insulin rises during amino acid solutions, as you would expect. When an amino acid solution enters the blood, blood glucose concentration actually rises temporarily. Which of the following substances is most likely the cause of this response?

   a. Gastrin
   b. GIP
   c. Glucagon
   d. Glucose
   e. Secretin

   ANALYSIS

   a. Gastrin secretion is stimulated by amino acids, although in the stomach rather than in the blood, but gastrin does not affect blood glucose. It stimulates gastric motility and acid secretion.
   b. GIP stimulates insulin secretion and would be expected to lower, rather than raise, blood glucose. Furthermore, the stimulus to secrete GIP is glucose, not amino acids.
   c. Glucagon antagonizes insulin and raises blood glucose. If it was released in response to elevated amino acid concentration in the blood, it could produce the observed response. In fact, glucagon is

released when amino acids enter the blood. A meal of protein without carbohydrate would raise blood amino acid levels, would stimulate insulin, and could dangerously lower blood glucose. Glucagon prevents this hypoglycemia and is thus an especially important hormone for carnivores.

d. Glucose is not infused in this experiment. The rise in blood glucose is a response that we want to explain, not the stimulus.

e. Secretin stimulates insulin secretion and would be expected to cause an even greater drop in blood glucose, not an increase.

3. Tadpoles, which are herbivorous, have a much longer intestine than adult frogs, which are carnivorous. Which of the following predictions would be most likely from this observation?

a. Tadpoles require less energy than adult frogs.
b. Cellulose-digesting enzymes work more slowly than proteases.
c. The intestinal epithelium is more permeable in tadpoles than in adult frogs.
d. Tadpoles are smaller than frogs.
e. The caloric content of cellulose is greater than that of protein.

ANALYSIS

a. If tadpoles require less energy, then, assuming that the diets have about the same energy content per gram, the tadpole would have to digest less food. This would lead to the prediction that the tadpole gut would be shorter, not longer.

b. This would mean that vegetable food would have to be exposed to the appropriate enzymes for a longer time than the same amount of meat. Assuming that food moves through the gut at the same speed in both cases, the herbivore (the tadpole) would be expected to have a longer gut than the carnivore.

c. If the epithelium is more permeable, the necessary amount of digested food could be absorbed more quickly and thus would require a shorter gut. In addition, most of the monomers are absorbed by carrier-mediated processes, not by simple diffusion, so permeability is not a factor.

d. Having a longer gut would lead to the prediction of a larger body. However, there is no reason to assume a constant proportion of body size devoted to any single organ. A larger gut could be accommodated even in a smaller body.

e. If cellulose in plant cell walls had more calories per gram than meat, an herbivore would require fewer total grams of food and would be expected to have a smaller gut.

Answers

Answers to Interactive Exercises

TYPES OF DIGESTIVE SYSTEMS AND THEIR FUNCTIONS (37-I)
1. particles; 2. molecules; 3. absorbed; 4. discontinuous; 5. continuous; 6. incomplete; 7. circulatory; 8. complete; 9. opening; 10. Motility; 11. secretion.

HUMAN DIGESTIVE SYSTEM: AN OVERVIEW/ INTO THE MOUTH, DOWN THE TUBE (37-II)
1. a. Mouth; b. Salivary glands; c. Stomach; d. Small intestine; e. Pancreas; f. Liver; g. Gallbladder; h. Large intestine; i. Rectum; 2. amylase; 3. epiglottis; 4. esophagus; 5. peristalsis; 6. stomach; 7. small intestine; 8. Pepsin; 9. Bile; 10. small intestine; 11. Amylase (Lipase, Trypsin, or Chymotrypsin); 12. salivary glands; 13. oral cavity; 14. liver; 15. stomach; 16. gallbladder; 17. small intestine; 18. large intestine; 19. anus; 20. pancreas; 21. esophagus; 22. pharynx; 23. T; 24. F; 25. T; 26. T; 27. F.

HUMAN NUTRITIONAL REQUIREMENTS (37-III)
1. a. 2,070; b. 2,900; c. 1,230; 2. 25; 3. Complex carbohydrates; 4. 50 to 60; 5. Phospholipids; 6. energy reserves; 7. 30; 8. essential fatty acids; 9. Proteins; 10. essential; 11. milk (eggs); 12. eggs (milk); 13. Vitamins; 14. Minerals.

NUTRITION AND METABOLISM (37-IV)
1. constructing hormones, nucleotides, proteins, and enzymes; 2. monosaccharides, free fatty acids, and glycerol; 3. The three uses are (a) to construct components of cells and storage forms (such as glycogen) and specialized derivatives such as steroids and acetylcholine; (b) to convert to amino acids as needed; and (c) to serve as a source of energy; 4. ammonia; 5. urea; 6. insulin; 7. glucagon; 8. glycogen; 9. epinephrine (norepinephrine); 10. norepinephrine (epinephrine); 11. glycogen; 12. glucose; 13. glucose.

Answers to Self-Quiz

1. b; 2. b; 3. a; 4. e; 5. d; 6. c; 7. e; 8. b; 9. e; 10. d; 11. c.

38

CIRCULATION

Interactive Exercises

CIRCULATORY SYSTEM: AN OVERVIEW / CHARACTERISTICS OF BLOOD (38-I, pp. 659–664)

1. Complete the following table, which describes the components of blood.

Components	Relative Amounts	Functions
Plasma Portion *(50%–60% of total volume):*		
Water	91%–92% of plasma volume	Solvent
a._____ _____ (albumin, globulins, fibrinogen, etc.)	7%–8%	Defense, clotting, lipid transport, roles in extracellular fluid volume, etc.
Ions, sugars, lipids, amino acids, hormones, vitamins, dissolved gases	1%–2%	Roles in extracellular fluid volume, pH, etc.

Components	Relative Amounts	Functions
Cellular Portion *(40%–50% of total volume):*		
b._____ _____ _____	4,500,000–5,500,000 per microliter	O_2, CO_2 transport
White blood cells:		
c._____	3,000–6,750	Phagocytosis
d._____	1,000–2,700	Immunity
Monocytes (macrophages)	150–720	Phagocytosis
Eosinophils	100–360	Roles in inflammatory response, immunity
Basophils	25–90	Roles in inflammatory response, anticlotting
e._____	250,000–300,000	Roles in clotting

Fill-in-the-Blanks

Blood is a highly specialized fluid (2)_____ tissue that helps stabilize internal (3)_____ and equalize internal temperature throughout an animal's body. Organisms with (4)_____ circulatory systems generally also have a supplementary (5)_____ _____ _____ that recovers and purifies interstitial fluid and returns it to the major blood vessels. Oxygen binds with the (6)_____ atom in a hemoglobin molecule. The red blood (7)_____ _____ in males is about 5.4 million cells per microliter of blood; in females, it is 4.8 million per microliter. When oxygen levels in tissues are low (as they would be if you moved from Phoenix, Arizona, to the higher altitude of Santa Fe, New Mexico), the kidneys secrete a substance that combines with a plasma protein, converting it into the hormone known as (8)_____ , which stimulates the production of erythrocytes in the (9)_____ _____ _____.
(10)_____ _____ are immature cells not yet fully differentiated. (11)_____ and monocytes are highly mobile and phagocytic; they chemically detect, ingest, and destroy bacteria, foreign matter, and dead cells. (12)_____ (thrombocytes) are cell fragments that aid in forming blood clots.

In humans, red blood cells lack their (13)_____ , but they contain enough to sustain them for about (14)_____ months. Platelets also have no (15)_____ , but they last a maximum of (16)_____ days in the human bloodstream.

Label-Match

Identify the numbered cell types in the illustration below. Complete the exercise by matching and entering the letter of the appropriate function in the parentheses following the given cell types. A letter may be used more than once.

17. _____

18. _____ _____ ( )

19. _____ ( )

20. _____ ( )

21. _____ ( )

22. _____ ( )

23. _____

24. _____ ( )

A. Phagocytosis
B. Plays a role in the inflammatory response
C. Plays a role in clotting
D. Immunity
E. O_2, CO_2 transport

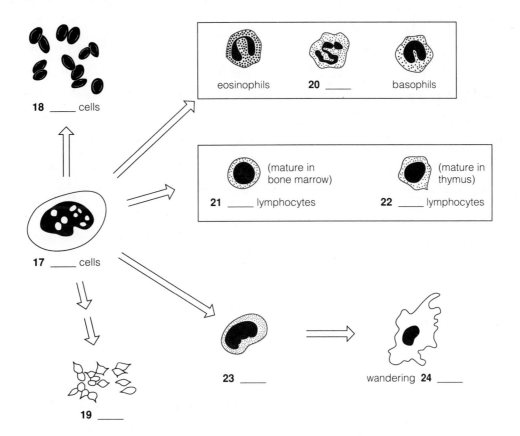

CARDIOVASCULAR SYSTEM OF VERTEBRATES (38-II, pp. 664–675)

Fill-in-the-Blanks

A receiving zone of a vertebrate heart is called a(n) (1)_____ ; a departure zone is called a(n) (2)_____. Each contraction period is called (3)_____ ; each relaxation period is (4)_____. The heart is a pumping station for two major blood transport routes: the (5)_____ circulation to and from the lungs, and the (6)_____ circulation to and from the rest of the body.

In the pulmonary circuit, the heart pumps (7)_____-poor blood to the lungs; then the (8)_____-enriched blood flows back to the (9)_____. The (10)_____ _____ is the cardiac pacemaker. During a cardiac cycle, contraction of the (11)_____ is the driving force for blood circulation; (12)_____ contraction helps fill the ventricles.

A(n) (13)_____ carries blood away from the heart. A(n) (14)_____ is a blood vessel with such a small diameter that red blood cells must flow through it single-file; its wall consists of no more than a single layer of (15)_____ cells resting on a basement membrane. In each (16)_____ _____ , small molecules move between the bloodstream and the (17)_____ fluid. (18)_____ are in the walls of veins and prevent backwashing. Both (19)_____ and (20)_____ serve as temporary reservoirs for blood volume.

(21)_____ are pressure reservoirs that keep blood flowing away from the heart while the (22)_____ are relaxing. (23)_____ are control points where adjustments can be made in the volume of blood flow to be delivered to different capillary beds. They offer great resistance to flow, so there is a major drop in (24)_____ in these tubes.

One cause of (25)_____ is the rupture of one or more blood vessels in the brain. A (26)_____ _____ blocks a coronary artery. (27)_____ is a term for a formation that can include cholesterol, calcium salts, and fibrous tissue. It is not healthful to have a high concentration of (28)_____-density lipoproteins in the bloodstream. The (29)_____ _____ is the principal organ that controls blood pressure in the entire cardiovascular system. Bleeding is stopped by several mechanisms that are referred to as (30)_____ ; the mechanisms include blood vessel spasm, (31)_____ _____ _____ , and blood (32)_____. Once the platelets reach a damaged vessel, through chemical recognition they adhere to exposed (33)_____ fibers in damaged vessel walls. This response is called the (34)_____ _____ _____. Reactions form the enzyme thrombin, which causes rod-shaped proteins (fibrinogen) to assemble into long, insoluble (35)_____ fibers. These trap blood cells and components of plasma. Under normal conditions, a clot eventually forms at the damaged site.

If you are blood type (36)_____ , you have no antibodies against A or B markers on red blood cell surfaces. If you are type (37)_____ , you have antibodies against A and B markers. (38)_____ is a response in which antibodies act against "foreign" cells bearing specific markers and cause them to clump together.

True-False

If false, explain why.

___ 39. The pulse rate is the difference between the systolic and the diastolic pressure readings.

___ 40. Because the total volume of blood remains constant in the human body, blood pressure must also remain constant throughout the circuit.

Labeling

Identify each indicated part of the accompanying illustrations.

41. _____

42. _____ _____ _____

43. _____ _____

44. _____ _____

45. _____ _____ _____

46. _____ _____

47. _____ _____ _____

48. _____ _____ _____

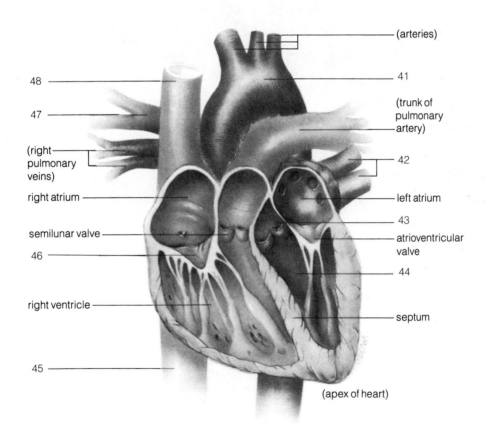

49. _____

50. _____

51. _____

52. _____

53. _____ _____ , _____ _____

54. _____

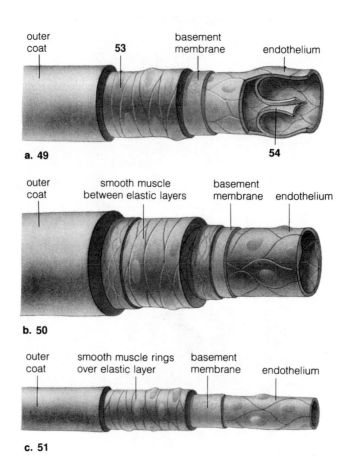

outer
coat   **53**   basement
membrane   endothelium

a. 49

54

outer
coat   smooth muscle
between elastic layers   basement
membrane   endothelium

b. 50

outer
coat   smooth muscle rings
over elastic layer   basement
membrane   endothelium

c. 51

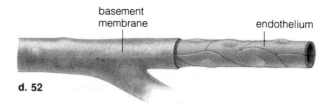

basement
membrane   endothelium

d. 52

LYMPHATIC SYSTEM (38-III, p. 676)

Labeling

Identify each indicated part of the accompanying illustration.

1. _____

2. _____ _____ _____

3. _____

4. _____ _____

5. _____

6. _____ _____

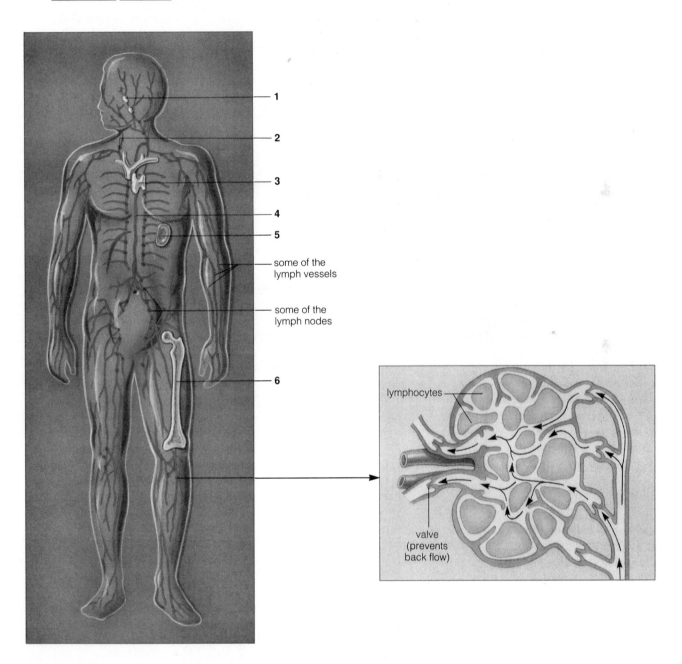

some of the
lymph vessels

some of the
lymph nodes

lymphocytes

valve
(prevents
back flow)

Fill-in-the-Blanks

(7)_____ vessels reclaim fluid lost from the bloodstream, purify the blood of microorganisms, and transport (8)_____ from the (9)_____ _____ to the bloodstream.

Chapter Terms

The following page-referenced terms are the important terms in the chapter. Refer to the instructions given in Chapter 1, p. 4.

circulatory system (659)	platelets (664)	arteries (667)	vasoconstriction (673)
blood (660)	pulmonary circuit (665)	arterioles (668)	hemostasis (674)
heart (660)	systemic circuit (665)	capillary (668)	antibodies (674)
blood vessels (660)	atrium, -ia (665)	venules (670)	agglutination (675)
lymphatic system (660)	ventricle (665)	veins (670)	lymphatic system (676)
red blood cells (662)	cardiac cycle (666)	hypertension (670)	lymph (676)
plasma portion (662)	cardiac conduction	atherosclerosis (670)	lymph vascular system (676)
cellular portion (662)	system (666)	arrhythmia (672)	lymphoid organs (676)
white blood cells (663)	blood pressure (667)	vasodilation (673)	

Self-Quiz

____ 1. Most of the oxygen in human blood is transported by _____.
  a. plasma
  b. serum
  c. platelets
  d. hemoglobin
  e. leukocytes

____ 2. When the oxygen level in human tissues is low, erythropoietin production is stimulated by enzymes secreted by the _____ ; the red blood cell count then increases.
  a. spleen
  b. lungs
  c. liver
  d. pancreas
  e. kidneys

____ 3. Open circulatory systems generally lack _____.
  a. a heart
  b. arterioles
  c. capillaries
  d. veins
  e. arteries

____ 4. Red blood cells originate in the _____.
  a. liver
  b. spleen
  c. yellow bone marrow
  d. thymus gland
  e. red bone marrow

____ 5. Hemoglobin contains _____.
  a. copper
  b. magnesium
  c. sodium
  d. calcium
  e. iron

____ 6. The pacemaker of the human heart is the _____.
  a. sinoatrial node
  b. semilunar valve
  c. inferior vena cava
  d. superior vena cava
  e. atrioventricular node

___ 7. During systole, _____.
   a. oxygen-rich blood is pumped to the lungs
   b. the heart muscle tissues contract
   c. the atrioventricular valves suddenly open
   d. oxygen-poor blood from all parts of the human body, except the lungs, flows toward the right atrium
   e. none of the above

___ 8. _____ are reservoirs of blood pressure in which resistance to flow is low.
   a. Arteries
   b. Arterioles
   c. Capillaries
   d. Venules
   e. Veins

___ 9. Which of the following is the proper sequence in clotting?
   a. prothrombin, fibrinogen, fibrin, thrombin, clot
   b. thrombin, prothrombin, fibrin, fibrinogen, clot
   c. prothrombin, fibrinogen, thrombin, fibrin, clot
   d. prothrombin, thrombin, fibrinogen, fibrin, clot
   e. prothrombin, fibrin, fibrinogen, thrombin, clot

___ 10. The lymphatic system is the principal avenue in the human body for transporting _____.
   a. fats
   b. wastes
   c. carbon dioxide
   d. amino acids
   e. interstitial fluids

Chapter Objectives/Review Questions

This section lists general and detailed chapter objectives that can be used as review questions. Answer them on a separate sheet of paper. To check for accuracy, compare your answers with information given in the chapter or glossary.

Page   *Objectives/Questions*

(660)   1. Distinguish between open and closed circulatory systems.
(660–664) 2. Describe the composition and functions of blood.
(662)   3. Describe the composition of human blood, using percentages of volume.
(662, 663) 4. Distinguish the five types of leukocytes from each other in terms of structure and functions.
(662–664) 5. State where erythrocytes, leukocytes, and platelets are produced.
(665)   6. Trace the path of blood in the human body. Begin with the aorta and name all major components of the circulatory system through which the blood passes before it returns to the aorta.
(666–667) 7. Explain what causes a heart to beat. Then describe how the rate of heartbeat can be slowed down or speeded up.
(666–671) 8. List the factors that cause blood to leave the heart and the factors that cooperate to return blood to the heart.
(667–668) 9. Explain what causes high pressure and low pressure in any animal circulatory system. Then relate this to the human circulatory system by explaining where major drops in blood pressure occur.
(667)   10. Describe how the structures of arteries, capillaries, and veins differ.
(670)   11. Explain how veins and venules can act as reservoirs of blood volume.
(670, 672) 12. Distinguish a stroke from a coronary occlusion.
(671)   13. Describe how hypertension develops, how it is detected, and whether it can be corrected.

Page	*Objectives/Questions*
(672)	14. State the significance of high- and low-density lipoproteins to cardiovascular disorders.
(674)	15. List in sequence the events that occur in the formation of a blood clot.
(674–675)	16. Describe how blood is typed for the ABO blood group and for the Rh factor.
(676)	17. Describe the composition and function of the lymphatic system.

Integrating and Applying Key Concepts

You observe that some people appear as though fluid had accumulated in their lower legs and feet. Their lower extremities resemble those of elephants. You inquire about what is wrong and are told that the condition is caused by the bite of a mosquito that is active at night. Construct a testable hypothesis that would explain (1) why the fluid was not being returned to the torso, as normal, and (2) what the mosquito did to its victims.

Critical Thinking Exercises

1. In the mammalian circulatory system, blood passes through four cardiac chambers in a series. Which of the following would be the most likely measurements of oxygen concentration in each of the four chambers?

	Right Atrium	*Left Atrium*	*Right Ventricle*	*Left Ventricle*
a.	40	40	40	40
b.	100	100	40	40
c.	100	40	100	40
d.	40	100	40	100
e.	40	40	100	100

ANALYSIS

Blood flows through the right side of the heart and then to the lungs, where oxygen concentration rises. Then blood returns to the left side of the heart and flows through it to the systemic organs, which consume oxygen, lowering the oxygen concentration in the blood. Thus, blood flowing through the two chambers of one side of the heart has the same oxygen concentration in both chambers. Furthermore, the blood in the left side of the heart has high oxygen concentration, and the blood in the right side has low oxygen concentration. The only choice that meets these criteria is (d).

2. Consider three flow patterns found in animal circulatory systems as shown in the diagrams below. Assume that in all three systems the rate of blood flow is the same at point B and that the supply and consumption of oxygen are the same in all three. Which of the following statements about the volume flow rate of blood in these systems would most likely be valid?

   a. Flows are equal at A, B, and C in system 3.
   b. Flows are equal at B and C in system 2.
   c. Flows are equal at A in all three systems.
   d. Flows are equal at A and C in system 2.
   e. Flows are equal at C in all three systems.

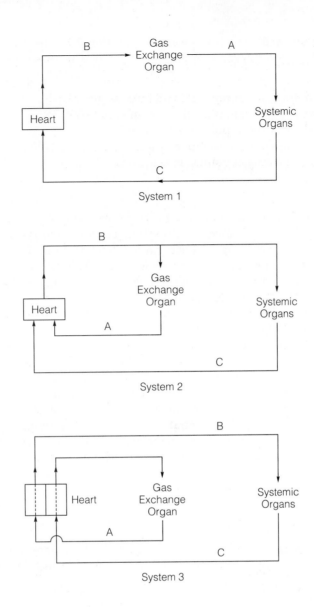

System 1

System 2

System 3

ANALYSIS

a. System 3 is the mammalian system and can be thought of as a single continuous loop. If blood flow, on the average, is different at any two points, blood will accumulate in one part of the system and be depleted in another. Thus, on the average, blood flows at all three of these positions will be expected to be equal.

b. In system 2, the flow from position B branches, with part of it going to A and part to C. Thus, the flow at C will be less than the flow at B.

c. The question assumes that flows at B are equal in all three systems. Systems 1 and 2 are single loops, and flow at any point must equal flow at all other points. However, flow at B in system 2 divides into two parts; hence, in this system, flow at A is less than that at B and less than that at A in the other two systems.

d. Because positions A and C in system 2 are on parallel loops, flow past them could be equal but does not have to be.

e. The argument in (c) also applies to this choice.

3. Use the circulatory systems and the assumptions of question 2 to answer this question: Which of the following statements about the oxygen concentration in blood in these systems would most likely be valid?

   a. The concentrations are the same at point B in all three systems.
   b. The concentrations are the same at points B and C in system 3.
   c. The concentrations are the same at point A in systems 1 and 3.
   d. The concentrations are the same at point B in systems 1 and 2.
   e. The concentrations are the same at point C in systems 2 and 3.

ANALYSIS

   a. System 1 pumps used blood through point B to the gas exchange organ. This blood has a low oxygen concentration. System 2 pumps a mixture of oxygenated and deoxygenated blood through point B. This blood has an intermediate oxygen concentration. System 3 pumps fully oxygenated blood from the gas exchange organ past point B to the systemic organs. This blood has high oxygen concentration.
   b. In system 3, the blood flows from point B through the systemic organs to point C. The systemic organs consume oxygen, lowering the oxygen concentration in the blood.
   c. In both systems, point A contains blood that has just come from the gas exchange organ and contains the highest concentration of oxygen. Given the assumptions stipulated in the question, the oxygen concentration should be the same at both of these points (and at point A in system 2 as well).
   d. The argument in (a) establishes why these concentrations should not be equal.
   e. The blood at both of these points has just left the systemic organs and has a low oxygen concentration. Because the problem stipulates that oxygen consumption rate is the same in all three systems, you might conclude that the oxygen concentration should be the same at both of these points. However, both the flow rate and the oxygen concentration entering the systemic organs are lower in system 2 than in system 3; thus, the supply of oxygen to the organs is lower in system 2. If the supply is less and the consumption is the same, the concentration left in the blood leaving the organs will be less.

Answers

Answers to Interactive Exercises

CIRCULATORY SYSTEM: AN OVERVIEW/ CHARACTERISTICS OF BLOOD (38-I)
1. a. Plasma proteins; b. Red blood cells; c. Neutrophils; d. Lymphocytes; e. Platelets; 2. connective; 3. pH; 4. closed; 5. lymph vascular system; 6. iron; 7. cell count; 8. erythropoietin; 9. red bone marrow; 10. Stem cells; 11. Neutrophils; 12. Platelets; 13. nucleus; 14. four; 15. nucleus; 16. nine; 17. stem; 18. red blood (E); 19. platelets (C); 20. neutrophils (A); 21. B (D); 22. T (D); 23. monocytes; 24. macrophages (A).

CARDIOVASCULAR SYSTEM OF VERTEBRATES (38-II)
1. atrium; 2. ventricle; 3. systole; 4. diastole; 5. pulmonary; 6. systemic; 7. oxygen; 8. oxygen; 9. heart; 10. SA node; 11. ventricles; 12. atrial; 13. artery; 14. capillary; 15. endothelial; 16. capillary bed (diffusion zone); 17. interstitial; 18. Valves; 19. veins (venules); 20. venules (veins); 21. Arteries; 22. ventricles; 23. Arterioles; 24. pressure; 25. stroke; 26. coronary occlusion; 27. Plaque; 28. low; 29. medulla oblongata; 30. hemostasis; 31. platelet plug formation; 32. coagulation; 33. collagen; 34. intrinsic clotting mechanism; 35. fibrin; 36. AB; 37. O; 38. Agglutination; 39. F; 40. F; 41. aorta; 42. left pulmonary veins; 43. semilunar valve; 44. left ventricle; 45. inferior vena cava; 46. atrioventricular valve; 47. right pulmonary artery; 48. superior vena cava; 49. vein; 50. artery; 51. arteriole; 52. capillary; 53. smooth muscle, elastic fibers; 54. valve.

LYMPHATIC SYSTEM (38-III)
1. tonsils; 2. right lymphatic duct; 3. thymus; 4. thoracic duct; 5. spleen; 6. bone marrow; 7. Lymph; 8. fats; 9. small intestine.

Answers to Self-Quiz

1. d; 2. e; 3. c; 4. e; 5. e; 6. a; 7. b; 8. a; 9. d; 10. a.

39

IMMUNITY

Interactive Exercises

NONSPECIFIC DEFENSE RESPONSES / SPECIFIC DEFENSE RESPONSES: THE IMMUNE SYSTEM (39-I, pp. 682–685)

Fill-in-the-Blanks

Among single-celled organisms, (1)_____ was a means of ingesting food that probably proved adaptive in defense. Twenty plasma proteins that are activated one after another in a "cascade" of reactions to help you destroy invading microorganisms are collectively referred to as the (2)_____ _____. (3)_____ are Y-shaped proteins that lock onto specific foreign targets and thereby tag them for destruction by phagocytes or by activating the complement system.

Several barriers prevent pathogens from crossing the boundaries of your body. Intact skin and (4)_____ membranes are effective barriers. (5)_____ is an enzyme that destroys the cell wall of many bacteria. (6)_____ fluid destroys many food-borne pathogens in the gut. Normal (7)_____ residents of the skin, gut, and vagina outcompete pathogens for resources and help keep their numbers under control.

The inflammatory response engages in battle both specific and nonspecific invaders. When the complement system is activated, circulating basophils and mast cells in tissues release (8)_____ , which dilates (9)_____ and makes them "leaky," so fluid seeps out. (10)_____ mechanisms help wall off infected or damaged tissues and work to keep blood vessels intact.

Labeling

Identify each numbered part in the accompanying illustration.

11. _____   15. _____

12. _____   16. _____

13. _____ _____ _____   17. _____

14. _____   18. _____

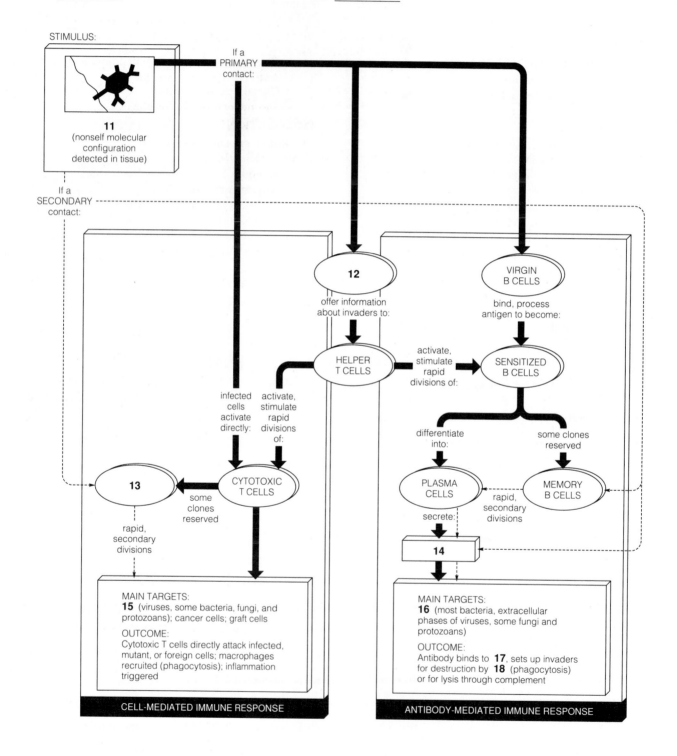

Matching

Match each of the white blood cell types with the appropriate function description.

19. _____ B cells

20. _____ cytotoxic T cells

21. _____ helper T cells

22. _____ macrophages

23. _____ memory cells

24. _____ suppressor T cells

A. Lymphocytes that directly destroy body cells already infected by certain viruses or by parasitic fungi

B. Lymphocytes that serve as master switches of the immune system; stimulate the rapid division of B cells and cytotoxic T cells

C. Lymphocytes that slow down or prevent immune responses

D. Nonlymphocytic white blood cells that develop from monocytes, engulf anything perceived as foreign, and alert helper T cells to the presence of *specific* foreign agents

E. Lymphocytes and their progeny that produce antibodies

F. A portion of B and T cell populations that were set aside as a result of a first encounter, now circulate freely and respond rapidly to any later attacks by the same type of invader

SPECIFIC DEFENSE RESPONSES: THE IMMUNE SYSTEM (cont.) / IMMUNIZATION (39-II, pp. 686–692)

Fill-in-the-Blanks

The (1)_____ _____ _____ explains how an individual has immunological memory, which is the basis of a secondary immune response; the theory also explains in part how *self* cells are distinguished from (2)_____ cells in the vertebrate immune response. Two distinct (3)_____ populations carry out specific immune responses: (4)_____ _____ and (5)_____ _____. Both kinds arise from (6)_____ _____ in bone marrow. Some of these progeny move to the thymus, where they acquire specific (7)_____ on their cell surfaces; in doing so, they become (8)_____ _____. (9)_____ _____ clones indirectly attack their targets through the production of antibodies. The (10)_____ _____ _____ is the route taken during a first-time contact with an antigen. Deliberately provoking the production of memory lymphocytes is known as (11)_____. (12)_____ refers to cells that have lost controls over cell division. Milstein and Kohler developed a means of producing large amounts of (13)_____ _____. (14)_____ drug therapy hooks up anticancer drug molecules with monoclonal antibodies, which locate precisely where the cancer cells are and deliver the drug to them only.

ABNORMAL OR DEFICIENT IMMUNE RESPONSES (39-III, pp. 692–695)

Fill-in-the-Blanks

(1)_____ is an altered secondary response to a normally harmless substance that may actually cause injury to tissues. (2)_____ _____ is a disorder in which the body mobilizes its forces against certain of its own tissues. AIDS is a constellation of disorders that follow infection by the (3)_____ _____ _____. In the United States, transmission has occurred most often among intravenous drug abusers who share needles and among (4)_____ _____. HIV is a (5)_____ ; its genetic

material is RNA rather than DNA, and it has several copies of an enzyme (6)_____ _____ ,
which uses the viral RNA as a template for making DNA, which is then inserted into a host chromosome.
Aroused macrophages war against any invaders, including cancer cells. (7)_____ are a group of small
proteins that perform antiviral activity and induce resistance to a wide range of viruses.

Matching

Match each letter with its best mate.

8. ___ allergy

9. ___ antibody

10. ___ antigen

11. ___ macrophage

12. ___ clone

13. ___ complement

14. ___ histamine

15. ___ MHC marker

16. ___ plasma cell

17. ___ T cell

A. Begins its development in bone marrow, but matures in the thymus gland
B. Cell that has directly or indirectly descended from the same parent cell
C. A potent chemical that causes blood vessels to dilate and let protein pass through the vessel walls
D. Y-shaped immunoglobulin
E. A nonself marker
F. The progeny of turned-on B cells
G. A group of about fifteen proteins that participate in the inflammatory response
H. An altered secondary immune response to a substance that is normally harmless to other people
I. The basis for self-recognition at the cell surface
J. Principal perpetrator of phagocytosis

Chapter Terms

The following page-referenced terms are the important terms in the chapter. Refer to the instructions given in Chapter 1, p. 4.

pathogens (682)
phagocytes (682)
complement system (682)
lysis (682)
inflammatory response (682)
histamine (683)
immune system (685)
macrophages (685)
B cells (685)
cytotoxic T cells (685)
helper T cells (685)

suppressor T cells (685)
memory cells (685)
MHC markers (685)
antigen (685)
primary immune
    response (686)
antibody-mediated
    response (686)
antibody (686)
plasma cells (686)

interleukin (686)
immunoglobulins (687)
perforins (688)
natural killer (NK) cells (688)
cancer (688)
immune therapy (688)
interferons (688)
monoclonal antibodies (688)
clonal selection theory (691)
clone (691)

secondary immune
    response (691)
memory cell (691)
immunization (692)
vaccine (692)
passive immunity (692)
allergies (692)
autoimmune response (693)
AIDS, acquired immune
    deficiency syndrome (693)

Self-Quiz

_____ 1. All the body's phagocytes are derived from stem cells in the _____.
   a. spleen
   b. liver
   c. thymus
   d. bone marrow
   e. thyroid

_____ 2. The plasma proteins that are activated when they contact a bacterial cell are collectively known as the _____ system.
   a. shield
   b. complement
   c. Ig G
   d. MHC
   e. HIV

_____ 3. _____ are divided into two groups: T cells and B cells.
   a. Macrophages
   b. Lymphocytes
   c. Platelets
   d. Complement cells
   e. Cancer cells

_____ 4. _____ produce and secrete antibodies that set up bacterial invaders for subsequent destruction by macrophages.
   a. B cells
   b. Phagocytes
   c. T cells
   d. Bacteriophages
   e. Thymus cells

_____ 5. Antibodies are shaped like the letter _____.
   a. Y
   b. W
   c. Z
   d. H
   e. E

_____ 6. The markers for every cell in the human body are referred to by the letters _____.

   a. HIV
   b. MBC
   c. RNA
   d. DNA
   e. MHC

_____ 7. Plasma cells _____.
   a. die within a week of being produced
   b. develop from B cells
   c. manufacture and secrete antibodies
   d. do not divide and form clones
   e. all of the above

_____ 8. Clones of B or T cells are _____.
   a. being produced continually
   b. sometimes known as memory cells if they keep circulating in the bloodstream
   c. only produced when their surface proteins recognize other specific proteins previously encountered
   d. produced and mature in the bone marrow
   e. both (b) and (c)

_____ 9. Whenever the body is reexposed to a specific sensitizing agent, IgE antibodies cause _____.
   a. prostaglandins and histamine to be produced
   b. clonal cells to be produced
   c. histamine to be released
   d. the immune response to be suppressed
   e. none of the above

_____ 10. The leading cause of death among transplant patients is _____.
   a. failure of the MHC in plasma cells to do its work
   b. pneumocystis infections
   c. loss of a vital organ when the transplant fails
   d. an excessive number of antigens being released into the bloodstream
   e. a transplant reaction similar to that which causes blood transfusion deaths

Chapter Objectives/Review Questions

This section lists general and detailed chapter objectives that can be used as review questions. You can make maximum use of these items by writing answers on a separate sheet of paper. Fill in answers where blanks are provided. To check for accuracy, compare your answers with information given in the chapter or glossary.

Page   *Objectives/Questions*

(682)   1. Describe typical external barriers that organisms present to invading organisms.
(682–684) 2. List and discuss four nonspecific defense responses that serve to exclude microbes from the body.
(682–683) 3. Explain how the complement system is related to an inflammatory response.
(684–685) 4. Distinguish the roles of T cells from the roles of B cells.
(685)   5. List the three general types of cells that form the basis of the vertebrate immune system.
(685)   6. Understand how vertebrates (especially mammals) recognize and discriminate between self and nonself tissues.
(685–686) 7. Describe how recognition proteins and antibodies are made. State how they are used in immunity.
(686–688) 8. Distinguish between the antibody-mediated response pattern and the cell-mediated response pattern.
(686; 691) 9. Explain what is meant by primary immune pathway as contrasted with secondary immune pathway.
(688–689) 10. Explain what monoclonal antibodies are and tell how they are currently being used in passive immunization and cancer treatment.
(690–691) 11. Describe the clonal selection theory and tell what it helps to explain.
(692)   12. Describe two ways that people can be immunized against specific diseases.
(692–693) 13. Distinguish allergy from autoimmune disease.
(693)   14. Describe some examples of immune failures and identify as specifically as you can which weapons in the immunity arsenal failed in each case.
(692–695) 15. Describe how AIDS specifically interferes with the human immune system.

Integrating and Applying Key Concepts

Suppose you wanted to get rid of forty-seven warts that you have on your hands by treating them with monoclonal antibodies. Outline the steps you would have to take.

Critical Thinking Exercises

1. An experiment was designed to test the effectiveness of lymphocytes from various organs in the immune response. The immune system of a mouse was suppressed by radiation, which kills all the lymphocyte precursor cells. The irradiated mouse was then injected with cells from organs of an intact mouse along with an antigen. The production of antibodies to that antigen was measured in the recipient mouse. The results follow.

Source of Injected Cells	Relative Level of Response
None	7.4
Thymus	12.3
Bone marrow	1.6
Both thymus and marrow	70.7

Reference: Claman et al. (1966) *Proc. Soc. Exp. Biol. Med.* 122:1167.

The conclusion was that the lymphocytes from thymus and bone marrow interacted. One cell type stimulated the other to produce antibodies so that the response of the combined cells was greater than the sum of the responses of the two types tested alone. What pattern of responses would have resulted from each of the following mechanisms?

a. Bone marrow cells produce antibodies independently, and thymus cells do not participate.
b. Thymus cells inhibit the production of antibodies by bone marrow cells.
c. Both bone marrow cells and thymus cells independently produce antibodies.
d. Both bone marrow cells and thymus cells produce antibodies, and both require stimulation from the other.
e. The same combination of cell types is found in both organs and produces antibodies.

ANALYSIS

Mechanism

Source of Injected Cells	a	b	c	d	e
None	low	low	low	low	low
Thymus	low	low	X	low	X
Bone marrow	X	X	X	low	X
Both thymus and marrow	X	low	2X	high	2X

a. Bone marrow cells produce the same amount of antibody with or without thymus cells.
b. Bone marrow cells produce antibody when they are injected alone but are inhibited by thymus cells.
c. Both cell types produce the same amount of antibody alone or in combination. When both cell types are injected, the response is the sum of the two single responses.
d. The cell types synergize in combination. The combined response is greater than the sum of the single responses.
e. As in (c), cells from both sources produce the same amount of antibody alone or in combination, so the combined response is the sum of the single responses.

2. Mice of an inbred strain are all almost genetically identical; that is, they are homozygous for the same alleles of almost all their genes. A mouse of inbred strain 1 is infected with virus X. A few days later, activated Tc lymphocytes are isolated from this mouse. Which of the following kinds of cells will these lymphocytes be able to kill?

a. Normal cells from strain 1 mice
b. Strain 1 cells infected with virus X
c. Strain 1 cells infected with virus Y
d. Normal cells from strain 2 mice
e. Strain 2 cells infected with virus X
f. Strain 2 cells infected with virus Y

ANALYSIS

These Tc lymphocytes are activated to recognize their own strain 1 MHC antigens in combination with virus X antigens. They will respond only to that combination of antigens. They will not kill cells from strain 2 at all. They will not kill strain 1 cells that are uninfected or infected with virus Y. They are specific for the cells of choice (b).

3. Among adult humans, 1 to 5 percent show an allergic response to penicillin. The response is caused by the presence of a specific protein antibody in their blood. One hypothesis is that penicillin causes a mutation in each individual and the mutant allele codes for the antibody. Which of the following is the best alternative explanation?

a. The allele for the antibody is dominant.
b. Penicillin treatment exerts a strong selective pressure against the mutant allele.
c. Allergic responses lead to penicillin resistance.
d. The allele for the antibody is already present in some individuals before penicillin exposure.
e. Penicillin exposure destroys the allele for the antibody.

ANALYSIS

a. The allele would be dominant, because even in an individual heterozygous for an inactive protein the active antibody would be present. However, this is not an alternative explanation for the source of the allele.
b. Because allergic reactions to penicillin can be lethal, this is a valid statement, but again it is not an alternative explanation.
c. This statement is meaningless, because the allergic response is a phenomenon in humans and penicillin resistance is a property of some bacteria.
d. This would explain why some individuals are allergic and others are not. The ultimate source of the allele would have to have been mutation some time in the past, not caused by penicillin exposure.
e. If penicillin destroyed the allele for the antibody, penicillin exposure would prevent, not trigger, the allergic response.

Note: Please see authors' note on page 371.

Answers

Answers to Interactive Exercises

NONSPECIFIC DEFENSE RESPONSES/ SPECIFIC DEFENSE RESPONSES: THE IMMUNE SYSTEM (39-I)
1. phagocytosis; 2. complement system; 3. Antibodies; 4. mucous; 5. Lysozyme; 6. Gastric; 7. bacterial; 8. histamine; 9. capillaries; 10. Clotting; 11. antigen; 12. macrophages; 13. memory T cells; 14. antibodies; 15. intracellular; 16. extracellular; 17. antigen; 18. macrophages; 19. E; 20. A; 21. B; 22. D; 23. F; 24. C.

SPECIFIC DEFENSE RESPONSES: THE IMMUNE SYSTEM/IMMUNIZATION (39-II)
1. clonal selection theory; 2. nonself; 3. lymphocyte; 4. B cells (T cells); 5. T cells (B cells); 6. stem cells; 7. markers; 8. T cells; 9. B cell; 10. primary immune response; 11. immunization; 12. Cancer; 13. monoclonal (pure) antibodies; 14. Targeted.

ABNORMAL OR DEFICIENT IMMUNE RESPONSES (39-III)
1. Allergy; 2. Autoimmune disease; 3. human immunodeficiency virus; 4. male homosexuals; 5. retrovirus; 6. reverse transcriptase; 7. Interferons; 8. H; 9. D; 10. E; 11. J; 12. B; 13. G; 14. C; 15. I; 16. F; 17. A.

Answers to Self-Quiz

1. d; 2. b; 3. b; 4. a; 5. a; 6. e; 7. e; 8. e; 9. a; 10. b.

40

RESPIRATION

Interactive Exercises

THE NATURE OF RESPIRATORY SYSTEMS (40-I, pp. 700–704)

Fill-in-the-Blanks

A (1)_____ is an outfolded, thin, moist membrane endowed with blood vessels. Gas transfer is enhanced by (2)_____ _____ , in which water flows past the bloodstream in the opposite direction. Insects have (3)_____ (chitin-lined air tubes leading from the body surface to the interior). (4)_____ are openings at the body surface of land arthropods. At sea level, atmospheric pressure is approximately (5)_____ mm Hg, and oxygen represents about (6)_____ percent of the total volume. Lungs originated as pockets off the front of the (7)_____. In some fishes, lung sacs became modified into (8)_____ _____ (buoyancy devices that help keep the fish from sinking).

    The energy to drive animal activities comes mainly from (9)_____ _____ , which uses (10)_____ and produces (11)_____ _____ wastes. In a process called (12)_____ , animals move (10) into their internal environment and give up (11) to the external environment.

All respiratory systems make use of the tendency of any gas to diffuse down its (13)_____

_____. Such a (13) exists between (10) in the atmosphere [(14)_____ pressure] and the

metabolically active cells in body tissues [where (10) is used rapidly; pressure is (15) [ ] highest [ ] lowest

here]. Another (13) exists between (11) in body tissues (16) [[ ] high [ ] low] pressure and the

atmosphere, with its (17) [ ] higher [ ] lower amount of (11).

According to (18)_____ _____ , the amount of a gas diffusing across a respiratory surface

in a given time depends on the (19)_____ _____ of the membrane and the differences in

(20) _____ _____ across it. (21)_____ is an important transport pigment, which can bind

loosely with as many as four O_2 molecules in the lungs.

A(n) (22)_____ is an internal respiratory surface in the shape of a cavity or sac. In all lungs,

(23)_____ carry gas to and from one side of the respiratory surface, and (24)_____ _____

carry gas to and from the other side.

Labeling

Identify the numbered parts of the accompanying illustration, which shows the respiratory system of many
fishes.

25. _____

26. _____

27. _____-_____ _____

28. _____-_____ _____

29. _____

30. _____

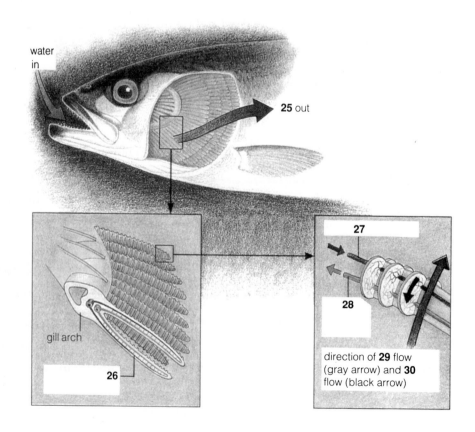

water
in

25 out

27

28

direction of **29** flow
(gray arrow) and **30**
flow (black arrow)

gill arch

26

HUMAN RESPIRATORY SYSTEM / AIR PRESSURE
CHANGES IN THE LUNGS (40-II, pp. 704–708)

Fill-in-the-Blanks

During inhalation, the (1)_____ moves downward and flattens, and the (2)_____ _____

moves outward and upward; when these things happen, the chest cavity volume (3) [choose one] ( )

increases, ( ) decreases, and the internal pressure (4) [choose one] ( ) rises, ( ) drops, ( ) stays the same.

Every time you take a breath, you are (5)_____ the respiratory surfaces of your lungs. The

(6)_____ _____ surrounds each lung. In succession, air passes through the nasal cavities,

pharynx, and (7)_____ , past the epiglottis into the (8)_____ (the space between the true vocal

cords), into the trachea, and then to the (9)_____ , (10)_____ , and alveolar ducts. Exchange of

gases occurs across the epithelium of the (11)_____ .

Labeling

Identify each indicated part of the accompanying illustration.

12. _____ _____   19. _____

13. _____   20. _____ _____

14. _____   21. _____ _____

15. _____   22. _____ _____

16. _____ _____   23. _____

17. _____   24. _____ (sing), _____ (plural)

18. _____   25. _____

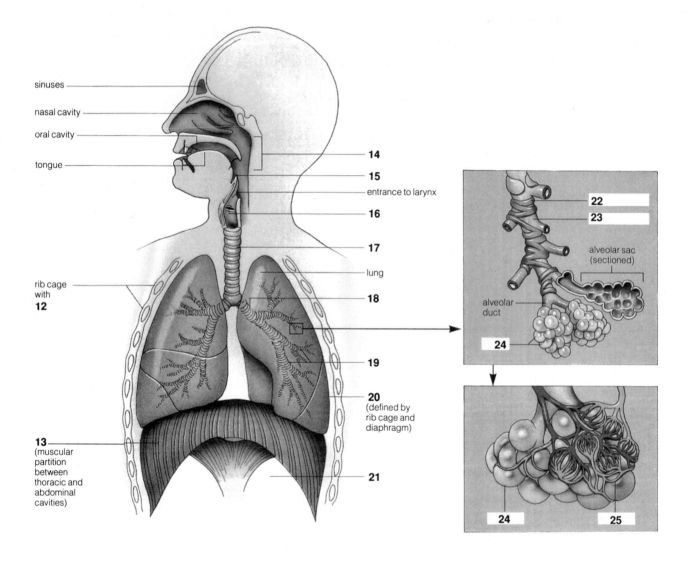

sinuses

nasal cavity

oral cavity

tongue

14

15

entrance to larynx

16

17

lung

18

rib cage with **12**

19

20
(defined by rib cage and diaphragm)

13
(muscular partition between thoracic and abdominal cavities)

21

22

23

alveolar sac (sectioned)

alveolar duct

24

24

25

GAS EXCHANGE AND TRANSPORT / RESPIRATION
IN UNUSUAL ENVIRONMENTS (40-III, pp. 708–715)

Fill-in-the-Blanks

Oxygen is said to exert a (1)_____ _____ of 760/21 or 160 mm Hg. (2)_____ alone moves
oxygen from the alveoli into the bloodstream, and it is enough to move (3)_____ _____ in the
reverse direction. (4)_____ is the medical name for oxygen deficiency; it is characterized by faster
breathing, faster heart rate, and anxiety at altitudes of 8,000 feet above sea level. About 70 percent of the
carbon dioxide in the blood is transported as (5)_____. Without (6)_____ , the plasma would be
able to carry only about 2 percent of the oxygen that whole blood carries. When oxygen-rich blood reaches
a (7)_____ tissue capillary bed, oxygen diffuses outward, and carbon dioxide moves from tissues into
the capillaries. With the assistance of (8)_____ _____ , carbonic acid dissociates to form water
and carbon dioxide. Clusters of cells in the (9)_____ _____ and (10)_____ _____
regulate contractions of the diaphragm and intercostal muscles associated with inhalation and exhalation.

(11)_____ is the distension of lungs and the loss of gas exchange efficiency such that running,
walking, and even exhaling are painful experiences. At least 90 percent of all (12)_____ _____
deaths are the result of cigarette smoking; only about 10 percent of afflicted individuals will survive.

When a diver ascends, (13)_____ tends to move out of the tissues and into the bloodstream. If the
ascent is too rapid, many bubbles of nitrogen gas collect at the (14)_____ , hence the common name,
"the bends," for what is otherwise known as (15)_____ sickness.

The rate of breathing is governed by a respiratory center in the (16)_____ _____ , which
monitors signals coming in from arterial walls, from blood vessels, and from other brain regions.

Chapter Terms

The following page-referenced terms are the important terms in the chapter. Refer to the instructions given
in Chapter 1, p. 4.

respiration (700)	integumentary	lung (703)	alveolus, -oli (707)
respiratory surface (700)	exchange (702)	pharynx (704)	alveolar sac (707)
Fick's law (700)	tracheas (702)	larynx (704)	oxyhemoglobin (708)
ventilating (701)	gill (702)	bronchus, -chi (706)	emphysema (710)
	countercurrent flow (702)	respiratory bronchioles (707)	lung cancer (711)

Self-Quiz

____ 1. Most forms of life depend on _____
to obtain oxygen and eliminate carbon
dioxide.
   a. active transport
   b. bulk flow
   c. diffusion
   d. osmosis
   e. muscular contractions

____ 2. _____ is the most abundant gas in
Earth's atmosphere.
   a. Water vapor
   b. Oxygen
   c. Carbon dioxide
   d. Hydrogen
   e. Nitrogen

___ 3. With respect to respiratory systems, countercurrent flow is a mechanism that explains how _____.
   a. oxygen uptake by blood capillaries in the lamellae of fish gills occurs
   b. ventilation occurs
   c. intrapleural pressure is established
   d. sounds originating in the vocal cords of the larynx are formed
   e. all of the above

___ 4. _____ have the most efficient respiratory system.
   a. Amphibians
   b. Reptiles
   c. Birds
   d. Mammals
   e. Humans

___ 5. Immediately before reaching the alveoli, air passes through the _____.
   a. bronchioles
   b. glottis
   c. larynx
   d. pharynx
   e. trachea

___ 6. During inhalation, _____.
   a. the pressure in the thoracic cavity is less than the pressure within the lungs
   b. the pressure in the chest cavity is greater than the pressure within the lungs
   c. the diaphragm moves upward and becomes more curved
   d. the thoracic cavity volume decreases
   e. all of the above

___ 7. Hemoglobin _____.

   a. releases oxygen more readily in active tissues
   b. tends to release oxygen in places where the temperature is lower
   c. tends to hold on to oxygen when the pH of the blood drops
   d. tends to give up oxygen in regions where partial pressure of oxygen exceeds that in the lungs
   e. all of the above

___ 8. Oxygen moves from alveoli to the bloodstream _____.
   a. whenever the concentration of oxygen is greater in alveoli than in the blood
   b. by means of active transport
   c. by using the assistance of carbaminohemoglobin
   d. principally due to the activity of carbonic anhydrase in the red blood cells
   e. by all of the above

___ 9. Oxyhemoglobin releases O_2 when _____.
   a. carbon dioxide concentrations are high
   b. body temperature is lowered
   c. pH values are high
   d. CO_2 concentrations are low
   e. all of the above occur

___ 10. Nonsmokers live an average of _____ longer than people in their mid-twenties who smoke two packs of cigarettes each day.
   a. 6 months
   b. 1–2 years
   c. 3–5 years
   d. 7–9 years
   e. over 12 years

Chapter Objectives/Review Questions

This section lists general and detailed chapter objectives that can be used as review questions. You can make maximum use of these items by writing answers on a separate sheet of paper. Fill in answers where blanks are provided. To check for accuracy, compare your answers with information given in the chapter or glossary.

Page   *Objectives/Questions*

(700–702) 1. Understand the behavior of gases and the types of respiratory surfaces that participate in gas exchange.

(704–712, 716) 2. Understand how the human respiratory system is related to the circulatory system, to cellular respiration, and to the nervous system.

(702)   3. Define countercurrent and explain how it works. State where such a mechanism is found.

Page		Objectives/Questions

Page *Objectives/Questions*

(702, 704) 4. Describe how incoming oxygen is distributed to the tissues of insects and contrast this process with the process that occurs in mammals.

(704–707) 5. List all the principal parts of the human respiratory system and explain how each structure contributes to transporting oxygen from the external world to the bloodstream.

(706–707) 6. Describe the relationship of the human lung to the pleural sac and to the thoracic cavity.

(709) 7. Explain why oxygen diffuses from the bloodstream into the tissues far from the lungs. Then explain why carbon dioxide diffuses into the bloodstream from the same tissues.

(708–709) 8. Explain why oxygen diffuses from alveolar air spaces, through interstitial fluid, and across capillary epithelium. Then explain why carbon dioxide diffuses in the reverse direction.

(709) 9. Describe what happens to carbon dioxide when it dissolves in water under conditions normally present in the human body.

(712) 10. List the structures involved in detecting carbon dioxide levels in the blood and in regulating the rate of breathing. Name the location of each structure.

(710–711) 11. Distinguish bronchitis from emphysema. Then explain how lung cancer differs from emphysema.

(710–715) 12. List some of the things that go awry with the respiratory system and describe the characteristics of the breakdown.

Integrating and Applying Key Concepts

Consider the amphibians—animals that generally have aquatic larval forms (tadpoles) and terrestrial adults. Outline the respiratory changes that you think might occur as an aquatic tadpole metamorphoses into a land-going juvenile.

Critical Thinking Exercises

1. Consider the following measurements of carbon dioxide concentrations:

Inhaled air	35
Exhaled air	28
Pulmonary arterial blood	47

Why would you most likely suspect these data to be in error?

a. The concentration of carbon dioxide in blood should be higher than in exhaled air.
b. The concentration of carbon dioxide in exhaled air should be higher than in the inhaled air.
c. The concentration of carbon dioxide in blood should be lower than in inhaled air.
d. The concentration of carbon dioxide in inhaled air should be higher than in exhaled air.
e. The concentration of carbon dioxide should be equal in all three compartments.

ANALYSIS

There is net movement of carbon dioxide out of the blood through the lungs. This means that the blood entering the lungs should contain the highest concentration of carbon dioxide. At most, the concentration of carbon dioxide in exhaled air could be equal to that in blood, if the two compartments are in contact long enough to reach equilibrium. If diffusion does not reach equilibrium, the blood concentration will still be larger than that in the exhaled air. Inhaled air should have the lowest concentration. This will allow net diffusion of carbon dioxide from blood into the alveolar lumen and raise the carbon dioxide concentration there. Choice (b) is the only one that fits this set of predictions.

2. Respiration was studied in a shark of the genus *Heterodontus*. The hypothesis was that the gills of this shark functioned by a countercurrent mechanism. Which of the following sets of measurements of oxygen concentration would most strongly support this hypothesis?

	Arterial Blood Leaving the Gills	Inhaled Water	Exhaled Water
a.	150	137	145
b.	100	137	85
c.	68	137	68
d.	60	137	77
e.	40	137	137

ANALYSIS

The gill, an organ of gas exchange, functions as a location of transfer of oxygen from inhaled water into blood. Thus, the concentration of oxygen in exhaled water must be lower than in inhaled water. This is not true in choices (a) and (e), so they can be eliminated. With a concurrent flow mechanism, the maximum transfer of oxygen from water to blood occurs when the two compartments reach equilibrium. Concurrent flow cannot result in a blood oxygen concentration higher than the concentration in exhaled water. Countercurrent flow can. This pattern is shown in choice (b), so this choice provides the greatest support for the hypothesis. Choices (c) and (d) are compatible with either countercurrent or concurrent flow patterns.

Reference: Grigg and Read (1971) *Z. vergl. Physiol.* 73:439.

3. When a whale comes to the surface of the water, it exhales a stream of gas through the blowhole in the top of its head. Consider the hypothesis that the whale brain sends action potentials to the muscles of exhalation only when the front part of the whale's head is out of the water. Which of the following assumptions must you make in order to accept this hypothesis?

a. Whales can perceive light underwater so that they know their distance from the surface.
b. When exposed to air, receptors in the skin of the whale's head send action potentials to the brain.
c. Receptors in the whale's flipper can respond to a decrease in pressure as the whale nears the water surface.
d. Whales inhale through their mouths and exhale through their blowholes.
e. When exposed to air, the whale's muscles contract automatically.

ANALYSIS

a. Assuming this might refine the hypothesis by adding an early warning mechanism to anticipate emergence from the water, but this assumption is not necessary to the hypothesis that the muscles are activated only when a certain part of the whale is out of the water.
b. The hypothesis requires some mechanism for perceiving when the head emerges from the water. Receptors like this would be such a mechanism.
c. This choice is answered by the same argument as choice (a).
d. This is the observation for which the hypothesis proposes a mechanism. It is not an assumption underlying the hypothesis.
e. This assumption would contradict the hypothesis. If this were true, there would be no requirement for action potentials from the brain in order to trigger exhalation. You might also argue that the muscles themselves are never exposed to air, because they are inside the skin. However, this would not be arguing that the assumption was not made in order to make the hypothesis. It would argue that the assumption was not valid and lead to rejection of the hypothesis.

Answers

Answers to Interactive Exercises

THE NATURE OF RESPIRATORY SYSTEMS (40-I)
1. gill; 2. countercurrent flow; 3. tracheas; 4. Spiracles; 5. 760; 6. 21; 7. gut; 8. swim bladders; 9. aerobic metabolism; 10. O_2 (oxygen); 11. carbon dioxide; 12. respiration; 13. pressure gradient; 14. high; 15. lowest; 16. high; 17. lower; 18. Fick's law; 19. surface area; 20. partial pressure; 21. Hemoglobin; 22. lung; 23. airways (blood vessels); 24. blood vessels (airways); 25. water; 26. blood vessel in gill filament; 27. oxygen-poor blood; 28. oxygen-rich blood; 29. water; 30. blood.

HUMAN RESPIRATORY SYSTEM/AIR PRESSURE CHANGES IN THE LUNGS (40-II)
1. diaphragm; 2. rib cage; 3. increases; 4. drops; 5. ventilating; 6. pleural sac; 7. larynx; 8. glottis; 9. bronchi; 10. bronchioles; 11. alveoli; 12. intercostal muscles; 13. diaphragm; 14. pharynx; 15. epiglottis; 16. vocal cords; 17. trachea; 18. bronchus; 19. bronchioles; 20. thoracic cavity; 21. abdominal cavity; 22. smooth muscle; 23. bronchiole; 24. alveolus, alveoli; 25. capillary.

GAS EXCHANGE AND TRANSPORT/ RESPIRATION IN UNUSUAL ENVIRONMENTS (40-III)
1. partial pressure; 2. Diffusion; 3. carbon dioxide; 4. Hypoxia; 5. bicarbonate; 6. oxyhemoglobin (hemoglobin); 7. systemic (low-pressure); 8. carbonic anhydrase; 9. reticular formation (medulla oblongata); 10. medulla oblongata (reticular formation); 11. Emphysema; 12. lung cancer; 13. N_2 (nitrogen gas); 14. joints; 15. decompression; 16. reticular formation (medulla oblongata).

Answers to Self-Quiz

1. c; 2. e; 3. a; 4. c; 5. a; 6. a; 7. a; 8. a; 9. a; 10. d.

41

SALT-WATER BALANCE AND TEMPERATURE CONTROL

Interactive Exercises

MAINTAINING THE EXTRACELLULAR FLUID (41-I, pp. 720–722)

Fill-in-the-Blanks

The body gains water by absorbing water from the slurry in the lumen of the small intestine and from

(1)_____ during condensation reactions. The mammalian body loses water mostly by excretion of

(2)_____ , evaporation through the skin and (3)_____ _____ , elimination of feces from

the gut, and (4)_____ as the body is cooled. (5)_____ behavior, in which the brain compels the

individual to seek liquids, influences the gain of water.

    The body gains solutes by absorption of substances from the gut, by the secretion of hormones and

other substances, and by (6)_____ , which produces CO_2 and other waste products of degradative

reactions. Besides CO_2, there are several major metabolic wastes that must be eliminated: (7)_____ ,

formed by deamination reactions; (8)_____ , which is produced in the liver during reactions that link two ammonia molecules to CO_2 and release a molecule of water; and (9)_____ _____ , which is formed in reactions that break down nucleic acids.

Labeling

Identify each indicated part of the accompanying illustrations.

10. _____

11. _____

12. _____ _____

13. _____

14. _____

15. _____

16. _____

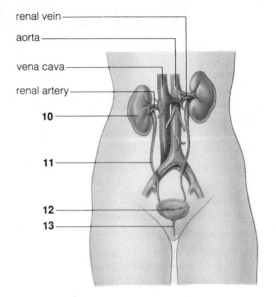

renal vein
aorta
vena cava
renal artery
10
11
12
13

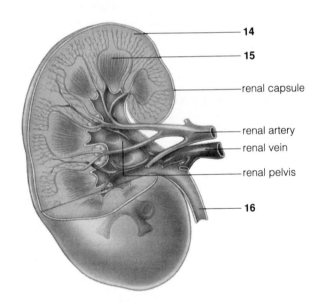

14
15
renal capsule
renal artery
renal vein
renal pelvis
16

URINE FORMATION (41-II, pp. 722–728)

True-False

If false, explain why.

___ 1. When the body rids itself of excess water, urine becomes more dilute.

___ 2. Water reabsorption into capillaries is achieved by diffusion and active transport.

Fill-in-the-Blanks

In mammals, urine formation occurs in a pair of (3)_____. Each contains about a million tubelike blood-filtering units called (4)_____. The function of (3) depends on intimate links between the (4) and the (5)_____. In every (4) blood flows from a(n) (6)_____ into a set of capillaries inside the (7)_____ _____ , then into a second set of capillaries that thread around the tubular parts of the nephron, then back to the bloodstream, leaving the kidney. Urine composition and volume depend on three processes: *filtration* of blood at the (8)_____ of a nephron, with (9)_____ _____ providing the force for filtration; *reabsorption,* in which water and (10)_____ move out of tubular parts of the nephron and back into adjacent (11)_____ _____ ; and (12)_____ , in which excess ions and a few foreign substances move out of those capillaries and back into the nephron so that they are disposed of in the urine. (13)_____ carry urine away from the kidney to the (14)_____ _____ , where it is stored until it is released via a tube called the (15)_____ , which carries urine to the outside.

Two hormones, ADH and (16)_____ , adjust the reabsorption of water and (17)_____ along the distal tubules and collecting ducts. An increase in the secretion of aldosterone causes (18) [choose one] ( ) more, ( ) less sodium to be excreted in the urine. When the body cannot rid itself of excess sodium, it inevitably retains excess water, and this leads to a rise in (19)_____ _____. Abnormally high blood pressure is called (20)_____ ; it can damage the kidneys, vascular system, and brain. One way to control hypertension is to restrict the intake of (21)_____ _____. Increased secretion of (22)_____ enhances water reabsorption at distal tubules and collecting ducts when the body must conserve water. When excess water must be excreted, ADH secretion is (23) [choose one] ( ) stimulated, ( ) inhibited.

Labeling

Identify each indicated part of the accompanying illustration.

24. _____ _____

25. _____ _____

26. _____ _____

27. _____ _____

28. _____ _____

29. _____ _____

30. _____ _____ _____

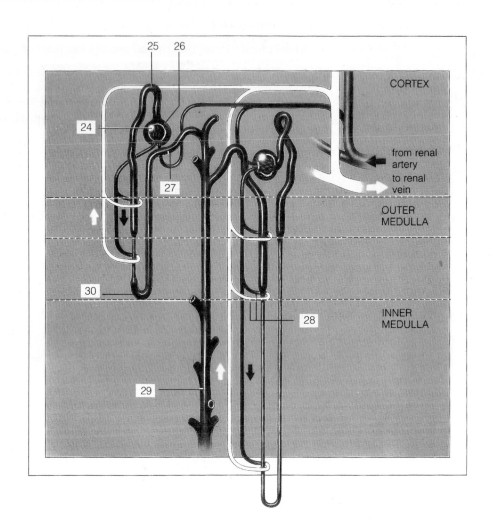

MAINTAINING BODY TEMPERATURE (41-III, pp. 728–734)

Fill-in-the-Blanks

In mammals, the (1)_____ is the seat of temperature control. Thermoreceptors located deep in the body are called (2)_____ thermoreceptors. The (3)_____ _____ contains smooth muscles that erect hairs or feathers and create an insulative layer of still air that helps prevent heat loss.

(4)_____ _____ is a response to cold stress in which the bloodstream's convective delivery of heat to the body's surface is reduced. A drop in body temperature below tolerance levels is referred to as (5)_____.

True-False

If false, explain why.

___ 6. Jackrabbits are endotherms.

___ 7. When the core temperature of the human body is about 86° F, consciousness is lost and heart muscle action becomes irregular.

___ 8. When the core temperature of the human body reaches 77° F, ventricular fibrillation sets in and death soon follows.

Matching

Choose the most appropriate answer to match with the following terms.

9. _____ conduction

10. _____ convection

11. _____ ectotherm

12. _____ endotherm

13. _____ evaporation

14. _____ heterotherm

15. _____ radiation

A. Body temperature determined more by heat exchange with the environment than by metabolic heat

B. Heat transfer by air or water heat-bearing currents away from or toward a body

C. Body temperature determined largely by metabolic activity and by precise controls over heat produced and heat lost

D. Direct transfer of heat energy between two objects in direct contact with each other

E. The emission of energy in the form of infrared or other wavelengths that are converted to heat by the absorbing body

F. Body temperature fluctuating at some times and heat balance controlled at other times

G. In changing from the liquid state to the gaseous state, the energy required is supplied by the heat content of the liquid

Chapter Terms

The following page-referenced terms are important; they were in boldface in the chapter. Refer to the instructions given in Chapter 1, p. 4.

urinary excretion (720)
kidneys (721)
urine (721)
nephrons (721)
urinary system (721)
urination (721)
glomerulus (722)
proximal tubule (722)
loop of Henle (722)

distal tubule (722)
filtration (722)
reabsorption (722)
secretion (723)
aldosterone (725)
ADH (antidiuretic hormone) (725)
thirst center (726)
core temperature (728)

radiation (729)
conduction (729)
convection (729)
evaporation (729)
ectotherms (730)
behavioral temperature regulation (730)
endotherms (730)

heterotherms (730)
peripheral vasoconstriction (731)
pilomotor response (732)
shivering (732)
nonshivering heat production (732)
peripheral vasodilation (733)

Self-Quiz

___ 1. The most toxic waste product of metabolism is _____.
   a. water
   b. uric acid
   c. urea
   d. ammonia
   e. carbon dioxide

___ 2. An entire subunit of a kidney that purifies blood and restores solute and water balance is called a _____.
   a. glomerulus
   b. loop of Henle
   c. nephron
   d. ureter
   e. none of the above

___ 3. In humans, the thirst center is located in the _____.
   a. adrenal cortex
   b. thymus
   c. heart
   d. adrenal medulla
   e. hypothalamus

___ 4. The longer the _____, the greater an animal's capacity to conserve water and to concentrate solutes to be excreted in the urine.
   a. loop of Henle
   b. proximal tubule
   c. ureter
   d. Bowman's capsule
   e. collecting tubule

___ 5. During reabsorption, sodium ions cross the proximal tubule walls into the interstitial fluid principally by means of _____.
   a. phagocytosis
   b. countercurrent multiplication
   c. bulk flow
   d. active transport
   e. all of the above

___ 6. Filtration of the blood in the kidney takes place in the _____.
   a. loop of Henle
   b. proximal tubule
   c. distal tubule
   d. Bowman's capsule
   e. all of the above

___ 7. _____ primarily controls the concentration of solutes in urine.
   a. Insulin
   b. Glucagon
   c. Antidiuretic hormone
   d. Aldosterone
   e. Epinephrine

___ 8. Hormonal control over excretion primarily affects _____.
   a. Bowman's capsules
   b. distal tubules
   c. proximal tubules
   d. the urinary bladder
   e. loops of Henle

___ 9. The last portion of the excretory system passed by urine before it is eliminated from the body is the _____.
   a. renal pelvis
   b. bladder
   c. ureter
   d. collecting ducts
   e. urethra

___ 10. Desert animals excrete _____ as their principal nitrogenous waste in a highly concentrated urine.
   a. urea
   b. uric acid
   c. ammonia
   d. amino acids
   e. ADH

Chapter Objectives/Review Questions

This section lists general and detailed chapter objectives that can be used as review questions. You can make maximum use of these items by writing answers on a separate sheet of paper. Fill in answers where blanks are provided. To check for accuracy, compare your answers with information given in the chapter or glossary.

Integrating and Applying Key Concepts

The hemodialysis machine used in hospitals is expensive and time-consuming. So far, artificial kidneys capable of allowing people who have nonfunctional kidneys to purify their blood by themselves, without having to go to a hospital or clinic, have not been developed. Which aspects of the hemodialysis procedure do you think have presented the most problems in development of a method of home self-care? If you had an unlimited budget and were appointed head of a team to develop such a procedure and its instrumentation, what strategy would you pursue?

Critical Thinking Exercises

1. One type of evidence for the existence of renal secretion comes from experiments with dye molecules. Slices of kidney tissue are incubated in a medium that contains the dye. After incubation, the concentration of dye in the tissue slice is higher than in the medium. In order to conclude that this accumulation of dye is evidence for renal secretion, which of the following assumptions must be made?

   a. The dye is chemically changed by the cells in the slice.
   b. The dye is accumulated in the lumen of the renal tubule.
   c. The dye is present in the filtrate formed during the incubation.
   d. The dye is not filtered during the incubation.
   e. Cell respiration does not occur during the incubation.

ANALYSIS

   a. Because the observations and conclusions concerned the location of the unchanged dye itself, this assumption would be irrelevant to the conclusion.
   b. Secretion is a process of active transport of material into the tubular lumen. The observation does not show where the dye is localized within the tissue. If the dye was accumulated inside the epithelial cells or some other compartment, it would not be evidence of secretion. This assumption is necessary for the conclusion.

c. This assumption would not be made, because the tissue is a slice removed from the animal; hence, it has no blood pressure or flow and no filtration.

d. This statement is a valid conclusion, not an assumption, because without blood circulation there can be no filtration. It is also a necessary condition for the interpretation, because if there was filtration the accumulation of dye could result from filtration of dye and water and reabsorption of water, leaving a concentrated dye solution.

e. The opposite assumption must be made. If respiration stops, ATP becomes depleted, and secretion by active transport cannot continue.

2. In the mammalian kidney, the rate of production of urine is directly affected by changes in blood pressure. Elevated blood pressure stimulates urine production; lower blood pressure results in decreased urine production. If you propose that this response is part of feedback regulation of blood pressure, which of the following statements must also be true?

   a. Decreased blood volume causes decreased blood pressure.
   b. Increased urine output causes increased blood pressure.
   c. Increased urine output causes decreased blood volume.
   d. Increased urine output causes decreased blood pressure.
   e. Decreased blood pressure causes decreased urine output.

ANALYSIS

Your observations demonstrate an effect of blood pressure on urine production. Your hypothesis that this is part of a feedback control system requires completion of the loop—there must be a negative effect of urine output on blood pressure. This is described in choice (d). Choice (b) also closes the loop, but this would be a positive feedback mechanism and would exacerbate an increase in blood pressure. Choice (e) is the observation given in the question. Choices (a) and (c) are elaborations of the mechanism of (d). If (a) and (c) are valid, they combine logically to give (d). These two choices might be regarded as hypotheses based on the assumption that (d) is valid.

3. Consider a mammalian kidney producing filtrate at 100 ml/min and excreting urine at 5 ml/min. Glucose, inulin, and PAH are measured in blood plasma and in urine. Inulin and PAH are water-soluble substances, harmless to the body at the concentrations used, that can be injected into the blood and easily measured in body fluids. Glucose can also be infused, and unusually high concentrations can be maintained during an experiment. The data follow (all concentrations expressed in mg/ml).

	Plasma Concentration	Urine Concentration
Glucose	12	40
Inulin	5	100
PAH	0.1	12

For each substance, decide which of the following processes are involved in its passage through the kidney.

   a. Filtration only
   b. Reabsorption only
   c. Secretion only
   d. Filtration and reabsorption
   e. Filtration and secretion

ANALYSIS

Substances that are either reabsorbed or secreted are also filtered, because they are small and soluble—glomerular filtration presents a permeability barrier only to molecules that are too large to be carried through a cell membrane. This eliminates choices (b) and (c) for all substances. Note also that no substances are both reabsorbed and secreted; that combination of processes would consume ATP and would result in no net transfer of the substance.

To decide which renal processes are applied to a substance, it is not sufficient in most cases to look at the concentration in blood and urine. Only if the urine concentration is lower than the plasma concentration is this conclusive; it shows that the substance must be reabsorbed. None of the substances in this problem have that pattern of concentration. For all of these substances, it is necessary to compare the *amount* filtered to the *amount* excreted in the urine. If more is excreted than filtered, the excess must have been secreted into the filtrate. If less is excreted than filtered, the missing amount must have been reabsorbed. If both amounts are the same, there are no carrier proteins that move the substance. The amount of the substance equals the product of the concentration and the volume.

	Amount Filtered	*Amount Excreted*
Glucose	1,200 mg/min	200 mg/min
Inulin	500 mg/min	500 mg/min
PAH	10 mg/min	60 mg/min

Although the concentration of glucose is higher in urine than in plasma, the total amount of urine is low enough, because of reabsorption of water, that less glucose is excreted than filtered. Thus, glucose is reabsorbed.

Inulin is excreted at the same rate as it is filtered, so it is neither reabsorbed nor secreted. As with glucose, the concentration of inulin rises in the urine because water is reabsorbed from the solution.

PAH must be secreted, because the amount excreted is greater than the amount filtered, even though its concentration in urine is the lowest of the three substances.

Answers

Answers to Interactive Exercises

MAINTAINING THE EXTRACELLULAR FLUID (41-I)
1. metabolism; 2. urine; 3. respiratory surfaces;
4. sweating; 5. Thirst; 6. metabolism; 7. ammonia;
8. urea; 9. uric acid; 10. kidney; 11. ureter; 12. urinary bladder; 13. urethra; 14. cortex; 15. medulla; 16. ureter.

URINE FORMATION (41-II)
1. T; 2. T; 3. kidneys; 4. nephrons; 5. bloodstream;
6. arteriole; 7. Bowman's capsule; 8. glomerulus;
9. blood pressure; 10. solutes; 11. blood capillaries (peritubular capillaries); 12. secretion; 13. Ureters;
14. urinary bladder; 15. urethra; 16. aldosterone;
17. sodium; 18. less; 19. blood pressure;
20. hypertension; 21. table salt (sodium chloride);
22. ADH; 23. inhibited; 24. glomerular capillaries;
25. proximal tubule; 26. Bowman's capsule; 27. distal tubule; 28. peritubular capillaries; 29. collecting duct;
30. loop of Henle.

MAINTAINING BODY TEMPERATURE (41-III)
1. hypothalamus; 2. core; 3. pilomotor response;
4. Peripheral vasoconstriction; 5. hypothermia; 6. T; 7. T;
8. T; 9. D; 10. B; 11. A; 12. C; 13. G; 14. F; 15. E.

Answers to Self-Quiz

1. d; 2. c; 3. e; 4. a; 5. d; 6. d; 7. d; 8. b; 9. e; 10. b.

42

PRINCIPLES OF REPRODUCTION AND DEVELOPMENT

THE BEGINNING: REPRODUCTIVE MODES
STAGES OF DEVELOPMENT
PATTERNS OF DEVELOPMENT
    Key Mechanisms: An Overview
        Cytoplasmic Localization
        Embryonic Induction
    Developmental Information
    in the Egg
    Fertilization
    Cleavage
    Gastrulation
CELL DIFFERENTIATION

MORPHOGENESIS
    Cell Migrations
    Changes in Cell Size and Shape
    Localized Growth and Cell Death
    Pattern Formation
        Vertebrate Eye Formation
        Chick Wing Formation
        Pattern Formation in *Drosophila*
        Inducer Signals
POST-EMBRYONIC DEVELOPMENT
AGING AND DEATH
    Commentary: Death in the Open

Interactive Exercises

THE BEGINNING: REPRODUCTIVE MODES / STAGES OF DEVELOPMENT (42-I, pp. 737–741)

Fill-in-the-Blanks

Asexual processes of reproduction include (1)_____ , which is common in animals such as *Hydra* and other cnidarians. (2)_____ _____ is considered the first stage of animal development. Rich stores of substances become assembled in localized regions of the (3)_____ cytoplasm. When sperm and egg unite and their DNA mingles and is reorganized, the process is referred to as (4)_____. At the end of fertilization, a (5)_____ is formed. (6)_____ includes the repeated mitotic divisions of a zygote that segregate the egg cytoplasm into a cluster of cells; the entire cluster is known as a (7)_____.
(8)_____ is the process that arranges cells into three germ layers. Ectoderm eventually will give rise to skin epidermis and the (9)_____ system; endoderm forms the inner lining of the (10)_____ and associated digestive glands. Mesoderm forms the circulatory system, the (11)_____ , and the muscles.
(12)_____ _____ activated at fertilization direct the initial stages of development until gastrulation occurs.

Copperheads show (13)_____: fertilization is internal; the fertilized eggs develop inside the mother's body without additional nourishment, and the young are born live. Birds show (14)_____: eggs with large yolk reserves are released from and develop outside the mother's body.

True-False

If false, explain why.

___ 15. In complex eukaryotes, development until gastrulation is governed by DNA in the nucleus of the zygote.

___ 16. Sperm penetration into the cytoplasm of the egg brings about specific structural changes and chemical reactions.

___ 17. Most animals reproduce sexually.

___ 18. Sexual reproduction is less advantageous in predictable environments; asexual reproduction is more advantageous in predictable environments.

___ 19. In a developing chick embryo, the heart begins to beat at some time between 30 and 36 hours after fertilization.

___ 20. Gastrulation precedes organ formation.

PATTERNS OF DEVELOPMENT / CELL DIFFERENTIATION (42-II, pp. 742–748)

True-False

If false, explain why.

___ 1. During gastrulation, maternal controls over gene activity are activated and begin the process of differentiation in each cell's nucleus.

Fill-in-the-Blanks

The third stage of animal development, (2)_____ , is characterized by the subdividing and compartmentalizing of the zygote; no growth occurs at this stage, and usually a hollow ball of cells, the (3)_____ , is formed. The fourth stage, (4)_____ , is concerned with the formation of ectoderm, mesoderm, and endoderm, the (5)_____ layers of the embryo; at the end of this stage, the (6)_____ is formed.

Much of the information that determines how structures will be spatially organized in the embryo begins with the distribution of (7)_____ _____ in the oocyte. As cleavage membranes divide up the cytoplasm, cytoplasmic determinants become localized in different daughter cells; this process, called (8)_____ _____ , helps seal the developmental fate of the descendants of those cells.

As the embryo develops, one group of cells may produce a substance (say, a growth factor) that diffuses to another group of cells and turns on protein synthesis in those cells. Such interaction among embryonic cells is called (9)_____ _____.

In amphibian eggs, sperm penetration on one side of an egg causes pigment granules on the opposite side of the egg to flow toward the (10)_____ _____. A lightly pigmented area called the

(11)_____ _____ results. It is a visible marker of the site where the (12)_____ _____ will be established and where gastrulation will begin.

Cleavage of a frog's zygote is total, because there is so little yolk that cleavage membranes can subdivide the entire cytoplasmic mass; in the chick, however, there is so much yolk that cleavage membranes cannot subdivide the entire mass. Cleavage is therefore said to be incomplete, and the chick grows from a primitive streak on the surface of the (13)_____ mass into a chick embryo complete with wing and leg buds and beating heart during the first (14)_____ days.

Through (15)_____ _____ , a single fertilized egg gives rise to an assortment of different types of specialized cells; these differentiated cells have the same number and same kinds of (16)_____ because they are all descended by mitosis from the same zygote. Through gene controls, however, (17)_____ are placed on which genes may be expressed (translated) in a given cell.

MORPHOGENESIS (42-III, pp. 748–753)

Fill-in-the-Blanks

In (1)_____ _____ _____ , cells move about by means of (2)_____ , which are temporary projections from the main cell body. They move in response to chemical gradients, a behavior called (3)_____ . These gradients are created when different cells release specific substances. Cells also move in response to (4)_____ cues provided by recognition proteins on other cell surfaces.

(5)_____ involves the growth, shaping, and spatial coordination necessary to form functional body units. Sometimes entire organs (such as testes in human males) change position in the developing organism, but the inward or outward folding of (6)_____ _____ is seen more often. Spemann demonstrated that the process known as (7)_____ _____ occurs in salamander embryos, where one body part differentiates because of signals it receives from an adjacent body part. Morphogenesis depends on (8)_____ _____ , which contributes to changes in the sizes, shapes, and proportions of body parts. A kitten's eyes after birth are an example of (9)_____ _____ _____ in action. (10)_____ mutations affect regulatory genes that activate *sets* of genes concerned with development; they are generally disadvantageous and cause mistakes in the developmental program of an organism.

POST-EMBRYONIC DEVELOPMENT / AGING AND DEATH (42-IV, pp. 754–756)

True-False

If false, explain why.

____ 1. The aging and death of a cell may be coded in large part in its DNA; external signals activate those DNA messages and tell the cell that it is time to die.

____ 2. A process of predictable cellular deterioration is built into the life cycle of all organisms that consist of differentiated cells that show considerable specialization.

____ 3. Body parts become folded, tubes become hollowed out, and eyelids, lips, noses, and ears all become slit or perforated by controlled cell death.

Fill-in-the-Blanks

In many animals, the embryo develops into a motile, independent (4)_____ , which extends the food supply and range of the population. A larva necessarily must undergo (5)_____ in order to become a juvenile. Any developmental program that includes an adaptive larval stage is called (6)_____ development. Normal cell types have a (7)_____ _____ _____ , and mitosis is scheduled to quit after so many cell divisions.

Chapter Terms

The following page-referenced terms are the important terms in the chapter. Refer to the instructions given in Chapter 1, p. 4.

sexual reproduction (737)	ectoderm (739)	gray crescent (743)	pattern formation (750)
asexual reproduction (737)	organ formation (741)	blastula (746)	homeotic mutation (753)
yolk (739)	growth and tissue	cell differentiation (748)	inducer signals (753)
gamete formation (739)	specialization (741)	morphogenesis (748)	adult (754)
fertilization (739)	cell differentiation (742)	active cell migration (748)	larva (754)
cleavage (739)	morphogenesis (742)	neural plate (749)	metamorphosis (754)
gastrulation (739)	cytoplasmic	neural tube (749)	regeneration (754)
endoderm (739)	localization (742)	localized growth (749)	aging (755)
mesoderm (739)	embryonic induction (742)	controlled cell death (750)	limited division potential (756)

Self-Quiz

____ 1. Animals such as birds lay eggs with large amounts of yolk; embryonic development happens within the egg covering outside the mother's body. This developmental strategy is called _____.
   a. ovoviviparity
   b. viviparity
   c. oviparity
   d. parthenogenesis
   e. none of the above

____ 2. The process of cleavage most commonly produces a(n) _____.
   a. zygote
   b. blastula
   c. gastrula
   d. third germ layer
   e. organ

____ 3. Imaginal disks are characteristic of the embryonic development of _____.
   a. frogs
   b. fruit flies
   c. chickens
   d. sea urchins
   e. humans

____ 4. Which of the following forms undergoes indirect development?
   a. frogs
   b. snakes
   c. whales
   d. horses
   e. hawks

____ 5. The differentiation of a body part in response to signals from an adjacent body part is _____.
   a. contact inhibition
   b. ooplasmic localization
   c. embryonic induction
   d. pattern formation
   e. none of the above

____ 6. A homeotic mutation _____.
   a. may cause a leg to develop on the head where an antenna should grow
   b. affects the expression of imaginal disks
   c. affects morphogenesis
   d. may alter the path of development
   e. all of the above

___ 7. Shortly after fertilization, the zygote is subdivided into a multicelled embryo during a process known as _____.
    a. meiosis
    b. parthenogenesis
    c. embryonic induction
    d. cleavage
    e. invagination

___ 8. Muscles differentiate from _____ tissue.
    a. ectoderm
    b. mesoderm
    c. endoderm
    d. parthenogenetic
    e. yolky

___ 9. The gray crescent is _____.
    a. formed where the sperm penetrates the egg
    b. next to the dorsal lip of the blastopore
    c. the yolky region of the egg
    d. where the first mitotic division begins
    e. formed opposite from where the sperm enters the egg

___ 10. The nervous system differentiates from _____ tissue.
    a. ectoderm
    b. mesoderm
    c. endoderm
    d. yolky
    e. parthenogenetic

Chapter Objectives/Review Questions

This section lists general and detailed chapter objectives that can be used as review questions. You can make maximum use of these items by writing answers on a separate sheet of paper. Fill in answers where blanks are provided. To check for accuracy, compare your answers with information given in the chapter or glossary.

Page   *Objectives/Questions*

(737–738) 1. Understand how asexual reproduction differs from sexual reproduction. Know the advantages and problems associated with having separate sexes.

(738–739) 2. Explain why evolutionary trends in many groups of organisms tend toward developing more complex, sexual strategies rather than retaining simpler, asexual strategies.

(739, 743, 746) 3. Explain how the amount of yolk in an ovum can influence an animal's cleavage pattern.

(738) 4. Define oviparous, viviparous, and ovoviviparous. For each of the three developmental strategies, cite an example of an animal that goes through it.

(739) 5. Name each of the three embryonic tissue layers and the organs formed from each.

(739–741) 6. Describe early embryonic development and distinguish among the following: oogenesis, fertilization, cleavage, gastrulation, and organ formation.

(740–741, 746–747) 7. Compare the early stages of frog and chick development (see Figures 42.5 and 42.8) with respect to egg size and type of cleavage pattern (incomplete or complete).

(742, 743) 8. Explain what causes polarity to occur during oocyte maturation in the mother and state how polarity influences later development.

(747) 9. Define gastrulation and state what process begins at this stage that did not happen during cleavage.

(748) 10. Define differentiation and give two examples of cells in a multicellular organism that have undergone differentiation.

(754) 11. Distinguish direct from indirect development.

(754) 12. Define what is meant by larva. Distinguish metamorphosis from morphogenesis.

(748–752) 13. Explain why the differentiation of cells in a multicellular organism goes hand in hand with the division of labor and the integration of life processes.

(748–752) 14. Explain how a spherical zygote becomes a multicellular adult with arms and legs.

Integrating and Applying Key Concepts

If embryonic induction did not occur in a human embryo, how would the eye region appear? What would happen to the forebrain and epidermis? If controlled cell death did not happen in a human embryo, how would its hands appear? its face?

Critical Thinking Exercises

1. The first sentence of a recent paper in a scientific journal says, "Reproduction in many species is confined to a time of year when the probability of survival for both the adults and offspring is maximum." When you start reading this paper, you would be most justified in saying that this statement is which of the following?

   a. A fact
   b. An observation
   c. An interpretation based on observations
   d. An assumption based on controlled experiments
   e. False

ANALYSIS

   a. This statement can never be regarded as a fact, because its major components cannot be observed. You could observe reproduction at a certain time of year, but you could not be sure that it did not also occur at other times of year and you just missed it. You could observe the actual survival rates of some sample at various seasons of some number of years, but you cannot observe probability. You can only infer it from its outcomes with a level of confidence that depends on the number of observations.
   b. An observation and a fact are the same thing, and the argument in (a) applies here as well.
   c. The argument in (a) supports this choice. You can observe how many animals of various ages are present at the beginning of some time period and how many are still present at the end. Those are the facts. You add some assumptions, such as that the missing ones died instead of just migrating. Then you can infer the probability of survival and extrapolate it to the species as a whole in all years.
   d. Assumptions are taken as true *without* experimental evidence. This choice contradicts itself.
   e. As a reader of scientific texts, you must always consider the possibility that the writer's conclusion is false. It is your job to evaluate the evidence and reasoning that led to the conclusion and decide for yourself.

2. As described in the text, the nucleus from a cell in the intestine of a frog was transplanted into an enucleated frog egg and directed the development of a normal frog. The investigator concluded that differentiated cells contain a full set of genes but express only a fraction of them. Which of the following observations would be the strongest criticism of this conclusion?

   a. Differentiation is a process of selective expression of genes.
   b. Differentiation is a process of selective transmission of genes.
   c. A nucleus transplanted from a brain cell also directs the full developmental program.
   d. Some of the cells in the intestine are not differentiated but in living tissue look just like the differentiated epithelium.
   e. Most of the attempts to perform this experiment fail because of damage to the transplanted nucleus or because the original nucleus is not completely removed.

ANALYSIS

   a. This is another wording of the hypothesis, not an observation.
   b. This is an alternative hypothesis, not an observation.
   c. This observation would support, not criticize, the hypothesis. It would be another example of a differentiated cell with a full set of genes.

d. This would be a serious criticism. It would point to the possibility that the transplanted nucleus had not undergone differentiation and thus still had its original complement of genes, even though differentiation occurs because different genes are transmitted to different cells.

e. This would support the hypothesis. It would explain the failures that the alternative hypothesis would say are evidence that differentiated nuclei simply don't have a full set of genes.

3. An intestinal cell nucleus from frog B is transplanted into an enucleated egg taken from frog A. Cells of frog A contain protein X, but the cells of frog B do not. The egg with the transplanted nucleus divides several times; then the cluster of cells is homogenized and analyzed. It contains protein X. Which of the following is the best interpretation?

a. The transplanted nucleus contains information for protein X.

b. Some mRNA, transcribed on the egg cell DNA, was left in the cytoplasm when the nucleus was removed.

c. Some tRNA, transcribed on the egg cell DNA, was left in the cytoplasm when the nucleus was removed.

d. Some DNA from the egg cell nucleus was left in the cytoplasm when the nucleus was removed.

e. Protein X is not synthesized in intestine cells.

ANALYSIS

a. This would explain the presence of protein X in the cells, but it would not explain its absence from the cells of the donor, frog B. The fact that the protein is present in the cells of frog A shows that the gene for this protein is expressible in a normal frog. What, except for the absence of the gene, would account for the protein's absence from frog B?

b. Frog A clearly has the gene for the protein. The probability that the gene will be transmitted to any specific egg by meiosis is 100 percent if the frog is homozygous, 50 percent if it is heterozygous. The gene was probably in the nucleus that was removed. If it had been transcribed before enucleation, mRNA would have been left in the cytoplasm that could be translated in the hybrid cell—any ribosomes can translate any mRNA—and would account for the presence of the protein.

c. Some tRNA from the original nucleus probably was left in the cytoplasm, but that is not specific for any protein. Molecules of tRNA participate readily in translation of all mRNA molecules. The protein product is determined by the mRNA only.

d. If DNA was left, it might have encoded the information for protein X, but this hypothesis would require the assumption of some very unusual and unlikely conditions to get the DNA into the transplanted nucleus, incorporated into a chromosome, and transcribed.

e. This statement would imply that the gene for protein X was absent from the transplanted nucleus, if you accepted the alternative hypothesis of differentiation. But that does not explain how the information for the protein got into the cell with the transplanted nucleus.

Answers

Answers to Interactive Exercises

THE BEGINNING: REPRODUCTIVE MODES/ STAGES OF DEVELOPMENT (42-I)

1. budding; 2. Gamete formation; 3. egg; 4. fertilization; 5. zygote; 6. Cleavage; 7. blastula; 8. Gastrulation; 9. nervous; 10. gut; 11. skeleton; 12. RNA transcripts (maternal messages); 13. ovoviviparity; 14. oviparity; 15. F; 16. T; 17. T; 18. T; 19. T; 20. T.

PATTERNS OF DEVELOPMENT/ CELL DIFFERENTIATION (42-II)

1. F; 2. cleavage; 3. blastula; 4. gastrulation; 5. germ; 6. gastrula; 7. cytoplasmic determinants; 8. cytoplasmic localization; 9. embryonic induction; 10. animal pole; 11. gray crescent; 12. body axis; 13. yolk; 14. three; 15. cell differentiation; 16. genes; 17. restrictions.

MORPHOGENESIS (42-III)
1. active cell migration; 2. pseudopods; 3. chemotaxis;
4. adhesive; 5. Morphogenesis; 6. ectodermal sheets;
7. embryonic induction; 8. localized growth;
9. controlled cell death; 10. Homeotic.

POST-EMBRYONIC DEVELOPMENT/
AGING AND DEATH (42-IV)
1. T; 2. T; 3. T; 4. larva; 5. metamorphosis; 6. indirect;
7. limited division potential.

Answers to Self-Quiz
1. c; 2. b; 3. b; 4. a; 5. c; 6. e; 7. d; 8. b; 9. e; 10. a.

43

HUMAN REPRODUCTION AND DEVELOPMENT

Interactive Exercises

HUMAN REPRODUCTIVE SYSTEM (43-I, pp. 759–769)

Fill-in-the-Blanks

Spermatogenesis occurs in the (1)_____ _____ , but sperm become somewhat motile in the (2)_____ of the male. The interstitial cells of the testis produce (3)_____. Accessory reproductive organs of the human male include the (4)_____ _____ , which either stores most of the sperm or, during sexual activity, moves them along by peristaltic action. The seminal vesicles and (5)_____ contribute secretions that make up most of the seminal fluid. A cap over most of the head of each sperm contains (6)_____ _____ that function in egg penetration. The (7)_____ of each sperm contains mitochondria, which provide the energy necessary for motility. (8)_____ secreted by the prostate gland into semen stimulate uterine contractions in the female.

Ovaries produce the important sex hormones (9)_____ and (10)_____. In the female, (11)_____ are passageways that channel ova from the ovary into the (12)_____ , which houses the embryo during pregnancy. In most mammals, the (13)_____ cycle is a predictably recurring time when the female becomes sexually receptive to the male. During ejaculation, a (14)_____ closes off the bladder so that urine cannot mix with semen. Each oocyte is contained in a spherical chamber called a (15)_____. At (16)_____ , the follicle ruptures, and the ovum escapes from the ovary and is swept by ciliary action into the (17)_____ , the road to the uterus. The ruptured follicle now changes into secretory tissue called the (18)_____ _____. Estrogen causes the (19)_____ (epithelial uterine lining) to thicken. (20)_____ stimulates glands in the thickened tissues to secrete various substances. The midcycle peak of (21)_____ , the level of which overshoots (22)_____ , triggers ovulation. The decrease of all the hormones that regulate the monthly cycle brings about (23)_____.

Labeling

Identify each indicated part of the accompanying illustrations.

24. _____ _____
25. _____ _____
26. _____ _____
27. _____ _____
28. _____ _____
29. _____

30. _____
31. _____
32. _____
33. _____ _____
34. _____ _____
35. _____

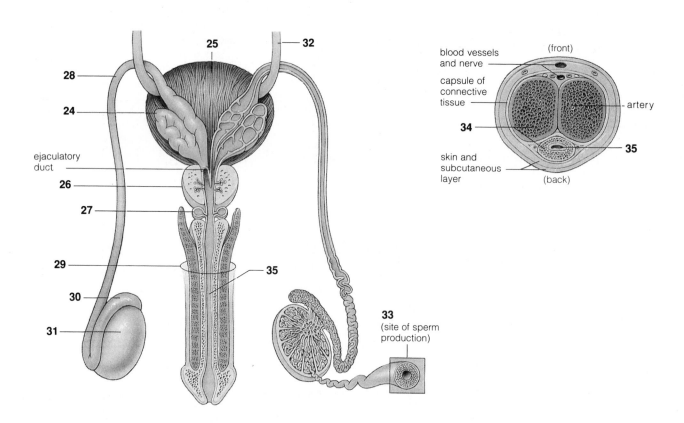

36. _____

37. _____ _____

38. _____ _____

39. _____ _____

40. _____

41. _____

42. _____

43. _____ _____

44. _____

45. _____

46. _____ _____

47. _____ _____

48. _____

49. _____

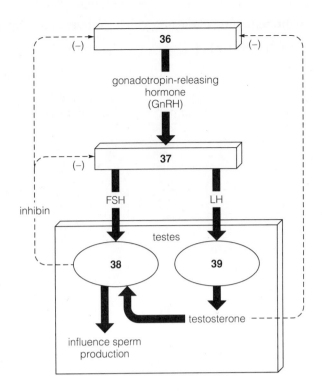

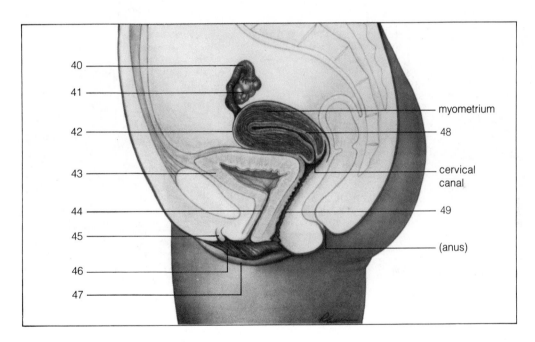

FROM FERTILIZATION TO BIRTH (43-II, pp. 769–777)

Fill-in-the-Blanks

Fertilization generally takes place in the (1)_____ ; five or six days after conception, (2)_____
begins as the blastocyst sinks into inside the endometrium. Extensions from the chorion fuse with the
endometrium of the uterus to form a (3)_____ , the organ of interchange between mother and fetus.

By the beginning of the (4)_____ trimester, all major organs have formed; the offspring is now referred to as a(n) (5)_____.

Labeling

Identify each indicated part of the accompanying illustrations.

6. _____ _____

7. _____ _____

8. _____ _____

9. _____ _____

10. _____ _____

11. _____

12. _____

13. _____

14. _____ _____

15. _____ _____

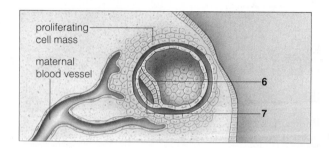

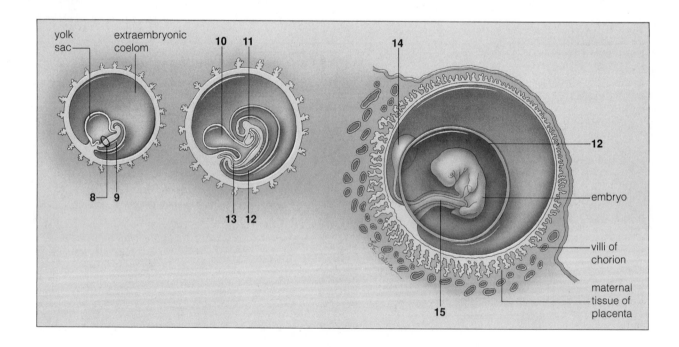

16. _____

17. _____ _____

18. _____

19. _____

20. _____

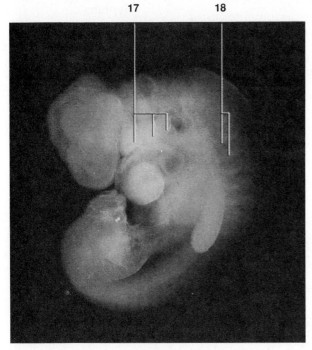

A human embryo at **16** weeks after conception.

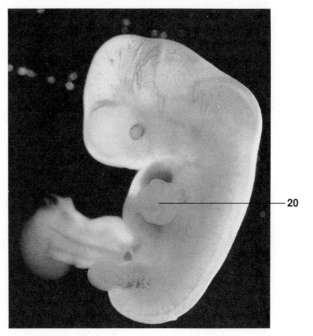

A human embryo at **19** weeks after conception.

POSTNATAL DEVELOPMENT, AGING, AND DEATH / CONTROL OF HUMAN FERTILITY (43-III, pp. 777–783)

Fill-in-the-Blanks

On Earth each week, (1)_____ more babies are born than people die. Each year in the United States, we still have about (2)_____ unwed teenage mothers and (3)_____ abortions. The most effective method of preventing conception is complete (4)_____. (5)_____ are about 85–93 percent reliable and help prevent venereal disease. A (6)_____ is a flexible, dome-shaped disk, used with a spermicidal foam or jelly, that is placed over the cervix. In the United States, the most widely used contraceptive is the Pill—an oral contraceptive of synthetic (7)_____ and (8)_____ that suppress the release of (9)_____ from the pituitary and thereby prevent the cyclic maturation and release of eggs. Two forms of surgical sterilization are vasectomy and (10)_____ _____.

Matching

Match each of the following with *all* applicable diseases.

11. _____ can damage the brain and spinal cord in ways leading to various forms of insanity and paralysis

12. _____ has no cure

13. _____ can cause violent cramps, fever, vomiting, and sterility due to scarring and blocking of the oviducts

14. _____ caused by a motile, corkscrew-shaped bacterium, *Treponema pallidum*

15. _____ affects about 14 million women a year, causing scarred oviducts, abnormal pregnancies, and sterility

16. _____ infected women typically have miscarriages, stillbirths, or sickly infants

17. _____ caused by a bacterium with pili, *Neisseria gonorrhoeae*

18. _____ caused by direct contact with the viral agent; about 5–20 million people in the United States infected by it

19. _____ produces a chancre (localized ulcer) 1–8 weeks following infection

20. _____ chronic infections by this can lead to cervical cancer

21. _____ caused by an intracellular parasite that infects 3–10 million Americans per year, especially college students

22. _____ can lead to lesions in the eyes that cause blindness in babies born to mothers with this

23. _____ acyclovir decreases the healing time and may also decrease the pain and viral shedding from the blisters

24. _____ following infection, the parasites migrate to lymph nodes, which become enlarged and tender; may lead to pronounced tissue swelling

25. _____ may be cured by antibiotics but can be infected again

26. _____ can be treated with tetracycline and sulfonamides

27. _____ generally preventable by correct condom usage

A. AIDS
B. Chlamydial infection
C. Genital herpes
D. Gonorrhea
E. Pelvic inflammatory disease
F. Syphilis

Chapter Terms

The following page-referenced terms are important; most were in boldface in the chapter. Refer to the instructions given in Chapter 1, p. 4.

testis, testes (759)
ovary, -ies (759)
seminiferous tubules (761)
sperm (762)
testosterone (762)
LH, luteinizing hormone (762)
FSH, follicle-stimulating hormone (762)
GnRH (763)

estrogen (763)
progesterone (763)
oviduct (763)
uterus (763)
endometrium (764)
cervix (764)
vagina (764)
menstrual cycle (764)
follicle (764)
primary oocyte (764)

secondary oocyte (764)
polar body (764)
ovulation (764)
corpus luteum (764)
hypothalamus (766)
anterior pituitary (766)
fertilization (769)
ovum (769)
implantation (769)
embryonic disk (769)

yolk sac (770)
allantois (770)
amnion (770)
chorion (770)
umbilical cord (770)
placenta (771)
fetus (775)
adult (777)
aging (777)

Self-Quiz

For questions 1–5, choose from the following answers:

    a. AIDS
    b. Chlamydial infection
    c. Genital herpes
    d. Gonorrhea
    e. Syphilis

___ 1. _____ is a disease caused by a spherical bacterium (*Neisseria*) with pili; it is curable by prompt diagnosis and treatment.

___ 2. _____ is a disease caused by a spiral bacterium (*Treponema*) that produces a localized ulcer (a chancre).

___ 3. _____ is an incurable disease caused by a retrovirus (an RNA-based virus).

___ 4. _____ is a disease caused by an obligate, intracellular parasite that migrates to regional lymph nodes, which swell and become tender.

___ 5. _____ is an extremely contagious viral infection (DNA-based) that causes sores on the facial area and reproductive tract; it is also incurable.

For questions 6–8, choose from the following answers:

    a. blastocyst
    b. allantois
    c. yolk sac
    d. oviduct
    e. cervix

___ 6. The _____ lies between the uterus and the vagina.

___ 7. The _____ is a pathway from the ovary to the uterus.

___ 8. The _____ results from the process known as cleavage.

For questions 9–12, choose from the following answers:

    a. interstitial cells
    b. seminiferous tubules
    c. vas deferens
    d. epididymis
    e. prostate

___ 9. The _____ connects a structure on the surface of the testis with the ejaculatory duct.

___ 10. Testosterone is produced by the _____.

___ 11. Meiosis occurs in the _____.

___ 12. Sperm mature and become motile in the _____.

Chapter Objectives/Review Questions

This section lists general and detailed chapter objectives that can be used as review questions. You can make maximum use of these items by writing answers on a separate sheet of paper. Fill in answers where blanks are provided. To check for accuracy, compare your answers with information given in the chapter or glossary.

Page   *Objectives/Questions*

(759–761) 1. Distinguish between primary and secondary sexual traits and between gonads and accessory reproductive organs.

(760–762) 2. Follow the path of a mature sperm from the seminiferous tubules to the urethral exit. List every structure encountered along the path and state the contribution to the nurture of the sperm.

(761–762) 3. Diagram the structure of a sperm, label its components, and state the function of each.

(761, 764) 4. Compare the function of the Leydig cells of the testis with the function of the ovarian follicle and the corpus luteum.

(761)   5. List in order the stages that compose spermatogenesis.

(762–763) 6. Name the four hormones that directly or indirectly control male reproductive function. Diagram the negative feedback mechanisms that link the hypothalamus, anterior pituitary, and testes in controlling gonadal function.

(764–767) 7. Distinguish the ovarian cycle from the menstrual cycle and explain how the two cycles are synchronized by hormones from the anterior pituitary, hypothalamus, and ovaries.

(764)   8. Indicate the ways in which the human menstrual cycle differs from estrus as it occurs in most mammals.

(764–766) 9. State which hormonal event brings about ovulation and which other hormonal events bring about the onset and finish of menstruation.

(766, 769) 10. List the physiological factors that bring about erection of the penis during sexual stimulation and the factors that bring about ejaculation.

(769)   11. List the similar events that occur in both male and female orgasm.

(763, 768–769) 12. Trace the path of a sperm from the urethral exit to the place where fertilization normally occurs. Mention in correct sequence all major structures of the female reproductive tract that are passed along the way and state the principal function of each structure.

(769–771, 774–775) 13. Describe the events that occur during the first month of human development. State how much time cleavage and gastrulation require, when organogenesis begins, and what is involved in implantation and placenta formation.

(772–775) 14. Explain why the mother must be particularly careful of her diet, health habits, and life-style during the first trimester after fertilization (especially during the first six weeks).

(775)   15. State when the embryo begins to be referred to as a fetus and at what point at least 10 percent of births result in survival.

(776–777) 16. Describe how a woman examines herself for breast cancer and how a man examines himself for testicular cancer.

(778–779) 17. Identify the factors that encourage and discourage methods of human birth control.

(779)   18. Identify the three most effective birth control methods used in the United States and the four least effective birth control methods.

(778)   19. State which birth control methods help prevent venereal disease.

(779)   20. Describe two different types of sterilization.

(783)   21. State the physiological circumstances that would prompt a couple to try in vitro fertilization.

(780–782) 22. For each STD described in the Commentary, know the causative organism and the symptoms of the disease.

Integrating and Applying Key Concepts

What rewards do you think a society should give a woman who has at most two children during her lifetime? In the absence of rewards or punishments, how can a society encourage women not to have abortions and yet ensure that the human birth rate does not continue to increase?

Critical Thinking Exercises

1. Spermatogenesis can fail because of loss or failure to form of any one of several cell types in the human testis. Each type of failure results in a characteristic pattern of hormonal changes. In each case below, state whether the levels of the following hormones would be elevated, normal, or depressed: FSH, LH, testosterone.
   a. Interstitial cells missing
   b. Sertoli cells missing
   c. Spermatogonia missing

ANALYSIS

a. Interstitial cells produce testosterone, so missing interstitial cells would cause very depressed levels of this hormone. Without the negative feedback from testosterone, the hypothalamus and pituitary would produce elevated levels of LH. The Sertoli cells, on the other hand, would continue to produce inhibin, and FSH levels would be expected to be normal.

b. Without Sertoli cells, inhibin would be absent, and FSH levels would rise. The testosterone-LH feedback loop would be intact, and their levels would be expected to be normal.

c. Spermatogonia are responsive to hormone levels but do not affect them by feedback mechanisms. All three hormone levels in this case would be normal.

2. The text points out that chromosomal abnormalities such as Down syndrome can result from unusual timing of hormonal cycles in older women that result in unusual hormone concentrations during oogenesis. At which of the following stages of meiosis is something most likely to happen to produce a gamete that would result in a Down syndrome zygote?

a. Prophase I
b. Anaphase I
c. Prophase II
d. Anaphase II
e. Telophase II

ANALYSIS

Down syndrome results from the presence of three chromosomes, rather than the usual two, in homologous pair #21. This means that an ovum was formed that had both members of this homologous pair and was fertilized by a normal haploid spermatozoon. Movement of both chromosomes of any homologous pair to the same daughter cell would have to occur during anaphase I, when the homologs are normally separated. Disruption of meiosis II would produce abnormal gametes but not with two homologs from the same pair. It is conceivable that an event in prophase I might prevent the later separation of the homologs, but remember that all oocytes are arrested in prophase I from birth until they commence development. Thus, this phase of meiosis is normally subjected to all the concentrations of hormones that occur during many cycles. This phase is therefore probably not sensitive to hormonal conditions.

3. Parts of the human reproductive tract develop from an embryonic structure called the urogenital sinus (UGS), which is identical in male and female embryos. The ectodermal epithelium and the mesodermal connective tissue components of the UGS can be surgically separated from each other, and combinations of ectoderm and mesoderm can be recombined. Experiments of this design have led to the conclusion that the ectoderm differentiates into female or male organs depending on the hormonal environment, rather than on the genome. To differentiate at all requires inducer from UGS mesoderm; in the presence of any other organ's mesoderm, UGS ectoderm grows but does not differentiate. If testosterone is present in the medium along with UGS mesoderm, both genetically male *and* genetically female UGS ectoderm differentiates into male organs. Conversely, if testosterone is absent, both genetically male *and* genetically female UGS ectoderm differentiates into female organs. The genetic sex of the mesoderm is irrelevant. Based on these observations, which gender's organs would you expect to form in each of the following cases?

a. A genetically male embryo from which the testes were removed before sexual differentiation occurred
b. A genetically female embryo from which the ovaries were removed before sexual differentiation occurred
c. A genetically male embryo with the chromosome abnormality XXXY
d. Male UGS ectoderm cultured with pancreas mesoderm plus testosterone
e. Male UGS ectoderm cultured in the presence of testosterone in medium in which male UGS mesoderm had been cultured for several days

ANALYSIS

a. This embryo would develop female organs. UGS ectoderm develops independent of genetic sex, in a direction determined by the presence of testosterone. Without testes, there will be no testosterone, and female development will result.

b. This embryo will also develop female organs. Female development occurs whenever testosterone is absent.

c. This embryo will develop male organs. The Y chromosome will cause the development of testes, which will produce testosterone, and if testosterone levels are sufficient, male organs will develop.

d. This ectoderm will fail to develop at all because it will not receive the UGS inducer. The UGS ectoderm has apparently already differentiated to the point that it is committed to one of only two developmental choices. It cannot respond to any other inducer.

e. Your answer to this question depends on your assumptions. The mesoderm would be expected to release the appropriate inducer into the medium, unless you assume that it requires a positive feedback signal from ectoderm. Then you must assume the presence or absence of the inducer in the medium. If your assumptions predict that the inducer will be present in the medium and still active, then you predict male development. If not, you predict no development at all. The answer also depends on your assumptions about when the testosterone was added to the medium and whether ectoderm, endoderm, or both require testosterone. In fact, the testosterone is required only by the mesoderm. Apparently, the mesoderm, depending on the presence or absence of testosterone, can release either of two inducers that have opposite effects on the ectoderm.

Answers

Answers to Interactive Exercises

HUMAN REPRODUCTIVE SYSTEM (43-I)
1. seminiferous tubules; 2. epididymis; 3. testosterone; 4. vas deferens; 5. prostate; 6. lytic (digestive) enzymes; 7. midpiece; 8. Prostaglandins; 9. estrogen (progesterone); 10. progesterone (estrogen); 11. oviducts; 12. uterus; 13. estrous; 14. sphincter; 15. follicle; 16. ovulation; 17. oviduct; 18. corpus luteum; 19. endometrium; 20. Progesterone; 21. LH; 22. FSH; 23. menstruation; 24. seminal vesicle; 25. urinary bladder; 26. prostate gland; 27. bulbourethral glands; 28. vas deferens; 29. penis; 30. epididymis; 31. testis; 32. ureter; 33. seminiferous tubule; 34. erectile tissues (corpora); 35. urethra; 36. hypothalamus; 37. anterior pituitary; 38. Sertoli cells; 39. Leydig cells (interstitial cells); 40. oviduct; 41. ovary; 42. uterus; 43. urinary bladder; 44. urethra; 45. clitoris; 46. labium minor; 47. labium major; 48. endometrium; 49. vagina.

FROM FERTILIZATION TO BIRTH (43-II)
1. oviduct; 2. implantation; 3. placenta; 4. second; 5. fetus; 6. embryonic disk; 7. amniotic cavity; 8. embryonic disk; 9. amniotic cavity; 10. yolk sac; 11. embryo; 12. amnion; 13. allantois; 14. yolk sac; 15. umbilical cord; 16. four; 17. gill arches; 18. somites; 19. five; 20. forelimb.

POSTNATAL DEVELOPMENT, AGING, AND DEATH/CONTROL OF HUMAN FERTILITY (43-III)
1. 1,700,000; 2. 200,000; 3. 1,500,000; 4. abstinence; 5. Condoms; 6. diaphragm; 7. estrogens (progesterones); 8. progesterones (estrogens); 9. gonadotropins; 10. tubal ligation; 11. A, F; 12. A, C; 13. D, E; 14. F; 15. E; 16. F; 17. D; 18. C; 19. F; 20. C; 21. B; 22. C, F; 23. C; 24. B, F; 25. D, F; 26. B; 27. A, B, C, D, E, F.

Answers to Self-Quiz

1. d; 2. e; 3. a; 4. b; 5. c; 6. e; 7. d; 8. a; 9. c; 10. a; 11. b; 12. d.

44

POPULATION ECOLOGY

Interactive Exercises

FROM POPULATIONS TO THE BIOSPHERE / POPULATION DYNAMICS (44-I, pp. 788–793)

For exercises 1–3, consider the equation $G = rN$, where G = the population growth rate, r = the net reproduction per individual, and N = the number of individuals in the population.

1. Assume that r remains constant at 0.2.

   a. As the value of G increases, what happens to the value of N?_____

   b. If the value of G decreases, what happens to the value of N?_____

   c. If the net reproduction per individual stays the same and the population grows faster, then what

      must happen to the number of individuals in the population?_____

2. If a society decides it is necessary to lower its value of N through reproductive means because supportive resources are dwindling, it must either lower its net reproduction per individual or its

   _____

3. The equation $G = rN$ expresses direct relationships with r and/or N. If G remains constant and N

   increases, what must the value of r do? (In this situation, r varies *inversely* with N.)_____

4. Look at line (a) in the graph below. After seven hours have elapsed, approximately how many individuals are in the population?_____

5. Look at line (b) in the graph below.

    a. After 24 hours have elapsed, approximately how many individuals are in the population?_____

    b. After 28 hours have elapsed, approximately how many individuals are in the population?_____

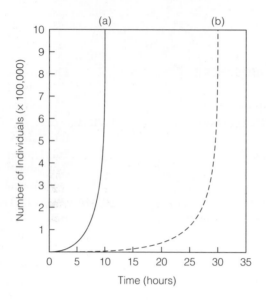

Fill-in-the-Blanks

A group of individuals of the same species occupying a given area at a specific time is a(n) (6)_____.
The place where a population lives is its (7)_____ . (8)_____ is the study of the ways in which
organisms interact with their physical and chemical environment. A(n) (9)_____ includes the
populations of all species occupying a habitat *and* their environment. (10)_____ _____ is the
number of individuals per unit of area or volume. Examples of nonrandom distribution are uniform
dispersion and (11)_____ dispersion. N = (natality + (12)_____) – (mortality + (13)_____).
Those members of a population that are in their offspring-producing time period represent the
(14)_____ _____ of the population.

  Any population that is not restricted in some way will show a pattern of (15)_____ growth,
because any increase in population size enlarges the (16)_____ base. When the course of such
growth is plotted on a graph, a (17)_____ -shaped curve is obtained. If (18)_____ factors
(essential resources in short supply) act on a population, population growth tapers off. In the equation
$G = r_{max} N (K - N / K)$, as the value of N approaches the value of K, and K and r_{max} remain constant, the
value of G (19) [choose one] ( ) increases, ( ) decreases, ( ) cannot be determined by humans, even if they

know algebra. As the value of r_{max} increases and G and N remain constant, the value of K (the carrying capacity) (20) [choose one] ( ) increases, ( ) decreases, ( ) cannot be determined by humans, even if they know algebra. (21)_____ _____ is a feature of the environment that is defined for one or more populations living in that environment and serves as a brake on runaway growth. The ability to produce the maximum possible number of new individuals is called the population's (22)_____ _____.
Sigmoid curves are characteristic of (23)_____ _____.

LIFE HISTORY PATTERNS (44-II, pp. 794–797)

Fill-in-the-Blanks

(1)_____ _____ use information summarized from life tables, which show trends in mortality and life expectancy. Type (2)_____ populations have low survivorship early in life. Food availability is a density-(3)_____ factor that works to cut back population size when it approaches the environment's (4)_____ _____. Environmental disruptions such as forest fires and floods are density-(5)_____ factors that may push a population above or below its tolerance range for a given variable.

Matching

Choose the most appropriate answer for each. A letter may be used more than once, and a blank may contain more than one letter.

6. _____ cohort
7. _____ density-independent factors
8. _____ density-dependent factors
9. _____ Reznick and Endler

A. Drought, floods, earthquakes
B. Reindeer in the Pribilof Islands in 1935
C. Food availability
D. Adrenal enlargement in wild rabbits in response to crowding
E. All of the 1987 human babies of New York City
F. Life history patterns of Trinidadian guppies

HUMAN POPULATION GROWTH (44-III, pp. 797–802)

1. Graph the following data in the space provided on page 507.

Year	Estimated World Population
1650	500,000,000
1850	1,000,000,000
1930	2,000,000,000
1975	4,000,000,000
1986	5,000,000,000
1991	5,400,000,000

a. Estimate the year that the world contained 3 billion humans._____

b. Estimate the year that Earth will house 8 billion humans._____

c. Do you expect Earth to house 8 billion humans within your lifetime?_____

Fill-in-the-Blanks

The (2)_____ _____ of a population shows how individuals are distributed in each age group.

The number of infants born each year per 1,000 women between the ages of 15 and 44 is known as the

(3)_____ _____. A world average of (4)_____ children per family is the estimated rate

that would bring us to zero population growth; if that were achieved, it would still require at least

(5)_____ years before the human population would stop growing, because (6)_____

_____. A simple way to slow things down would

be to encourage (7)_____ _____. In 1991, the human population on Earth reached

(8)_____ billion. In that year, almost (9)_____ million more individuals were added.

Chapter Terms

The following page-referenced terms are important; most were in boldface in the chapter. Refer to the instructions given in Chapter 1, p. 4.

ecology (788)
population (788)
community (788)
ecosystem (788)
biosphere (788)
zero population growth (790)

r, net reproduction per individual (790)
G, population growth rate (790)
N, number of individuals (790)
exponential growth (790)

J-shaped curve (790)
biotic potential (790)
carrying capacity (792)
logistic growth (792)
S-shaped curve (793)
density-dependent controls (793)

density-independent controls (793)
life history patterns (794)
life tables (794)
survivorship curves (795)
demographic transition model (799)

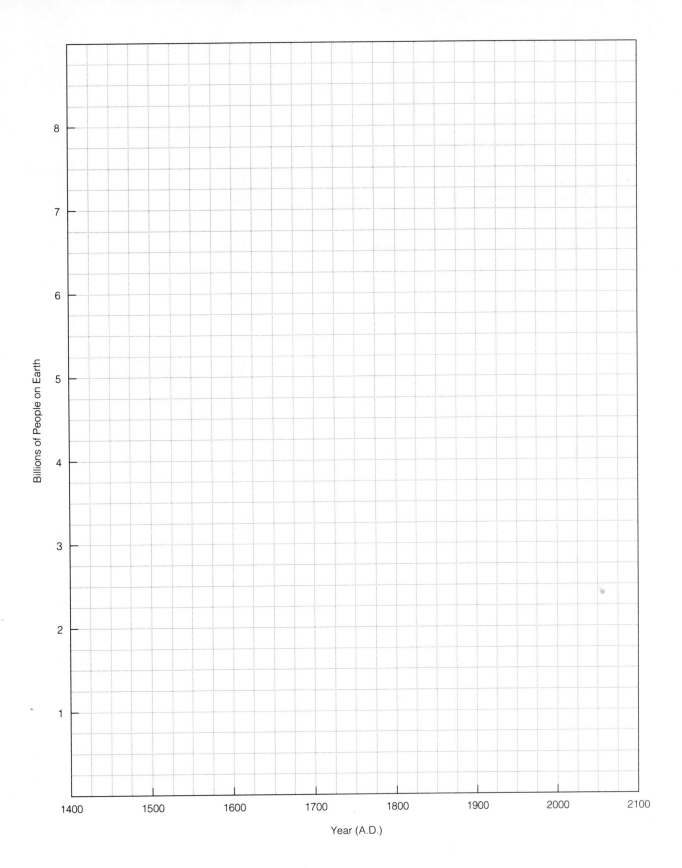

Billions of People on Earth

Year (A.D.)

Self-Quiz

____ 1. The total number of individuals
of the same species that occupy a
given area at a given time is
_____.
   a. the population density
   b. the population growth
   c. the population birth rate
   d. the population size

____ 2. The average number of individuals of the
same species per unit area at a given time
is _____.
   a. the population density
   b. the population growth
   c. the population birth rate
   d. the population size

____ 3. A population that is growing
exponentially in the absence of limiting
factors can be illustrated accurately by
a(n) _____.
   a. S-shaped curve
   b. J-shaped curve
   c. curve that terminates in a plateau
      phase
   d. tolerance curve

____ 4. If reproduction occurs early in the life
cycle, _____.
   a. higher population levels tend to result
   b. it represents an extrinsic factor that
      limits population size
   c. it represents a density-dependent
      factor that limits population size
   d. all of the above

____ 5. A situation in which the birth rate plus
immigration over the long term equal the
death rate plus emigration is called
_____.
   a. an intrinsic limiting factor
   b. exponential growth
   c. saturation
   d. zero population growth

____ 6. The rate of increase for a population (*r*)
refers to the _____ the birth rate and
death rate plus immigration and minus
emigration.

   a. sum of
   b. product of
   c. doubling time between
   d. difference between

____ 7. _____ is a way to express the growth
rate of a given population.
   a. Doubling time
   b. Population density
   c. Population size
   d. Carrying capacity

____ 8. Any population that is not restricted in
some way will grow exponentially
_____.
   a. except in the case of bacteria
   b. irrespective of doubling time
   c. if the death rate is even
      slightly greater than the birth
      rate
   d. all of the above

____ 9. Interaction between resource availability
and a population's tolerance to prevailing
environmental conditions defines
_____.
   a. the carrying capacity of the environment
   b. exponential growth
   c. the doubling time of a population
   d. density-independent factors

____ 10. In natural communities, some feedback
mechanisms operate whenever populations
change in size; they are _____.
   a. density-dependent factors
   b. density-independent factors
   c. always intrinsic to the individuals of
      the community
   d. always extrinsic to the individuals of
      the community

____ 11. Which of the following is *not* an intrinsic
factor that can influence population size?
   a. behavior
   b. metabolism
   c. predation
   d. fertility

Chapter Objectives/Review Questions

This section lists general and detailed chapter objectives that can be used as review questions. You can make maximum use of these items by writing answers on a separate sheet of paper. Fill in answers where blanks are provided. To check for accuracy, compare your answers with information given in the chapter or glossary.

Page   *Objectives/Questions*

(788–789)   1. List the three ways that individuals can be distributed in space and provide an example of each.
(790)   2. Calculate a population growth rate (*G*); use values for birth, death, and number of individuals (*N*) that seem appropriate.
(790)   3. Define zero population growth and describe how achieving it would affect the human population of the United States.
(790)   4. State how increasing the death rate of a population affects its doubling time.
(791–793)   5. Contrast the conditions that promote J-shaped curves with those that promote S-shaped curves in populations.
(792–793)   6. Understand the meaning of the logistic growth equation and know how to calculate values of *G* by using the logistic growth equation. Understand the meaning of r_{max} and *K*.
(791–793)   7. Define limiting factors and tell how they influence population curves.
(793)   8. Define density-dependent controls, give two examples, and indicate how density-dependent factors act on populations.
(793)   9. Define density-independent controls, give two examples, and indicate how such controls affect populations.
(794–795) 10. Understand the significance and use of life tables; be able to interpret survivorship curves.
(794–795) 11. Explain how the construction of life tables and survivorship curves can be useful to humans in managing the distribution of scarce resources.
(797–800) 12. List three possible reasons why the human growth rate has accelerated.
(800–801) 13. Define age structure and explain why this is the principal reason it would be 70 to 100 years before the world population would stabilize even if the world average became 2.5 children per family.
(800–801) 14. Explain how timing of reproduction can affect the degree of intraspecific competition for available resources.

Integrating and Applying Key Concepts

Assume that the world has reached zero population growth. The year is 2110, and there are 10.5 billion individuals of *Homo pollutans* on Earth. You have seen stories on the community television screen about how people used to live 120 years ago. List the ways that life has changed and comment on the events that no longer happen because of the enormous human population.

Critical Thinking Exercises

1. To estimate the number of individuals in a population of animals, the mark-recapture technique is often used. In this method, a sample of animals is captured, marked, and released. After a short time period, another sample is captured. The percentage of marked animals (recaptures) in the second sample is taken as an estimate of the percentage of marked animals in the whole population. Because the total number of marked animals is known—it is the number released from the first sample—the total number of animals in the population can be calculated. Which of the following assumptions does this method require?

   a. There is no immigration into the population during the period between the samples.
   b. There is no emigration from the population during the period between the samples.
   c. There are no births in the population during the period between the samples.
   d. There are no deaths in the population during the period between the samples.
   e. The marked individuals mix freely into the whole population during the period between the samples.

ANALYSIS

a. This assumption is not required as long as it can be assumed that any immigrants mix freely with the indigenous population and have the same probability of being caught as native individuals.

b. Emigration does not cause error in this technique as long as no marked individuals emigrate. If any marked individuals leave the population, the number used in the calculation will be in error.

c. Births would cause the estimate of population to be too low, because newborn individuals are not as mobile and hence are less likely to be caught than adults.

d. The assumption about deaths is the same as the assumption about emigration.

e. This assumption underlies the whole method. If the released animals all stay near the trapping area and are more likely to be trapped than others, the percentage of marked individuals in the second sample will be too large, and the estimate will be low. On the other hand, if the marked animals leave the area and become less likely to be caught, the error will be in the opposite direction.

2. The text says that the number of offspring per couple needed to achieve zero population growth is a little larger than 2.0, because some females die before they reach reproductive age. The text does not apply the same argument to prereproductive deaths of males. Which of the following assumptions are they most likely making?

   a. Mate choice is random.
   b. All matings are monogamous.
   c. Males may mate with more than one female.
   d. No females fail to mate.
   e. No males fail to mate.

ANALYSIS

a. This assumption is not necessary. The conclusion concerns only the overall rate of reproduction, not the success of any particular segment of the population.

b. If all matings were monogamous, then loss of a male would lower the overall birthrate the same as loss of a female.

c. This means that if one male is lost before he reproduces, the female he would have impregnated will be mated by another male, and the total production of offspring will be unchanged. If the loss of unbred males is to have no effect on population dynamics, this must be assumed.

d. and e. These assumptions need not be made. The calculation requires only that a constant percentage of each sex fails to mate. However, if some do not reproduce, the number of offspring per couple must rise to compensate.

3. Use the data presented in Table 44.1 to answer the following questions.

   a. Why is the death rate (column 4) higher in the third row than in the first, though the number of deaths in the third row is only about one-third as many as in the first row?
   b. Calculate the total number of offspring (seeds) produced in the population in each age interval.
   c. Suppose these seeds follow the same life table. Would the size of the population increase, decrease, or remain unchanged?
   d. Why would you not expect a population of mammals to have a life table like this?

ANALYSIS

a. The death rate is the number of deaths divided by the number of individuals in the population. In the third age interval, the population size has decreased by a greater factor since the first interval than has the number of deaths. Hence, the ratio has increased.

b. The number of seeds is the birth rate (column 5) times the population size.

Age Interval	Number of Seeds
0–63	0
63–124	0
124–184	0
184–215	0
215–264	0
264–278	0
278–292	0
292–306	52
306–320	482
320–334	798
334–348	972
348–362	95
362–	0

c. Your answer depends on your assumptions. If you assume that all the seeds produce individuals in the next generation, then you simply add the total number of seeds produced and, because that number is greater than the beginning population size in the table, predict that even more offspring will be produced every generation and that the population will continue to grow. On the other hand, you do not know what stage of development the plants were in during the first age interval in the table. If they were seeds, then the first assumption is valid. If they were young plants, seeds after germination, then considerable losses could have occurred before the seeds reached the stage of development represented by that first age interval. In that case, you would predict that the population was growing, stable, or shrinking, depending on your estimate of the magnitude of the losses. A stable population would begin each generation with the same number of individuals as the population in the table, produce the same total number of offspring, and suffer the same number of losses between seed set and the first age interval.

d. Young mammals require considerable parental care. This life table shows parents dying very soon after the offspring are "born." Mammalian offspring would die, too, if their parents died at an equivalent age.

Answers

Answers to Interactive Exercises

FROM POPULATIONS TO THE BIOSPHERE/ POPULATION DYNAMICS (44-I)
1. a. It increases; b. It decreases; c. It must increase; 2. population growth rate; 3. It must decrease; 4. 100,000; 5. a. 100,000; b. 300,000; 6. population; 7. habitat; 8. Ecology; 9. ecosystem; 10. Population density; 11. clumped; 12. immigration; 13. emigration; 14. reproductive base; 15. exponential; 16. reproductive; 17. J; 18. limiting; 19. decreases; 20. decreases; 21. Carrying capacity; 22. biotic potential; 23. logistic growth.

LIFE HISTORY PATTERNS (44-II)
1. Insurance companies; 2. III; 3. dependent; 4. carrying capacity; 5. independent; 6. [B], E; 7. A; 8. B, C, D; 9. F.

HUMAN POPULATION GROWTH (44-III)
1. a. 1962–63; b. 2,025 or sooner; c. depends on the age and optimism of the reader; 2. age structure; 3. fertility rate; 4. 2.4–2.5; 5. sixty; 6. people in the prereproductive base will be moving into their reproductive years, and people still in the reproductive years can still produce children; 7. delayed reproduction; 8. 5.4; 9. 92.

Answers to Self-Quiz

1. d; 2. a; 3. b; 4. a; 5. d; 6. d; 7. a; 8. b; 9. a; 10. a; 11. c.

45

COMMUNITY INTERACTIONS

Interactive Exercises

CHARACTERISTICS OF COMMUNITIES / MUTUALLY BENEFICIAL INTERACTIONS (45-I, pp. 805–807)

Fill-in-the-Blanks

A community is characterized by the kinds and diversity of species, as well as by the numbers and (1)_____ of their individuals throughout the habitat. A(n) (2)_____ is limited to all the different populations that occupy and are adapted to a defined area. The (3)_____ of a population is the sort of place where it is typically located; in contrast, the (4)_____ of a population is defined by its role in a community, including all the ecological requirements and interactions that influence that population in its community. Robins and fruit flies are (5)_____ with human populations. The flowering plants and their pollinators form (6)_____ relationships from which both participating populations benefit. (7)_____ _____ is the number of species present in a community. (8)_____ _____ is an expression of the number of individuals of each kind of organism present.

COMPETITIVE INTERACTIONS / CONSUMER-VICTIM INTERACTIONS (45-II, pp. 807–816)

Fill-in-the-Blanks

In (1)_____ , one population or individual exploits the same limited resources as another or intervenes with another sufficiently to keep it from gaining access to the resources. According to the concept of (2)_____ _____ , when two species are competing for the same resource, one tends to exclude the other from the area of niche overlap. To a greater or lesser extent, one would have the advantage; the other would be forced to modify its (3)_____. When one population denies another species access to a limited resource (usually by aggressive behavior), the tactic is called (4)_____ _____. When two or more populations share resources in different ways, in different areas, or at different times, coexistence is also possible through (5)_____ _____. Populations of the Canadian lynx and snowshoe hare undergo (6) _____ _____ , providing support for the Lotka-Volterra model of predator-prey interactions. Wolves and dogs establish (7)_____ to avoid direct competition.

(8)_____ often occurs through reciprocal selection pressures operating on two ecologically interacting populations. Prey populations attempt to avoid predation by adaptations for flight, hiding, fighting, and/or (9)_____. (10)_____ refers to adaptations in form, patterning, color, or behavior that enable an organism to blend with its background and escape detection. In (11)_____ mimicry, harmful species resemble each other.

Matching

Match each of the following with the most appropriate answer. The same letter may be used more than once. Use only one letter per blank.

12. ___ American bittern
13. ___ blood flukes (schistosomes) and humans
14. ___ blue whale and krill
15. ___ bombardier beetle and grasshopper mouse
16. ___ brown-headed cowbird, Kirtland's warbler, and botflies
17. ___ Canadian lynx and snowshoe hare
18. ___ fungal mycelia and plant root hairs
19. ___ jackpine sawfly larvae and other parasitoid larvae
20. ___ katydid
21. ___ killer whale and seals and dolphins
22. ___ *Lithiops*
23. ___ short-eared owl
24. ___ tiger and tall-stalked golden grasses
25. ___ yucca moth and yucca plant

A. Parasitism
B. Mutualism
C. Predator-prey relationship
D. Camouflage
E. Startle display

COMMUNITY ORGANIZATION, DEVELOPMENT, AND DIVERSITY (45-III, pp. 816–824)

1. There are two islands (B and C) of the same size and topography equidistant from the African coast (A), as shown in the illustration below. Which will have the higher species diversity values?_____

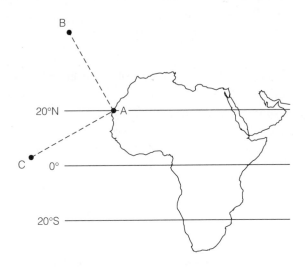

2. Name two factors responsible for creating the higher species diversity values.

   a. _____

   b. _____

Fill-in-the-Blanks

(3)_____ minimizes the intensity of competition among prey species. Two different species with similar niche requirements can occupy that niche if they are able to (4)_____ that resource—that is, use it at different times, in different areas, or in different ways. (5)_____ _____ occurs after a disturbance wipes out one or more communities; in the years that follow, earlier community stages become reestablished. (6)_____ _____ is the ratio of the total number of different species in an area to the total number of all individuals in the same area; its value (7) [choose one] ( ) increases, ( ) decreases, ( ) stays the same as one moves from Earth's poles toward the equator.

Sequence

Arrange the following in correct sequence of their appearance in the primary succession of glaciated regions of Alaska, from first to last.

8. ___   A. alders
              B. cottonwoods and willows
9. ___   C. mountain avens (*Dryas*)
10. ___   D. Sitka spruce and western hemlock
11. ___

True-False

If false, explain why.

___ 12. If two islands are equivalent distances from source areas, larger islands tend to support more species than smaller islands.

___ 13. The closer an island is to a source of potential colonists, the more different species it supports.

Chapter Terms

The following page-referenced terms are important; most were in boldface in the chapter. Refer to the instructions given in Chapter 1, p. 4.

community (805)
habitat (805)
species diversity (806)
niche (806)
species interactions (806)
symbiosis (807)
interspecific competition
  (807)

competitive exclusion (808)
predator (810)
parasite (810)
predator-prey
  interactions (810–811)
coevolution (812)
mimicry (813)

moment-of-truth
  defenses (813)
Müllerian mimicry (813)
Batesian mimicry (813)
aggressive mimicry (813)
speed mimicry (813)
camouflage (814)
true parasites (815)

parasitoids (815)
social parasites (816)
resource partitioning (816)
succession (820)
climax community (820)
primary succession (820)
secondary succession (820)

Self-Quiz

___ 1. All the populations of different species that occupy and are adapted to a given area are referred to as a(n) _____.
  a. biosphere
  b. community
  c. ecosystem
  d. niche

___ 2. The range of all factors that influence whether a species can obtain resources essential for survival and reproduction is called the _____ of a species.
  a. habitat
  b. realized niche
  c. carrying capacity
  d. ecosystem

___ 3. A one-way relationship in which one species benefits at the expense of another is called _____.
  a. commensalism
  b. competitive exclusion
  c. parasitism
  d. an obligatory relationship

___ 4. The weakest symbiotic attachment, in which one species simply lives in the presence of another species, is _____.
  a. commensalism
  b. competitive exclusion
  c. mutualism
  d. an obligate relationship

___ 5. A symbiotic relationship in which both species benefit is best described as _____.
  a. commensalism
  b. mutualism
  c. predation
  d. parasitism

___ 6. The relationship between the brown-headed cowbird and the Kirtland warbler is an example of _____.
  a. commensalism
  b. competitive exclusion
  c. mutualism
  d. parasitism

7. During the process of community succession, _____.
   a. the total biomass remains constant
   b. there are increasing possibilities for niche partitioning
   c. the pioneer community gives way quickly to the climax community, followed by a succession of more diverse arrays of organisms
   d. all of the above

8. In a community, friction between two competing populations might be minimized by _____.
   a. partitioning the niches in time
   b. partitioning the niches spatially
   c. fights to the death
   d. both a and b

9. Robins probably have a(n) _____ relationship with humans.
   a. parasitic
   b. mutualistic
   c. obligate
   d. commensal

10. The relationship between an insect and the plants it pollinates (for example, apple blossoms, dandelions, and honeysuckle) is best described as _____.
    a. mutualism
    b. competitive exclusion
    c. parasitism
    d. commensalism

11. The relationship between the yucca plant and the yucca moth that pollinates it is best described as _____.
    a. camouflage
    b. commensalism
    c. competitive exclusion
    d. an obligate relationship

12. In 1934, G. F. Gause utilized two species of *Paramecium* in a study that described _____.
    a. interspecific competition and competitive exclusion
    b. resource partitioning
    c. the establishment of territories
    d. coevolved mutualism

Chapter Objectives/Review Questions

This section lists general and detailed chapter objectives that can be used as review questions. You can make maximum use of these items by writing answers on a separate sheet of paper. Fill in answers where blanks are provided. To check for accuracy, compare your answers with information given in the chapter or glossary.

Page   Objectives/Questions

(805–806)   1. Define and distinguish between habitat and niche.
(806–807)   2. Define the following ecological terms: community, symbiosis, competition, predation, parasitism, mutualism.
(806–808)   3. Explain how the concept of competitive exclusion is related to the concept of a niche.
(806–807)   4. Define commensalism and mutualism and give one example of each.
(808)   5. Describe a study that demonstrates interference competition.
(808–809)   6. Describe a study that demonstrates laboratory evidence in support of the competitive exclusion concept.
(810–812)   7. Define predator and tell how predators benefit from one another.
(810–812, 816)   8. Discuss the positive and negative aspects of predation on prey populations.
(812)   9. Define coevolution and suggest why coevolving might serve the interests of the populations concerned.
(812–813) 10. Identify an outstanding vulnerability shared by coevolved populations.
(813–814) 11. Name an animal that utilizes a startle display behavior when cornered.
(812–814) 12. Distinguish mimicry from camouflage.
(813)   13. Contrast Batesian mimicry and Müllerian mimicry.
(816–818) 14. Discuss (by giving examples) the effects of resource partitioning and predation on interspecific competition and niche relationships.

(818–820) 15. Explain how the introduction of nonnative species can disrupt succession. List five specific examples of species introductions into the United States that have had adverse results.

(820–822) 16. Describe the sequence of communities that might occur if the climax community nearest to you were burned to the ground.

(822–823) 17. Estimate qualitatively the differences in species diversity and abundance of organisms likely to exist on two islands with the following characteristics: Island A has an area of 6,000 square miles, and Island B has an area of 60 square miles; both islands lie at 10° N latitude and are equidistant from the same source area of colonizers.

Integrating and Applying Key Concepts

If you were Ruler of All People on Earth, how would you organize industry and human populations in an effort to solve our most pressing pollution problems?

Is there a *fundamental niche* that is occupied by humans? If you think so, describe the minimal abiotic and biotic conditions required by populations of humans in order to live and reproduce. (Note that *thrive* and *be happy* are not criteria.) If you do not think so, state why.

These minimal niche conditions can be viewed as resource categories that must be protected by populations if they are to survive. Do you believe that the cold war between the United States and the Soviet Union primarily involved protection of minimal niche conditions, or do you believe that the cold war was based on other, more (or less) important factors?

a. If the former, how do you think *minimal* niche conditions might have been guaranteed for all humans willing and able to accept certain responsibilities as their contribution toward enabling this guarantee to be met?

b. If the latter, identify what you think those factors are and explain why you consider them more (or less) important than minimal niche conditions.

Critical Thinking Exercises

1. Not long after feral dogs were introduced to Australia, the native doglike marsupial disappeared. One hypothesis proposed to explain this observation was that the dogs were stronger competitors for food than the marsupials. Which of the following assumptions was most likely made in proposing this hypothesis?

a. Dogs produce larger litters of pups than marsupials.
b. Marsupials could not switch to alternative food sources.
c. Dogs introduced diseases that are lethal to marsupials.
d. The marsupials prey on puppies of the feral dogs.
e. Feral dogs and coyotes compete for food in America.

ANALYSIS

a. This could make competition more intense, but it is not a necessary assumption. Even producing smaller litters, but with higher survival rates, dogs could be more effective competitors for food species.

b. If the marsupials switched to preying on alternative food species, they would escape the competition, and both dogs and marsupials could coexist. The assumption of inability to partition resources is necessary in order to conclude that competition operates between two species.

c. This is an alternative hypothesis that would also explain the disappearance without competition.

d. This assumption would tend to lead to the opposite outcome, extinction of the dogs, and therefore does not contribute to an explanation of the observations.

e. This would show only that dogs can compete for food with some species under some circumstances. It is not a necessary assumption in order to conclude that they do compete in this situation.

2. Examine the graphs of Figure 45.18. Which of the following is the best explanation of why curve *a* is broader and flatter than curve *b*?

   a. Ocean temperature is more variable than land temperature.
   b. Tidal flow exposes molluscs to air for part of each day.
   c. Ants construct protective nests.
   d. Ants exploit a wider variety of food than molluscs.
   e. Ants are terrestrial.

ANALYSIS

   a. Ocean temperature is *less* variable than land temperature; thus, more species can exist in the ocean at high latitudes than on land where the temperature is extremely low for most of the year.
   b. This is true only for shoreline molluscs, not for the majority that live in deeper water or burrow in the substrate. Furthermore, exposure would tend to *limit*, not expand, species number by necessitating special adaptations for resistance.
   c. This would tend to *increase*, not limit, species number. Species that could protect themselves with nests would be able to invade otherwise hostile environments.
   d. Again, this explanation would be an argument for greater, not less, species diversity in ants.
   e. The land environment is much more hostile in high latitudes than the marine environment because of the very low temperatures and the snow and ice cover. Furthermore, there simply is less land area and more ocean in high latitudes, which means less niche diversity and fewer species, as on islands.

3. Beach grass is commonly planted on coastal sand dunes to stabilize them and prevent sand movement. Fewer species and smaller populations of burrowing arthropods are found in dunes with beach grass than in uncovered dunes. One hypothesis to explain this observation is that the dune grass roots form a dense mat that prevents movement by burrowing arthropods. Which of the following observations would give the strongest support to this hypothesis?

   a. Burrowing arthropods eat beach grass.
   b. Burrowing arthropods die when exposed to sand in which beach grass was grown.
   c. Burrowing arthropods are extremely rare in open sand with no plant growth at all.
   d. Burrowing arthropods are rare in dunes with any kind of plant growth that produces matted roots.
   e. The humidity among matted plant roots is much higher than in open sand.

ANALYSIS

   a. This observation would tend to contradict the hypothesis. It would lead to the prediction that arthropod populations would be higher when beach grass was present, especially if they ate the roots.
   b. This observation would support an alternative hypothesis that beach grass releases some substance toxic to burrowing arthropods.
   c. This contradicts the observations. If it were true, it would contradict the hypothesis that arthropods are excluded from stabilized dunes by matted roots, because they would already be excluded from unstabilized dunes.
   d. This observation would support the hypothesis by showing a correlation between low arthropod populations and matted roots without regard to the species of plant. This would tend to eliminate alternative hypotheses specific to the plant species, such as choice (b). Even stronger evidence could be provided by embedding artificial mats of some neutral material in dunes and demonstrating that arthropods are thereby excluded.
   e. This observation would suggest an alternative hypothesis that is also not species-specific. It could be that the arthropods are excluded by the humidity rather than by the mechanical obstruction of the roots.

Answers

Answers to Interactive Exercises

CHARACTERISTICS OF COMMUNITIES/ MUTUALLY BENEFICIAL INTERACTIONS (45-I)

1. dispersion; 2. community; 3. habitat; 4. niche;
5. commensal (symbiotic); 6. mutualistic; 7. Species richness; 8. Relative abundance.

COMPETITIVE INTERACTIONS/CONSUMER-VICTIM INTERACTIONS (45-II)

1. interference; 2. competitive exclusion; 3. niche;
4. interference competition; 5. resource partitioning;
6. cyclic replacement; 7. territories; 8. Coevolution;
9. camouflage; 10. Camouflage; 11. Müllerian; 12. D;
13. A; 14. C; 15. E (C); 16. A; 17. C; 18. B; 19. C; 20. D;
21. C; 22. D; 23. E; 24. D; 25. B.

COMMUNITY ORGANIZATION, DEVELOPMENT, AND DIVERSITY (45-III)

1. C; 2. a. Greater annual amount of sunlight promotes greater resource availability; b. Species diversity is self-reinforcing because more plant species can support a high diversity of herbivores, which can support more carnivores; 3. Predation; 4. partition; 5. Secondary succession; 6. Species diversity; 7. increases; 8. C; 9. A;
10. B; 11. D; 12. T; 13. T.

Answers to Self-Quiz

1. b; 2. b; 3. c; 4. a; 5. b; 6. d; 7. b; 8. d; 9. d; 10. a; 11. d;
12. a.

46

ECOSYSTEMS

Interactive Exercises

CHARACTERISTICS OF ECOSYSTEMS / STRUCTURE
OF ECOSYSTEMS (46-I, pp. 827–830)

Fill-in-the-Blanks

Photosynthetic (1)_____ capture (2)_____ and concentrate (3)_____ for ecosystems. All organisms in a community that are the same number of energy transfers away from the initial energy input into their ecosystem constitute a (4)_____ _____. All (5)_____ are primary consumers.

(6)_____ obtain their energy from the organic remains and products of other organisms.

(7)_____ are invertebrates that feed on partially decomposed bits of organic matter. Energy flows

(8)_____ through an ecosystem; materials are (9)_____ through an ecosystem. During every

transfer of energy, (10)_____ is released as a by-product.

Matching

Match each organism in the Antarctic food chain given below with the principal trophic level it occupies. Some letters may not be needed. (See Figure 46.3 in the text.)

11. _____ emperor penguins
12. _____ krill
13. _____ blue whale
14. _____ diatoms
15. _____ leopard seal
16. _____ fishes, small squids
17. _____ killer whale

A. Chemosynthetic autotrophs
B. Herbivores
C. Photosynthetic autotrophs
D. Secondary consumers
E. Tertiary consumers

ENERGY FLOW THROUGH ECOSYSTEMS (46-II, pp. 830–832)

Fill-in-the-Blanks

The energy remaining after respiration and contained in plant biomass is its (1)_____ _____

_____. (2)_____ typically is expressed as grams (dry weight) of organic matter per unit of area.

In a (3)_____ _____ _____ , decomposers feed on organic wastes, dead tissues, and

decomposed organic matter. Earthworms, millipedes, and fly and beetle larvae are examples of

(4)_____ feeders. Only about (5)_____ to (6)_____ percent of the energy entering one

trophic level becomes available to organisms at the next level.

True-False

____ 7. Energy pyramids are always "right-side up" with a large energy base at the bottom.

____ 8. The amount of useful energy flowing through the consumer trophic levels of an ecosystem increases at each energy transfer as metabolically generated heat is lost and as energy tied up in the bonds of food molecules is shunted into organic wastes.

BIOGEOCHEMICAL CYCLES (46-III, pp. 833–843)

Fill-in-the-Blanks

In the (1)_____ _____ , Earth's surface is warmed by sunlight and radiates (2)_____ (infrared

wavelengths) to the atmosphere and space. Greenhouse gases such as (3)_____ , which are used as

refrigerants, solvents and plastic foams, and (4)_____ _____ , which is released from burning

fossil fuels, deforestation, car exhaust, and factory emissions, will together be responsible for about 75

percent of the global warming trend by 2020. The (5)_____ _____ studies demonstrated that

stripping the land of vegetation disrupts nutrient retention by an entire ecosystem for a long time; in the

watershed, (6)_____ _____ efficiently mined the soil for calcium, moving it up into the growing

plant parts. (7)_____ and weathering of rocks brought calcium replacements back into the watershed.

In (8)_____ cycles, the element does not have a gaseous phase. In most major ecosystems, the amount of a nutrient that is (9)_____ within the ecosystem is greater than the amount entering or leaving in a given year. Water molecules stay in the atmosphere for approximately (10)_____ days.

(11)_____ can assimilate nitrogen from the air in the process known as (12)_____ _____. In (13)_____ , either ammonia or ammonium ions are stripped of electrons, and (14)_____ (NO_2^-) is released as a product of the reactions. Under some conditions, nitrate is converted into (15)_____ _____ by denitrifying bacteria. DDT was sprayed in Borneo to control (16)_____ responsible for transmitting the organisms that cause (17)_____. Fill in the blank for this food web:

detrital particles → bugs, worms → songbirds → peregrine falcons

DDT sprayings

→ eggs with (18)_____ _____

dead insects → salmon → osprey

Because of its stability, DDT is a prime candidate for (19)_____ _____—the increasing concentration of a nondegradable substance as it moves up through trophic levels. (20)_____ _____ is a method of combining crucial bits of information about an ecosystem through computer programs and models in order to predict the outcome of the next disturbance.

Chapter Terms

The following page-referenced terms are important; most were in boldface in the chapter. Refer to the instructions given in Chapter 1, p. 4.

primary producers (827)
consumers (828)
detritivores (828)
decomposers (828)
ecosystem (828)
trophic levels (828)
photosynthetic autotroph (828)
chemosynthetic autotroph (828)
food chain (829)
food webs (829)

primary productivity (830)
gross primary productivity (830)
net primary productivity (830)
grazing food webs (830)
detrital food webs (830)
ecological pyramid (831)
pyramid of biomass (831)
energy pyramid (832)
biogeochemical cycles (833)

hydrologic cycle (833)
atmospheric cycles (833)
sedimentary cycles (833)
watersheds (835)
transpiration (835)
precipitation (835)
evaporation (835)
carbon cycle (836)
greenhouse effect (838)
carbon dioxide (839)
chlorofluorocarbons (839)

methane (839)
nitrous oxide (839)
nitrogen cycle (840)
nitrogen fixation (840)
ammonification (840)
nitrification (840)
denitrification (840)
phosphorus cycle (842)
biological magnification (843)
ecosystem modeling (843)

Self-Quiz

____ 1. A network of interactions that involve the cycling of materials and the flow of energy between a community and its physical environment is a(n) _____.

a. population
b. community
c. ecosystem
d. biosphere

____ 2. One square meter of an ecosystem receives 1,600 kilocalories of light energy each day. How much energy is likely to be passed on to the ecosystem's herbivores from that square meter?

a. 160 to 480 kilocalories
b. 16 to 48 kilocalories
c. 1.6 to 4.8 kilocalories
d. 0.16 to 0.48 kilocalories

_____ 3. In the Antarctic, blue whales feed mainly on _____.
  a. petrels
  b. krill
  c. seals
  d. fish and small squids

_____ 4. _____ is a process in which nitrogenous waste products or organic remains of organisms are decomposed by soil bacteria and fungi that use the amino acids being released for their own growth and release the excess as NH_3 or NH_4^+.
  a. Nitrification
  b. Ammonification
  c. Denitrification
  d. Nitrogen fixation

_____ 5. In a natural community, the primary consumers are _____.
  a. herbivores
  b. carnivores
  c. scavengers
  d. decomposers

_____ 6. A growing animal must take in _____ of photosynthetically derived food energy to produce every 1 kilocalorie of energy stored in its body.
  a. 1 kilocalorie
  b. 10 kilocalories
  c. 100 kilocalories
  d. 2,000 kilocalories

_____ 7. Which of the following is a primary consumer?
  a. cow
  b. dog
  c. hawk
  d. all of the above

_____ 8. Of the 1,700,000 kilocalories of solar energy that entered an aquatic ecosystem in Silver Springs, Florida, H. T. Odum determined that about _____ percent was trapped during photosynthesis and used in generating plant biomass.
  a. 1
  b. 10
  c. 25
  d. 74

_____ 9. Water that passes from a plant to its surrounding air arrives there by _____.
  a. evaporation
  b. transpiration
  c. detention
  d. precipitation

_____ 10. Chemosynthesizers utilize as their direct energy source _____.
  a. energy released from degrading the organic remains and products of all other organisms
  b. energy from sunlight
  c. energy released during dehydration synthesis (condensation) reactions
  d. energy released from the oxidation of inorganic substances

Chapter Objectives/Review Questions

This section lists general and detailed chapter objectives that can be used as review questions. You can make maximum use of these items by writing answers on a separate sheet of paper. Fill in answers where blanks are provided. To check for accuracy, compare your answers with information given in the chapter or glossary.

Page	Objectives/Questions
(828)	1. List the principal trophic levels in an ecosystem of your choice; state the source of energy for each trophic level and give one or two examples of organisms associated with each trophic level.
(828, 830–831)	2. Explain why nutrients can be completely recycled but energy cannot. Compare grazing food webs with detrital food webs. Present an example of each.
(830)	3. Distinguish between *net* and *gross* primary production.
(830–831)	4. Understand how materials and energy enter, pass through, and exit an ecosystem.
(831–832)	5. Describe an important study that determined the annual pattern of energy flow in an aquatic ecosystem.
(832)	6. Explain how one goes about preparing an ecosystem analysis such as the one prepared by H. T. Odum.

Integrating and Applying Key Concepts

In 1971, *Diet for a Small Planet* was published. Frances Moore Lappé, the author, felt that people in the United States of America wasted protein and ate too much meat. She said, "we have created a national consumption pattern in which the majority, who can pay, overconsume the most inefficient livestock products [cattle] well beyond their biological needs (even to the point of jeopardizing their health), while the minority, who can not pay, are inadequately fed, even to the point of malnutrition." Cases of *marasmus* (a nutritional disease caused by prolonged lack of food calories) and *kwashiorkor* (caused by severe, long-term protein deficiency) have been found in Nashville, Tennessee, and on an Indian reservation in Arizona, respectively. Lappé's partial solution to the problem was to encourage people to get as much of their protein as possible directly from plants and to supplement that with less meat from the more efficient converters of grain to protein (chickens, turkeys, and hogs) and with seafood and dairy products. Most of us realize that feeding the hungry people of the world is not just a matter of distributing the abundance that exists—that it is being prevented in part by political, economic, and cultural factors. Devise two full days of breakfasts, lunches, and dinners that would enable you to exploit the lowest acceptable trophic levels to sustain yourself healthfully.

Critical Thinking Exercises

1. A sealed chamber was prepared containing five different organisms under optimum conditions for extended survival. One organism was a plant containing starch labeled with radioactive carbon. Another was a carnivore. If you predict that radioactive carbon would eventually be found in all the organisms, which of the following are you most likely assuming?

   a. The other three organisms were carnivores.
   b. At least one of the other organisms was an herbivore.
   c. The plant would oxidize some of its own starch to carbon dioxide.
   d. At least two of the other organisms were also plants.
   e. The carnivore respired more rapidly than the plant.

ANALYSIS

Your answer depends on your assumption about the radioactive carbon to be found in the organisms. If you assume that this radioactive carbon could simply be in some carbon dioxide molecules in the animals' body fluids, then you are most likely assuming (c). Oxidation of its own starch would be one way to release radiocarbon from the plant into the atmosphere, from where it could diffuse into the blood of any animal present. However, if one of the other organisms was an herbivore, it could eat the plant and release some radioactive carbon dioxide, or it could incorporate radiocarbon into its own molecules and then be eaten by the carnivore. The latter point illustrates the other interpretation of the question. If you assume that the radioactive carbon must be found in the organisms' own molecules, then you must be assuming that at least one of the animals is an herbivore. Only via an herbivore can radioactive carbon move from plant starch into a carnivore's molecules. This also assumes that the carnivore is a strict carnivore and never eats a little plant material.

2. Algae were grown in two aquaria, one with and one without herbivores. At the beginning and at day six, the biomasses of algae and herbivores were measured in the two aquaria. The data are shown in the table below.

		Biomass (gm)	
		Algae	Herbivores
Aquarium 1	Day 0	0.5	0
	Day 6	0.7	0
Aquarium 2	Day 0	0.5	0.1
	Day 6	0.6	0.2

Which of the following is the best interpretation of these observations?

a. The total growth of algae was less in aquarium 1.
b. The total growth of algae was the same in both aquaria.
c. The total growth of algae was greater in aquarium 1.
d. There were too many herbivores in aquarium 2.
e. If a carnivore species had been included, net algal growth would have been greater.

ANALYSIS

a. This statement clearly denies the data, which show an observable growth of 0.2 gram of algae in aquarium 1 and 0.1 gram in aquarium 2.
b. This statement would assume direct transfer of 0.1 gram of algal biomass to the herbivores. However, when biomass is transferred from one trophic level to another, inevitably there is significant loss. More than 0.1 gram of algae would be required to produce 0.1 gram of herbivore growth.
c. This statement is in accord with the analysis for choice (b). In order to grow by 0.1 gram, the herbivores would have had to eat more than 0.1 gram of algae. This means that the total growth of algae was more than 0.2 gram.
d. What defines "too many"? The net growth of algae was reduced, but because there was net positive growth the population of algae was apparently not endangered by the herbivores. As argued in choices (b) and (c), the presence of the herbivores actually stimulated total growth of algae.
e. This is a prediction that might be based on these data, but it does not interpret the observations. You might argue that the presence of a carnivore would reduce the population of herbivores and thus allow greater net growth of algae. But, if the effect of carnivores on herbivores is like that of herbivores on algae—that is, to stimulate total growth—the carnivores might result in greater consumption of algae by herbivores.

3. The following data on calcium contained in several compartments of forests were gathered in North and Central America. Note: ha = hectare = 10,000 sq. meters.

	Calcium Content (kg/ha)			
	Leaves	Litter	Soil	Wood
Northeastern hardwood forest	40	1,740	690	530
Northwestern fir forest	73	137	741	260
Tropical rainforest	55	11	176	380

Data like these are typical for elements found in forests and are often used to account for the very slow and difficult regrowth of deforested tropical rainforest. In using the data to draw such a conclusion, which of the following assumptions is most likely made?

a. Tropical trees grow more slowly than trees in other forests.
b. Tropical trees have a higher calcium content than trees in other forests.
c. There is a very low influx of calcium into tropical forests from surrounding regions.
d. Calcium is a limiting nutrient for growth of tropical forests.
e. There is a high turnover rate of calcium in tropical forests.

ANALYSIS

a. This assumption, if made, would account by itself for the slow process of reforestation without invoking mineral distribution.
b. This assumption denies the data, which show that tropical trees contain less calcium per unit area of forest (380 kg/ha) than hardwood forests (530 kg/ha).
c. This may be one cause of the observed pattern, but it does not lead from the data to the conclusion.
d. If calcium is not required for growth and calcium supply does not limit the growth rate of tropical trees, the low concentration of calcium in tropical forest soil cannot account for the low rate of reforestation growth. This assumption must be made in order to reach the conclusion.
e. This is another conclusion that can be reached from the data. Because there is almost no calcium in the litter, the calcium contained in any plant must be reabsorbed into living material almost immediately when the plant dies. This is rapid turnover of calcium. However, this is not a necessary assumption for the conclusion in the problem.

Reference: Jordan, C. F.; Kline, J. R.; Sasscer, D. S. (1972) *Amer. Naturalist* 106:237–253.

Answers

Answers to Interactive Exercises

CHARACTERISTICS OF ECOSYSTEMS/ STRUCTURE OF ECOSYSTEMS (46-I)
1. autotrophs; 2. sunlight; 3. energy; 4. trophic level; 5. herbivores; 6. Decomposers; 7. Detritivores (crabs, nematodes or earthworms); 8. one-way; 9. cycled; 10. heat; 11. E; 12. B; 13. D; 14. C; 15. E; 16. D; 17. E.

ENERGY FLOW THROUGH ECOSYSTEMS (46-II)
1. net primary production; 2. Biomass; 3. detrital food web; 4. detrital; 5. 6; 6. 16; 7. T; 8. F.

BIOGEOCHEMICAL CYCLES (46-III)
1. greenhouse effect; 2. heat; 3. CFCs; 4. carbon dioxide; 5. Hubbard Brook; 6. tree roots; 7. Erosion (Rainfall); 8. sedimentary; 9. cycled; 10. ten; 11. Bacteria (*Rhizobium*); 12. nitrogen fixation; 13. nitrification; 14. nitrite; 15. nitrous oxide; 16. mosquitoes; 17. malaria; 18. thin shells (brittle shells); 19. biological magnification; 20. Ecosystem modeling.

Answers to Self-Quiz
1. c; 2. c; 3. b; 4. b; 5. a; 6. b; 7. a; 8. a; 9. b; 10. d.

47

THE BIOSPHERE

Interactive Exercises

CHARACTERISTICS OF THE BIOSPHERE (47-I, pp. 847–850)

Matching

Choose the single most appropriate letter for each.

1. ___ climate
2. ___ doldrums
3. ___ greenhouse effect
4. ___ gyres
5. ___ lithosphere
6. ___ westerly
7. ___ ozone
8. ___ rain shadow
9. ___ trade winds
10. ___ upwellings
11. ___ Benguela current
12. ___ Canary current
13. ___ California current
14. ___ Gulf Stream
15. ___ equatorial countercurrent
16. ___ Humboldt current
17. ___ Japan current
18. ___ Labrador current
19. ___ North Atlantic drift
20. ___ west wind drift

A. Absorbs ultraviolet wavelengths of incoming solar radiation
B. Located between 0° and ±30° north and south
C. Regions of rising, warmed air spreading northward and southward
D. Caused by lack of abundant precipitation and intervening mountains
E. Regions of rising, nutrient-laden water
F. Prevailing weather conditions that are caused by four principal factors
G. Caused by molecules of the lower atmosphere absorbing some heat and then reradiating it back to Earth
H. A wind that blows in an eastward direction
I. Rocks, soils, sediments, and outer portions of crust
J. Circular movements of large water masses
K. A cold water mass that maintains the temperate rain forest associated with Washington and Oregon
L. A southward-flowing current that runs into the Gulf Stream and creates eddies
M. Feeds into the Gulf of Mexico
N. Southward off Spain and West Africa
O. An eastward-flowing warm-water current in the Pacific Ocean
P. Splits to give rise to a northeastward drift and a southward current
Q. Off the west coast of Peru
R. An eastward-flowing cold-water current of the southern hemisphere
S. Flows past the northern coasts of Great Britain and Sweden
T. Merges with the California current in the northern Pacific gyre

Fill-in-the-Blanks

Besides being the primary energy source for ecosystems, (21)_____ _____ influences their distribution on Earth's surface. (22)_____ energy derived from the sun warms the atmosphere and drives Earth's weather systems. Global air circulation patterns, ocean currents, and topographic features interact to produce regional variations in temperature and (23)_____ , which influence the composition of (24)_____ and sediments, which, in turn, influence the growth and distribution of the primary (25)_____ and, through them, the distribution of (26)_____ and ecosystems. When descending dry air warms, it picks up moisture from the land and leaves it very dry so that (27)_____ are created; most of these occur at about 30° latitudes.

THE WORLD'S BIOMES (47-II, pp. 851–861)

Fill-in-the-Blanks

(1)_____ is formed as the products of living and dead organisms (humus) are mixed with weathered and eroded loose rock. As rainwater percolates down through the topsoil, minerals dissolve in it (that is, are (2)_____) and are carried along with particles to the (3)_____. Tree roots penetrate far below the soil surface, breaking up rocks as they go and absorbing minerals in solution. Arid regions that support little primary production are known as (4)_____. Scrubby plants are dominant primary producers in (5)_____ _____ , and cacti, yuccas, and other drought-resistant plants dominate (6)_____ _____. At one time, (7)_____ extended westward from the Mississippi River region to the Rocky Mountains, from Canada through Texas. (8)_____ _____ were converted into vast fields of corn and wheat, and (9)_____ _____ were converted into ranges for imported cattle. Overgrazing and ill-advised attempts to farm the region led to massive erosion, which crippled the primary productivity of much of this land.

Warm temperatures combined with uniform, abundant rainfall encourage the growth of (10)_____ _____ _____ , which contain (11)_____ communities dominated by trees spreading their (12)_____ far above the forest floor. This type of biome exists where the annual mean temperature is about (13)_____°C and where heavy, regular rainfall coincides with (14)_____ of 80 percent or more and produces more litter than any other forest biome; however, (15)_____ and mineral cycling are rapid in the hot, humid climate, and its soils are not a good reservoir of nutrients. In (16)_____ _____ forests, most plants lose their leaves during part of the year as an adaptation to the dry seasons that alternate with the wet monsoons. In (17)_____ _____ forests, most of the dominant trees drop their leaves—not in response to dry seasons, but as protection against (18)_____ _____. At lower elevations of the montane coniferous forest, (19)_____ and Douglas fir predominate. (20)_____ is synonymous with boreal forest. Just beneath the surface of the (21)_____ is the permafrost. The main limiting factor in the tundra is the low level of (22)_____ _____.

THE WATER PROVINCES (47-III, pp. 862–871)

Fill-in-the-Blanks

Water is densest at 4°C; at this temperature, it sinks to the bottom of its basin, displacing the nutrient-rich bottom water upward and giving rise to spring and fall (1)_____. Portions of a lake where light penetrates to the bottom compose the (2)_____ zone, where rooted vegetation and decomposers are abundant. The (3)_____ zone includes areas of a lake *below* the depth where sufficient light penetration balances the respiration and photosynthesis of the primary producers; anaerobic bacterial decomposers are the principal organisms here. (4)_____ lakes are nutrient-poor. In bodies of water that are sufficiently deep, there is a depth at which the temperature of the water decreases rapidly with a small increase in depth; this region is known as the (5)_____. A region where freshwater mixes with saltwater is a(n) (6)_____. Organisms that depend on currents and drifts to distribute them are collectively referred to as (7)_____. Most marine ecosystems fall within the shallow (8)_____ _____ (between the high and low tide marks) and the (9)_____ _____ (between 2,000 and 4,000 meters beneath the ocean's surface). The (10)_____ _____ is a geothermal ecosystem 2,500 meters beneath the ocean's surface, where sulfur-oxidizing bacteria are the primary producers.

(11)_____ is a marsh grass that is the dominant primary producer in New England salt marshes.

Chapter Terms

The following page-referenced terms are important; most were in boldface in the chapter. Refer to the instructions given in Chapter 1, p. 4.

biosphere (847)	biogeographic realms (851)	spring overturn (862)	benthic province (869)
hydrosphere (847)	biome (851)	fall overturn (862)	hydrothermal vents (870)
atmosphere (847)	soil (852)	streams (864)	upwelling (870)
climate (848)	permafrost (861)	intertidal zone (866)	El Niño Southern Oscillation,
rain shadow (850)	plankton (862)	pelagic province (867)	ENSO (870–871)

Self-Quiz

___ 1. The Galápagos Rift, a geothermal ecosystem 2,500 meters beneath the ocean's surface, has _____ as its primary producers.
   a. blue-green algae
   b. protistans
   c. nitrogen-fixing organisms
   d. chemosynthetic organisms

___ 2. In a(n)_____ , water draining from the land mixes with seawater carried in on tides.
   a. abyssal zone
   b. rift zone
   c. upwelling
   d. estuary

___ 3. A biome with grasses as primary producers and scattered trees adapted to prolonged dry spells is known as a _____.
   a. warm desert
   b. savanna
   c. tundra
   d. taiga

___ 4. Monocrops of fast-growing softwood trees _____.

   a. are replacing climax forests in parts of the national forest system
   b. are less vulnerable to insect predators than constituents of a mixed deciduous climax forest
   c. use fewer nutrients than climax forests
   d. are desirable for their nitrogen-fixing ability

___ 5. In tropical rain forests, _____.
   a. competition for available sunlight is intense
   b. diversity is limited because the tall forest canopy shuts out most of the incoming light
   c. conditions are extremely favorable for growing luxuriant crops
   d. all of the above

___ 6. In a lake, the open sunlit water with its suspended phytoplankton is referred to as its _____ zone.
   a. epipelagic
   b. limnetic
   c. littoral
   d. profundal

___ 7. Differences in _____ determine the distribution of producer organisms in marine ecosystems.
   a. the intensity of incoming solar radiation
   b. surface water salinity
   c. the availability of nutrients
   d. all of the above

___ 8. _____ is least influential in determining the distribution of biomes on land.
   a. Light intensity
   b. Rainfall
   c. Salinity
   d. Temperature

___ 9. A wind system that influences large climatic regions and reverses direction seasonally, producing dry and wet seasons, is referred to as a(n)_____.
   a. geothermal ecosystem
   b. upwelling
   c. taiga
   d. monsoon

___ 10. _____ *least* affects the amount of incoming light that strikes an area.
   a. Latitude
   b. Temperature
   c. The amount of recurring cloud cover
   d. The degree that a slope is exposed to incoming light

Chapter Objectives/Review Questions

This section lists general and detailed chapter objectives that can be used as review questions. You can make maximum use of these items by writing answers on a separate sheet of paper. Fill in answers where blanks are provided. To check for accuracy, compare your answers with information given in the chapter or glossary.

Page		Objectives/Questions
(847, 851)	1.	Describe the ways in which climate affects the biomes of Earth and influences how organisms are shaped and how they behave.
(848–850)	2.	Explain what causes the prevailing air currents. Compare the prevailing air currents with the prevailing ocean currents and state whether or not they are similar.
(849, 854)	3.	State why land masses heat and cool more rapidly than oceans.
(852–853)	4.	Describe the typical layered structure of soils and state your understanding of how each layer was formed from its original parent material—solid rock.
(852–861)	5.	State the relationship between temperature and rainfall (on the one hand) and the abundance and diversity of producers (on the other hand) in the different biomes.
(856–857)	6.	List the factors that influence the distribution of biomes on land. Then consider the prairie biome and indicate the factors that can bring about more specialized ecosystems within the prairie biome.
(858–859)	7.	List the factors that encourage the development of broadleaf forests.
(858–859)	8.	Describe a broadleaf deciduous forest and construct a typical food web for this biome.
(860–861)	9.	List the factors that encourage the development of evergreen coniferous forests.
(858)	10.	Describe a monsoon forest and explain what causes a monsoon forest to be deciduous rather than to be a rain forest.
(861)	11.	Describe the physical and biotic features of the tundra ecosystem. Distinguish between alpine and arctic tundra.
(862–863)	12.	Describe the causes of temperature stratification in bodies of water.
(862)	13.	Discuss how spring and fall overturns can occur in freshwater ecosystems.
(862–864; 865–870)	14.	Contrast life in lake ecosystems with that in oceans and estuaries.
(865–870)	15.	List the variables that determine how producer organisms (and the consumer organisms that depend on them) are distributed in marine environments.
(870)	16.	Explain how a hydrothermal vent ecosystem operates.

Integrating and Applying Key Concepts

One species, *Homo sapiens*, uses about 40 percent of all of Earth's productivity, and its representatives have invaded every biome, either by living there or by dumping waste products there. Many of Earth's residents are being denied the minimal resources they need to survive, while human populations continue to increase exponentially. Can you suggest a better way of keeping Earth's biomes healthy while providing at least the minimal needs of *all* Earth's residents (not just humans)? If so, outline the requirements of such a system and devise a way in which it could be established.

Critical Thinking Exercises

1. Bags made of fine nylon mesh are used in studies of the rate of breakdown of litter in forests. Fresh litter is placed in a bag, weighed, and replaced on the forest floor. At intervals, the bagged litter is again weighed. Weight loss is counted as breakdown. Which of the following assumptions must be made in order to use the technique?

   a. Detritivores can freely enter and leave the bag.
   b. Detritivores are insignificant contributors to breakdown of forest litter.
   c. The weight of detritivores is an insignificant fraction of the weight of litter.
   d. No undecomposed fragments of litter can fall out of the bag.
   e. The water content of the bagged litter is constant.

   ANALYSIS

   a. This assumption may or may not be made. The choice, however, affects the other assumptions that are necessary.
   b. If you assume that the bag excludes detritivores, then you must assume that detritivores do not contribute significantly to the process you are studying. You want to infer the rate of litter breakdown on the open forest floor from data taken on litter inside the bag. You have to assume that conditions inside the bag are not significantly different from conditions outside the bag.
   c. If you assume that detritivores freely move in and out of the bag, then they will contribute something to the weight of the bag. Unless you can calculate that contribution and correct for it, you have to assume that it is insignificant relative to the weight of the litter.
   d. If any undecomposed fragments of litter fall out through the mesh, they will represent a weight loss not due to decomposition and will lead to a falsely high estimate of breakdown rate.
   e. Water is a dense substance and contributes a great deal to the weight of organic matter, dead or alive. If you weigh the whole bag and ascribe weight loss over time to decomposition, you must be assuming constant water concentration. Alternatively, you could take a sample of the material inside the bag, dry it in the laboratory, weigh it, and calculate the dry weight of the bag's contents. These weights would be comparable at all times.

2. Eutrophic lakes are characterized by enormous growth of photosynthetic algae that release oxygen. Why, then, do these lakes undergo severe oxygen depletion?

   a. The algae exhaust the available supply of phosphorus and stop photosynthesis.
   b. Fish in the lakes die as a result of the low oxygen concentration.
   c. The rate of decomposition of dead algae becomes greater than the rate of photosynthesis by living algae.
   d. Oxygen diffuses into the lake water from the atmosphere.
   e. Oxygen diffuses out of the lake water into the atmosphere.

ANALYSIS

a. Even if the production of new algal cells stops because of exhaustion of phosphorus, photosynthesis will continue in the cells already there. Phosphorus is a required nutrient for ATP production, nucleic acid synthesis, and cell reproduction, but not for photosynthesis.

b. This is a common consequence of eutrophic oxygen depletion, but it does not explain the depletion.

c. Decomposition is respiration of decomposers, which consumes oxygen. The rate of change of oxygen concentration is the difference between processes that add oxygen and processes that remove oxygen. As the amount of dead algae becomes larger, the population of decomposers also grows, the rate of oxygen consumption by decomposers can exceed the rate of oxygen production by the living algae, and the oxygen concentration will decrease.

d. This would tend to reverse the depletion, not account for it.

e. Diffusion can only equalize concentrations of a substance between two compartments. This case is development of an inequality. Diffusion cannot accomplish that.

3. The experiment in Figure 47.25 is interpreted to indicate that phosphorus, not carbon, is the growth-limiting nutrient in lakes. However, plant tissue contains a much higher concentration of carbon than of phosphorus. Furthermore, the concentration of carbon in lake water is also very low. Which of the following statements contribute to an explanation of this apparent discrepancy?

a. Photosynthetic enzymes function at low carbon dioxide concentrations.

b. Carbonate ion is constantly dissolved into the lake water from the inorganic substrate.

c. Carbon dioxide constantly diffuses into the lake water from the atmosphere.

d. Carbon dioxide is constantly released into the lake water by decomposers.

e. Nucleotides are continually broken down.

ANALYSIS

The problem here is to account for the movement of large amounts of carbon out of a reservoir that contains only a little carbon. It is analogous to asking how a rubber hose that contains only a little water can deliver large volumes of water onto a lawn in a short time. The answer is that a lot of water flows into the hose from the faucet as the water in the hose is delivered out the other end. In this problem, some sources of carbon are identified in choices (b), (c), and (d). The resolution of the conflict is that these sources deliver enough carbon into the lake to support growth, whereas the sources of phosphorus are more limited, and growth can continue only as rapidly as phosphorus is delivered. Choice (a) is also part of the answer; enzymes that can utilize carbon at a low concentration allow growth to consume the entering carbon as fast as it becomes available without requiring accumulation of a supply. Choice (e), on the other hand, would tend to provide a continuous phosphorus supply because phosphorus is found in nucleotides. This would lead to the argument that phosphorus should not be the limiting nutrient.

Answers

Answers to Interactive Exercises

CHARACTERISTICS OF THE BIOSPHERE (47-I)
1. F; 2. C; 3. G; 4. J; 5. I; 6. H; 7. A; 8. D; 9. B; 10. E; 11. M; 12. N; 13. K; 14. P; 15. O; 16. Q; 17. T; 18. L; 19. S; 20. R; 21. solar radiation; 22. Heat; 23. rainfall; 24. soils; 25. producers; 26. biomes; 27. deserts (rain shadows).

THE WORLD'S BIOMES (47-II)
1. Soil; 2. leached; 3. subsoil; 4. deserts; 5. dry shrublands (cold deserts); 6. warm deserts; 7. prairies (grasslands); 8. Tallgrass prairies; 9. shortgrass prairies; 10. tropical rain forests; 11. vertical (stratified); 12. canopies; 13. 25; 14. humidity; 15. decomposition; 16. tropical deciduous; 17. temperate deciduous; 18. freezing weather; 19. pine; 20. Taiga; 21. tundra; 22. soil nutrients (solar radiation).

THE WATER PROVINCES (47-III)
1. overturns; 2. littoral; 3. profundal; 4. Oligotrophic; 5. thermocline; 6. estuary; 7. plankton; 8. intertidal zone; 9. abyssal zone (bathypelagic); 10. hydrothermal vent; 11. *Spartina*.

Answers to Self-Quiz

1. d; 2. d; 3. b; 4. a; 5. a; 6. b; 7. d; 8. c; 9. d; 10. b.

48

HUMAN IMPACT ON
THE BIOSPHERE

ENVIRONMENTAL EFFECTS OF HUMAN
POPULATION GROWTH

CHANGES IN THE ATMOSPHERE
    Local Air Pollution
    Acid Deposition
    Damage to the Ozone Layer

CHANGES IN THE HYDROSPHERE
    Consequences of Large-Scale
    Irrigation
    Maintaining Water Quality

CHANGES IN THE LAND
    Solid Wastes
    Conversion of Marginal Lands for Agriculture
    Deforestation
    Desertification

A QUESTION OF ENERGY INPUTS
    Fossil Fuels
    Nuclear Energy
    Commentary: Biological Principles
    and the Human Imperative

Interactive Exercises

ENVIRONMENTAL EFFECTS OF HUMAN POPULATION GROWTH /
CHANGES IN THE ATMOSPHERE (48-I, pp. 875–880)

Fill-in-the-Blanks

(1)_____ _____ _____ attack marble, metals, mortar, rubber, and plastic; they also form

droplets of (2)_____ _____ that create holes in nylon stockings. When a layer of dense, cool air

gets trapped beneath a layer of warm air, the situation is known as a(n) (3)_____ _____. When

(4)_____ and nitrogen dioxide are exposed to sunlight, they are converted into poisonous substances

collectively called (5)_____ smog. (6)_____ _____ kills fish in northeastern U.S. lakes and

streams, while greenhouse gases such as (7)_____ destroy the ozone layer and raise the (8)_____

_____ _____ on Earth, melting the ice caps and flooding coastal lands. The human population

has undergone rapid exponential growth since the (9) mid-_____ century. (10)_____ are

substances with which ecosystems have had no prior evolutionary experience and that therefore have no

mechanisms for dealing with them. The precipitation in much of eastern North America is (11) [choose

one] ( ) 30, ( ) 50, ( ) 75, ( ) 100 times more acidic than it was several decades ago, and croplands and

(12)_____ are suffering. The (13)_____ layer in the lower stratosphere has been thinning, due

mostly to odorless, invisible (14)_____.

CHANGES IN THE HYDROSPHERE / CHANGES IN THE LAND (48-II, pp. 880–886)

Fill-in-the-Blanks

In a generalized resource recovery system, (1)_____ could be used to extract steel and iron, and (2)_____ _____ could send plastic and paper to different recovery chambers. In (3)_____ waste-water treatment, mechanical screens and sedimentation tanks are used to force coarse suspended solids out of the water to become a sludge. (4)_____ treatment involves advanced methods of precipitation of suspended solids and phosphate compounds.

(5)_____ on steep slopes leads to soil erosion and watershed disruption. The buildup of salt in irrigated soils is called (6)_____. Farmers take so much water from the (7)_____ _____ that the overdraft nearly equals the annual flow of the Colorado River. (8)_____ accounts for almost two-thirds of the human population's annual use of freshwater; (9)_____ may not ever be cost-effective enough to supply the needs of agriculture. Mechanized agriculture is based on massive inputs of fertilization, pesticides, and ample irrigation to sustain (10)_____-_____ crops. (11)_____ _____ energy instead of animal energy is used to drive farm machines, and crop yields are (12) [choose one] ( ) 4, ( ) 8, ( ) 20 times as high in green revolution countries, but the modern practices use up (13) [choose one] ( ) 5, ( ) 25, ( ) 50, ( ) 100 times more energy and mineral resources than does subsistence agriculture. Today, (14)_____ on marginal lands is the main cause of large-scale desertification.

A QUESTION OF ENERGY INPUTS (48-III, pp. 887–890)

Fill-in-the-Blanks

The Chernobyl power station experienced a (1)_____ , and (2)_____ was released into the atmosphere as the plant's containment structures were breached. Oil shale is buried rock that contains (3)_____ , a hydrocarbon compound. Nuclear-powered electricity-generating plants produce (4)_____ percent of the United States' electrical energy. Only (5)_____ has good long-term, intermediate, and short-term availability as an energy option for the United States. Unlike conventional nuclear-fission reactors, (6)_____ reactors could potentially explode like atomic bombs. A nuclear exchange involving about one-third of the existing American and Soviet arsenals would probably kill between 40 and 65 percent of the human population and most other forms of life; the detonations would create a (7)_____ _____ , in which a huge dark cloud of soot and smoke would block out the sun. (8)_____ _____ is the energy left over after subtracting the energy used to locate, extract, transport, store, and deliver energy to consumers. Energy supplies in the form of (9)_____ _____ are nonrenewable and dwindling, and their extraction and use come at high environmental cost. (10)_____ _____ normally does not pollute the environment as much as fossil fuels do, but the costs and risks associated with fuel (11)_____ and with storage of (12)_____ wastes are enormous.

Chapter Terms

The following page-referenced terms are important; most were in boldface in the chapter. Refer to the instructions given in Chapter 1, p. 4.

pollutants (876)
thermal inversion (876)
industrial smog (876)
photochemical smog (877)
dry acid deposition (877)

wet acid deposition (877)
chlorofluorocarbons (880)
salination (880)
desalination (880)
green revolution (882)

subsistance
 agriculture (883)
deforestation (883)
shifting cultivation (883)
desertification (886)

net energy (887)
fossil fuels (887)
nuclear energy (887)
meltdown (888)

Self-Quiz

___ 1. Which of the following processes is not generally considered a component of tertiary waste-water treatment?
   a. microbial action
   b. precipitation of suspended solids
   c. reverse osmosis
   d. adsorption of dissolved organic compounds

___ 2. When fossil-fuel burning gives off particulates and sulfur oxides, we have _____.
   a. photochemical smog
   b. industrial smog
   c. a thermal inversion
   d. both (a) and (c)

___ 3. _____ result(s) when nitrogen dioxide and hydrocarbons react in the presence of sunlight.
   a. Photochemical smog
   b. Industrial smog
   c. A thermal inversion
   d. Both (a) and (c)

___ 4. Between _____ of urban wastes consists of paper products.
   a. 10 and 20 percent
   b. 20 and 40 percent
   c. 50 and 65 percent
   d. 75 and 90 percent

___ 5. About _____ of the waste water in the United States is not even receiving primary treatment.
   a. 20 percent
   b. 30 percent
   c. 50 percent
   d. 65 percent

___ 6. The wastes from a nuclear power reactor must be kept out of the environment for _____ years before they are safe.
   a. 25
   b. 250
   c. 1 million
   d. 250,000

___ 7. What proportion of the world's human population does *not* have enough pure water?
   a. one-fourth
   b. one-half
   c. three-fourths
   d. two-thirds

___ 8. For every million gallons of water in the world, only about _____ gallons are in a form that can be used for human consumption or agriculture.
   a. 6
   b. 60
   c. 600
   d. 6,000

___ 9. Each day in the United States, _____ metric tons of pollutants are discharged into the atmosphere.
   a. 1,000
   b. 100,000
   c. 700,000
   d. 5,000,000

___ 10. The most abundant fossil fuel in the United States is _____.
   a. carbon monoxide
   b. oil
   c. natural gas
   d. coal

Chapter Objectives/Review Questions

This section lists general and detailed chapter objectives that can be used as review questions. You can make maximum use of these items by writing answers on a separate sheet of paper. Fill in answers where blanks are provided. To check for accuracy, compare your answers with information given in the chapter or glossary.

Page	Objectives/Questions
(876)	1. Identify the principal air pollutants, their sources, their effects, and the possible methods for controlling each pollutant.
(876–877)	2. Define thermal inversion and indicate its cause.
(876–877)	3. Distinguish photochemical smog from industrial smog.
(877–878, 879–880)	4. Explain what acid rain does to an ecosystem. Contrast those effects with the action of CFCs.
(880, 882–886)	5. Examine the effects that modern agriculture has wrought on desert and grassland ecosystems.
(881–882)	6. Define primary, secondary, and tertiary waste-water treatment and list some of the methods used in each of the three types of treatment.
(882)	7. Distinguish a recycling system for solid wastes from an ecologically based system. State the conditions under which an ecologically based system for handling solid wastes would benefit humans more than a recycling system would.
(883–885)	8. Explain the repercussions of deforestation that are evident in soils, water quality, and genetic diversity in general.
(887)	9. State when the U.S. petroleum reserves are expected to run out if present consumption rates are maintained. State when the U.S. natural gas reserves are expected to run out.
(887)	10. Explain why exploiting oil shale deposits may not be worth doing.
(887–890)	11. Describe how our use of fossil fuels, solar energy, and nuclear energy affects ecosystems.
(888–889)	12. Compare the functions of a conventional nuclear power plant reactor with those of a breeder reactor and assess the principal benefits and risks of each.
(888)	13. Explain what a meltdown is.
(890)	14. Understand the magnitude of pollution problems in the United States.
(890)	15. List five ways in which you could become personally involved in ensuring that institutions serve the public interest in a long-term, ecologically sound way.

Integrating and Applying Key Concepts

If you were Ruler of All People on Earth, how would you encourage people to depopulate the cities and adopt a way of life by which they could supply their own resources from the land and dispose of their own waste products safely on their own land?

Explain why some biologists believe that the endangered species list now includes all species.

Critical Thinking Exercises

1. Assessment of hazards due to chemicals or radiation is based in part on dose-effect curves, in which some measure of the toxic or damaging effect is plotted as a function of some expression of the amount of substance or radiation applied. Which of the following dose-effect curves would lead to the conclusion that there is no safe level of radiation?

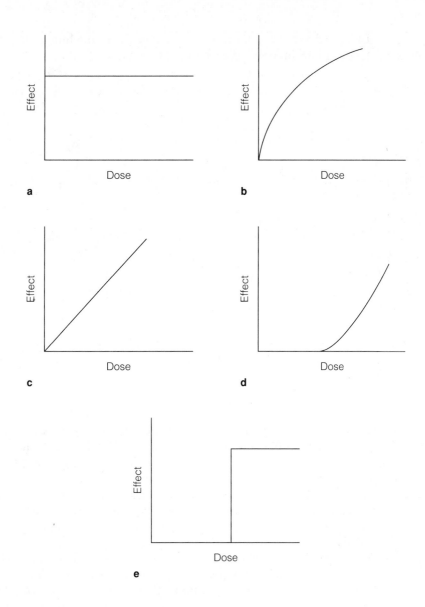

ANALYSIS

   a. This curve shows an effect that is independent of dose. No matter what the exposure, the effect is the same. This would indicate that any exposure is equally as dangerous as any other and that there is no safe dose. However, this is not a realistic situation. The allergic reaction to a substance is largely independent of dose; a full response is triggered by a very small exposure. But ordinary toxic effects mediated by chemical reactions or radiation damage are dose-dependent. Furthermore, as drawn, the curve shows a maximum effect even at zero dose, which is nonsense.

   b. This curve shows rapidly increasing effects at low doses and a lower slope at higher doses, possibly leading to a saturation effect. Because a detectable effect results from all nonzero doses, there is no absolutely safe dose. A decision has to be made about what risk level is acceptable in order to determine what level of exposure is tolerable.

   c. This is a linear response such as is often assumed to be the case for radiation exposure. Again, there is no dose level that causes no damage, and a decision of tolerable risk must be made.

   d. This substance shows a threshold behavior. Below a certain dose, it is not effective, and doses in this range could be regarded as safe if the substance does not accumulate in an exposed organism.

   e. This pattern shows an all-or-none response with a threshold. As for (d), low doses could be deemed safe if the substance does not accumulate.

2. The two curves below represent two models for the long-term future of the human population. From the following list, select the assumptions inherent in each of the two models.

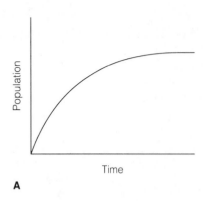

A

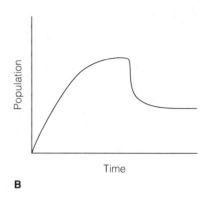

B

a. The environment imposes a carrying capacity.
b. All essential resources are renewable or inexhaustible.
c. The environment is stable with only minor fluctuations around a mean condition.
d. The carrying capacity can be enlarged indefinitely by technology.
e. The carrying capacity can be enlarged temporarily by technology.

ANALYSIS

a. Both models assume a carrying capacity. In A, the population rises to the capacity and remains stable thereafter. In B, the population rises beyond the carrying capacity, and then crashes and stabilizes.
b. Both models assume that no essential resources become exhausted. If any required resources did disappear, the population would become extinct.
c. Model A makes this assumption. Major environmental changes would reset the carrying capacity at a new level or cause temporary swings in population. Model B assumes environmental changes that trigger the crash. Some essential resource becomes temporarily reduced to a level insufficient to sustain the inflated population.
d. Model A may or may not make this assumption. The graph does not indicate whether the stable population is at a natural or a technologically inflated carrying capacity.
e. Model B makes this assumption. The final stable population is at the natural carrying capacity. The precrash population was artificially inflated by technology until an unmanageable environmental change occurred.

3. a. Figure 48.13 in the text shows the current world consumption of various forms of energy. Suppose the world population decided to convert entirely to the use of renewable energy. By what factor would the production of renewable energy have to increase?
b. The United States has about 6 percent of the world population, and United States residents consume about 230,000 kilocalories of energy per person per day. In contrast, the worldwide average energy consumption is about 40,000 kilocalories per person per day. (Only about 2,000 to 3,000 kilocalories per person per day is directly consumed as food.) In order to bring the entire world to the United States' level of energy consumption, by what factor would the total worldwide energy production have to increase?
c. To accomplish this rise in world standard of living with only renewable energy, by what factor would the production of renewable energy have to increase?
d. In industrial England around 1875, the energy consumption was 70,000 kilocalories per person per day. To achieve this standard of living worldwide with renewable energy sources would require what level of increase in renewable energy production?
e. The figure shows only the biosphere and hydropower as renewable energy sources, but the biosphere is actually shrinking, and almost all the suitable dam sites have been developed. What other sources of renewable energy are there?

f. What major assumption was made in the above calculations? What other assumptions must be made in order to consider any of the multiplications of renewable power production feasible?

ANALYSIS

a. About 20 percent of current use is of renewable sources. To make that 100 percent, multiply by 5.

b. Total consumption now = $0.06 \times 230,000 + 0.94 \times 40,000$
   Total goal consumption = $1.00 \times 230,000$

   Ratio of goal to now = 4.5

c. $5 \times 4.5 = 22.5$

d. To bring the world to 70,000 kilocalories per person per day is a 1.75-fold increase of the present 40,000. This must also be multiplied by 5 to convert to renewable sources: 8.75

e. Renewable sources include solar power, fusion and other hypothetical technological advances, and minor sources such as geothermal power, windmills, and tidal basins.

f. The most important assumption by far in the calculations was that population is constant. All factors increase as population increases.

In order to accomplish the conversion, adequate supplies of raw materials must be available for the new equipment, the project must not produce intolerable amounts of chemical wastes, and there must be sufficient educated personnel, industrial plants, and transportation systems. All of these are assumptions. There are doubtless many more.

Answers

Answers to Interactive Exercises

ENVIRONMENTAL EFFECTS OF HUMAN POPULATION GROWTH/CHANGES IN THE ATMOSPHERE (48-I)

1. Wet acid depositions; 2. sulfuric acid (nitric acid); 3. thermal inversion; 4. hydrocarbons; 5. photochemical; 6. Acid rain; 7. CFCs (chlorofluorocarbons); 8. average annual temperature; 9. eighteenth; 10. Pollutants; 11. 30; 12. forests; 13. ozone; 14. CFCs.

CHANGES IN THE HYDROSPHERE/ CHANGES IN THE LAND (48-II)

1. magnets; 2. conveyor belts; 3. primary; 4. Tertiary; 5. Deforestation; 6. salination; 7. Ogallala aquifer; 8. Agriculture; 9. desalination; 10. high-yield; 11. Fossil fuel; 12. 4; 13. 100; 14. overgrazing.

A QUESTION OF ENERGY INPUTS (48-III)

1. meltdown; 2. radiation; 3. kerogen; 4. 8; 5. coal; 6. breeder; 7. nuclear winter; 8. Net energy; 9. fossil fuels; 10. Nuclear energy; 11. containment; 12. radioactive.

Answers to Self-Quiz

1. a; 2. b; 3. a; 4. c; 5. b; 6. d; 7. c; 8. a; 9. c; 10. d.

49

BEHAVIORAL RESPONSES TO THE ENVIRONMENT

MECHANISMS UNDERLYING BEHAVIOR
    Genetic Effects on Behavior
    Hormonal Effects on Behavior
FORMS OF BEHAVIOR
    Instinct and Learning
    Forms of Learning

    Imprinting
    Song Learning
THE ADAPTIVE VALUE OF BEHAVIOR
    Feeding Behavior
    Mating Behavior

Interactive Exercises

MECHANISMS UNDERLYING BEHAVIOR / FORMS OF BEHAVIOR (49-I, pp. 893–899)

Fill-in-the-Blanks

(1)_____ is a collection of observable responses by muscles and glands to environmental stimuli.

(2)_____ and endocrine systems signal the muscles and glands after evaluating information

transmitted from (3)_____ _____ that were disturbed by the environmental stimuli. Starlings

deck their nests with wild carrot in order to control populations of (4)_____. A (5)_____

_____ _____ is a sequence of motor outputs determined by the central nervous system that results

in a coordinated movement. It is genetically determined—that is, (6)_____—and predictably runs its

course after some stimulus sets it in motion. Any stimulus that activates a fixed action pattern is well

defined and (7)_____. (8)_____ is a change in behavior that involves an enduring potential for

adapting future responses as a result of past experience. In (9)_____ environments, rigid, genetically

determined motor patterns are common. The work of Ivan Pavlov centered on a form of associative

learning called (10)_____ _____ , in which new responses called (11)_____ _____

were established in the behavior of dogs. In another form of associative learning, (12)_____

conditioning, an animal learns to associate a voluntary activity with the consequences that follow *after* a

response is given, and learning occurs by (13)_____ _____ _____. If a response has no

positive or negative consequences, an animal learns through experience not to respond—a learning process

called (14)_____. (15)_____ _____ has been demonstrated to occur only in some primates;

it is a trial-and-error process that goes on in the brain before an appropriate response is made. For many animals, learning certain forms of behavior occurs only during limited time spans called (16)_____ _____ of development; preferential behavior toward a stimulus acquired during such a time is called (17)_____.

Male and female songbirds differ in the structure and size of their brain region, the (18)_____ _____ , that controls the muscles of their vocal organ, the syrinx. Adult male zebra finches sing, but adult females normally don't. In adult male zebra finches, levels of the male hormone (19)_____ are higher than in adult females, but in very young males, levels of the female hormone (20)_____ are higher than in nesting females, and it is this hormone and another, progesterone, that establish and *organize* brain regions that deal with singing.

Singing behavior depends indirectly on a different hormone, melatonin, which is secreted by the (21)_____ gland in the head. High levels of melatonin suppress secretion of gonadal hormones. Increased amounts of (22)_____ suppress the production of melatonin and free the gonads to secrete their hormones. Adult male songbirds in spring produce high levels of (23)_____ , which activates the song system and prepares the bird to sing the song of his species that he was required to hear during the (24)_____ period of his youth.

True-False

If false, explain why.

___ 25. Instincts and learned behavior are regulated by different kinds of neural and hormonal systems.

___ 26. Both genes and the environment contribute to the mechanisms that underly instinctive behavior.

___ 27. Both genes and the environment contribute to the mechanisms that underly learned behavior.

THE ADAPTIVE VALUE OF BEHAVIOR (49-II, pp. 899–903)

Fill-in-the-Blanks

Behavior has a(n) (1)_____ basis in that response to a stimulus depends on the organization of (2)_____ in a nervous system, and that is determined by the creature's genetic endowment, as is the sensitivity of its receptors. Those animals whose behavioral responses enable them to survive and (3)_____ will pass those genetic assortments on to their offspring. If we assume that behavioral mechanisms have evolved as a result of (4)_____ _____ , then those mechanisms must have promoted the (5)_____ _____ of the individual; in other words, they helped the individual pass on its genes to its offspring.

Matching

Match each example or definition below with the correct term.

6. ___ adaptive behavior

7. ___ altruistic behavior

8. ___ natural selection

9. ___ reproductive success

10. ___ selfish behavior

11. ___ sexual selection

A. A male cat eats a female cat's newborn kittens and then mates with her

B. Survival and reproduction of offspring

C. Activities dictated by genes the frequency of which increases in successive generations

D. Scenario in which the "fittest" survive and reproduce; the less "fit" survive and reproduce to a lesser extent

E. The captain of a ship shouting "Women and children first!" as the boat begins to sink and the lifeboats are launched

F. Competition for mates and selectivity among potential mates

True-False

If false, explain why.

___ 12. In developing a working hypothesis to explain some behavioral trait, it will usually be more profitable to use an approach based on benefit to the species rather than an approach based on natural (individual) selection.

___ 13. Wandering ravens give carcass-advertising calls, but territorial residents do not.

___ 14. Members of the same species often create obstacles to reproductive success for each other, either through competition for access to mates or through selective mate choice.

___ 15. A male white-crowned sparrow reared in soundproof isolation sings the typical adult's song, which indicates that the song system is genetically based.

___ 16. Lemmings disperse because they are self-sacrificing and altruistic, not because they are searching for less crowded habitats where they can reproduce successfully.

Chapter Terms

The following page-referenced terms are important; most were in boldface in the chapter. Refer to the instructions given in Chapter 1, p. 4.

behavior (893)
song system (895)

instinctive behavior (896)
learned behavior (896)

sexual selection (900)

Self-Quiz

___ 1. Any stimulus that activates a fixed action pattern, or even some part of it, is

_____.

a. well defined and simple
b. hormonal in nature
c. genetically determined
d. an example of selfish behavior

___ 2. Many animals learn through experience not to give a specific response to a situation if the response elicits no positive or negative consequences; this form of learning is _____.

a. insight learning
b. imprinting
c. latent learning
d. habituation

_____ 3. Newly hatched goslings follow any large moving objects to which they are exposed shortly after hatching; this is an example of _____.
  a. homing behavior
  b. imprinting
  c. piloting
  d. migration

_____ 4. The principal difference between classical Pavlovian conditioning and operant conditioning is _____.
  a. one uses a reinforcing stimulus, and the other doesn't
  b. one presents the reinforcing stimulus before the response and the other one after
  c. one uses a bell as a reinforcing stimulus, and the other uses an instrument as the reinforcing stimulus
  d. the pineal gland is involved in one, and an operator center is part of the other

_____ 5. The example used to demonstrate that Darwinian individual selection explained some behavioral traits better than did "good of the group" selection was _____.

  a. the dilution effect in wildebeest and zebra populations
  b. siblicide among egrets
  c. courtship behavior in albatrosses
  d. the dispersal of Norwegian lemmings when population densities became extremely high

_____ 6. A cat explores all the rooms of a new home even though such exploration is not rewarded. Later, when the cat begins to feel chilled, it goes directly to the warmest room. This is an example of _____.
  a. classical conditioning
  b. echolocation
  c. spatial or latent learning
  d. imprinting

_____ 7. Motor activity and metabolic rates associated with biological clocks are coordinated by _____.
  a. thyroxin secreted by the thyroid gland
  b. melatonin secreted by the pineal gland
  c. a magnetic sense that is attuned to variations in Earth's magnetic field
  d. pheromones released by the dominant animal of the group

Chapter Objectives/Review Questions

This section lists general and detailed chapter objectives that can be used as review questions. You can make maximum use of these items by writing answers on a separate sheet of paper. Fill in answers where blanks are provided. To check for accuracy, compare your answers with information given in the chapter or glossary.

Page   *Objectives/Questions*

(893–894)   1. Define behavior and name four factors that produce it.
(894)   2. Describe Steven Arnold's studies that suggest that feeding behavior in certain California garter snakes has some genetic basis.
(893–895)   3. Understand the components of behavior that have a genetic and/or hormonal basis.
(896)   4. Define learning and distinguish it from innate behavior.
(896)   5. Explain what a fixed action pattern is and state its relationship to innate behavior.
(897)   6. Explain what is meant by classical Pavlovian conditioning and distinguish it from operant conditioning by describing an example of each process. Explain why both forms of conditioning are examples of associative learning.
(897)   7. Define insight and give an example.
(899–900)   8. Know the aspects of behavior that have an adaptive value.
(899–900)   9. Contrast altruistic behavior with selfish behavior and cite examples of each in animal populations.
(900)   10. State whether Gary Larson's lemming cartoon supports or undermines group selection.

Integrating and Applying Key Concepts

Explain whether or not you think humans have any critical periods for establishing the ability to learn certain kinds of knowledge. State whether or not you think humans undergo imprinting. Do you think humans employ resource-defense behavior? Female-defense behavior? If so, can you cite an example?

Critical Thinking Exercises

1. Two species of butterfly are almost identical. There are no visible differences between the females of the two species. The males of one species have four red spots on the ventral side of the forewing that are lacking in the other species. One aspect of reproductive isolation between these two species lies in mating behavior. The females have several axons in the optic nerve that transmit action potentials only when the eyes are illuminated with red light. These nerves may be the sensory basis for the females' decisions to accept or reject courting males who display their wings as they solicit mating. Which of the following behaviors would be expected of the males?

    a. Court only females whose red dots match their own.
    b. Wait passively for females to make their choice.
    c. Court all observed females.
    d. Court all objects of appropriate size and color.
    e. Dispense with courtship and mate with all observed females.

ANALYSIS

    a. This strategy is impossible, because females of both species are identical.
    b. This would not be an effective strategy, because females make their choices only after the males perform courtship behavior.
    c. This strategy would ensure the maximum probability of success, even though some energy would be wasted courting nonreceptive females. Because there is apparently no detectable difference between the females of the two species, this is the only available strategy.
    d. This is an extension of the strategy in (c). It would not be expected, because it would increase the waste of energy devoted to inappropriate objects without increasing the probability of successful courting. However, it is a strategy used by some insects, such as the periodic cicadas. The resolution is that there are not very many objects in the environment that look at all like female cicadas.
    e. This strategy would likely not result in successful mating, because the females avoid males that do not exhibit the specific courtship behavior and signals of the species. Furthermore, some percentage of the successful copulations would be with a female of the other species and would not result in successful reproduction—a maladaptive waste of gametes.

2. Many birds dustbathe. They scratch and flap in dry soil, tossing clouds of dust over their bodies. One hypothesis is that this behavior controls ectoparasites. To test this hypothesis, you watch a flock of birds and collect several immediately after their dustbathing and a control group that did not dustbathe during the observation period. All the birds are free of ectoparasites. In order to evaluate the hypothesis, which of the following observations do you also have to make?

    a. How much time elapses before the control birds bathe
    b. How many ectoparasites were on the birds before they bathed
    c. Whether uninfested birds ever dustbathe
    d. The temperature of the dust
    e. The chemical composition of the dust

ANALYSIS

    a. Time between bathing might be relevant if the experiment was set up to show a correlation between level of infestation by ectoparasites and frequency of bathing, but this experiment is a different design.
    b. The hypothesis predicts that dustbathing will reduce the number of ectoparasites on the birds. To show this requires measurements both before and after bathing.

c. An elaboration of the hypothesis might predict that infestation triggers the behavior and would predict less frequent bathing in more lightly infested birds, perhaps even no bathing in uninfested birds. On the other hand, the behavior might be a form of preventive maintenance that all birds of the species do. However, the simple hypothesis that the behavior controls parasites does not predict the frequency of bathing, and this observation is thus irrelevant to testing of the present hypothesis.

d. This information would be needed to test an alternative hypothesis—that dustbathing is thermoregulatory behavior.

e. Once it is established that dustbathing controls ectoparasites, this observation would be potentially useful in testing hypotheses about the mechanism of the effect, but at this stage it is not needed.

3. To test the hypothesis that crabs of the genus *Callinectes* actively avoid direct light, twenty crabs were placed in an aquarium with twenty shelters, ten of clear glass and ten of stone. After an hour, ten crabs were found in glass shelters and ten in stone shelters. In order to reject the hypothesis on the basis of this evidence, which of the following assumptions must be made?

a. Crabs are willing to enter shelters already occupied by another crab.
b. Crabs prefer contact with rough surfaces to contact with smooth surfaces.
c. Crabs can locate shelters within one hour.
d. Crabs hide at low tide and forage in the open at high tide.
e. All the crabs in the experiment were of the same sex.

ANALYSIS

a. Unless this assumption is valid, the evidence is inconclusive. If crabs have a stronger preference for sole occupancy of their space than they have for dark shelters, the first ten crabs will occupy the preferred shelters, and the second ten will have no choice. Either there must be twenty shelters of each type, or this assumption must be made.

b. This is an alternative hypothesis that predicts the same outcome as the hypothesis being tested. If this is assumed, the original hypothesis would be rejected, but the evidence has the same bearing on this statement as it does on the original hypothesis.

c. This does not have to be assumed. The observations show that the crabs did find shelters within an hour.

d. This is an unrelated hypothesis about crab behavior that does not affect the evaluation of the present evidence. You might argue that this assumption would govern the choice of whether to fill the aquarium with water. However, the problem does not stipulate what was done about this condition.

e. If you assumed that the behavior of the two sexes was different with respect to light and dark shelters, you would have to make this assumption, too, in order to evaluate the evidence. But making unnecessary assumptions weakens the experiment.

Answers

Answers to Interactive Exercises

MECHANISMS OF UNDERLYING BEHAVIOR/ FORMS OF BEHAVIOR (49-I)

1. Behavior; 2. Nervous; 3. sensory receptors; 4. mites; 5. fixed action pattern; 6. instinctive (innate); 7. simple; 8. Learning; 9. predictable (stable); 10. classical conditioning; 11. conditioned responses; 12. operant; 13. trial and error; 14. habituation; 15. Insight learning; 16. sensitive periods; 17. imprinting; 18. song system; 19. testosterone; 20. estrogen; 21. pineal; 22. sunlight (daylight); 23. testosterone; 24. sensitive (imprinting); 25. T; 26. T; 27. T.

THE ADAPTIVE VALUE OF BEHAVIOR (49-II)

1. heritable (genetic); 2. neurons; 3. reproduce; 4. natural selection; 5. reproductive success; 6. C; 7. E; 8. D; 9. B; 10. A; 11. F; 12. F; 13. T; 14. T; 15. F; 16. F.

Answers to Self-Quiz

1. a; 2. d; 3. b; 4. b; 5. d; 6. c; 7. b.

50

SOCIAL INTERACTIONS

MECHANISMS OF SOCIAL LIFE
    Functions of Communication Signals
    Types of Communication Signals
        Chemical Signals
        Visual Signals
        Acoustical Signals
        Tactile Signals
COSTS AND BENEFITS OF SOCIAL LIFE
    Disadvantages to Sociality
    Advantages to Sociality
        Predator Avoidance
        The Selfish Herd

SOCIAL LIFE AND SELF-SACRIFICE
THE EVOLUTION OF ALTRUISM
    Parental Behavior and Caring for Relatives
    Indirect Selection and Social Insects
    Doing Science: Naked Mole-Rats and the
    Evolution of Self-Sacrificing Behavior
HUMAN SOCIAL BEHAVIOR

Interactive Exercises

MECHANISMS OF SOCIAL LIFE (50-I, pp. 907–911)

Fill-in-the-Blanks

Social interactions are based on (1)_____ _____ that have developed over evolutionary time; these are actions or cues that benefit both the sender and the receiver more than they cost the sender and the receiver. Natural selection tends to favor communication signals that promote the (2)_____ success of both the signaler and the receiver. (3)_____ signalers and receivers are members of one kind of population that exploit signals produced by members of a different species, so that benefits do not flow both ways.

Matching

Choose the single most appropriate letter for each.

4. ___ acoustical signal
5. ___ chemical signal
6. ___ tactile signal
7. ___ visual signal

  A. Pheromone
  B. Sam Peabody, Peabody, Peabody and whine plus "chuck"
  C. Bioluminescent messages
  D. Dance of the foraging honeybee

COSTS AND BENEFITS OF SOCIAL LIFE (50-II, pp. 911–913)

Fill-in-the-Blanks

The reason for diversity in social life becomes clearer if (1)_____ are considered, as well as benefits of social life, in terms of individual success with (2)_____. (3)_____ is by far the dominant selection pressure favoring the evolution of social behavior. Some animals live in groups simply for the purpose of using other individuals as living shields against (4)_____ ; such groups have been labeled (5)_____ _____.

True-False

If false, explain why.

____ 6. Insect-eating songbirds are more likely to consume solitary caterpillars than a cluster of them.

____ 7. Australian sawflies execute a "waggle dance" in response to a disturbance.

____ 8. The largest, most powerful bluegill males prepare nests in the center of the colony rather than on the periphery of the colony, in order to attract mates.

____ 9. Dominant animals advertise their status by means of threat displays.

SOCIAL LIFE AND SELF-SACRIFICE / THE EVOLUTION OF ALTRUISM / HUMAN SOCIAL BEHAVIOR (50-III, pp. 913–920)

Fill-in-the-Blanks

(1)_____-_____ behavior helps *other* individuals survive and reproduce at personal cost to oneself for the time being; however, patient individuals may survive longer by being associated with a group than by being solitary and may be able to (2)_____ later on if the dominant members of the society are displaced. Fighting results in establishing (3)_____ for one individual and submission by others. Ranking of members in a social group establishes a (4)_____ _____ that serves to lessen the frequency of damaging fighting but also establishes unequal opportunities to (5)_____ and differential access to food supplies and protection. Submission finds new expression in avoidance behavior and (6)_____ behavior as a further show of deference. The key point in the theory of (7)_____ _____ is that individuals can indirectly pass on their genes by helping their relatives survive and reproduce. Among the (8)_____ _____ , division of labor reaches the level that exists in a complex society; the (9)_____ becomes the reproductive "organ" for all, and the workers forage for food and repel invaders. (10)_____ are male members of the colony that have no stinger; some mate with the queen. Naked mole-rats show extreme examples of (11)_____ , because many individuals appear to spend their entire lives as (12)_____ helpers in the groups in which they live. The method of (13)_____ _____ established that nonbreeding helper mole-rats are very (14)_____ genetically to the reproducing members of their population. Even though helper mole-rats don't breed, their activities act to preserve genotypes very similar to their own, which dictate gene-based cooperative behaviors, so giving up reproduction under such circumstances does not condemn the individual's genes to (15)_____.

Chapter Terms

The following page-referenced terms are important; most were in boldface in the chapter. Refer to the instructions given in Chapter 1, p. 4.

social behavior (907)   visual signals (908)   selfish herd (913)   altruistic behavior (914)
communication signals (907)   acoustical signals (909)   self-sacrificing behavior (913) indirect selection (914)
chemical signals (908)   tactile signals (910)   dominance hierarchy (914)   siblicide (919)

Self-Quiz

___ 1. An animal that arranges its schedule of activities to minimize competition for essential resources with a dominant member of the same group is said to be engaging in _____ behavior.
   a. appeasement
   b. avoidance
   c. ritualized
   d. territorial

___ 2. The studies of siblicide among the egrets demonstrated that _____.
   a. sibling aggression promotes reproductive success of the parents
   b. parents will often intervene when siblings fight with the runt of the litter
   c. parent egrets incubate eggs so that they hatch synchronously
   d. the parents engage in complex rituals before mating

___ 3. The example used to demonstrate that competitive interactions lead to the formation of dominance hierarchies involved _____.
   a. albatrosses
   b. a honeybee colony
   c. baboon troops
   d. greylag geese

___ 4. Parental support of offspring is an example of _____.
   a. artificial selection
   b. kin selection
   c. natural selection
   d. negative selection

___ 5. A submissive animal that exposes its throat or genitals to a dominant member of the same group is said to be engaging in _____ behavior.
   a. appeasement
   b. avoidance
   c. ritualized
   d. dispersive

___ 6. In highly integrated insect societies, _____.
   a. natural selection favors individual behaviors that lead to greater diversity among members of the society
   b. there is scarcely any division of labor
   c. cooperative behavior predominates
   d. patterns of behavior are flexible, and learned behavior predominates
   e. all of the above

___ 7. Social behavior among insects depends on _____.
   a. diversity
   b. echolocation
   c. polymorphism
   d. genetic similarity
   e. communication

___ 8. The termite colony described in the text has _____.
   a. one queen that reproduces parthenogenetically
   b. one queen and a harem of males to fertilize her
   c. one king and one queen
   d. several queens operating at one time

Chapter Objectives/Review Questions

This section lists general and detailed chapter objectives that can be used as review questions. You can make maximum use of these items by writing answers on a separate sheet of paper. Fill in answers where blanks are provided. To check for accuracy, compare your answers with information given in the chapter or glossary.

Integrating and Applying Key Concepts

If you were Ruler of All People on Earth, what mechanisms would you employ to distribute essential resources so as to minimize deadly competition and wars? Would a person who believed in biological determinism be likely to advocate the equal distribution of essential resources among all members of a social group?

Critical Thinking Exercises

1. An animal, the sender, performed a set of movements, the signal. The signal was observed by three other animals, A, B, and C, each of which then performed a different set of movements. Animal A moved toward the sender, animal B moved away from the sender, and animal C did not move. Which of the following statements is the best interpretation?

   a. Either A or B misinterpreted the signal's intended meaning.
   b. The signal had different intended meanings for A and B.
   c. The signal communicated to A and B.
   d. The signal did not communicate to C.
   e. The signal communicated to all three animals.

ANALYSIS

It is not possible to interpret this single observation with much confidence. More information is needed about the characteristics of the animals and especially about the consistency with which they demonstrate this set of behaviors. Are all four animals of the same species? Are the three receivers all the same sex and age and in the same condition? Does each animal always make the same response to the signal? What responses do they make to other signals?

   a. This could be valid if A and B have the same critical characteristics. The sender may have perceived A and B as being identical individuals and may have intended to signal the same response from both. Because they gave different responses, one or the other could have misinterpreted or even completely missed the intent of the signal.
   b. A single signal might have different intent depending on the nature of the receiver. For example, the same gesture might signal "come play" to an immature male and "threat" to a mature male. If the receivers in this case were different, their opposite responses might both have been appropriate to the signal's intent.
   c. A sequence of behaviors by different animals might indicate that communication has been accomplished. However, in an isolated instance like this, it is not a strong conclusion. The signal might have been ignored or not perceived at all, and the behaviors might be inappropriate responses or irrelevant behaviors. A consistent pattern must be observed before behavior can be classed as communication.
   d. This statement is the strongest interpretation of a single observation such as this, although it is also weak. The absence of response gives no evidence of communication but does not give positive

evidence that communication has not occurred. The signal to animal C might have been a "Danger—freeze!" message, in which case communication *was* accomplished.

e. As discussed in (c) and (d), this is a possible interpretation but is far from conclusive in the absence of repeated observations of a consistent pattern of responses that depend on the characteristics of the responders and the conditions at the time of the interaction.

2. Ants commonly move between the nest and a food source along an indirect path and follow the same track on repeated trips. To test the hypothesis that ants communicate the location of food by laying a trail of chemical signals as they walk, you set up a good food source some distance from an ant nest on a table in the laboratory. When the ants have found the food and many ants are going to and from it along a single path, you rub your finger on the table across the path of ants. The ants begin to walk aimlessly in loops and can't find the food or the nest. You assume that your finger removed any chemicals from the table surface and conclude that your hypothesis is supported. Which of the following is the best criticism of this experiment?

   a. The experiment should have been performed with more than one species of ant.
   b. The effect of varying the direction of the light source should have been investigated.
   c. Chemicals from human skin might have disrupted the ants' behavior.
   d. The trail should have been disrupted by rubbing a dead ant across it.
   e. The experiment should have been performed on a natural surface of loose dirt.

ANALYSIS

   a. Results obtained with only one species cannot be generalized to all ants, but they can still lead to valid conclusions about that one species. Because an experiment cannot be repeated with all the almost 9,000 known species of ants, how many species must be used before the results *can* be generalized?

   b. This investigation would test an alternative hypothesis that ants navigate by the direction of the light source. However, this is a separate investigation, not a criticism of the experiment.

   c. This is a strong criticism of the experiment. The procedure not only may disrupt any chemical trails the ants have laid but also would deposit on the surface any chemicals present on the investigator's finger as a result of secretion or contamination. If these chemicals caused a response in the ants, the response would be an experimental artifact, not an interpretable result.

   d. This procedure would create an even stronger possibility of artifactual result. A dead ant would deposit all chemical signaling substances the individual carried, and the response to the mixed signals would be uninterpretable.

   e. In this design, the disorientation of the ants might be due to disruption of a chemical trail, but it might also be due to changes in topography. The experiment must be designed so that only one variable is changed at a time. The best design would be to allow the ants to establish their pathway across a movable surface such as a piece of paper, then change the orientation of the movable segment. If the ants' direction does not change, they are not navigating by a chemical trail. If they follow their old path in its new direction, they probably are using chemical communication.

3. In the discussion of termite communication, the text says that "volatile odors released from the gluey substance attract more soldiers to the spot." Which of the following observations would provide the strongest evidence in support of this conclusion?

   a. Termite soldiers congregate in defensive postures at breaks in the tunnel.
   b. Termite soldiers congregate at a spot in an experimental box where the gluey substance has been deposited.
   c. Termite soldiers congregate at a break in the tunnel, even when no gluey substance has been deposited.
   d. Termite soldiers released into the stem of a T-tube consistently move to the arm at the end of which the gluey substance has been deposited.
   e. Ants released into the stem of a T-tube consistently move away from the arm at the end of which the gluey substance has been deposited.

ANALYSIS

a. This is the original observation that is explained by the hypothesis. It is not independent evidence that supports the hypothesis.

b. This observation would not indicate that the gluey substance emits vapors that attract the soldiers from other locations. It could be that the soldiers wander randomly and stop when they encounter high concentrations of the gluey substance.

c. This observation would be contradictory evidence and would indicate that the congregation phenomenon was independent of the gluey substance.

d. This observation would support the hypothesis of attraction. At the time that the soldier makes the choice of which arms of the tube to follow, it is still some distance from the gluey substance. If it consistently chooses to go toward the substance, it must be able to perceive the substance at a distance by volatile odors and be attracted toward them. Note that the position (right or left) of the substance has to be changed randomly to eliminate the possibility that the soldier favors one direction for some other reason.

e. This observation would lead to conclusions about the defensive significance of the gluey substance but would not support the hypothesis that it is a communication signal.

Answers

Answers to Interactive Exercises

MECHANISMS OF SOCIAL LIFE (50-I)
1. communication signals; 2. reproductive;
3. Illegitimate; 4. B; 5. A; 6. D; 7. C.

COSTS AND BENEFITS OF SOCIAL LIFE (50-II)
1. costs; 2. reproduction; 3. Predation; 4. enemies;
5. selfish herds; 6. T; 7. F; 8. T; 9. T.

SOCIAL LIFE AND SELF-SACRIFICE/EVOLUTION OF ALTRUISM/HUMAN SOCIAL BEHAVIOR (50-III)
1. Self-sacrificing; 2. reproduce; 3. dominance;
4. dominance hierarchy; 5. reproduce; 6. appeasement;
7. indirect selection; 8. social insects; 9. queen;
10. Drones; 11. altruism; 12. nonbreeding; 13. DNA fingerprinting; 14. similar; 15. extinction.

Answers to Self-Quiz

1. b; 2. a; 3. c; 4. b; 5. a; 6. c; 7. c; 8. b.